5

Microbiology Monographs

Series Editor: Alexander Steinbüchel

Microbiology Monographs

Volumes published in the series

Inclusions in Prokaryotes
Jessup M. Shively (Editor)
Volume 1 (2006)

Complex Intracellular Structures in Prokaryotes
Jessup M. Shively (Editor)
Volume 2 (2006)

Magnetoreception and Magnetosomes in Bacteria
Dirk Schüler (Editor)
Volume 3 (2007)

Predatory Prokaryotes – Biology, Ecology and Evolution
Edouard Jurkevitch (Editor)
Volume 4 (2007)

Amino Acid Biosynthesis – Pathways, Regulation and Metabolic Engineering

Volume Editor: Volker F. Wendisch

With 64 Figures, 6 in Color

Springer

Volume Editor:

Prof. Dr. Volker F. Wendisch
Institute of Molecular Microbiology and Biotechnology
Westfalian Wilhelms University Münster
Corrensstr. 3
48149 Münster
Germany
e-mail: wendisch@uni-muenster.de

Series Editor:

Professor Dr. Alexander Steinbüchel
Institut für Molekulare Mikrobiologie und Biotechnologie
Westfälische Wilhelms-Universität
Corrensstraße 3
48149 Münster
Germany
e-mail: steinbu@uni-muenster.de

Library of Congress Control Number: 2006940356

ISSN 1862-5576
ISBN 978-3-540-48595-7 Springer Berlin Heidelberg New York
DOI 10.1007/978-3-540-48596-4

Springer is a part of Springer Science+Business Media

springer.com

Editor: Dr. Christina Eckey, Heidelberg, Germany
Desk Editor: Dr. Jutta Lindenborn, Heidelberg, Germany
Cover Design: WMXDesign GmbH, Heidelberg, Germany
Typesetting and Production: LE-TEX Jelonek, Schmidt & Vöckler GbR, Leipzig, Germany

Printed on acid-free paper SPIN 11768746 149/3100 YL – 5 4 3 2 1 0

Preface

Amino acids are simple organic compounds containing at least one amino and one carboxylic function. The L-α-amino acids and glycine can be considered building blocks of life since they constitute the biopolymers proteins in all organisms, but also D-isomers, e.g. D-alanine in bacterial cell walls, and β- and γ-amino acids like β-alanine as component of vitamin B5 and γ-aminobutyrate as neurotransmitter occur in nature. While mature proteins contain many different amino acids due to post-translational modifications, only 22 proteinogenic L-α-amino acids are genetically encoded, i.e. they are used for protein biosynthesis as amino acyl-tRNAs by the ribosome. Since the deciphering of the genetic code for 20 amino acids about 40 years ago, L-selenocysteine (Chambers et al. 1986; Zinoni et al. 1986) and L-pyrrolysine (Hao et al. 2002; Srinivasan et al. 2002) were discovered as the 21st and 22nd genetically encoded amino acids. The co-translational insertion of selenocysteine and pyrrolysine into nascent proteins requires unique *cis*- and *trans*-acting factors to recode UGA and UAG stop codons, respectively, as codons for selenocysteine and pyrrolysine, respectively. The existence of further widely distributed and genetically encoded amino acids is unlikely as suggested in a recent bioinformatic analysis of tRNAs encoded in about 150 bacterial and archaeal genomes (Lobanov et al. 2006). The essential L-amino acids isoleucine, leucine, lysine, methionine, phenylalanine, threonine, tryptophan, and valine cannot be synthesized by humans, but have to supplied in the diet, and infants require arginine and histidine in addition. Microorganisms differ considerably in their capabilities to synthesize amino acids de novo, for example *Leuconostoc mesenteroides* only grows, when 16 amino acids are supplied, while bacteria like *Escherichia coli, Bacillus subtilis* and *Corynebacterium glutamicum* are able to synthesize all amino acids de novo from ammonium.

This monograph deals with amino acid biosynthetic pathways and their genetic and biochemical regulation on one hand and with the use and metabolic engineering of microorganisms for biotechnological production of amino acids on the other. The current knowledge of amino acid metabolic pathways and transport systems specific for uptake or export of amino acids is covered.

The characterization of the control mechanisms of amino acid biosynthesis have revealed fundamental insights into genetic and biochemical regulation. Feedback inhibition of biosynthetic enzymes by metabolic end-products (e.g.

of aspartate transcarbamoylase in the biosynthesis of arginine and pyrimidines by CTP), enzyme activity control by covalent modification (e.g. adenylation of glutamine synthetase), co-repressor dependent transcriptional repression (e.g. by the tryptophan-activated repressor of the *trp* operon), and attenuation control (e.g. translation-mediated attenuation of the tryptophan biosynthetic operon) are integral parts of biochemistry, genetics and microbiology textbooks. Recently, global gene expression analyses allowed determining the regulons of a number of transcriptional regulators of amino acid biosynthesis. Moreover, the discovery of a lysine-specific riboswitch (Sudarsan et al. 2003), a new class of regulatory elements, illustrates that the characterization of the regulation of amino acid metabolism continues to spur new discoveries.

Amino acids are used on the basis of their chemical characteristics, physiological activities, nutritional value and taste as chemical building blocks, as pharmaceuticals and, particularly, as food and feed additives. The discovery of *C. glutamicum* as suitable catalyst to produce the flavor enhancer monosodium glutamate 1957 in Japan marked the birth of the amino acid fermentation industry. Reports on the beginnings of microbial amino acid production (Yamada et al. 1972; Aida et al. 1986; Enei et al. 1989) and an up-to-date review of *C. glutamicum* (Eggeling & Bott 2005) are available. In this monograph, recent achievements to enable or to improve production of amino acids and of dipeptides by fermentation and enzyme catalysis are comprehensively reviewed with a particular focus on metabolic engineering, i.e. the rational improvement of a cell's metabolic functions using recombinant DNA technology. Genome sequencing and post-genomics approaches to strain improvement for amino acid production will be covered as their impact, although already quite visible, is expected to increase considerably in the future accelerating the development of new and more efficient biocatalysts for amino acid production.

We are grateful to the authors who contributed excellent chapters to the volume *Amino Acids – Pathways, Regulation and Metabolic Engineering*. We highly appreciate the expertise and enthusiasm devoted to their chapters. Despite their many other obligations and duties their comprehensive overview chapters were timely completed. Our thanks also go to Springer for publishing this monograph and especially to Christina Eckey and Jutta Lindenborn for their valuable suggestions and support.

Münster, January 2007

Volker F. Wendisch
Alexander Steinbüchel

References

Aida K, Chibata I, Nakayama K, Takinami K, Yamada H (1986) Biotechnology of Amino Acid Production. Progress in Industrial Microbiology. Kodansha Ltd, Tokyo and Elsevier, Amsterdam–Oxford–New York–Tokyo

Chambers I, Frampton J, Goldfarb P, Affara N, McBain W, Harrison PR (1986) The structure of the mouse glutathione peroxidase gene: the selenocysteine in the active site is encoded by the 'termination' codon, TGA. EMBO J 5:1221–1227

Eggeling L, Bott M (2005) Handbook of Corynebacterium glutamicum. CRC Press (Taylor Francis Group), Boca Raton

Enei H, Yokozeki K, Akashi K (1989) Recent progress in microbial production of amino acids. Gordon and Breach, Amsterdam

Hao B, Gong W, Ferguson TK, James CM, Krzycki JA, Chan MK (2002) A new UAG-encoded residue in the structure of a methanogen methyltransferase. Science 296:1462–1466

Lobanov AV, Kryikov GV, Hatfield DF, Gladyshev VN (2006) Is there a twenty third amino acid in the genetic code? Trends Genet 22:357–360

Srinivasan G, James CM, Krzycki JA (2002) Pyrrolysine encoded by UAG in Archaea: charging of a UAG-decoding specialized tRNA. Science 296:1459–1462

Sudarsan N, Wickiser JK, Nakamura S, Ebert MS, Breaker RR (2003) An mRNA structure in bacteria that controls gene expression by binding lysine. Genes Dev 17:2688–2697

Yamada K, Kinoshita S, Tsunoda T, Aida K (1972) The microbial production of amino acid. Kodansha Ltd, Tokyo

Zinoni F, Birkmann A, Stadtman TC, Bock A (1986) Nucleotide sequence and expression of the selenocysteine-containing polypeptide of formate dehydrogenase (formate-hydrogen-lyase-linked) from *Escherichia coli*. Proc Natl Acad Sci USA 83:4650–4654

Contents

Microbiol Monogr (5)
V. F. Wendisch: Amino Acid Biosynthesis
DOI 10.1007/7171_2006_064/Published online: 9 December 2006

Production of Glutamate and Glutamate-Related Amino Acids: Molecular Mechanism Analysis and Metabolic Engineering

Hiroshi Shimizu[1] (✉) · Takashi Hirasawa[2]

[1]Department of Bioinformatic Engineering,
Graduate School of Information Science and Technology, Osaka University,
2-1 Yamadaoka, Suita, Osaka 565-0871, Japan
shimizu@ist.osaka-u.ac.jp

[2]Institute for Advanced Biosciences, Keio University, 403-1 Nipponkoku, Daihoji,
Tsuruoka, Yamagata 997-0017, Japan

Abstract Glutamate production is a typical success in industrial fermentation. Annual production of glutamate by *Corynebacterium glutamicum* is over 1.5 million tons per year worldwide. It is well known that there are some triggers of glutamate overproduction by *C. glutamicum*: depletion of biotin, which is required for cell growth; addition of detergent; addition of β-lactam antibiotics such as penicillin; and addition of ethambutol or cerulenin. A marked change in metabolic pathways occurs after glutamate overproduction is triggered. In this chapter, recent studies on the molecular mechanisms of glutamate production are described with a particular focus on triggering mechanisms, changes in key enzyme activities, and secretion of glutamate. Recent advances in genome-wide studies, including genomics, proteomics, metabolomics, and on metabolic flux analysis of flux redistribution during glutamate overproduction are discussed as well. The biosynthesis of the related amino acids glutamine and proline and strategies for their overproduction are also described.

Keywords Glutamate · Glutamine · Proline · Metabolic engineering · Triggering of glutamate overproduction

1 Glutamate

In this section, recent studies on the molecular mechanisms of glutamate production and metabolic engineering analysis of flux redistribution in the central and anaplerotic pathways are described.

1.1 General Introduction and History

Monosodium glutamate is a substance that exhibits the specific taste of "Umami", well-known in Japanese cuisine. It was discovered and isolated from hydrolyzates of "konbu", seaweed, in 1908 by Dr. Kikunae Ikeda. In the presentation at the 8th International Congress of Applied Chemistry, Chicago, 1912 he reported that: "*An attentive taster will find out something common in the complicated taste of asparagus, tomatoes, cheese and meat, which is quite peculiar and cannot be classified under any of the well defined four taste qualities, sweet, sour, salty and bitter*". He succeeded in identification of this substance as monosodium glutamate. Ajinomoto was the first company to produce monosodium glutamate on an industrial scale by extraction from wheat hydrolyzates. A coryneform bacterium, *Corynebacterium glutamicum* (former name *Micrococcus glutamicus*), was isolated from a soil sample in the 1950s by Japanese researchers of Kyowa Hakko Kogyo (Kinoshita et al. 1957; Udaka 1960). They developed a fermentation method to produce glutamate directly from cheap sugar and ammonia, thus reducing production costs of glutamate considerably. Currently, *C. glutamicum* is widely used as a producer of amino acids such as lysine, arginine, histidine, valine, and so forth (Eggeling and Sahm 1999, Kimura 2003, Ikeda 2003).

The market of amino acids is increasing with an annual growth rate of 5–7% (Leuchtenberger et al. 2005). With the exploitation of the wide spectrum of the uses of amino acids as food additives, pharmaceuticals, feed supplements, cosmetics, and polymer precursors, the demand for amino acids has increased. The biggest market among the amino acids is that of glutamate, and the main use of this amino acid is as a flavor enhancer. Recently, the annual production of glutamate was more than 1.5 million tons per year worldwide (Ajinomoto 2006).

C. glutamicum is a facultatively anaerobic, non-spore-forming, Gram-positive bacterium. This microorganism has the ability to produce large amounts of amino acids such as glutamate and lysine. Recently, the genome sequencing of *C. glutamicum* was completed (Ikeda and Nakagawa 2003, Kalinowski et al. 2003). A related species, *Corynebacterium efficiens*, was found to be another glutamate producer (Fudou et al. 2002) and the genome sequencing of this strain was also performed (Nishio et al. 2003).

There are some triggers of glutamate overproduction: depletion of biotin, which is required for cell growth: addition of detergent; addition of β-lactam antibiotics such as penicillin; addition of ethambutol or cerulenin; and for temperature-sensitive mutants a temperature shock. A marked change in metabolic flux typically occurred after these triggers.

1.2
Brief Summary of Glutamate Production by *C. glutamicum*

Glutamate is synthesized from 2-oxogluratrate by a one-step reaction catalyzed by glutamate dehydrogenase (GDH) (Börmann et al. 1992), which is the main pathway for glutamate formation when the ammonium concentration is sufficiently high (Ertan 1992a,b; Börmann-El Kholy et al. 1993). 2-Oxoglutarate is a member of the tricarboxylic acid cycle and, thus, of the central carbon metabolism, including the glycolytic pathway, the pentose phosphate pathway, and the anaplerotic pathways, which should be discussed for overproduction of glutamate.

C. glutamicum requires biotin for cell growth. Glutamate is not produced by wild-type *C. glutamicum* when excess biotin is present in the culture medium. However, a significant production of glutamate occurs when biotin is depleted (Shiio et al. 1962). Even with excess biotin, the addition of a detergent compound such as polyoxyethylene sorbitan monopalmitate (Tween 40) or polyoxyethylene sorbitan monostearate (Tween 60) causes *C. glutamicum* to produce significant glutamate (Takinami et al. 1965). Addition of penicillin, one of the β-lactam antibiotics, also enhances the overproduction of glutamate by *C. glutamicum* (Nara et al. 1964). Similarly, the addition of ethambutol (Radmacher et al. 2005), a chemotherapeutic used to treat tuberculosis, or the addition of cerulenin (Hashimoto et al. 2006) trigger glutamate production by wild-type *C. glutamicum*. Temperature-sensitive mutants produce glutamate after a heat-shock (Momose and Takagi, 1978; Delaunay et al. 1999, 2002), although this process is not stable (Uy et al. 2003).

Since biotin is a cofactor of acetyl-CoA carboxylase, which is necessary for fatty acid synthesis, it was thought that the cell membrane permeability increased when biotin was depleted in the culture medium. Similarly, addition of detergent, penicillin, ethambutol, or cerulenin alters the composition of the cell membrane or cell wall of this microorganism. Consequently, it was thought that the permeability of the cell membrane and cell wall should change due to these operations. This explanation was called the "leak model" of glutamate production. However, from the viewpoint of the material balance of intracellular and extracellular glutamate, it is difficult to explain the high accumulation of glutamate to more than 60–80 g/L by this model alone.

Very early on, it was proposed that *C. glutamicum* produces glutamate because it might lack 2-oxoglutarate dehydrogenase complex (ODHC) and all flux from isocitrate to 2-oxoglutarate is channeled to glutamate. However, about 40 years ago *C. glutamicum* was shown to possess ODHC activity (Shingu and Terui 1971; Shiio and Ujigawa-Takeda 1980) and the genes encoding the subunits of ODHC were later characterized (Usuda et al. 1996; Schwinde et al. 2001). The proposal that the activity control of ODHC might be critical for glutamate production (Kinoshita 1985) was substantiated in recent studies (Kawahara et al. 1997) and the signal transduction pathway for regulation of ODHC, which involves protein kinase PknG and the specific inhibitory protein OdhI, were unraveled in 2006 (Niebisch et al. 2006).

1.3 Recent Advances in Analysis of the Mechanism of Glutamate Overproduction by *C. glutamicum*

ODHC is a branch-point enzyme complex between the tricarboxylic acid (TCA) cycle and glutamate biosynthesis and catalyzes the conversion of 2-oxoglutarate to succinyl-CoA. ODHC consists of three subunits, 2-oxoglutarate dehydrogenase (E1o), dihydrolipoamide *S*-succinyltransferase (E2o) and dihydrolipoamide dehydrogenase (E3). The *C. glutamicum odhA* gene encoding the E1o subunit shows high homology with the gene encoding E1o subunit from other bacteria, but the N-terminal extension in *C. glutamicum* OdhA cannot be found in other reported E1os (Usuda et al. 1996).

The *lpd* gene encoding the E3 subunit was characterized by Schwinde et al. (2001) and the *sucB* gene encoding the E2o subunit was found during the genome sequencing project (Ikeda and Nakagawa 2003; Kalinowski et al. 2003).

In 1971, Shingu and Terui reported that the decrease in ODHC activity of *Brevibacterium* sp. was observed under biotin-limiting conditions rather than in biotin-rich conditions. To further understand the relationship between the enzyme activity and glutamate production by *C. glutamicum*, Kawahara et al. (1997) measured activities of ODHC as well as isocitrate dehydrogenase (ICDH) and glutamate dehydrogenase (GDH) of *C. glutamicum*, both of which are located at the 2-oxoglutarate branch point, during glutamate production under biotin limitation, Tween 40 addition, and penicillin addition conditions. The activities of ICDH and GDH did not change significantly, but that of ODHC decreased during glutamate production under these three conditions. These results indicate that glutamate overproduction by *C. glutamicum* is correlated with the change in metabolic flows from the TCA cycle to glutamate production due to the decrease in ODHC activity. The reduction of the ODHC activity is one of the key factors in glutamate overproduction by *C. glutamicum*.

1.4 Analysis of the Molecular Mechanisms of Glutamate Overproduction by *C. glutamicum*

Genetic experiments revealed the role of ODHC, DtsR1, and of the signal transduction proteins PknG and OdhI for glutamate production by *C. glutamicum*.

1.4.1 Effect of *odhA* Mutations on Glutamate Production

Firstly, Nakazawa et al. (1994) obtained mutant strains of *C. glutamicum* that obviously showed lower ODHC activity than the wild-type strain and investigated glutamate production by these mutants. The mutants were obtained after treatment with the mutagen *N*-methyl-*N*′-nitro-*N*-nitrosoguanidine. The obtained mutants showed 10 to 1000 times less ODHC activity than the parent strain and produced high amounts of glutamate in the presence of excess biotin.

Asakura et al. (1995) constructed *C. glutamicum* strains in which the *odhA* gene encoding the E1o subunit of ODHC was either overexpressed or deleted, and observed the effects on glutamate overproduction. ODHC activity increased by overexpressing *odhA* and the increase in ODHC activity suppressed Tween 40-triggered and penicillin-triggered glutamate overproduction. However, the effect of the increase in ODHC activity by *odhA* overexpression on Tween 40-triggered glutamate production was larger than that on penicillin-triggered glutamate production. Moreover, the *odhA* knockout strain had an ability to produce high amounts of glutamate in the presence of excess biotin. But, as the *odhA* knockout strain shows a severe growth defect, this strain is not suitable for industrial production of glutamate. To improve the growth defect of the *odhA* knockout strain, other *odhA* mutants were constructed using polymerase chain reaction-based in vitro mutagenesis (Nakamura et al. 2006). The *odhA* mutants showed a leaky phenotype with less than half of the ODHC activity of the wild type and only a small growth defect was observed. As expected, the *odhA* mutants produced high amounts of glutamate in the presence of excess biotin.

These genetic experiments established that the change in ODHC activity actually affects the glutamate production by *C. glutamicum.*

1.4.2 Role of DtsR1 for Induction of Glutamate Production

The *dtsR1* gene was cloned as the gene that suppresses the Tween 40-sensitivity of a Tween 40-sensitive mutant of *C. glutamicum* (Kimura et al. 1996, 1997, 1999). The *dtsR1* gene product shows high homology to the β subunits of some biotin-containing acyl-CoA carboxylase complexes, but does not have any biotin-binding motif. DtsR1 is thought to form a complex with AccBC, the biotin-carboxyl-carrier protein/biotin carboxylase protein, which is a subunit of acyl-CoA carboxylases (Jäger et al. 1996) required for fatty acid biosynthesis. The *dtsR2* (*accD2*) gene shows high sequence homology to *dtsR1* and is located downstream of *dtsR1*. Recently, further homologs of *dtsR1* and *dtsR2* (*accD2*), namely *accD3* and *accD4*, were identified from the genome sequence of *C. glutamicum* (Gande et al. 2004; Portevin et al. 2005). The *dtsR2*, *accD3* and *accD4* genes also encode β subunits of acyl-CoA carboxylase, and their products form complexes with AccBC and are involved in the biosynthesis of mycolic acids, fatty acids found in the cell wall of *C. glutamicum* and its related species such as *Mycobaterium* and *Rhodococcus.*

The *dtsR1*-disruptant showed auxotrophy of oleic acid, oleate ester (Tween 80) or laurate ester (Tween 20) and produces high amounts of glutamate in the presence of excess biotin (Kimura et al. 1997). Overproduction of glutamate induced by Tween 40 addition is completely suppressed, while glutamate overproduction induced by biotin limitation and penicillin addition is partially suppressed by *dtsR1* disruption (Kimura et al. 1997). The expression level of DtsR1 protein decreased by addition of Tween 40 and biotin limitation, but not by penicillin addition (Kimura et al. 1999). ODHC activity decreased by *dtsR1* gene disruption and was suppressed by overexpression of *dtsR1* gene (Kimura et al. 1999). While a link between expression levels of *dtsR1*, the level of ODHC activity, and glutamate overproduction induced by biotin limitation and Tween 40 addition is clear, the reason why penicillin-induced glutamate overproduction is inhibited by *dtsR1* overexpression still remains obscure.

Recently, the transcriptional regulator of *dtsR1* gene, DRP (*dtsR1*-regulator protein) was identified (Hirano et al. 2001; Kimura 2002). DRP, identical to NCgl0291, was proposed to have a helix-turn-helix motif conserved in cAMP receptor proteins (CRPs), which was speculated to bind a putative CRP binding sequence, TGTGA-N_6-TCACA, in the upstream region of *dtsR1*. Further investigations of the relationship between the regulation of *dtsR1* gene expression by DRP and glutamate overproduction are required.

1.4.3 Involvement of Protein Kinase G and OdhI Protein in Regulation of ODHC Enzyme Activity

Very recently, one of the regulation mechanisms of ODHC enzyme activity in *C. glutamicum* was found. Niebisch et al. (2006) found a novel protein kinase, PknG, regulating ODHC activity via the phosphorylation of OdhI protein. A *C. glutamicum pknG* mutant could not utilize glutamine as sole carbon and nitrogen source and the intracellular level of glutamate was twice as high in the Δ*pknG* mutant than in the wild type, indicating that the Δ*pknG* mutant has a defect in glutamate catabolism (Niebisch et al. 2006). Comparative proteome analyses between the wild-type, Δ*pknG* mutant and *pknG*-complemented Δ*pknG* mutant strains revealed that the in vivo substrate of PknG kinase is the protein encoded by *cg1630*, designated OdhI for ODHC inhibitory protein (Niebisch et al. 2006). *C. glutamicum* OdhI protein shows high homology with GarA of *Mycobacterium smegmatis* (Belanger and Hatfull 1999) and has a forkhead-associated domain, which binds phosphothreonine epitopes in protein and mediates phosphorylation-dependent protein–protein interaction (Pallen et al. 2002). OdhI might be phosphorylated by other kinases such as PknB (a homolog of *M. smegmatis* PknB phosphorylating GarA; Villarino et al. 2005) as well as by PknG. The defect of glutamine utilization in the Δ*pknG* mutant could be suppressed by additional mutation of Δ*odhI* (Niebisch et al. 2006). Comparison of amino acid sequences in *C. glutamicum* PknG with its homologous proteins and functional analysis of OdhI mutant proteins revealed that Thr14 in OdhI is phosphorylated by PknG kinase (Niebisch et al. 2006).

The OdhA protein (E1o subunit of ODHC) was copurified with unphosphorylated OdhI mutant protein expressed in the Δ*odhI* mutant and only a low amount of OdhA was copurified with the wild-type OdhI protein, which was predominantly phosphorylated (Niebisch et al. 2006). This suggested that unphosphorylated OdhI protein interacts with OdhA protein. Considering these results, as well as phenotypes related to the glutamine utilization in Δ*pknG* and Δ*odhI* mutants, unphosphorylated OdhI protein can function as an inhibitor of ODHC. This was confirmed by an in vitro enzyme as-

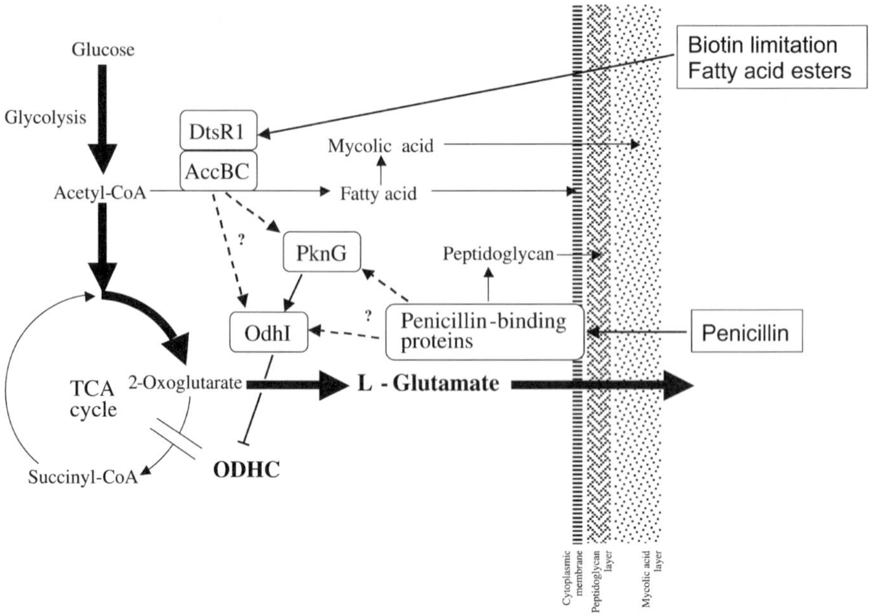

Fig. 1 Possible molecular mechanism of glutamate overproduction by *C. glutamicum*

say (Niebisch et al. 2006). Furthermore, OdhA was also copurified with SucB (E2o subunit of ODHC), Lpd [E3 subunit of ODHC and pyruvate dehydrogenase complex (PDHC)], and AceE (E1 subunit of PDHC), suggesting that mixed complexes are formed by OdhA, AceE, SucB and Lpd having ODHC and PDHC activities (Niebisch et al. 2006). Surprisingly, OdhA or AceE, copurified with DtsR1, DtsR2 and AccBC, are involved in fatty acid biosynthesis, indicating that OdhA, unphosphorylated OdhI, and DtsR1 proteins might be included in the same protein complex. Interaction of these proteins might be involved in glutamate overproduction by *C. glutamicum.*

The decrease in ODHC activity is necessary for glutamate overproduction by *C. glutamicum* as described above. The regulatory mechanism in ODHC activity by the phosphorylation state of OdhI establishes one clue for understanding the mechanism of glutamate overproduction by *C. glutamicum.* The current possible mechanism of glutamate overproduction by *C. glutamicum,* including OdhI and PknG, is shown in Fig. 1.

1.5 Analysis of Glutamate Secretion Mechanisms of *C. glutamicum*

After isolating *C. glutamicum* as a glutamate overproducing bacterium (Kinoshita et al. 1957; Udaka 1960), many researchers have attempted to understand the mechanisms of glutamate overproduction and secretion by *C. glu-*

tamicum. In the 1960s, it was thought that glutamate overproduction was achieved by passive leakage through a membrane due to alteration of the membrane state (Shiio et al. 1963; Shibukawa and Ohsawa 1966; Demain and Birnbaum 1968; Shibukawa et al. 1970). In the 1980s and 1990s, other hypotheses of glutamate secretion mechanism were proposed. Y. Clement and coworkers suggested that glutamate secretion is catalyzed by uncoupled glutamate uptake (Clement et al. 1984; Clément and Lanéele 1986). Recently, GltS, the sodium-coupled glutamate uptake system was identified (Trötschel et al. 2003). Characterization of *gltS* deletion strain revealed that GltS transporter has a significant impact on glutamate production induced by biotin limitation. Hoischen and Krämer (1989) proposed that the efflux carrier system is involved in glutamate secretion induced by biotin limitation, based on kinetic analyses. Detailed analysis showed that glutamate secretion is not coupled with the movement of protons or potassium or chloride ions, and ATP or other high energy metabolites are thought to be involved in the activity of glutamate secretion system (Gutmann et al. 1992).

1.6 Relationship between Cell Surface Structure and Glutamate Production by *C. glutamicum*

Glutamate overproduction by *C. glutamicum* is induced by biotin limitation (Shiio et al. 1962), fatty acid ester surfactant addition (Takinami et al. 1965), and penicillin addition (Nara et al. 1964), as described above. Biotin is a coenzyme for fatty acid biosynthesis enzymes, and fatty acid esters also affect the fatty acid biosynthesis. Penicillin is a β-lactam antibiotic that binds to the penicillin-binding proteins and inhibits the peptidoglycan biosynthesis. These treatments affect the cell surface integrity of *C. glutamicum*, thus, it had been thought that glutamate leaked passively through a membrane. As described above, changes in metabolic flow toward glutamate biosynthesis as well as changes in cell surface structure are among the key factors of glutamate overproduction by *C. glutamicum*. There might be some relationships between the alteration of the cell surface state and glutamate overproduction. Some researchers have attempted to analyze the mechanisms of glutamate overproduction through change in cell surface structure of *C. glutamicum*.

1.6.1 Cell Surface Structure of *C. glutamicum*

C. glutamicum belongs to the mycolic acid-containing actinomycetes, and is phylogenetically related to *Mycobacterium*, *Nocardia*, and *Rhodococcus*. Mycobacteria have highly hydrophobic, lipid-bound cell walls. The cell wall lipids mainly consist of long chain fatty acids, called mycolic acids. Mycolic

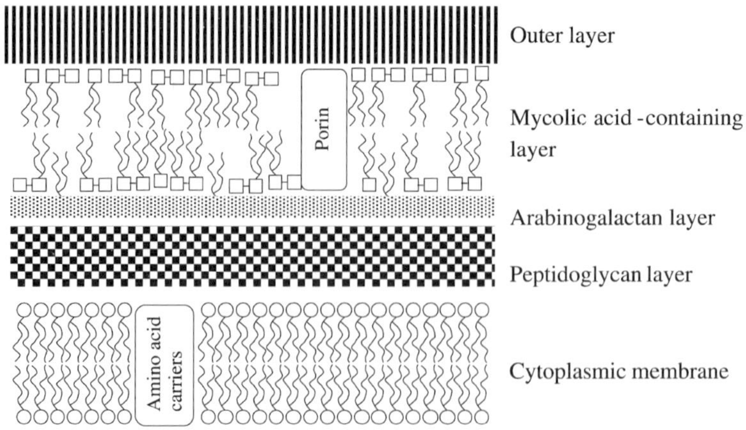

Fig. 2 Cell surface organization of *C. glutamicum*. The figure is a modification of that by Eggeling and Sahm (2001). See same reference for detailed explanation

acid is an α-alkyl-β-hydroxylated fatty acid $R_1 - CH(OH) - CH(R_2) - COOH$ (R_1 and R_2 represent alkyl chains). It is known that *C. glutamicum* has two major types of mycolic acids, $C_{32:0}$3OH and $C_{34:1}$, and other types, $C_{34:0}$, $C_{36:0}$, $C_{36:1}$, $C_{36:2}$ (Collins et al. 1982; Jang et al. 1997). The chain lengths of mycolic acids found in *C. glutamicum* are shorter than those found in other related species (50–56 carbons for *Mycobacterium* and 34–48 for *Rhodococcus*) (Barry III et al. 1998).

C. glutamicum has a similar cell wall structure and composition to mycobacteria (Fig. 2): the cytoplasmic membrane is covered with a thick peptidoglycan layer, and the arabinogalactan layer surrounds the peptidoglycan. The composition of peptidoglycan of *C. glutamicum* is the same as that of *E. coli*, consisting of β-1,4 linked *N*-acetylglucosamine and *N*-acetylmuramic acid-containing side chains of the peptide, L-alanyl-D-glutamyl-*meso*-diaminopimelyl-D-alanine (Schleifer and Kandler, 1972). The D-alanine residue in one peptide side chain is covalently linked to the carboxyl-group of amidated *meso*-diaminopimelate in the other side chain. As in mycobateria, mycolic acid is covalently bound to the terminal hexa-arabinosyl motif of arabinogalactan and forms a layer in the most exterior part. This is because the outer layer, consisting of arabinogalactan and mycolic acid, probably functions as a permeability barrier.

1.6.2
Relationship between Lipid and Fatty Acid Compositions and Glutamate Production

Soon after isolating *C. glutamicum* as a glutamate overproducing bacterium, since *C. glutamicum* showed auxotrophy of biotin (a coenzyme for the enzymes related to fatty acid biosynthesis) some researchers became interested in the relationship between membrane permeability (probably due to the changes in lipid or fatty acid composition) and glutamate overproduction by biotin limitation (Shiio et al. 1963).

In 1990, Hoischen and Krämer reported the relationship between alteration of membrane state and glutamate overproduction by *C. glutamicum* in detail. The total amount of lipids or fatty acids, as well as phospholipids was decreased and the ratio of saturated/unsaturated fatty acids (decreased level in oleic acid and increased level in palmitic acid) was changed under biotin-limiting conditions. Moreover, the total content of phospholipids was decreased, but the distribution of the phospholipids species was not changed. The time courses of the change in fatty acid or lipid composition were not correlated to the change in glutamate secretion, and specific fatty acid or phospholipids compounds involved in glutamate overproduction could not be found. It was concluded that alteration of the membrane state was required but was not sufficient for glutamate overproduction.

Recently, the effects of overexpression or deletion of genes related to lipid or fatty acid biosynthesis on glutamate overproduction by *C. glutamicum* were investigated (Nampoothiri et al. 2002). The changes in expression of the genes related to lipid or fatty acid biosynthesis caused severe alteration of phospholipid composition and temperature-sensitive growth. The alteration of phospholipid composition was obvious with overexpression of *fadD15* (encoding acyl-CoA ligase), *pgsA2* (phosphatidyl glycerophosphate synthase) and *cdsA* (CDP-diacylglycerol synthase) genes, respectively. The mutants of *cls* gene encoding caldiolipin synthase most significantly showed temperature sensitivity. Not only changes in phospholipid composition and growth phenotype but also changes in glutamate efflux were observed by changing the expression of the phospholipid or fatty acid biosynthesis genes. Overexpression of the *acp* gene encoding acyl carrier protein dramatically reduced glutamate accumulation triggered by addition of Tween 60. In contrast, Tween 60-triggered glutamate accumulation was enhanced by overexpression of *cma* (cyclopropane mycolic acid synthase), *cls*, or *plsC* (acylglycerol-3-phosphate acyltransferase). The *cls* and *cma* mutants exhibited increased Tween 60-sensitivity and accumulated less glutamate than the wild type. Furthermore, the strains overexpressing *acp*, *cls*, *pgsA2*, and *cdsA* genes, respectively, showed higher amounts of glutamate than the wild type without Tween 60 addition. However, no correlation between phospholipid composition and glutamate efflux was found.

Further investigation of the relationship between phospholipid or fatty acid biosynthesis and other factors, such as decrease in ODHC activity, involvement of glutamate carrier protein, and membrane permeability, might be necessary.

1.6.3 Studies on Glutamate Production by Lysozyme-Sensitive Mutants of *C. glutamicum*

C. glutamicum shows tolerance to the lytic enzyme lysozyme, probably because of its mycolic acid-containing outer layer of the peptidoglycan layer. Thus, the lysozyme-sensitive mutants of *C. glutamicum* are thought to have defect(s) in cell surface structure. Hirasawa et al. (2000) analyzed lysozyme-sensitive mutants of *C. glutamicum* and glutamate production by these mutants.

The *C. glutamicum* mutant strain KY9714, originally isolated as a lysozyme-sensitive mutant (Katsumata et al. 1991), cannot grow at 37 °C. The wild-type strain KY9611 showed a normal rod shape at both 30 and 37 °C. On the other hand, the KY9714 mutant strain became little fat rods or club-shaped rods even at the permissive temperature (30 °C), and they became swollen at the restrictive temperature (37 °C) (Fig. 3). These morphologies are typical of temperature-sensitive mutants of *C. glutamicum* (Kijima et al. 1998). The *ltsA* gene was cloned as the gene that complements both temperature- and lysozyme-sensitivity of the KY9714 mutant. The *ltsA* gene product is highly homologous with PurF-type glutamine-dependent asparagine synthetases (belonging to the glutamine-dependent amidotransferases) of various organisms, but the *ltsA* gene cannot complement the asparagine auxotrophy of an *E. coli asnA asnB* double mutant. The *ltsA* mutants have an ability to produce glutamate at elevated growth temperatures, as shown in Table 1 (Hirasawa et al. 2001), that is, the relationship between a defect in cell surface structure due to the *ltsA* mutations and glutamate overproduction was found. The lysozyme-sensitive phenotype and mutational

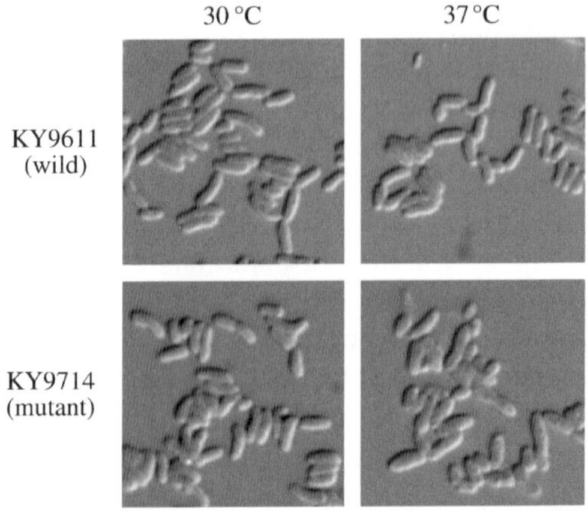

Fig. 3 Cell morphology of *C. glutamicum* wild-type and *ltsA* mutant strains at 30 °C and 37 °C. Differential interference contrast microphotographs are shown

Table 1 Glutamate production by *C. glutamicum ltsA* mutants

Stain	Glutamate production (g/L) 30 °C	35 °C	37 °C
KY9611 (wild-type)	0.07 ± 0.01	0.22 ± 0.01	0.29 ± 0.02
KY9714 (*ltsA9714*)	0.10 ± 0.01	0.82 ± 0.12	0.62 ± 0.08
KY9611 *ltsA::kan* (*ltsA::kan*)	0.47 ± 0.02	1.10 ± 0.13	2.45 ± 0.18

Culture condition and descriptions of the mutant alleles are described by Hirasawa et al. (2001)

analysis of *ltsA* mutants isolated by mutagen treatment reveals that the *ltsA* gene product is a novel glutamine-dependent amidotransferase involved in the formation of rigid cell surface structure and a defect due to the *ltsA* mutations induced glutamate overproduction by *C. glutamicum*. Glutamine-dependent amidotransferases catalyze the transfer of the amido residue of glutamine to substrate. The acceptor of amido residue of *C. glutamicum* LtsA protein has not been identified yet, but it is thought that the LtsA protein might be involved in formation of the mycolic acid-containing layer.

Recently, the LtsA protein of *Rhodococcus erythropolis*, which is related to *C. glutamicum*, was characterized by Mitani et al. (2005). The lysozyme-sensitivity of a *R. erythropolis ltsA* mutant was suppressed by introduction of *C. glutamicum ltsA* and *B. subtilis asnB* (one of the homologs of *C. glutamiucm ltsA*; Yoshida et al. 1999). Moreover, the lysozyme-sensitivity of *R. erythropolis ltsA* mutant was suppressed by addition of NH_4Cl to the culture medium. The glutaminase activity of the mutant LtsA protein inactivated by site-directed mutagenesis was also restored by NH_4Cl, suggesting that NH_3 as well as glutamine can be used as an amido donor for LtsA protein in *R. erythropolis*. However, an amido residue acceptor for LtsA protein was not identified. Further investigations of the reaction mechanism of *C. glutamicum* LtsA protein and its homologs are expected to lead to an understanding of cell surface formation and glutamate overproduction by *C. glutamicum*.

1.6.4 Role of the Structures of Mycolic Acid-Containing Layer in Glutamate Overproduction and Secretion by *C. glutamicum*

As described above, *C. glutamicum* and its related bacteria have an outer cell surface layer of peptidoglycan. This outer layer mainly consists of mycolic acids and plays a role as a permeability barrier, leading to resistance to some antibiotics.

Hashimoto et al. (2006) investigated the relationship between the formation of the mycolic acid layer and glutamate overproduction by *C. glutamicum*. The major mycolic acids of *C. glutamicum* were C_{30}, C_{32}, and C_{34} under normal growth conditions. C_{32} mycolic acid is the most abundant and forms about 70% of total mycolic acid. C_{32} mycolic acid was composed of two C_{16} fatty acids (palmitate, one of the abundant fatty acids in *C. glutamicum*). Another abundant fatty acid, oleic acid ($C_{18:1}$) was hardly found in the mycolic acid layer. The cellular content of mycolic acid decreased under biotin limitation, Tween 40 addition, penicillin treatment, and cerulenin addition (cerulenin is an antibiotic that inhibits fatty acid biosynthesis and this treatment also induces glutamate overproduction). Moreover, the content of short chain length mycolic acids increased with biotin limitation and cerulenin addition. These indicate that defects in the mycolic acid layer are caused by treatments inducing glutamate overproduction. It is thought that the mycolic acid layer might function as a permeability barrier to glutamate secretion. These results support the old hypothesis that the increase in cell surface permeability is also one possible key factor for glutamate overproduction.

Radmacher et al. (2005) also reported the relationship between mycolic acid biosynthesis and glutamate overproduction. They showed that the treatment of ethambutol, which inhibits formation of the mycolic acid layer and is one of the drugs used to treat tuberculosis, induces glutamate overproduction by *C. glutamicum*. In *M. tuberculosis*, the target of ethambutol is arabinosyltransferases encoded by *embCAB* genes (Telenti et al. 1997; Escuyer et al. 2001). *C. glutamicum* also has a homolog of *embC* gene, namely *emb*. The *emb* gene expression correlates with glutamate efflux, that is, decrease in *emb* gene expression enhances glutamate overproduction and the glutamate overproduction was suppressed by overexpression of *emb* gene. Ethambutol treatment caused less arabinan deposition in cell wall arabinogalactan and the reduction of content of cell wall-bounded mycolic acid. These results suggest that the alteration of the mycolic acid layer triggers glutamate efflux.

Further investigation is expected and necessary for understanding the mechanism of the reduction of the content of mycolic acid in the mycolic acid layer by the treatments triggering glutamate overproduction. Recently, some genes whose products are involved in mycolic acid biosynthesis were identified from the genome sequence of *C. glutamicum* (Gande et al. 2004; Portevin et al. 2005). It is expected that some relationship between the mycolic acid biosynthesis genes and glutamate overproduction will be found. Moreover, Kumagai et al. (2005) reported that fluorescent phospholipid analogs can be used as a microscopic probe for detection of the mycolic acid layer. These phospholipid analogs can detect alteration of the mycolic acid layer by ethambutol treatment. The tools might help to understand the mechanism of glutamate overproduction as well as to develop novel drugs against tuberculosis.

1.7 Genome-Wide Analysis of Glutamate Production by *C. glutamicum*

In 2003, the genome sequence of *C. glutamicum* ATCC 13032 was reported by individual two research groups in Japan and Europe (Ikeda and Nakagawa 2003; Kalinowski et al. 2003). Before completing the genome sequencing, genome-wide studies of *C. glutamicum* had been performed.

The first DNA microarray experiment of *C. glutamicum* to monitor expression of 52 genes was reported by Loos et al. (2001). Hayashi et al. (2002) also reported the transcriptome analysis of about 200 genes for central metabolism and amino acid biosynthesis in acetate metabolism. Soon after that, whole-genome DNA microarrays for *C. glutamicum* were developed enabling genome-wide gene expression analysis (Wendisch 2003; Hüser et al. 2003; Radmacher et al. 2005).

Recently, the gene expression profiles of *C. glutamicum* under glutamate overproducing conditions due to biotin limitation, Tween 40 addition, and penicillin addition using DNA microarray were reported by Kataoka et al. (2006). The expression of most of the genes related to the central metabolism (glycolysis, pentose phosphate pathway, anaplerotic pathway, and TCA cycle) were down-regulated under glutamate overproducing conditions. The expression of *odhA* and *sucB* genes encoding E1o and E2o subunits of ODHC, respectively, were also down-regulated, but the expression of the *lpd* encoding E3 subunit of ODHC was not down-regulated. Expression of *dtsR1* and *accBC* genes, encoding the subunits of acyl-CoA carboxylase and related to glutamate overproduction induced by biotin limitation and Tween 40 addition, was down-regulated in the presence of Tween 40 and penicillin. Upon biotin limitation, the *dtsR1* expression was not changed, but the *accBC* expression was severely down-regulated. The genes highly up-regulated under at least one condition triggering glutamate overproduction were *NCgl0917*, *NCgl2944*, *NCgl2945*, *NCgl2946*, and *NCgl2975*. *NCgl2944*, *NCgl2945*, and *NCgl2946* showed high homology with each other, but no proteins homologous to these gene products were found in the public database. *NCgl0917* gene product showed homology with the hypothetical protein of *C. glutamicum*, NCgl2252. NCgl2975 protein was homologous to metal-binding proteins. Transcriptome analysis was performed during temperature-triggered glutamate production by Stansen et al. (2005). Characterization of these genes and their products is one of the clues for understanding the mechanism of glutamate overproduction by *C. glutamicum*.

Other genome-wide analysis systems of *C. glutamicum*, such as proteome, metabolome and fluxome, have been also established by some researchers. For proteome analysis, the methods for 2-dimensional gel electrophoresis and protein identification by mass spectrometry for *C. glutamicum* have been developed (Hermann et al. 2000, 2001; Schaffer et al. 2001). In the proteome analysis, it is difficult to separate membrane proteins by 2-dimensional gel electrophoresis. Recently, Schlusener et al. (2005) and Fischer et al. (2006) reported that a novel method to separate and identify membrane proteins of *C. glutamicum* was established. For metabolome analysis of *C. glutamicum*, the analytical method based on gas chromatography-mass spectrometry (GC-MS) was established (Strelkov et al. 2004). In this metabolome analysis, more than 1000 metabolites were detected and 164 compounds (121 different metabolites) were identified per single experiment. The fluxome analysis, i.e., determination of intracellular metabolic fluxes in the stoichiometric metabolic network, is generally carried out based on ^{13}C-labeling experiments (Szyperski 1995; Marx et al. 1996; Christensen and Nielsen 1999; Dauner and Sauer 2000; Shirai et al. 2006). Calculation of intracellular fluxes based on ^{13}C-labeling experiments is performed using determined extracellular reaction rates and the data of ^{13}C enrichment measured by nuclear magnetic resonance spectroscopy and/or GC-MS analysis in the proteinogenic amino acids.

It has been about 50 years since *C. glutamicum* was isolated, but the clear mechanism of glutamate overproduction by *C. glutamicum* has not been found yet. For understanding the physiology and the mechanism of glutamate overproduction in *C. glutamicum*, the systems biology approach might be necessary (Wendisch et al. 2006). Integration of the quantitative data obtained by the genome-wide analysis systems is expected to encourage understanding of the mechanism of glutamate overproduction by *C. glutamicum*.

1.8 Metabolic Engineering of Glutamate Production

Metabolic engineering is a valuable tool widely used in biotechnology. Metabolic engineering was also used to improve glutamate-producing strains.

1.8.1 Introduction to Metabolic Engineering

Metabolic engineering is a methodology for a targeted improvement of metabolite formation or cellular properties through modification of specific biochemical reactions in complex metabolic networks (Bailey 1991; Stephanopoulos and Vallino 1991). To enhance the target product in bioprocesses, both genetic improvement of metabolic pathways and process operation strategies are very important. Recently, various methodologies in metabolic engineering have been developed: metabolic flux analysis based on measurements of extracellular metabolite reaction rates, ^{13}C isotope labeling and enrichment measurements of metabolites (Marx et al. 1996) and analysis of control mechanism in the complex metabolic networks [metabolic control analysis (MCA)] (Small and Kacser 1993: Stephanopoulos and Simpson

1997), and application of metabolic flux distribution analysis to operation and control of bioprocesses (Takiguchi et al. 1997)

In metabolic flux distribution analysis, metabolic fluxes are determined by intracellular and extracellular metabolite measurements. Balance equations for the intracellular and extracellular metabolites included in the metabolic reactions given above can be represented as:

$$Ar_c = r_m\,, \tag{1}$$

where r_c, r_m, and A are the calculated reaction rate vector (n-dimensional flux vector), measured reaction rates (m-dimensional accumulation reaction vector), and (m x n) matrix of stoichiometric coefficients, respectively (Vallino and Stephanopoulos 1993; Stephanopoulos et al. 1998; Shimizu 2002). The r_m involves intracellular and extracellular metabolite production rates. A pseudo-steady state approximation is used for intracellular metabolites, namely accumulation rates of intracellular metabolites is taken to be zero.

If the number of measured values is greater than the number of unknown reaction rates, the system has redundancy to be determined, and metabolic fluxes are determined by measured values using the least square methods:

$$r_c = \left[A^T A\right]^{-1} A^T r_m\,. \tag{2}$$

In this case, the consistency of the metabolic reaction model can be checked by comparison of the consistency index of the developed model, h, with χ^2 statistics value of the degree of redundancy of the system (Vallino and Stephanopoulos 1993; Stephanopoulos et al. 1998).

1.8.2
Effect of Changes in Enzyme Activities on Metabolic Fluxes during Glutamate Production

As shown in Sect. 1.2, there are some triggering operations for glutamate production by *C. glutamicum*: depletion of the biotin that is required for cell growth, addition of a detergent such as polyoxyethylene sorbitan monopalmitate (Tween 40), and addition of a β-lactam antibiotic such as penicillin. It was reported that a decrease in activity of the ODHC and enhancement of glutamate production were observed after these triggering operations (Kawahara et al. 1997; Kimura 2002). In this section, the degree of impact of changes in enzyme activities around a key branched point, i.e., 2-oxoglutarate, on a target flux of glutamate is quantitatively described

A metabolic reaction (MR) model, as shown in Fig. 4 for central carbon metabolism and glutamate synthetic pathways, was constructed. The metabolic flux distribution was analyzed in detail using the constructed MR model. In particular, the flux distribution at a key branch point, 2-oxo-

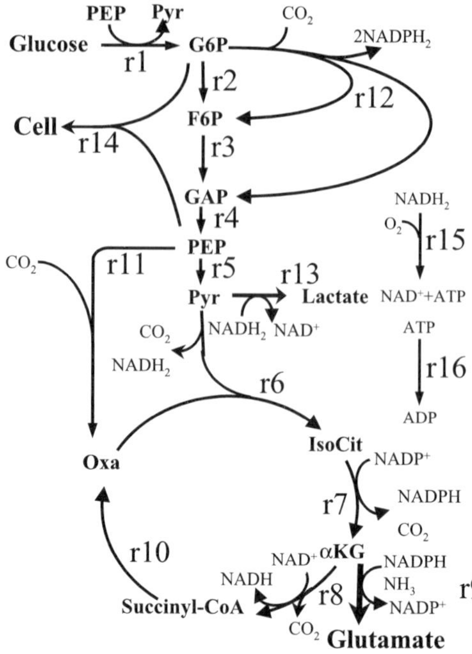

Fig. 4 Metabolic pathways of central carbon metabolism and glutamate production and metabolic reaction model

glutarate, was investigated in detail. Activities of ICDH and GDH at the branch point were changed by introducing the plasmids carrying *icd* (Eikmanns et al. 1995) and *gdh* (Bormann et al. 1992) genes, respectively, into the parent strain. ODHC activity was attenuated by biotin depletion. The mole flux distributions in these strains were calculated using the MR model, and the effects of the changes in the enzyme activities on the mole flux distributions were compared. Sensitivity of the glutamate production against change in enzyme activity was evaluated.

C. glutamicum AJ1511 (ATCC 13869) was used as a glutamate-producing wild-type. *C. glutamicum* AJ13678 and AJ13679 are genetically engineered strains that harbor plasmids with *gdh* and *icd* genes, respectively.

1.8.3 Metabolic Flux Distribution Analysis at the 2-Oxoglutarate Branch-Point in Perturbation Experiments

The metabolic flux distribution was analyzed in *C. glutamicum* AJ1511, AJ13678, and AJ13679. The flux distribution at the 2-oxoglutarate branch point is shown in Fig. 5. It was shown that enhancement of ICDH and GDH activities did not significantly affect the flux distribution at the 2-oxoglutarate

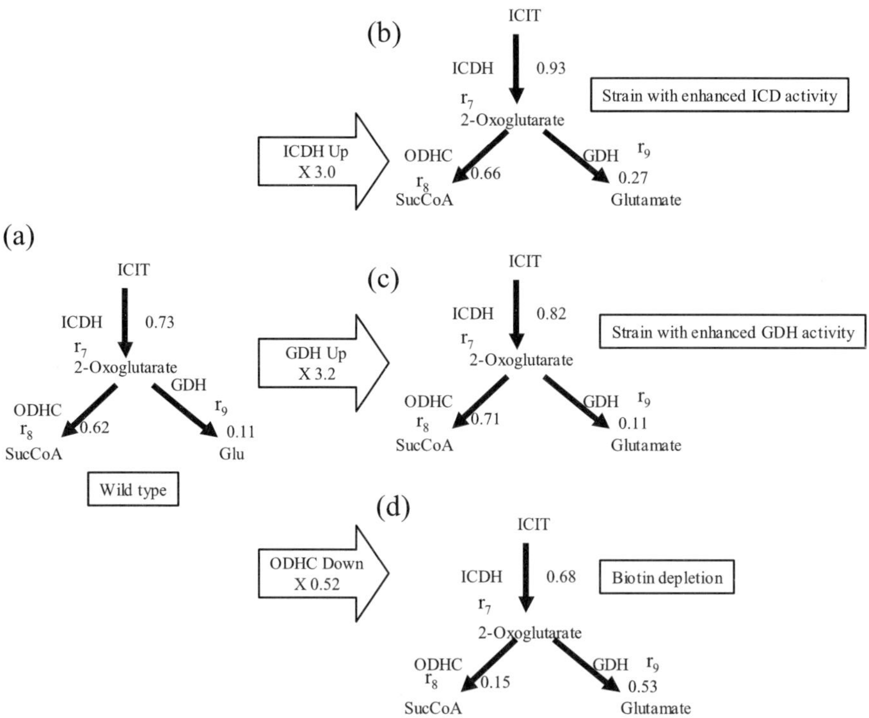

Fig. 5 Comparison of flux distribution at 2-oxoglutarate branch point in glutamate production. Flux distribution of wild-type strain (**a**) was perturbed by enhancement of *icd* gene (**b**), enhancement of *gdh* gene (**c**), and attenuation of ODHC by biotin depletion (**d**)

branch point. Even though ICDH and GDH activities were enhanced 3.0- and 3.2-fold, respectively, more than 70% carbon flux still flows into the TCA cycle. On the other hand, when ODHC activity decreased to about 52% after depletion of biotin, marked changes in fluxes of GDH and ODHC were observed. More than 75% carbon was used for glutamate production; this result shows that ODHC has the most significant responsibility for glutamate production quantitatively by metabolic engineering (Shimizu et al. 2003; Heijnen et al. 2004).

1.8.4 Comparative Study of Flux Redistribution of Metabolic Pathways in Glutamate Production by *C. glutamicum* and *C. efficiens*

C. efficiens can grow at a higher temperature than *C. glutamicum*, consequently, refrigeration to eliminate the fermentation heat is not necessary for its culture, and therefore, it is expected that the production cost would be less during industrial fermentation (Fudou et al. 2002). The complete *C. efficiens* genome was also sequenced, and its thermostability has been discussed by comparing the amino acid sequences of the enzymes of *C. efficiens* with those of *C. glutamicum* having the same function (Nishio et al. 2003). In this section, quantitative investigations of the difference in the amount of glutamate between *C. glutamicum* and *C. efficiens* are described under the conditions of varying fermentation temperature, biotin depletion, and Tween 40 addition by metabolic flux analysis. The flow control mechanism at the branch point is also mentioned by comparing flux distribution analysis results with those of enzyme activity measurements (Shirai et al. 2005).

Fermentations of *C. glutamicum* and *C. efficiens* were performed under biotin depletion condition at 31.5 °C and 37 °C, respectively. It was clear that glutamate production by *C. glutamicum* was higher at 31.5 °C than that at 37 °C. Cell growth and glutamate production of *C. glutamicum* at 37 °C were lower than those at 31.5 °C. However, the growth and glutamate production patterns of *C. efficiens* were almost the same at both temperatures. From these results, it is clear that *C. efficiens* is more tolerant than *C. glutamicum* to high temperature. However, the glutamate production by *C. glutamicum* was threefold larger than that by *C. efficiens* at both temperatures.

Using the MR model, metabolic flux was determined under conditions of biotin depletion and Tween 40 addition in *C. glutamicum* and *C. efficiens*. The flux distribution at the key 2-oxoglutarate branch point was investigated in detail, as shown in Fig. 6, before and after the triggering effect occurred. With both Tween 40 addition and biotin depletion, the flux distribution ratio from 2-oxoglutarate to succinyl-CoA in *C. glutamicum* markedly decreased and the flux to glutamate increased, as shown in Figs. 6a and 6c. The flux distribution ratio from 2-oxoglutarate to succinyl-CoA in *C. efficiens* also decreased and the flux to glutamate increased. However, the degree of flux redistribution of *C. efficiens* was much smaller than that of *C. glutamicum*, as shown in Figs. 6b and 6d.

Figure 7 shows time courses of the specific activities of ICDH and GDH. ICDH and GDH activities were almost constant in both *C. glutamicum* and *C. efficiens* or under both triggering operations (Tween 40 addition and biotin depletion). However, as shown in Fig. 8, the specific activity of ODHC significantly changed in both coryneform bacteria when both triggering effects were added. The specific activity of ODHC was significantly attenuated and glutamate was overproduced. The degree of the change in the ODHC activity in *C. efficiens* was smaller than that in *C. glutamicum*. In two strains, the increase in the flux distribution ratio from 2-oxoglutarate to glutamate showed positive correlation with the decrease in the specific activity of ODHC in both conditions, as shown in Fig. 9. It was concluded that the decrease in the ODHC activity is the main and direct factor behind the flux redistribution of 2-oxoglutarate in both species. The difference in glutamate production between both species was due to the difference in the decrease in the ODHC activity after the triggering operations.

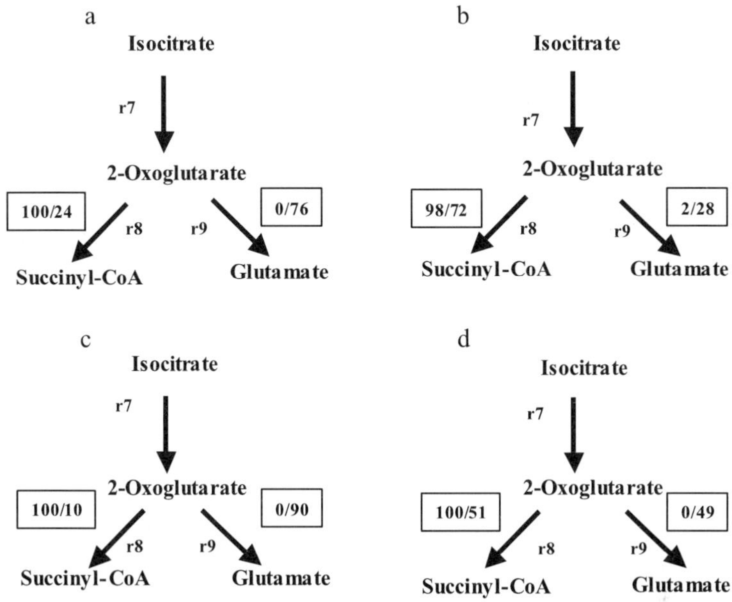

Fig. 6 Flux distribution ratios at α-KG branch point in *C. glutamicum* (**a** and **c**) and *C. efficiens* (**b** and **d**) under two conditions: addition of Tween 40 (**a** and **b**) and depletion of biotin (**c** and **d**). Numbers in the *left-hand boxes* indicate values of the flux distributions before glutamate production, and those on the *right* indicate values after production

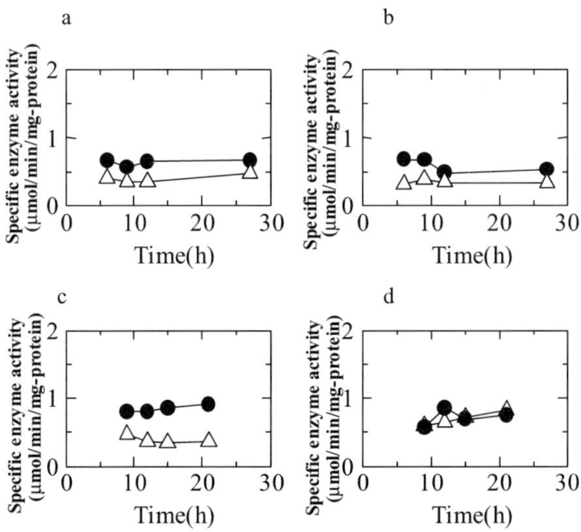

Fig. 7 Time courses of ICDH and GDH specific activities of *C. glutamicum* (**a** and **c**) and *C. efficiens* (**b** and **d**) under two conditions: addition of Tween 40 (**a** and **b**) and depletion of biotin (**c** and **d**). *Open triangles* ICDH specific activity, *closed circles* GDH specific activity

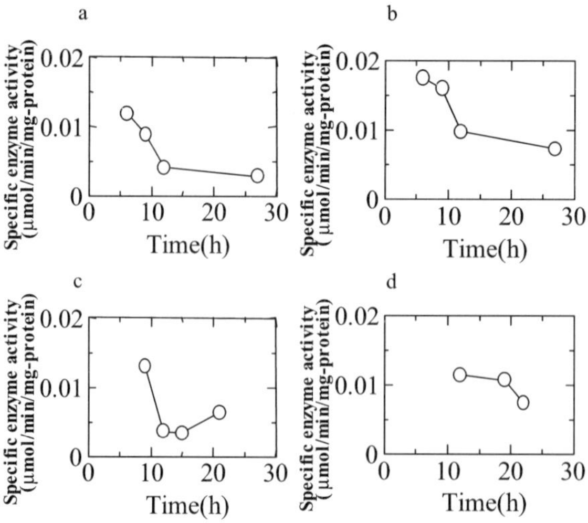

Fig. 8 Time courses of ODHC specific activity of *C. glutamicum* (**a** and **c**) and *C. efficiens* (**b** and **d**) under two conditions: addition of Tween 40 (**a** and **b**) and depletion of biotin (**c** and **d**)

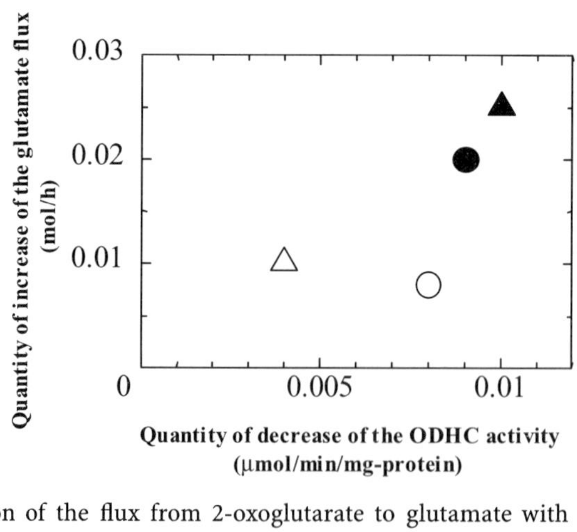

Fig. 9 Correlation of the flux from 2-oxoglutarate to glutamate with the ODHC activity. *Closed circle* and *open circle*, Tween 40 addition in *C. glutamicum* and *C. efficiens*, respectively. *Closed triangle* and *open triangle*, biotin depletion in *C. glutamicum* and *C. efficiens*, respectively

Furthermore, to clarify the enzyme-to-substrate affinity, the Michaelis–Menten constants (K_m) of the enzymes around 2-oxoglutarate: ICDH, GDH, and ODHC were determined on the basis of Lineweaver–Burk plots. Results are summarized in Table 2. In both coryneform bacteria, the K_m values of GDH were 50- to 100-fold higher than those of ICDH and ODHC. Therefore,

Table 2 Comparison of K_m values (mM) of enzymes around 2-oxoglutarate

Enzyme	*C. glutamicum*	*C. efficiens*
ICDH	0.03 (0.99)	0.03 (0.99)
ODHC	0.08 (0.98)	0.03 (0.99)
GDH	2 (0.99)	5 (0.99)

K_m of ICDH for isocitrate and those of ODHC and GDH for 2-oxoglutarate are shown. Bracket values for each enzyme indicate sum of residual square (r^2 values) for Lineweaver–Burk plot

the affinity of 2-oxoglutarate for GDH was much lower than that for ODHC. This is the reason why glutamate synthetic flux was not observed before the decrease in the specific activity of ODHC, even though that of GDH was determined to be sufficient. After the decrease in the specific activity of ODHC, the flux to glutamate occurred due to the increase in 2-oxoglutarate levels.

The glutamate production by *C. glutamicum* was higher than that by *C. efficiens*, even though the growth and glutamate production by *C. glutamicum* was lower at high temperatures. In both strains, the decrease in the ODHC specific activity is the one of the main factors. Since the relationship between the decrease in the ODHC specific activity and the change in flux was the same for both strains, an improvement in glutamate production is expected with the decrease in ODHC activity. Without the triggering effects, almost

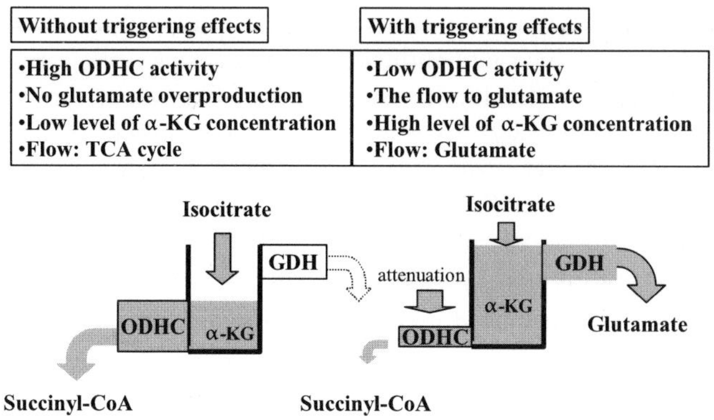

Fig. 10 Scheme of flux redistribution at 2-oxogluratate (α-KG) branch point in coryneform bacteria. Without the triggering effects, almost all 2-oxoglutarate was converted to succinyl-CoA catalyzed by ODHC and, consequently, glutamate was not overproduced. Although sufficient GDH-specific activity was observed, the flux catalyzed by GDH was very small because the K_m of GDH was much higher than that of ODHC. Once the ODHC specific activity was decreased after the triggering effects, 2-oxoglutarate accumulated and, consequently, glutamate was overproduced catalyzed by GDH

all 2-oxoglutarate was converted to succinyl-CoA as catalyzed by ODHC, and consequently, glutamate was not overproduced by either coryneform bacterium. Although sufficient GDH specific activity was observed, the flux catalyzed by GDH was very small because the K_m of GDH was much higher than that of ODHC. Once the ODHC specific activity was decreased after the triggering effects, 2-oxoglutarate accumulated, and consequently, glutamate was overproduced as catalyzed by GDH. Figure 10 summarizes the mechanism of flux redistribution at the 2-oxoglutarate branch point.

1.8.5 Studies on the Improvements of Glutamate Overproduction by Metabolic Engineering

Several studies have also been reported on improvement of productivity of glutamate. One of the important issues is the anaplerotic pathway of the PEP–pyruvate–oxaloacetate node (Sauer and Eikmanns 2005) because glutamate is synthesized from 2-oxoglutarate (a member of TCA cycle) and carbon is released as CO_2 in the TCA cycle. Peters-Wendisch et al. (2001) characterized the importance of the anaplerotic pathways of *C. glutamicum*. *C. glutamicum* possesses both phosphoenolpyruvate carboxylase (PPC) and pyruvate carboxylase (PYC) as anaplerotic enzymes for growth. To analyze the significance of PYC for the glutamate production by this organism, the wild-type *pyc* gene, encoding PYC, was used for the construction of defined *pyc*-inactive and *pyc*-overexpressing strains, and the synthesis of glutamate and other amino acids was tested. In a detergent-dependent glutamate production assay, the *pyc*-overexpressing strain showed more than sevenfold higher glutamate production, and the *pyc*-deficient strain about twofold lower, than the wild-type under the detergent-inducing conditions. As other improvement methods for glutamate production, a mutant of *C. glutamicum* ATCC 14067 with reduced H^+-ATPase activity was also reported (Aoki et al. 2005). A novel strategy of simultaneous production of lysine and glutamate was also reported (Shiratsuchi et al. 1995). Glutamate production was noted not to differ with the carbon source, when glutamate production from glucose, fructose, and sucrose were compared (Georgi et al. 2005).

2 Short Review of Glutamine and Proline Production

2.1 Glutamine

Glutamine is synthesized from glutamate by a one-step reaction catalyzed by glutamine synthetase and plays an important role in the regulation of nitrogen metabolism, especially the synthesis of amino acids, proteins, nucleic acids, and cell wall components. Glutamate and glutamine are key nitrogen or amino residue donors for amino acid synthesis.

In general, glutamate can be synthesized by two different pathways. Glutamate dehydrogenase (GDH) catalyzes the reductive amination of 2-oxoglutarate with concomitant oxidation of NADPH and is responsible for glutamate formation in the presence of a high concentration of ammonia (> 1 mM) (Ertan 1992a,b; Börmann-El Kholy et al. 1993). When the ammonia concentration is low, however, glutamate is synthesized by the coupled reactions of glutamine synthetase (GS) and glutamate synthase (glutamine: 2-oxoglutarate aminotransferase; GOGAT). Most of the studies on glutamine synthesis are concerned with ammonia uptake under conditions of low ammonia concentration. Burkovski (2003) reviewed mechanisms of nitrogen control with respect to ammonia uptake, its assimilation, and glutamate/glutamine metabolism in coryneform bacteria.

The *glnA* gene encoding glutamine synthetase I (GSI) and the *gltBD* genes encoding large and small subunits of GOGAT of *C. glutamicum* were isolated and characterized (Jakoby et al. 1997; Beckers et al. 2001). Under nitrogen starvation conditions, expression of *glnA* gene and *gltBD* operon occurred (Schulz et al. 2001). Deletion of *glnA* in *C. glutamicum* led to a glutamine auxotrophic phenotype and complete loss of glutamine synthetase activity. From this result, it seems that GSI has an important role in the synthesis of glutamine. Using in vivo flux analysis, the ammonium assimilatory pathways of *C. glutamicum* were studied (Tesch et al. 1999): unexpectedly, 28% of the $NH4^+$ was assimilated via the GS reaction in glutamine under high nitrogen surplus conditions, while 72% were assimilated by the reversible GDH reaction via glutamate. No GOGAT activity was found.

Signal transduction of nitrogen regulation via glutamine in *C. glutamicum* has been also studied. In general, GS activity is controlled by its adenylylation status. The adenylyltransferase for GS forms a complex with P_{II} protein and the activity of adenylyltransferase is regulated via the uridylylation of P_{II}. The complex of uridylylated P_{II} and adenylyltransferase functions as a GS-deadenylylating enzyme to change to active form of GS, while the complex of unuridylylated P_{II} and adenylyltransferase can adenylylate GS to decrease its activity. Sequence analysis revealed that GS of *C. glutamicum* also has a putative adenylation site, Tyr405 (Jakoby et al. 1997). The nitrogen-regulated signal transduction cascade of *C. glutamicum*, including uridilyltransferase encoded by *glnD*, P_{II}-type protein GlnK, and adenyltransferase encoded by *glnE*, is assumed (Nolden et al. 2001a,b). However, adenylyltransferase works independently on GlnK protein in *C. glutamicum*: the *glnK* or *glnD* deletion mutant cannot express the genes regulated by a transcriptional regulator AmtR in response to nitrogen depletion. The *glnD* gene expression is upregulated upon nitrogen limitation. The quantification of intracellular ammonium concentration showed a correlation between the decrease in the intracellular ammonium level responding nitrogen depletion and the transcriptional response. Recently, nitrogen-regulation via adenylylation or deadenylylation of GlnK protein was studied (Strösser et al. 2004). Under nitrogen starvation conditions, the GlnK protein is adenylated by the *glnD* gene product and deadenylylated responding to nitrogen supply. Intracellular localization of GlnK protein is changed depending on its modification status. Under low nitrogen condition, the adenylated GlnK protein localizes in the cytoplasm and regulates the AmtR-dependent gene expression. On the other hand, the deadenylated GlnK protein localizes to the cytoplasmic membrane under nitrogen-sufficient conditions and binds to the ammonium transporter AmtB, assuming that the unmodified GlnK protein interferes with the uptake of ammonium via AmtB. The binding of GlnK to AmtB leads to the interaction with FtsH protein, marking GlnK for proeolytic degradation by the Clp protease complex.

In *C. glutamicum*, a global repressor protein AmtR regulates the nitrogen depletion-dependent gene expression (Jakoby et al. 2000). Deletion of the *amtR* gene leads to the constitutive expression of the *amt* gene encoding the ammonium uptake system. AmtR regulates the gene expression of the *amtB-glnK-glnD* operon encoding another ammonium transporter AmtB, P_{II}-type protein GlnK and uridylyltransferase, *gltBD* encoding GOGAT and *glnA* encoding GS.

Yamada et al. (1979) reported glutamine formation by a penicillin-resistant mutant of *Flavobacterium rigense*, and the fermentation conditions using this microorganism were established by Nabe et al. (1980). Glutamine-producing mutants were selected from wild-type glutamate-producing coryneform bacteria. A sulfaguanidine-resistant mutant of *C. glutamicum* accumulates more than 40 g/L glutamine in 48 h from 100 g/L glucose (Tsuchida et al. 1987) Production yield with respect to glucose is greater than 40% (Yoshihara et al. 1992). A coupling method for enzymatic glutamine production by a bacterial GS reaction with energy transfer by baker's yeast or a cell-free extract was also developed (Tachiki et al. 1981a,b, 1983; Wakisaka et al. 1998).

2.2 Proline

In some bacteria, proline is synthesized in four reactions from glutamate via γ-glutamyl phosphate, γ-glutamic semialdehyde, and its spontaneous cyclization product Δ^1-pyrroline-5-carboxylate (Fig. 11). The enzymes involved in proline biosynthesis and their genes have been characterized. For example, in *E. coli* (for review, see Leisinger 1996), the gene product of *proB* (γ-glutamyl kinase) that forms a complex with the *proA* gene product (γ-glutamyl phosphate reductase) catalyzes the conversion of glutamate to γ-glutamyl phosphate. γ-Glutamyl phosphate is reduced by the product of *proA* to γ-glutamic semialdehyde, which is converted nonenzymatically

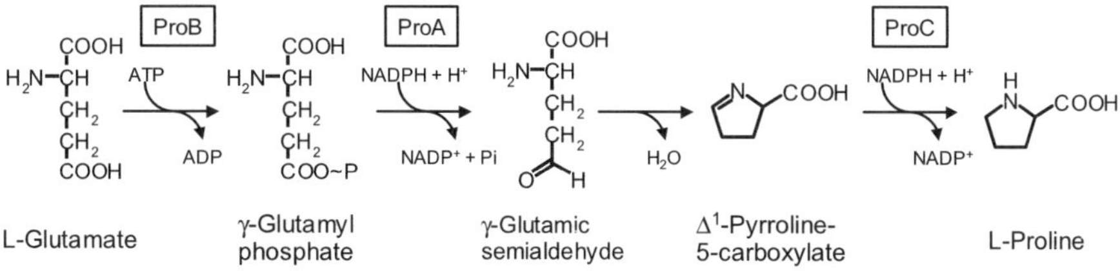

Fig. 11 Biosynthetic pathway of proline in typical bacteria

to Δ^1-pyrroline-5-carboxylate. Proline is synthesized from Δ^1-pyrroline-5-carboxylate by the *proC* gene product.

In 1966, Yoshinaga et al. first reported the direct fermentation of proline using an isoleucine auxotrophic mutant of *C. glutamicum*. The productivity of proline by this strain was improved by conferring sulfaguanidine resistance to increase the intracellular level of ATP (Tsuchida et al. 1986). Kato et al. (1968) reported that a serine auxotrophic mutant of *Kurthia catenaforma* produced high amounts of proline.

Proline biosynthesis is strictly controlled by proline-mediated feedback inhibition. Proline inhibits the reaction between glutamate and γ-glutamic semialdehyde. It is thought that escaping the proline-mediated feedback inhibition is one of the strategies for overproduction of proline. Proline analogs, such as azetidine-2-carboxylate, 3,4-dehydroproline, and thiazoline-4-carboxylate, inhibit γ-glutamyl kinase but support protein synthesis. Resistance to these analogs is based either on a deficiency in proline uptake or on proline overproduction due to (a) defect(s) in pathway regulation. Mutants carrying resistance of the latter type are thought of as proline overproducers, and the resistance is thought to be closely related to the mutations in the *proB* gene encoding γ-glutamyl kinase.

Baich and Pierson (1965) reported that an *E. coli* mutant showing resistance to the proline analog 3,4-dehydroproline excreted proline into the culture medium. Nakamori et al. (1982) also reported that the 3,4-dehydroproline- and sulfaguanidine-resistant mutant strains of *C. glutamicum* produced high amounts of proline. Sano et al. (1987) reported that the proline productivity of dehydroproline-resistant strains of *C. glutamicum* was improved by overexpression of the phosphoenolpyruvate carboxylase gene to supply oxaloacetate. Sugiura et al. (1985) obtained proline analog-resistant mutants of *Serratia marcescens* harboring mutations in the *putA* gene encoding proline oxidase, which is involved in proline degradation. Hyperproduction of proline was achieved by mutants showing 3,4-dehydroproline, thiazoline-4-carboxylate and azetidine-2-carboxylate resistance (Sugiura et al. 1985). Proline overproduction via the arginine biosynthetic pathway was also achieved by *S. marcescens* strains carrying the mutations of derepression of arginine biosynthesis, resistance to feedback inhibition of *N*-acetylglutamate synthase, and the defect in *N*-acetylornithine aminotransferase as well as proline analog resistance. However, the production of by-products, such as *N*-acetylglutamate-γ-semialdehyde and arginine was observed (Sugiura et al. 1986). Omori et al. (1991, 1992) cloned the wild-type *proBA* operon of *S. marcescens* and characterized this operon isolated from proline analog-resistant mutants. The γ-glutamyl kinase in the mutants was no longer susceptible to feedback inhibition by proline. DNA sequence analysis revealed that a proline hyperproducing mutant of *S. marcescens* showing resistance to three proline analogs had two base substitutions as well as *putA* mutation and an unknown mutation related to the increased level of glutamate dehydrogenase (Sugiura et al. 1985); one base exchange in *proB* rendering γ-glutamyl kinase 700-fold less sensitive to proline-mediated feedback inhibition than the wild-type enzyme; and one base exchange in the spacer region of the *proBA* promoter, resulting in the fourfold increase in the transcriptional activity of the *proBA* operon (Omori et al. 1992). Kosuge and Hoshino (1998) isolated the dehydroproline-resistant and proline degradation-deficient mutant of the extremely thermophilic bacterium *Thermus thermophilus* by site-directed mutagenesis of the *proB* gene and UV mutagenesis. This mutant strain overproduced proline at high temperatures (70 °C). However, proline productivity of the *T. thermophilus* mutant was lower than that of the *C. glutamicum* and *S. marcescens* mutants.

In bacteria, proline can accumulate intracellularly by increased net synthesis or by enhanced uptake from the medium under high osmotic pressure condition (reviewed in Csonka, 1989). Some Gram-positive bacteria are able to increase the intracellular level of proline under high osmotic pressure condition without exogenous proline, suggesting that the synthesis or degradation of proline in these bacteria is controlled by change in osmotic pressure. On the other hand, proteobacteria generally achieve high intracellular accumulation of proline during osmotic stress by enhancing proline transport. However, it was reported that the proline-overproducing strains of *E. coli*, *Salmonella typhimurium*, and *S. marcescens* exhibited tolerance to osmotic stress (Csonka 1981; Mahan and Csonka 1983; Smith 1985; Jakowec et al. 1985; Sugiura and Kisumi 1985; Csonka et al. 1988).

C. glutamicum also accumulates proline as a compatible solute under high osmotic pressure condition (Kawahara et al. 1989; Guilouet and Engasser 1995). From 1995 to 1996, the proline biosynthetic pathway of *C. glutamicum* was characterized (Serebrijski et al. 1995: Ankri et al. 1996). The *proA*, *proB*, and *proC* genes of *C. glutamicum* were identified by complementation analysis of proline auxotrophy of the *E. coli proA*, *proB*, and *proC* mutants, respectively, but the complementation of *E. coli proA* mutation by the *C. glutamicum proA* gene was optimal at high osmolarity. The *proB* and *proA* genes are separated by an unknown gene *unk* (Ankri et al. 1996), while the *proB* and *proA* genes comprise the same operon structure as in almost all bacteria. The null *proA* mutant of *C. glutamicum* can grow slowly under proline depletion conditions, suggesting that *C. glutamicum* has a ProA-independent bypass of conversion of γ-glutamyl phosphate to γ-glutamic semialdehyde. *C. glutamicum* and *E. coli proA* mutants were suppressed by a plasmid carrying *C. glutamicum asd* gene encoding aspartate β-semialdehyde dehydrogenase, indicating that aspartate β-semialdehyde dehydrogenase might play a role in this bypass. The osmotic regulation of gene expression and enzyme activity in proline biosynthesis of *C. glutamicum* has not been reported yet. However, the enzyme activity of ProC, but not ProB, was elevated threefold under conditions of high osmotic pressure (Kawahara et al. 1989).

Varela et al. (2003) reported the metabolic flux redistribution responding to a gradual increase in osmolarity in lysine-overproducing *C. glutamicum*. During increasing osmolarity, trehalose is first accumulated and then proline is produced at higher osmolarity. The fluxes for glycolysis and the TCA cycle increase over saline gradient, suggesting that the increase in net ATP production rate might support the higher energy requirement for cellular maintenance. On the other hand, the flux for the pentose phosphate pathway remains constant, because the reducing power (NADPH) is needed for production of proline, valine, and glutamate at higher osmolarity as well as for lysine production at lower osmolarity. The oxaloacetate node is initially rigid at lower osmolarity, but the TCA/aspartate ratio increases at higher osmolarity because of the high cellular demand for proline.

C. glutamicum normally prefers uptake of compatible solutes to their biosynthesis under high osmotic pressure conditions. *C. glutamicum* has five transporters for uptake of compatible solutes, PutP, BetP, ProP, EctP, and LcoP (Peter et al. 1997, 1998; Steger et al. 2004). Recently, DNA microarray analysis revealed that the MtrA-MtrB two-component signal transduction system of *C. glutamicum* is involved in the regulation of gene expression of the transpoter genes *betP*, *proP*, and *lcoP* (Möker et al. 2004). Among these transporters, PutP, ProP, and EctP are responsible for proline uptake. PutP is not osmoregulated in its activity and is involved in proline uptake for proline utilization (Peter et al. 1997). ProP and EctP are regulated in their activity according to the change in external osmotic pressure (Peter et al. 1998).

References

Ajinomoto (2006) Fact sheet: Feed-use amino acids business http://www.ajinomoto.com/ar/i_r/pdf/presentation/FY2005
data.pdf

Ankri S, Serebrijski I, Reyes O, Leblon G (1996) Mutations in the *Corynebacterium glutamicum* proline biosynthetic pathway: a natural bypass of the *proA* step. J Bacteriol 178:4412–4419

Aoki R, Wada M, Takesue N, Tanaka K, Yokota A (2005) Enhanced glutamic acid production by a H^+-ATPase-defective mutant of *Corynebacterium glutamicum*. Biosci Biochem Biotech 69:1466–1472

Asakura Y, Kimura E, Abe C, Kawahara Y, Nakamatsu T, Usuda Y, Tsujimoto Y, Kurahashi O (1995) Japan Patent H08-501926 (International Patent WO95/34672)

Baich A, Pierson DJ (1965) Control of proline synthesis in *Escherichia coli.* Biochim Biophys Acta 104:397–404

Bailey JE (1991) Towards a science of metabolic engineering. Science 252:1668–1675

Barry III CE, Lee RE, Mdlili K, Samson AE, Schroeder BG, Slayden RA, Yuan Y (1998) Mycolic acids: structure, biosynthesis and physiological functions. Prog Lipid Res 37:143–179

Beckers G, Nolden L, Burkovski A (2001) Glutamate synthase of Corynebacterium glutamicum is not essential for glutamate synthesis and is regulated by the nitrogen status. Microbiology 147:2961–2970

Belanger AE, Hatfull GF (1999) Exponential-phase glycogen recycling is essential for growth of *Mycobacterium smegmatis.* J Bacteriol 181:6670–6678

Bormann ER, Eikmanns EJ, Sahm H (1992) Molecular analysis of the *Corynebacterium glutamicum gdh* gene encoding glutamate dehydrogenase. Mol Microbiol 6:317–326

Börmann-El Kholy ER, Eikmanns BJ, Gutmann M, Sahm H (1993) Glutamate dehydrogenase is not essential for glutamate formation by *Corynebacterium glutamicum.* Appl Environ Microbiol 59:2329–2331

Burkovski A (2003) Ammonium assimilation and nitrogen control in *Corynebacterium glutamicum* and its relatives: an example for new regulatory mechanisms in actinomycetes. FEMS Microbiol Rev 27:617–628

Christensen B, Nielsen J (1999) Isotopomer analysis using GC-MS. Metab Eng 4:282–290

Clement Y, Escoffier B, Trombe MA, Lanéele G (1984) Is glutamate excreted by its uptake system in *Corynebacterium glutamicum*? A working hypothesis. J Gen Microbiol 130:2589–2594

Clément Y, Lanéele G (1986) Glutamate excretion mechanism in *Corynebacterium glutamicum*: triggering by biotin starvetion or by surfactant addition. J Gen Microbiol 132:925–929

Collins MD, Goodfellow M, Minnikin DE (1982) A survey of the structure of mycolic acids in *Corynebacterium* and related taxa. J Gen Microbiol 128:129–149

Csonka LN (1981) Proline over-production results in enhanced osmotolerance in *Salomonella typhimurium.* Mol Gen Genet 182:82–86

Csonka LN (1989) Physiological and genetic responses of bacteria to osmotic stress. Microbiol Rev 53:121–147

Csonka LN, Gelvin SB, Goodner BW, Orser CS, Siemieniak D, Slightom JL (1988) Nucleotide seqence of a mutation in the *proB* gene of *Escherichia coli* that confers proline overproduction and enhanced tolerance to osmotic stress. Gene 64:199–205

Dauner M, Sauer U (2000) GC-MS analysis of amino acids rapidly provides rich information for isotomoper balancing. Biotechnol Prog 16:642–649

Delaunay S, Gourdon O, Lapujade P, Mailly E, Oriol E, Engasser JM, Lindley ND, Goergen J-L (1999) An improved temperature-triggered process for glutamate production with *Corynebacterium glutamicum.* Enzyme Microbial Technol 25:762–768

Delaunay S, Lapujade P, Engasser JM, Goergen JL (2002) Flexibility of the metabolism of *Corynebacterium glutamicum* 2262, a glutamic acid-producing bacterium, in response to temperature upshocks. J Ind Microbiol Biotechnol 28:333–337

Demain AL, Birnbaum J (1968) Alternation of permeability for the release of metabolites from the microbial cell. Curr Top Microbiol Immunol 46:1–25

Eggeling L, Sham H (1999) L-Glutamate and L-lysine: traditional products with impetuous development. Appl Microbiol Biotechnol 52:146–153

Eggeling J, Sahm H (2001) The cell wall barrier of *Corynebacterium glutamicum* and amino acid efflux. J Biosci Bioeng 92:201–213

Eikmanns BJ, Rittmann D, Sahm H (1995) Cloning, sequence analysis, expression, and inactivation of the *Corynebacterium glutamicum icd* gene encoding isocitrate dehydrogenase and biochemical characterization of the enzyme. J Bacteriol 177:774 -782

Ertan H (1992a) Some properties of glutamate dehydrogenase, glutamine synthetase and glutamate synthase from *Corynebacerium callunae*. Arch Microbiol 158:35–41.

Ertan H (1992b) The effect of the various culture conditions on the levels of ammonia assimilatory enzymes of *Corynebacerium callunae*. Arch Microbiol 158:42–47

Escuyer VE, Lety MA, Torrelles JB, Khoo KH, Tang JB, Rithner CD, Frehel C, McNeil MR, Brennan PJ, Chatterjee D (2001) The role of the *embA* and *embB* gene products in the biosynthesis of the terminal hexaarabinofuranosyl motif of *Mycobacterium smegmatis* arabinogalactan. J Biol Chem 276:48854–48862

Fischer F, Wolters D, Rögner M, Poetsch A (2006) Toward the complete membrane proteome. High coverage of integral membrane proteins through transmembrane peptide detection. Mol Cell Proteomics 5:444–453

Fudou R, Jojima Y, Seto A, Yamada F, Kimura E, Nakamatsu T, Hiroishi A, Yamanaka S (2002) *Corynebacterium efficiens* sp. nov., a glutamic acid-producing species from soil and vegetables. Int J Syst Evol Microbiol 52:1127–1131

Gande R, Gibson KJ, Brown AK, Krumbach K, Dover LG, Sahm H, Shioyama S, Oikawa T, Besra GS, Eggeling L (2004) Acyl-CoA carboxylases (*accD2* and *accD3*), together with a unique polyketide synthase (*Cg-pks*), are key to mycolic acid biosynthesis in *Corynebacterianeae* such as *Corynebacterium glutamicum* and *Mycobacterium tuberculosis*. J Biol Chem 279:44847–44857

Georgi T, Rittman D, Wendisch VF (2005) Lysine and glutamate production of *Corynebacterium glutamicum* on glucose, fructose, and sucrose: roles of malic enzyme and fructose-1,6-bisphosphate. Metab Eng 7:291–301

Guillouet S, Engasser JM (1995) Sodium and proline accumulation in *Corynebacterium glutamicum* as a response to an osmotic saline upshock. Appl Microbiol Biotechnol 43:315–320

Gutmann M, Hoischen C, Kramer R (1992) Carrier-mediated glutamate secretion by *Corynebacterium glutamicum* under biotin limitation. Biochim Biophys Acta 1112:115–123

Hashimoto K, Kawasaki H, Akazawa K, Nakamura J, Asakura Y, Kudo T, Sakuradani E, Shimizu S, Nakamatsu T (2006) Changes in composition and content of mycolic acids in glutamate-overproducing *Corynebacterium glutamicum*. Biosci Biotechnol Biochem 70:22–30

Hayashi M, Mizoguchi H, Shiraishi N, Obayashi M, Nakagawa S, Imai J, Watanabe S, Ota T, Ikeda M (2002) Transcriptome analysis of acetate metabolism in *Corynebacterium glutamicum* using a newly developed metabolic array. Biosci Biotechnol Biochem 66:1337–1344

Heijnen JJ, van Gulik WM, Shimizu H, Stephanopoulos G (2004) Metabolic flux control analysis of branch points: an improved approach to obtain flux control coefficients from large perturbation data. Metab Eng 6:391–400

Hermann T, Finkemeier M, Pfefferle W, Wersch G, Krämer R, Burkovski A (2000) Two-dimensional electrophoretic analysis of *Corynebacterium glutamicum* membrane fraction and surface proteins. Electrophoresis 21:654–659

Hermann T, Pfefferle W, Baumann C, Busker E, Schaffer S, Bott M, Sahm H, Dusch N, Kalinowski J, Pühler A, Bendt AK, Krämer R, Burkovski A (2001) Proteome analysis of *Corynebacterium glutamicum*. Electrophoresis 22:1712–1723

Hirasawa T, Wachi M, Nagai K (2000) A mutation in the *Corynebacterium glutamicum ltsA* gene causes susceptibility to lysozyme, temperature-sensitive growth, and L-glutamate production. J Bacteriol 182:2696–2701

Hirasawa T, Wachi M, Nagai K (2001) L-Glutamate production by lysozyme-sensitive *Corynebacterium glutamicum ltsA* mutant strains. BMC Biotechnol 1:9

Hirano S, Kimura E, Sugimoto S (2001) Japan Patent 2001-258560

Hoischen C, Krämer R (1989) Evidence for an efflux carrier system involved in the secretion of glutamate by *Corynebacterium glutamicum*. Arch Microbiol 151:342–347

Hoischen C, Krämer R (1990) Membrane alteration is necessary but not sufficient for effective glutamate secretion in *Corynebacterium glutamicum*. J Bacteriol 172:3409–3416

Hüser AT, Becker A, Brune I, Dondrup M, Kalinowski J, Plassmeier J, Puhler A, Wiegräbe I, Tauch A (2003) Development of a *Corynebacterium glutamicum* DNA microarray and validation by genome-wide expression profiling during growth with propionate as carbon source. J Biotechnol 106:269–286

Ikeda M (2003) Amino acid production processes. In: Faurie R, Thommel J (eds) Adv Biochem Eng Biotech 79:1–36

Ikeda M, Nakagawa S (2003) The *Corynebacterium glutamicum* genome: features and impacts on biotechnological processes. Appl Microbiol Biotechnol 62:99–109

Jäger W, Peters-Wendisch PG, Kalinowski J, Pühler A (1996) A Corynebacterium glutamicum gene encoding a two-domain protein similar to biotin carboxylases and biotin-carboxy-carrier proteins. Arch Microbiol 166:76–82

Jakoby M, Nolden L, Meier-Wagner J, Krämer R, Burkovski A (2000) AmtR, a global repressor in the nitrogen regulation system of *Corynebacterium glutamicum*. Mol Microbiol 37:964–977

Jakoby M, Tesch M, Sahm H, Kramer R, Burkovski A (1997) Isolation of the *Corynebacterium glutamicum glnA* gene encoding glutamate synthetase I. FEMS Microbiol Lett 154:81–88

Jakowec MW, Smith LT, Dandekar AM (1985) Recombinant plasmid conferring proline overproduction and osmotic tolerance. Appl Environ Microbiol 50:441–446

Jang KH, Pierotti D, Kemp GW, Best GR, Britz ML (1997) Mycolic acid composition of *Corynebacterium glutamicum* surface mutants: effects of growth with glycine and isonicotinic acid hydrazide. Microbiology 143:3209–3221

Kalinowski J, Bathe B, Bartels D, Bischoff N, Bott M, Burkovski A, Dusch N, Eggeling L, Eikmanns BJ, Gaigalat L, Goesmann A, Hartmann M, Huthmacher K, Krämer R, Linke B, McHardy AC, Meyer F, Möckel B, Pfefferle W, Pühler A, Rey DA, Rückert C, Rupp O, Sahm H, Wendisch VF, Wiegräbe I, Tauch A (2003) The complete *Corynebacterium glutamicum* ATCC 13032 genome sequence and its impact on the production of L-aspartate-derived amino acids and vitamins. J Biotechnol 104:5–25

Kataoka M, Hashimoto KI, Yoshida M, Nakamatsu T, Horinouchi S, Kawasaki H (2006) Gene expression of Corynebacterium glutamicum in response to the conditions inducing glutamate overproduction. Lett Appl Microbiol 42:471–476

Kato J, Komatsubara S, Kisumi M, Chibata I (1968) Production of L-proline by *Kurthia catenaforma*. Appl Microbiol 16:1200–1206

Katsumata R, Oka T, Furuya A (1991) Japan Patent H01-3475

Kawahara Y, Ohsumi T, Yoshihara Y, Ikeda S (1989) Proline in osmoregulation of *Brevibacterium lactofermentum*. Agric Biol Chem 53:2475–2479

Kawahara Y, Takahashi-Fuke K, Shimizu E, Nakamatsu T, Nakamori S (1997) Relationship between the glutamate production and the activity of 2-oxoglutarate dehydrogenase in *Brevibacterium lactofermentum*. Biosci Biotechnol Biochem 61:1109–1112

Kijima N, Goyal D, Takada A, Wachi M, Nagai K (1998) Induction of only limited elongation instead of filamentation by inhibition of cell division in *Corynebacterium glutamicum*. Appl Microbiol Biotechnol 50:227–232

Kimura E (2002) Triggering mechanism of L-glutamate overproduction by DtsR1 in coryneform bacteria. J Biosci Bioeng 94:545–551

Kimura E (2003) Metabolic engineering of glutamate production. Adv Biochem Eng Biotech 79:37–58

Kimura E, Abe C, Kawahara Y, Nakamatsu T (1996) Molecular cloning of a novel gene, *dtsR*, which rescues the detergent sensitivity of a mutant derived from *Brevibacterium lactofermentum*. Biosci Biotechnol Biochem 60:1565–1570

Kimura E, Abe C, Kawahara Y, Nakamatsu T, Tokuda H (1997) A *dtsR* gene-disrupted mutant of *Brevibacterium lactofermentum* requires fatty acids for growth and efficiently produces L-glutamate in the presence of ac excess of biotin. Biochem Biophys Res Commun 234:157–161

Kimura E, Yagoshi C, Kawahara Y, Ohsumi T, Nakamatsu T, Tokuda H (1999) Glutamate overproduction in *Corynebacterium glutamicum* triggered by a decrease in the level of a complex comprising DtsR and a biotin-containing subunit. Biosci Biotechnol Biochem 63:1274–1278

Kinoshita S (1985) Glutamic acid bacteria. In: Demain AL, Solomon NA (eds) Biology of industrial microorganisms. Benjamin Cummings, London, pp 115–145

Kinoshita S, Udaka S, Shimono M (1957) Studies of the amino acid fermentation, Part I. Production of L-glutamic acid by various microorganisms. J Gen Appl Microbiol 3:193–205

Kosuge T, Hoshino T (1998) Construction of a proline-producing mutant of the extremely thermophilic eubacterium *Thermus thermophilus* HB27. Appl Environ Microbiol 64:4328–4332

Kumagai Y, Hirasawa T, Hayakawa K, Nagai K, Wachi M (2005) Fluorescent phospholipid analogs as microscopic probes for detection of the mycolic acid-containing layer in *Corynebacterium glutamicum*: detecting alterations in the mycolic acid-containing layer following ethambutol treatment. Biosci Biotechnol Biochem 69:2051–2056

Leisinger T (1996) Biosynthesis of proline. In: Nedihardt F, Curtis R III, Ingraham JI, Kin ECC, Low B, Magasanik B, Reznikoff WS, Riley M, Schaechter M, Umbarger E (eds) *Escherichia coli* and *Salmonella*, cellular and molecular biology, 2nd edn. ASM, Washington, DC, pp 434–441

Leuchtenberger W, Huthmacher K, Drauz K (2005) Biotechnological production of amino acids and derivatives: current status and prospects. Appl Microbiol Biotechnol 69:1–8

Loos A, Glanemann C, Willis LB, O'Brien XM, Lessard PA, Gerstmeir R, Guillouet S, Sinskey AJ (2001) Development and validation of *Corynebacterium* DNA microarrays. Appl Environ Microbiol 67:2310–2318

Mahn MJ, Csonka LN (1983) Genetic analysis of the *proBA* genes of *Salmonella typhimurium*: physical and genetic analyses of the cloned $proB^{+}A^{+}$ genes of *Escherichia coli* and of a mutant allele that confers proline overproduction and enhanced osmotolerance. J Bacteriol 156:1249–1262

Marx A, de Graaf AA, Wiechert A, Eggeling L, Sahm H (1996) Determination of the fluxes in the central metabolism of *Corynebacterium glutamicum* by nuclear magnetic resonance spectroscopy combined with metabolite balancing. Biotechnol Bioeng 49:111–129

Mitani Y, Meng X, Kamagata Y, Tamura T (2005) Characterization of LtsA from *Rhodococcus erythropolis*, an enzyme with glutamine amidotransferase activity. J Bacteriol 187:2582–2591

Möker N, Brocker M, Schaffer S, Krämer R, Morbach S, Bott M (2004) Deletion of the genes encoding the MtrA-MtrB two component system of *Corynebaacterium glutamicum* has a strong influence on cell morphology, antibiotics susceptibility and expression of genes involved in osmoprotection. Mol Microbiol 54:420–438

Momose H, Takagi T (1978) Glutamic acid production in biotin-rich media by temperature sensitive mutants of *Brevibacterium lactofermentum*, a novel fermentation process. Agric Biol Chem 42:1911–1917

Nabe K, Ujimaru T, Izuo N, Yamada S, Chibata I (1980) Production of L-glutamine by a penicillin-resistant mutant of *Flavobacterium rigense*. Appl Environ Microbiol 40:19–24

Nakamura J, Hirano S, Yamaguchi M, Ito H (2006) Japan Patent 2006-101875

Nakazawa E, Kawashima N, Koyama I, Ishii K, Kawahara Y (1994) Japan Patent H6-237779

Nakamori S, Morioka H, Yoshinaga F, Yamanaka S (1982) Fermentative production of L-proline by DL-3,4-dehydroproline resistant mutant of L-glutamate producing bacteria. Agric Biol Chem 46:487–491

Nampoothiri KM, Hoischen C, Bathe B, Mockel B, Pfefferle W, Krumbach K, Sahm H, Eggeling L (2002) Expression of genes of lipid synthesis and altered lipid composition modulates L-glutamate efflux of *Corynebacterium glutamicum*. Appl Microbiol Biotechnol 58:89–96

Nara T, Samejima H, Kinoshita S (1964) Effect of penicillin on amino acid fermentation. Agric Biol Chem 28:120–124

Niebisch A, Kabus A, Schults C, Weil B, Bott M (2006) Corynebacterial protein kinase G controls 2-oxoglutarate dehydrogenase activity via phosphorylation status of the OdhI protein. J Biol Chem 281:12300–12307

Nishio Y, Nakayama Y, Kawarabayasi Y, Usuda Y, Kimura E, Sugimoto S, Matsui K, Yamagishi A, Kikuchi H, Ikeo K, Gojobori T (2003) Comparative complete genome sequence analysis of the amino acid replacements responsible for the thermostability of *Corynebacterium efficiens*. Genome Res 13:1572–1579

Nolden L, Farwick M, Krämer R, Burkovski A (2001a) Glutamine synthetases of *Corynebacterium glutamicum*: transcriptional control and regulation of activity. FEMS Microbiol Lett 201:91–98

Nolden L, Ngouoto-Nkili C-E, Bendt AK, Krämer R, Burkovski A (2001b) Sensing nitrogen limitation in *Corynebacterium glutamicum*: the role of *glnK* and *glnD*. Mol Microbiol 42:1281–1295

Omori K, Suzuki S, Imai Y, Komatsubara S (1991) Analysis of the *Serratia marcescens proBA* operon and feedback control of proline biosynthesis. J Gen Microbiol 137:509–517

Omori K, Suzuki S, Imai Y, Komatsubara S (1992) Analysis of the mutant *proBA* operon from a proline-producing strain of *Serratia marcescens*. J Gen Microbiol 138:693–699

Pallen M, Chaudhuri R, Khan A (2002) Bacterial FHA domains: neglected players in the phospho-threonine signalling game? Trends Micobiol 10:556–563

Peter H, Bader A, Burkovski A, Lambert C, Krämer R (1997) Isolation of the *putP* gene of *Corynebacterium glutamicum* and characterization of a low-affinity uptake system for compatible solutes. Arch Microbiol 168:143–151

Peter H, Weil R, Burkovski A, Krämer R, Morbach S (1998) *Corynebacterium glutamicum* is equipped with four secondary carriers for compatible solutes: identification, sequencing, and characterization of the proline/ectoine uptake system, ProP, and the ectoine/proline/glycine betaine carrier, EctP. J Bacteriol 180:6005–6012

Peters-Wendisch PG, Schiel B, Wendisch VF, Katsoulidis E, Mockel B, Sahm H (2001) Pyruvate carboxylase is a major bottleneck for glutamate and lysine production by *Corynebacterium glutamicum*. J Mol Microbiol Biotechnol 3:295–300.

Portevin D, de Sousa-D'Auria C, Montrozier H, Houssin C, Stella A, Laneelle MA, Bardou F, Guilhot C, Daffé M (2005) The acyl-AMP ligase FadD32 and AccD4-containing acyl-CoA carboxylase are required for the synthesis of mycolic acids and essential for mycobacterial growth. Identification of the carboxylation product and determination of the acyl-CoA carboxylase components. J Biol Chem 280:8862–8874

Radmacher E, Stansen KC, Besra GS, Alderwick LJ, Maughan WN, Hollweg G, Sahm H, Wendisch VF, Eggeling L (2005) Ethambutol, a cell wall inhibitor of *Mycobacterium tuberculosis*, elicits L-glutamate efflux of *Corynebacterium glutamicum*. Microbiology 151:1359–1368

Sano K, Ito K, Miwa K, Nakamori S (1987) Amplification of the phosphoenol pyruvate carboxylase gene of *Brevibacterium lactofermentum* to improve amino acid production. Agric Biol Chem 51:597–599

Sauer U, Eikmanns BJ (2005) The PEP-pyruvate-oxaloacetate node as the switch point for carbon flux distribution in bacteria. FEMS Microbiol Rev 29:765–794

Schaffer S, Weil B, Nguyen VD, Dongmann G, Gunther K, Nickolaus M, Hermann T, Bott M (2001) A high-resolution reference map for cytoplasmic and membrane-associated proteins of *Corynebacterium glutamicum*. Electrophoresis 22:4404–4422

Schleifer KH, Kandler O (1972) Peptidoglycan types of bacterial cell walls and their taxonomic implications. Bacteriol Rev 36:407–477

Sculuesener D, Fischer F, Kruip J, Rögner M, Poetsch A (2005) Mapping the membrane proteome of *Corynebacterium glutamicum*. Proteomics 5:1317–1330

Schulz AA, Collet HJ, Reid SJ (2001) Nitrogen and carbon regulation of glutamine synthetase and glutamate synthase in *Corynebacterium glutamicum* ATCC 13032. FEMS Microbiol Lett 205:361–367

Schwinde JW, Hertz PF, Sahm H, Eikmanns BJ, Guyonvarch A (2001) Lipoamide dehydrogenase from *Corynebacterium glutamicum*: molecular and physiological analysis of the lpd gene and characterization of the enzyme. Microbiology 147:2223–2231

Serebrijski I, Wojcik F, Reyes O, Leblon G (1995) Multicopy suppression by *asd* gene and osmotic stress-dependent complementation by heterologous *proA* in *proA* mutants. J Bacteriol 177:7255–7260

Shibukawa M, Kurima M, Ohuchi S (1970) L-Glutamic acid fermentation with molasses. Part XII. Relationship between the kind of phospholipids and their fatty acid composition in the mechanism of extracellular accumulation of L-glutamate. Agric Biol Chem 34:1136–1141

Shibukawa M, Ohsawa T (1966) L-Glutamic acid fermentation with molasses. Part XII. Effect of saturated-unsaturated fatty acid ratio in the cell membrane fraction on the extracellular accumulation of L-glutamate. Agric Biol Chem 30:750–758

Shiio I, Otsuka S, Katsuya N (1963) Cellular permeability and extracellular formation of glutamic acid in *Brevibacterium flavum*. J Biochem 53:333–340

Shiio I, Otsuka S, Takahashi M (1962) Effect of biotin on the bacterial formation of glutamic acid. I. Glutamate formation and cellular permeability of amino acids. J Biochem 51:56–62

Shiio I, Ujigawa-Takeda K (1980) Presence and regulation of α-ketoglutarate dehydrogenase complex in a glutamate-producing bacterium, *Brevibacterium flavum*. Agric Biol Chem 44:1887–1890

Shimizu H (2002) Metabolic engineering – integrating methodologies of molecular breeding and bioprocess systems engineering. J Biosci Bioeng 94:563–573

Shimizu H, Tanaka H, Nakato A, Nagahisa K, Kimura E, Shioya S (2003) Effects of the changes in enzyme activities on metabolic flux redistribution around the 2-oxoglutarate branch in glutamate production by *Corynebacterium glutamicum*. Biopro Biosys Eng 25:291–298

Shingu H, Terui G (1971) Studies on the process of glutamic acid fermentation at the enzyme level: I. On the changes of a-ketoglutaric acid dehydrogenase in the course of culture. J Ferment Technol 49:400–405

Shirai T, Nakato A, Izutani N, Nagahisa K, Shioya S, Kimura E, Kawarabayasi Y, Yamagishi A, Gojobori T, Shimizu H (2005) Comparative study of flux redistribution of metabolic pathway in glutamate production by two coryneform bacteria. Metab Eng 7:59–69

Shirai T, Matsuzaki K, Kuzumoto M, Nagahisa K, Furusawa C, Shioya S, Shimizu H (2006) Precise metabolic flux analysis of coryneform bacteria with GC-MS measurement and verification of NMR analysis. J Biosci Bioeng 102:413–424

Shiratsuchi M, Kuronuma H, Kawahara Y, Yoshihara Y, Miwa H, Nakamori S (1995) Simultaneous and high fermentative production of L-lysine and L-glutamic acid using a strain of *Brevibacterium lactofermentum*. Biosci Biochem Biotechnol 59:83–86

Small JR, Kacser H (1993) Responce of metabolic systems to large changes in enzyme activities and effectors. 1. The linear treatment of unbranched chains. Eur J Biochem 213:613–624

Smith LT (1985) Characterization of a γ-glutamyl kinase from *Escherichia coli* that confers proline overproduction and osmotic tolerance. J Bacteriol 164:1088–1093

Stansen KC, Uy D, Delaunay S, Eggeling L, Goergen J-L, Wendisch VF (2005) Characterization of a *Corynebacterium glutamicum* lactate utilization operon induced during temperature-triggered glutamate production. Appl Environ Microbiol 71:5920–5928

Steger R, Weinand M, Krämer R, Morbach S (2004) LcoP, an osmoregulated betaine/ectoine uptake system from Corynebacterium glutamicum. FEBS Lett 573:155–160

Stephanopoulos G, Aristidou A, Nielsen J (1998) Metabolic engineering. Academic, San Diego, CA

Stephanopoulos G, Simpson T (1997) Flux amplification in complex metabolic network. Chem Eng Sci 52:2607–2627

Stephanopoulos G, Vallino JJ (1991) Network rigidity and metabolic engineering in metabolite over production. Science 252:1675–1681

Strelkov S, von Elstermann M, Schomburg D (2004) Comprehensive analysis of metabolites in *Corynebacterium glutamicum* by gas chromatography/mass spectrometry. Biol Chem 385:853–861

Strösser J, Lüdke A, Schaffer S, Krämer R, Burkovski A (2004) Regulation of GlnK activity: modification, membrane sequestration and proteolysis as regulatory principles in the network of nitrogen control in *Corynebacterium glutamicum*. Mol Microbiol 54:132–147

Sugiura M, Kisumi M (1985) Proline-hyperproducing strains of *Serratia marcescens*: enhancement of proline analog-mediated growth inhibition by increasing osmotic stress. Appl Environ Microbiol 49:782–786

Sugiura M, Suzuki S, Takagi T, Kisumi M (1986) Proline production via the arginine biosynthetic pathway: transfer of regulatory mutations of arginine biosynthesis into a proline-producing strain of *Serratia marcescens*. Appl Microbiol Biotechnol 24:153–158

Sugiura M, Takagi T, Kisumi M (1985) Proline production by regulatory mutants of *Serratia marcescens*. Appl Microbiol Biotechnol 21:213–219

Szyperski T (1995) Biosynthetically directed fractional ^{13}C-labeling of proteinogenic amino acids. An efficient analytical tool to investigate intermediary metabolism. Eur J Biochem 232:433–448

Tachiki T, Matsumoto H, Yano T, Tochikura T (1981a) Glutamine production by coupling with energy transfer: employing yeast cell-free extracts and *Gluconobacter* glutamine synthetase. Agric Biol Chem 45:705–710

Tachiki T, Matsumoto H, Yano T, Tochikura T (1981b) Glutamine production by coupling with energy transfer: employing yeast cells and *Gluconobacter* glutamine synthetase. Agric Biol Chem 45:1237–1241

Tachiki T, Suzuki H, Wakisaka S, Yano T, Tochikura T (1983) Glutamine production in high concentrations with energy transfer employing glutamine synthetase from *Micrococcus glutamicus*. J Gen Appl Microbiol 29:355–363

Takiguchi N, Shimizu H, Shioya S (1997) An on-line physiological state recognition system for the lysine fermentation process based on a metabolic reaction model. Biotechnol Bioeng 55:170–181

Takinami K, Yoshii H, Tsuri H, Okada H (1965) Biochemical effects of fatty acid and its derivatives on L-glutamic acid fermentation, Part III. Biotin-Tween 60 relationship in the accumulation of L-glutamic acid and the growth of *Brevibacterium lactofermentum*. Agric Biol Chem 29:351–359

Telenti A, Philipp WJ, Sreevatsan S, Bernasconi C, Stockbauer KE, Wieles B, Musser JM, Jacobs WR (1997) The *emb* operon, a gene cluster of *Mycobacterium tuberculosis* involved in resistance to ethambutol. Nat Med 3:567–570

Tesch M, de Graaf AA, Sahm H (1999) In vivo fluxes in ammonium assimilatory pathways in *Corynebacterium glutamicum* studied by 15N nuclear magnetic resonance. Appl Environ Microbiol 65:1099–1109

Trötschel C, Kandirali S, Diaz-Achirica P, Meinhardt A, Morbach S, Kramer R, Burkovski A (2003) GltS, the sodium-coupled L-glutamate uptake system of *Corynebacterium glutamicum*: identification of the corresponding gene and impact on L-glutamate production. Appl Microbiol Biotechnol 60:738–742

Tsuchida T, Kubota K, Yoshinaga F (1986) Improvement of L-proline production by sulfaguanidine resistant mutants derived from L-glutamic acid-producing bacteria. Agric Biol Chem 50:2201–2207

Tsushida T, Kubota K, Yoshihara Y, Kikuchi K, Yoshinaga F (1987) Fermentative production of L-glutamine by sulfaguanidine resistant mutants derived from L-glutamate producing-bacteria. Agric Biol Chem 51:2089–2094

Udaka S (1960) Screening method for microorganisms accumulating metabolites and its use in the isolation of *Micrococcus glutamicus*. J Bacteriol 79:754–55

Usuda Y, Tujimoto N, Abe C, Asakura Y, Kimura E, Kawahara Y, Kurahashi O, Matsui H (1996) Molecular cloning of the *Corynebacterium glutamicum* (*Brevibacterium lactofermentum* AJ12036) *odhA* gene encoding a novel type of 2-oxoglutarate dehydrogenase. Microbiology 142:3347–3354

Uy D, Delaunay S, Germain P, Engasser J-M, Goergen J-L (2003) Instability of glutamate production by *Corynebacterium glutamicum* 2262 in continuous culture using the temperature-triggered process. J Biotechnol 104:173–184

Vallino JJ, Stephanopoulos G (1993) Metabolic flux distributions in *Corynebacterium glutamicum* during growth and lysine over production. Biotechnol Bioeng 41:633–646

Varela C, Agostin E, Baez M, Klapa M, Stephanopoulos G (2003) Metabolic flux reditribution in *Corynebacterium glutamicum* in response to osmotic stress. Appl Microbiol Biotechnol 60:547–555

Villarino A, Duran R, Wehenkel A, Fernandez P, England P, Brodin P, Cole ST, Zimny-Arndt U, Jungblut PR, Cervenansky C, Alzari PM (2005) Proteomic identification of *M. tuberculosis* protein kinase substrates: PknB recruits GarA, a FHA domain-containing protein, through activation loop-mediated interactions. J Mol Biol 350:953–963

Wakisaka S, Ohshima Y, Ogawa M, Tochikura T, Tachiki T (1998) Characteristics and efficiency of glutamine production by coupling of a bacterial glutamine synthetase reaction with alcoholic fermentation system of baker's yeast. Appl Env Microbiol 64:2952–2957

Wendisch VF (2003) Genome-wide expression analysis in *Corynebacterium glutamicum* using DNA microarray. J Biotechnol 104:273–285

Wendisch VF, Bott M, Kalinowski J, Oldiges M, Wiechert W (2006) Emerging *Corynebacterium glutamicum* systems biology. J Biotechnol 124:74–92

Yamada S, Nabe K, Ujimaru T, Izuo N, Chibata I (1979) L-Glutamine formation by *Flavobacterium rigense*. Appl Environ Microbiol 37:1063–1066

Yoshida K, Fujita Y, Ehrlich D (1999) Three asparagines synthetase genes of Bacillus subtilis. J Bacteriol 181:6081–6091

Yoshihara Y, Kawahara Y, Yamada Y (1992) Process for producing L-amino acids by fermentation. US Patent 5 164 307

Yoshinaga F, Konishi S, Okumura S, Katsuya N (1966) Studies on the fermentative production of L-proline. I. Production of L-proline by an isoleucine auxotrophic mutant of *Brevibacterium flavum* 2247. J Gen Appl Microbiol 12:219–228

Microbiol Monogr (5)
V. F. Wendisch: Amino Acid Biosynthesis
DOI 10.1007/7171_2006_089/Published online: 24 February 2007

The L-Lysine Story: From Metabolic Pathways to Industrial Production

Christoph Wittmann (✉) · Judith Becker

Biochemical Engineering, Saarland University, Building A1.5, Im Stadtwald,
66041 Saarbrücken, Germany
c.wittmann@mx.uni-saarland.de

Abstract L-lysine is an essential amino acid required for nutrition of animals and humans. It has to be present in food and feed, which, in many cases, is realized by supplementation of the feed-stuffs with pure lysine. The high importance of lysine in nutrition has stimulated intensive research on the lysine biosynthetic pathways and their regulation and the search for microorganisms capable of over-producing this amino acid. As an important milestone, the glutamate producing soil bacterium *Corynebacterium glutamicum* was isolated in 1956 and soon received interest to be used for production of another amino acid stemming from the TCA cycle: lysine. Within a few years the first lysine producing strains were obtained. The past 50 years following the discovery of *C. glutamicum* were characterized by a huge progress towards understanding the physiology of this organism and developing and optimizing industrial production strains. This has resulted in effective biotechnological processes currently used for producing about 750 000 tons of lysine per year. Today, systems-oriented approaches aiming at investigating the link between the different components of cellular physiology such as transcriptome, fluxome and metabolome, provide a novel powerful platform that will surely drive future research towards holistic understanding of lysine over-producing microorganisms as well as the creation of superior production strains.

1
Introduction

The essential amino acid L-lysine is one of the most important amino acids applied as supplement in animal feed. Animal feed, which is typically based on corn, wheat or barley, is poor in lysine. The supplementation of such feed materials with a lysine rich source leads to optimized growth of e.g. pigs or chicken. The direct addition of lysine hereby has proven especially valuable. It does not cause an extra uptake and metabolization of other amino acids beyond their need so that superfluous formation of ammonia and environmental burden by increased nitrogen loads in the manure is avoided. The continuing development of an increased consumption of white meat in various countries of the western as well as the eastern world has led to an enormous market growth for lysine during the past decades (Fig. 1). Currently, the world-wide production of lysine is about 750 000 tons with a predicted market growth over the next years of about 8%. Since only the L-isomer of lysine is effective as feed supplement, all manufacturing processes utilize fermentative production (Leuchtenberger 1996). The cradle of biotechnological lysine production stands in Japan. Stimulated by the increased demand for amino acids, especially glutamate, a large screening program was initiated in Japan about 50 years ago which, in 1956, led to the discovery of the glutamate excreting microorganism *Corynebacterium glutamicum* (Kinoshita et al. 1957; Udaka 1960). Within a few years, the first lysine excreting mutants of *C. glutamicum* were available and applied for production in large scale (Kelle et al. 2005; Kinoshita et al. 1961). Since then a continuous optimization of producing strains and processes has led to efficient manufacturing of lysine from renewable resources covering the high demand for this compound required today (de Graaf et al. 2001; Eggeling and Sahm 1999; Hermann 2003; Pfefferle et al. 2003).

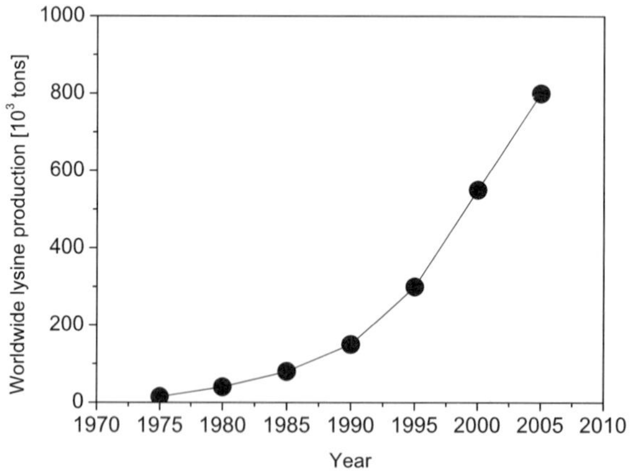

Fig. 1 Development of the world-wide annual biotechnological production of L-lysine

2
Lysine Biosynthetic Pathways

In microorganisms, lysine can be synthesized by two completely different routes: either from 2-oxoglutarate and acetyl-CoA via the α-aminoadipate route or from aspartate via the diaminopimelate route. Two variants of the α-aminoadipate route (for review see Velasco et al. 2002 and references therein) occur in higher fungi and in archaea like *Thermoproteus neutrophilus* on the one hand and in the bacterium *Thermus thermophilus* on the other hand. In five reactions that are catalyzed by homoisocitrate synthase (EC 4.1.3.21), homoaconitate hydratase/cis-homoaconitase (EC 4.2.1.36) and homoisocitrate dehydrogenase (EC 1.1.1.155), 2-oxoglutarate and acetyl-CoA are converted to α-aminoadipate. In one variant of the pathway, α-aminoadipate is converted to lysine by α-aminoadipate reductase (EC 1.2.1.32), saccharopine reductase (EC 1.5.1.10) and saccharopine dehydrogenase (EC 1.5.1.7 or 1.5.1.8). In another variant of the pathway, first described in *Thermus thermophilus*, the conversion of α-aminoadipate to lysine occurs via the acetylated intermediates N^2-acetyl-L-α-aminoadipate, N^2-acetyl-L-aminoadipyl-δ-phosphate, N^2-acetyl-L-α-aminoadipate semialdehyde and N^2-acetyl-L-lysine catalyzed by the gene products of *lysX*, *lysZ*, *lysY* and either *lysJ* or *argD* and either *lysK* or *argE*.

In bacteria and plants, lysine may be synthesized from aspartate by one or several of four variants of the diaminopimelate route. These pathway variants diverge at the common intermediate tetrahydrodipicolinate (Born and Blanchard 1999; Schrumpf et al. 1991; McCoy et al. 2006). As shown in Fig. 2, one of these pathways involves succinylated intermediates, while the acetylase pathway comprises acetylated intermediates and the so-called dehydrogenase pathway directly forms DL-diaminopimelate from tetrahydrodipicolinate (Schrumpf et al. 1991; Wehrmann et al. 1998). The aminotransferase pathway, recently described to operate in *Chlamydia*, converts tetrahydrodipicolinate to LL-diaminopimelate, which can then be epimerized to DL-diaminopimelate (McCoy et al. 2006). Most bacteria only comprise one of these pathways (Bartlett and White 1985; White 1983). Whereas the succinylase pathway is present both in gram negative and gram positive bacteria, the acetylase variant seems to be exclusively used by some *Bacillus* species (Bartlett and White 1985; Born and Blanchard 1999; Weinberger and Gilvarg 1970). Only in a few organisms like different species of the genera *Corynebacterium* and in *Bacillus macerans* two lysine biosynthetic pathways operate together (Bartlett and White 1985; Malumbres and Martin 1996; Schrumpf et al. 1991). A common feature involved in the different pathways concerns the regulation of pathway flux by feedback inhibition of aspartate kinase. In *C. glutamicum* only one isoenzyme of aspartate kinase (or aspartokinase) exists, which is encoded by two genes, *lysCα* and *lysCβ*, representing the coding sequences for the two subunits of the enzyme (Kalinowski et al. 1990). Its activity is bio-

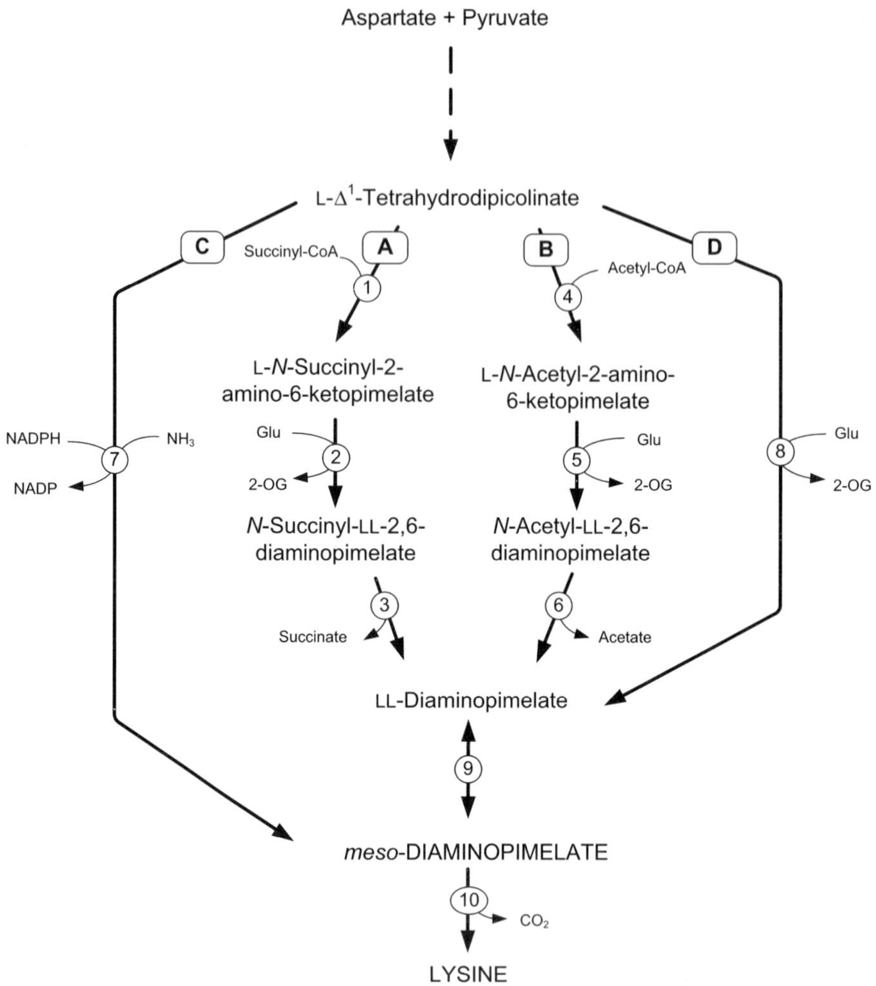

Fig. 2 The four pathways of D,L-diaminopimelate and lysine synthesis from aspartate in prokaryotes: succinylase pathway (**A**); acetylase pathway (**B**), dehydrogenase pathway (**C**), and aminotransferase pathway (**D**). Enzymes involved in the succinylase pathway are: (1) tetrahydrodipicolinate succinylase (DapD); (2) *N*-succinyl-aminoketopimelate aminotransferase (DapC); and (3) *N*-succinyl-diaminopimelate desuccinylase (DapE). Enzymes of the acetylase pathway are: (4) tetrahydrodipicolinate acetylase; (5) N-acetyl-aminoketopimelate aminotransferase; and (6) *N*-acetyl-diaminopimelate deacetylase. Via the aminotransferase pathway, L,L-diaminopimelate is formed by (8) tetrahydrodipicolinate aminotransferase. (9) Diaminopimelate epimerase (DapF) is common for these three pathways. The dehydrogenase pathway directly forms D,L-diaminopimelate via (7) diaminopimelate dehydrogenase (Ddh). Lysine is built from D,L-diaminopimelate by (10) diaminopimelate decarboxylase

chemically regulated through concerted feedback inhibition by lysine and threonine that bind at the regulatory β-subunits (Kalinowski et al. 1991; Malumbres and Martin 1996). A similar regulation has been described for lysine synthesis in *B. megaterium* which, however, has two different isoenzymes. One of the present isoenzymes is also inhibited by a concerted action of two amino acids, which are lysine and methionine, whereas the other isoenzyme is only subject to inhibition by threonine (Chatterjee and White 1982). Additionally to the modulation of enzyme activity by metabolites, some bacteria e.g. *E. coli* and *B. megaterium* also show a transcriptional regulation for enzymes involved in lysine biosynthesis (Chatterjee and White 1982; Cremer et al. 1988).

3
Corynebacterium glutamicum as a Production Organism

Corynebacterium glutamicum, including its subspecies *Brevibacterium flavum*, *Brevibacterium lactofermentum*, *Corynebacterium lilium*, *Corynebacterium efficiens* and *Brevibacterium divaricatum* (Liebl 2005; Liebl et al. 1991) is the most important organism for industrial lysine production. The only other species used for lysine production are recombinant *E. coli* strains (Imaizumi et al. 2005, 2006). The capability of *C. glutamicum* to secrete amino acids was discovered in the 1950s (Kinoshita et al. 1957; Udaka 1960). It is a gram-positive, rod-like, non-motile, aerobic bacterium (Fig. 3). Fur-

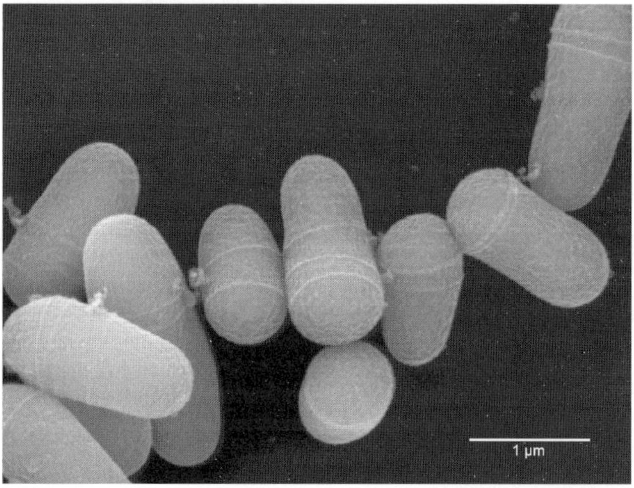

Fig. 3 Raster electron micrograph of *Corynebacterium glutamicum* ATCC 13032 cultivated on minimal glucose medium. Samples were taken during exponential growth (Krömer, Heinzle and Wittmann 2006, unpublished results)

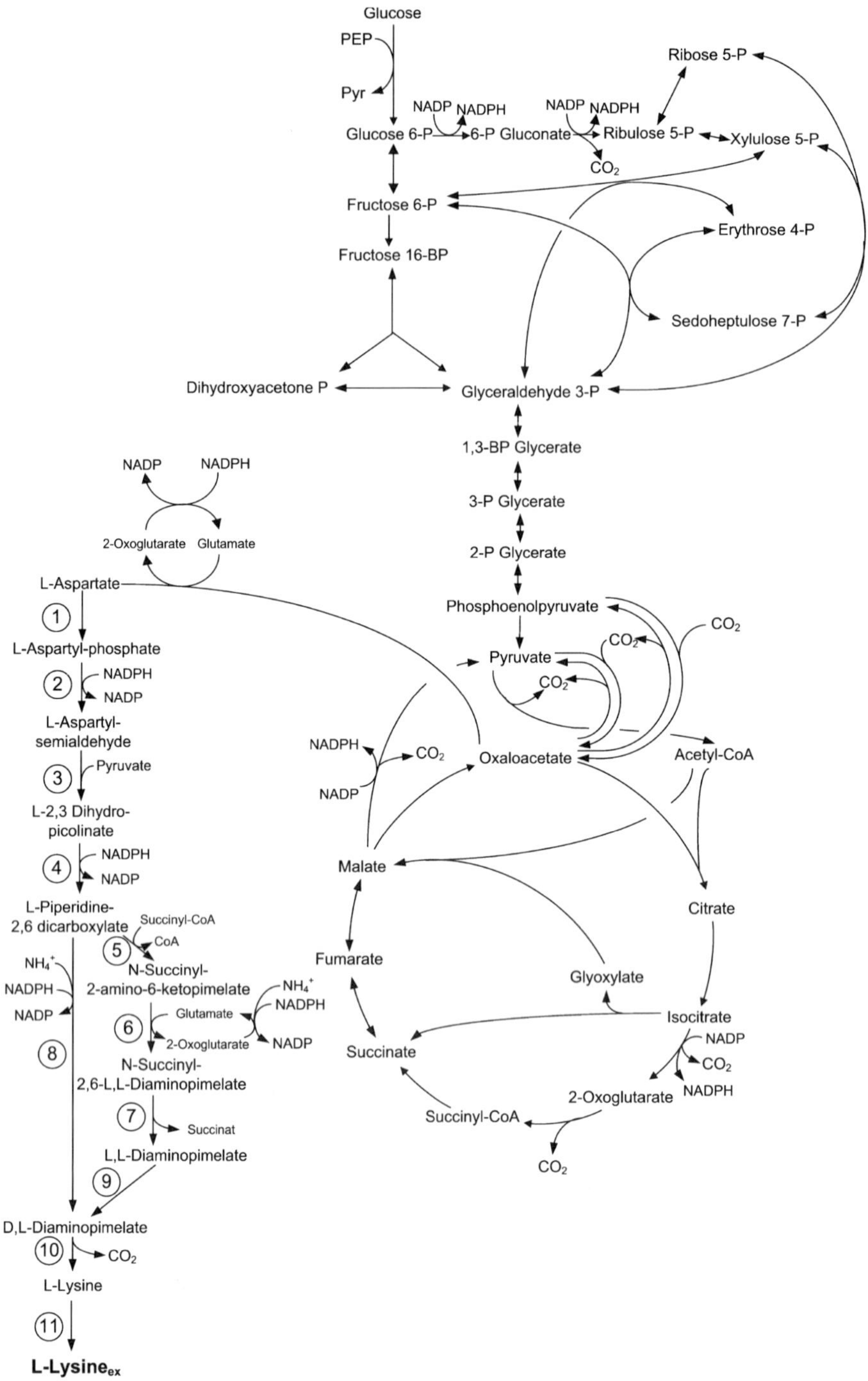
Glucose
PEP
Pyr
NADP
NADPH
Glucose 6-P
6-P Gluconate
Ribulose 5-P
Ribose 5-P
Xylulose 5-P
CO2
Fructose 6-P
Erythrose 4-P
Fructose 16-BP
Sedoheptulose 7-P
Dihydroxyacetone P
Glyceraldehyde 3-P
1,3-BP Glycerate
3-P Glycerate
2-P Glycerate
Phosphoenolpyruvate
Pyruvate
Oxaloacetate
Acetyl-CoA
Malate
Citrate
Fumarate
Glyoxylate
Isocitrate
Succinate
2-Oxoglutarate
Succinyl-CoA
2-Oxoglutarate
Glutamate
L-Aspartate
L-Aspartyl-phosphate
L-Aspartyl-semialdehyde
Pyruvate
L-2,3 Dihydro-picolinate
L-Piperidine-2,6 dicarboxylate
Succinyl-CoA
CoA
N-Succinyl-2-amino-6-ketopimelate
NH4+
Glutamate
2-Oxoglutarate
N-Succinyl-2,6-L,L-Diaminopimelate
Succinat
L,L-Diaminopimelate
D,L-Diaminopimelate
L-Lysine
L-Lysine ex
1
2
3
4
5
6
7
8
9
10
11

◀ **Fig. 4** Central metabolic pathways and lysine biosynthesis in *Corynebacterium glutamicum*. The numbers given for the reactions in the lysine biosynthesis pathway relate to Table 1, where further details, for example on the corresponding enzymes, the encoding genes, and the patents claimed by major industrial players in the field are given

ther characteristics comprise a cell wall with arabino-galactan and mycolic acids with 26 to 36 carbon atoms and a murein sacculus with peptido-glycan cross linked via meso-diaminopimelic acid (Goodfellow et al. 1976). The GC-content of its genome is 53.8% (Kalinowski et al. 2003). Extensive biochemical studies in the last decades as reviewed in a recent handbook on *Corynebacterium glutamicum* (Eggeling and Bott 2005) and the recent unravelling of the genome sequence (Haberhauer et al. 2001; Ikeda and Nakagawa 2003; Tauch et al. 2002; Kalinowski et al. 2003) have contributed to a detailed knowledge of the reactions of lysine biosynthesis and central metabolism in this microorganism (Fig. 4).

3.1 Lysine Biosynthesis

Lysine belongs to the aspartate amino acid family and in *C. glutamicum* is produced from pyruvate, oxaloacetate and two ammonia molecules involving the additional supply of four NADPH as reducing power (Michal 1999). Interestingly, the organism has a split pathway for the biosynthesis of lysine (Schrumpf et al. 1991; Sonntag et al. 1993). The two alternative branches give *C. glutamicum* an increased flexibility in response to changing environmental conditions, involving e.g. different ammonia levels (Sahm et al. 2000). DL-diaminopimelate as intermediate of the lysine pathway additionally is an essential building block for the synthesis of the murein sacculus (Wehrmann et al. 1998). Concerning regulation of lysine biosynthesis, aspartokinase (EC 2.7.2.4), catalyzing the formation of aspartyl phosphate from aspartate is the key enzyme. It is feedback regulated by concerted action of lysine and threonine (Kalinowski et al. 1991). Via the requirement for building blocks and cofactors lysine formation is closely linked to the central metabolism. The enzymes involved in lysine biosynthesis are summarized in Table 1.

3.2 Central Carbon Metabolism

The most relevant substrates for industrial lysine production are starch and molasses. They are based on glucose, fructose, and sucrose as major carbon sources. These compounds are taken up via phosphoenolpyruvate-dependent phosphotransferase systems (Dominguez and Lindley 1996; Dominguez et al. 1998; Malin and Bourd 1991; Moon et al. 2005). The carbon source has an in-

Table 1 Genes and enzymes involved in L-lysine biosynthesis in *C. glutamicum*. The information on patents and patent applications by the lysine producing companies Archer Daniels Midland (■), Ajinomoto (▲), BASF(•), Degussa (♦) and Kyowa Hakko Kogyo (▼) claiming an improvement of lysine production through modification of the lysine biosynthetic pathway in *C. glutamicum* is taken from Kelle et al. (2005). The reaction numbers relate to the corresponding metabolic reactions of lysine biosynthesis in the metabolic network shown in Fig. 4

Gene	Enzyme	EC number	Inhibitory ligands	Transcription unit	Transcriptional regulation	Reference	Reaction number	Patents
lysC	aspartate kinase	2.7.2.4	lysine threonine	lysC		(Kalinowski et al. 1991)	1	■ ▲ • ♦ ▼
asd	aspartate semialdehyde dehydrogenase	1.2.1.11	lysine threonine	asd		(Cremer et al. 1988)	2	■ • ▼
dapA	dihydrodipicolinate synthase	4.2.1.52		dapB-orf2-dapA-orf4		(Patek et al. 1997)	3	■ ▲ • ♦ ▼
dapB	dihydrodipicolinate reductase	1.3.1.26		dapB-orf2-dapA-orf4		(Patek et al. 1997)	4	■ ▲ • ♦ ▼
dapD	tetrahydrodipicolinate succinylase	2.3.1.117		dapD		(Wehrmann et al. 1998)	5	• ▼
dapC	succinyl-amino-keto-pimelate transaminase	2.6.1.17		dapC		(Hartmann et al. 2003)	6	• ♦ ▼
dapE	succinyl-diamino-pimelate desuccinylase	3.5.1.18		dapE		(Wehrmann et al. 1998)	7	• ▼
ddh	meso-diaminopimelate dehydrogenase	1.4.1.16		ddh		(Cremer et al. 1988)	8	■ ▲ • ♦ ▼
dapF	diaminopimelate epimerase	5.1.1.7		dapF		(Hartmann et al. 2003)	9	• ♦ ▼
lysA	diaminopimelate decarboxylase	4.1.1.20		lysA	inhibited by lysine	(Cremer et al. 1988)	10	■ ▲ • ▼
lysE	lysine permease			lysE	induced by lysine	(Vrljic et al. 1996)	11	• ♦ ▼

fluence on lysine production by *C. glutamicum*, which is related to the resulting different entry points into the central metabolism (Kiefer et al. 2002). The central catabolic network previously identified in *C. glutamicum* comprises the pathways of glycolysis, pentose phosphate pathway (PPP), tricarboxylic acid (TCA) cycle and glyoxylate cycle, while the Entner–Doudoroff pathway has not been detected (Eikmanns 2005; Kalinowski et al. 2003; Yokota and Lindley 2005). The oxidative part of the PPP comprises the NADPH dependent enzymes glucose 6-phosphate dehydrogenase (EC 1.1.1.49) (Ihnen and Demain 1969; Moritz et al. 2000) and 6-phophogluconate dehydrogenase (1.1.1.44) (Moritz et al. 2000; Ohnishi et al. 2005). Both enzymes were intensively studied regarding their kinetic properties (Moritz et al. 2000). *C. glutamicum* exhibits an extensive set of different enzymes for the interconversion of C_4 metabolites of the TCA cycle and C_3 metabolites of glycolysis (Fig. 4). For anaplerotic replenishment of the TCA cycle *C. glutamicum* possesses pyruvate carboxylase (EC 6.4.1.1) (Peters-Wendisch et al. 1998) and phosphoenolpyruvate carboxylase (EC 4.1.1.31). Malic enzyme (EC 1.1.1.40) (Gourdon et al. 2000), PEP carboxykinase (EC 4.1.1.32) (Jetten and Sinskey 1993; Riedel et al. 2001), oxaloacetate decarboxylase (EC 4.1.1.3) (Jetten and Sinskey 1995) and probably phosphoenolpyruvate synthetase (Jetten et al. 1994), of which the genes of oxaloacetate decarboxylase and phosphoenolpyruvate synthetase have not been found yet, catalyze decarboxylation reactions from the TCA cycle towards glycolysis. In selected cases, NADPH supply by malic enzyme supports growth and lysine production of *C. glutamicum* (Dominguez et al. 1998; Kim et al. 2006). However, deletion of *malE* does not affect growth of *C. glutamicum* on glucose or acetate, suggesting that the metabolic pathways involved in NADPH-production are highly flexible (Gourdon et al. 2000). It has been proposed that cyclic cooperation of the enzymes between pyruvate/phosphoenolpyruvate and malate/oxaloacetate is involved in the regeneration of excess ATP (Marx et al. 1996; Petersen et al. 2000; Riedel et al. 2001).

In addition to the central catabolic routes, also anabolic pathways have been studied in detail providing detailed information on the cellular composition and the precursor demand for growth (Wittmann and de Graaf 2005). Overall about (16.4 mmol NADPH) g^{-1} is required for biomass synthesis. It is obvious that the anabolic NADPH requirement competes with lysine production. Considering a biomass yield of 0.5 (g dry biomass) (g glucose)$^{-1}$, which is achieved by *C. glutamicum* under aerobic conditions this results in 1.7 mol NADPH (mol glucose)$^{-1}$ that have to be generated by the NADPH forming reactions in the PPP and the TCA cycle. It should be noted that the knowledge about the metabolic network is continuously being updated with novel findings such as the detection of pyruvate carboxylase as anaplerotic enzyme (Peters-Wendisch et al. 1998) or the consideration of new reactions for growth on specific substrates such as sucrose (Dominguez and Lindley 1996).

3.3 Maximal Lysine Production Capacity

The maximal capacity, i.e. the theoretical maximum yield, of a *C. glutamicum* cell for lysine production is an important characteristic, since it provides an estimate of the remaining optimization potential of a running industrial process and gives advice for process or genetic engineers. Previous stoichiometric calculation considering only the major pathways involved in lysine production have yielded a molar lysine yield on glucose of 75% (Hollander 1994). A more detailed insight can be gained via stoichiometric network analysis considering the full set of available pathways in the central metabolism with information on reversibility or irreversibility of the different reactions and additional assumptions and restrictions posed on the metabolic network. This approach, named elementary flux mode analysis (Schilling et al. 1999; Schuster et al. 2002), has been recently applied to elucidate the potential of *C. glutamicum* for methionine production (Krömer et al. 2006). The theoretical maximum molar yield of *C. glutamicum* for lysine production obtained by such an analysis is 82% (Fig. 5). The in silico pathway analysis also provides information on the reactions contributing to this theoretical optimum, i.e. the corresponding theoretical metabolic flux distribution. Shown for one of the optimal elementary flux modes, i.e. optimal lysine production, PPP and malic enzyme supply the required NADPH (Fig. 5). Hereby the concerted action of pyruvate carboxylase, malate dehydrogenase (EC 1.1.1.37) and malic enzyme form a transhydrogenase-like cycle converting NADH into NADPH and thus providing an additional amount of this important cofactor. This metabolic cycle is different from a previously described cyclic pathway involving PCx, PCK, and MDH as shown to operate in vivo (Petersen et al. 2000; Riedel et al. 2001). One should further notice that, in addition to the soluble malate dehydrogenase *C. glutamicum* possesses another enzyme for interconversion of oxaloacetate and malate, the membrane-bound malate:quinone oxidoreductase (MQO, EC 1.1.99.16). Both enzymes probably catalyze a cyclic reaction in vivo leading to a net oxidation of NADH (Molenaar et al. 2000). The dehydrogenase branch of lysine biosynthesis is the sole lysine producing route. Admittedly, the given scenario implies zero flux through the TCA cycle and no biomass formation, which can hardly be realized. The achievable optimum can be expected to be somewhat lower than the value calculated here, but should still be significantly higher than the yield achieved in practice, which is in the range of about 55% (Ikeda 2003; Kawahara et al. 1990).

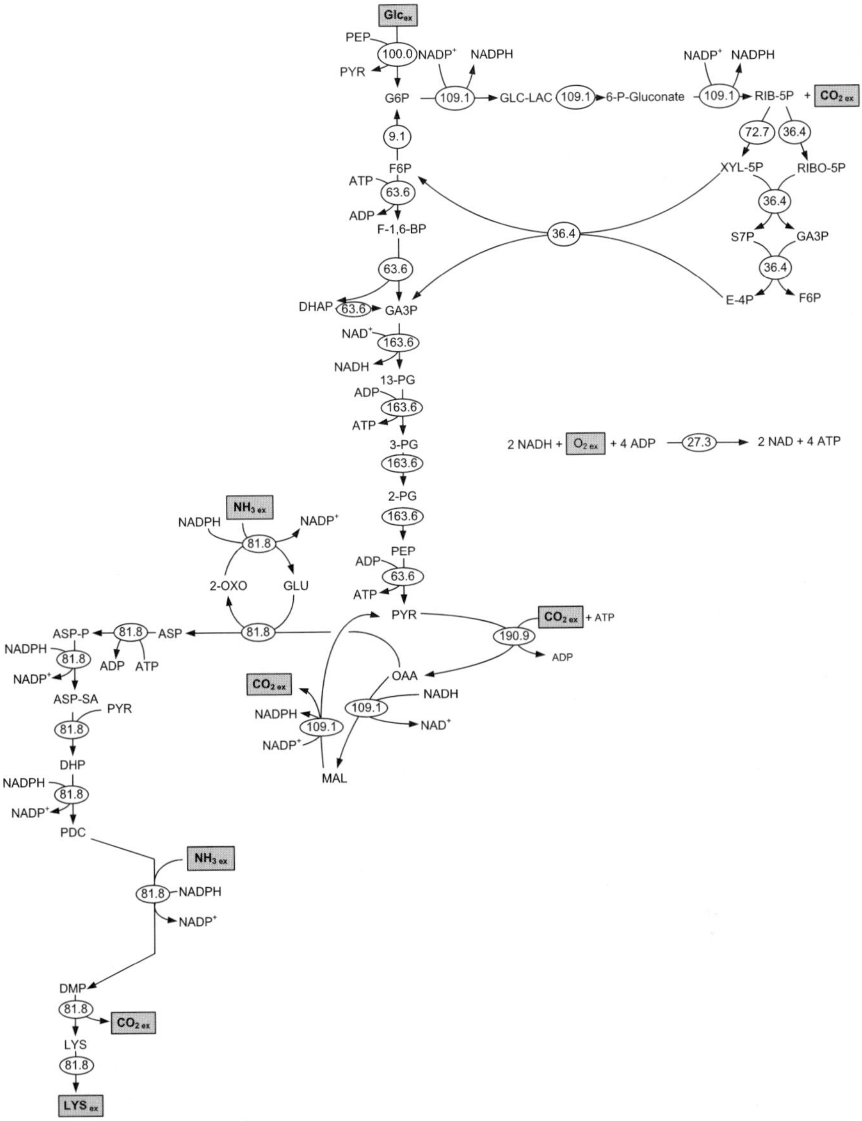

Fig. 5 Metabolic flux distribution of *Corynebacterium glutamicum* with maximal theoretical lysine carbon yield. All fluxes are given as relative molar fluxes to the glucose uptake in mol (mol)$^{-1}$ $\times$ 100 (Krömer, Heinzle and Wittmann 2006, unpublished). One should note that additionally another elementary mode with optimal production exists which, however, only slightly differs in reaction steps involved in cofactor metabolism

4 Strain Engineering

After the discovery of its ability to produce and excrete amino acids (Kinoshita et al. 1957), *C. glutamicum* was used to establish a biotechnological production process for several amino acids. Through the years various methods for strain engineering have been developed to create more efficient production strains.

4.1 Classical Engineering

The first production strains were created within a few years after the discovery of *C. glutamicum* using an iterative procedure of random mutagenesis with UV light or chemical mutagens and subsequent strain selection (Nakayama et al. 1978). The key to success in these days was the use of toxic lysine analogues, such as S-(2-aminoethyl) cysteine, to screen for feedback resistant strains (Nakayama and Araki 1973). These strains later all revealed point mutations in the aspartokinase gene, through which the encoded enzyme was released from feedback inhibition by lysine and threonine (Kalinowski et al. 1991; Thierbach et al. 1990). This modification displays one of the most important characteristics of lysine production strains at all. Consequently, also strains were developed which exhibited a weakened or even blocked biosynthesis of threonine, i.e. auxotrophy for threonine (Nakayama and Araki 1973). Through further cycles of mutagenesis and selection strains with different auxotrophies for other amino acids, vitamins and resistance to other anti-metabolites were obtained (Kelle et al. 2005). The subsequent mutants from such a strain genealogy exhibited a stepwise improvement of production (Schrumpf et al. 1992; Wittmann and Heinzle 2002). Remarkable production properties such as a conversion yield up to 50% and a lysine·HCl titre above 100 g L^{-1} were achieved with such classically derived strains (Ikeda 2003; Leuchtenberger 1996). The additional nutrient requirement and the weak stress tolerance, due to the large number of undesired mutations accumulated during strain development (Ohnishi et al. 2002), display, however, severe disadvantages of conventional production strains and stimulated targeted approaches for strain optimization. After the gene targets improving lysine production had been identified (see below), it became possible to introduce mutant alleles of these genes isolated from the classically obtained producer strains into the wild type to generate stable and stress-tolerant lysine producer strains without additional nutrient requirements (Georgi et al. 2005; Hayashi et al. 2006a; Ohnishi et al. 2002, 2005).

4.2
Metabolic Engineering of Lysine Biosynthesis

The possibility to perform targeted genetic modifications through developments of molecular biology and genetic engineering tools initiated a number of efforts towards rational optimization to *C. glutamicum* (Ohnishi et al. 2002, 2005; Sahm et al. 2000). Logically, many of these studies have focussed on the optimization of the flux through the lysine biosynthesis by directly modifying enzymes of the pathway (Fig. 5). The modification of three of the enzymes, i.e. aspartate kinase (LysC), dihydrodipicolinate synthase (DapA, EC 4.2.1.52) and the lysine exporter (LysE), was especially valuable with respect to improvement of lysine production (Fig. 6). Aspartate kinase is the key enzyme with regard to metabolic control of the lysine pathway as it is subject to a feedback inhibition by threonine and lysine (Kalinowski et al. 1991; Malumbres and Martin 1996). Different point mutations in the *lysC* gene, i.e. in the region coding for its regulatory β-subunit, have been shown to release the enzyme from feedback control and lead to enhanced lysine formation (Cremer et al. 1991; Follettie et al. 1993; Kalinowski et al. 1991; Sugimoto et al. 1997). Similarly, also overexpression of the aspartate kinase gene stimulated production (Jetten et al. 1995). Today, the release of aspartate kinase from feedback control is regarded as one of the most important features of industrial lysine producer strains. This is also underlined by the various patents claiming different amino acid exchanges in this enzyme (Table 1) (Kelle et al.

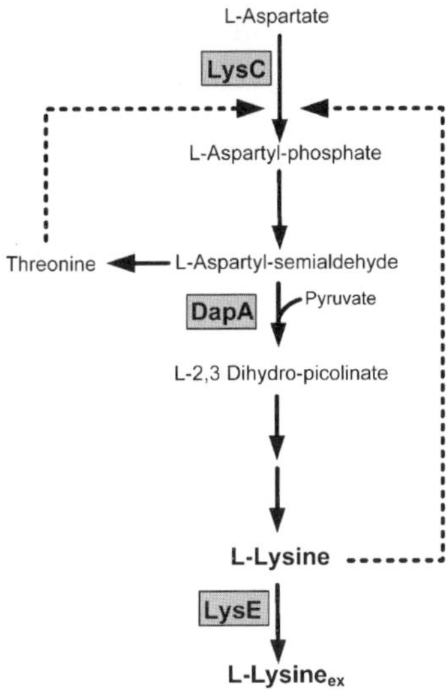

Fig. 6 Lysine biosynthetic pathway in *Corynebacterium glutamicum*. Metabolic regulation through feedback inhibition of asparate kinase by concerted action of lysine and threonine and important targets for optimization of lysine overproduction

2005). Plasmid encoded amplified expression of the *dapA* gene significantly increases lysine (Bonnassie et al. 1990; Cremer et al. 1991; Eggeling et al. 1998; Pisabarro et al. 1993). Amplification of *dapA* expression was further achieved through an extensive mutation of the promoter sequence (Vasicova et al. 1999), whereby a hot spot was discovered at the – 10 region (de Graaf et al. 2001). Also overproduction of diaminopimelate epimerase (DapF, EC 5.1.1.7) and succinyl-aminoketopimelate transaminase (DapC, EC 2.6.1.17), two enzymes of the succinylase branch, was beneficial for lysine formation (Kelle et al. 2005). A striking discovery with respect to lysine production was the discovery of the lysine exporter (LysE) and the subsequent overexpression of the *lysE* gene which resulted in an increased lysine secretion rate (Bellmann et al. 2001; Vrljic et al. 1996, 1999). The recently performed expression of *lysE* from *C. glutamicum* in a *Methylophilus methylotrophus* lysine producing strain was shown to also improve lysine production from methanol by this organism (Gunji and Yasueda 2006). Summarizing, the importance of engineering enzymes of the lysine pathway for efficient lysine production is underlined by the fact that today every single gene of the lysine biosynthetic pathway is covered with one or several patents by the major players in the field (Table 1).

4.3
Metabolic Engineering of NADPH Metabolism

NADPH is consumed by the lysine biosynthetic pathway in four steps either directly or indirectly through the assimilation of ammonium and its efficient supply appears crucial for lysine overproduction. A detailed insight into the NADPH metabolism of *C. glutamicum* has been obtained by ^{13}C metabolic flux studies under various physiological conditions (Becker et al. 2005; Kiefer et al. 2004; Marx et al. 1996, 1997; Sonntag et al. 1995; Wittmann and de Graaf 2005; Wittmann and Heinzle 2001a, 2002; Wittmann et al. 2004a). These allow establishing a NADPH balance. Glucose 6-phosphate dehydrogenase, 6-phosphogluconate dehydrogenase and isocitrate dehydrogenase (EC 1.1.1.42), which uses NADP as cofactor in *C. glutamicum* (Chen and Yang 2000; Eikmanns et al. 1995), are hereby considered as NADPH-generating reactions, whereas NADPH-consuming reactions comprise lysine production and growth with a stoichiometric demand of 16.4 mmol NADPH (g biomass)$^{-1}$ (Wittmann and de Graaf 2005). Figure 7A illustrates that the NADPH metabolism of *C. glutamicum* is highly flexible adjusting to the given overall physiological needs of the cell. NADPH supply and consumption vary depending on the physiological growth state (Marx et al. 1997), the applied carbon source (Dominguez et al. 1998; Wendisch et al. 2000; Wittmann et al. 2004a), or the genetic background (Marx et al. 1999, 2003). In most cases this results in an apparent NADPH excess which seems to increase with diminishing NADPH oxidation. In selected cases even an apparent limitation for NADPH is observed. Examples are phases of maximal lysine production dur-

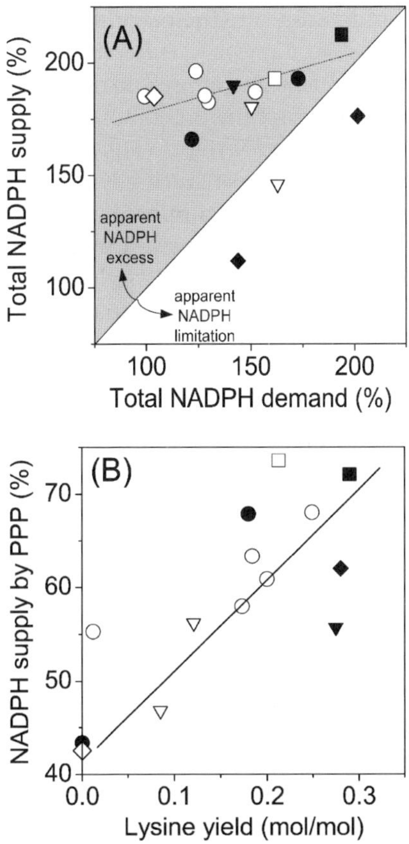

Fig. 7 NADPH metabolism of different *C. glutamicum* strains under various cultivation conditions as revealed by ^{13}C flux analysis: Overall flux through NADPH supply and demand (**A**), flux through the lysine pathway and the PPP (**B**). NADPH supply considers G6P dehydrogenase, 6PG dehydrogenase, and isocitrate dehydrogenase. Lysine production (4 NADPH/lysine) and anabolism are considered as NADPH-consuming reactions. Hereby, the NADPH demand for biomass formation, 16.4 mmol (g cell dry mass)$^{-1}$, was based on a detailed analysis of cellular composition (Wittmann and de Graaf 2005). Studies involved flux analysis of (i) non-producing *C. glutamicum* ATCC 13032 in batch culture on glucose (*open diamond*) (Sonntag et al. 1995), (ii) non-producing *C. glutamicum* ATCC 13032 and lysine producing *C. glutamicum* ATCC 13287, ATCC 21253, ATCC 21526, and ATCC 21543 in batch culture on glucose (*open circle*) (Wittmann and Heinzle 2002), (iii) lysine producing *C. glutamicum* ATCC 21253 during maximum production phase in batch culture on glucose (*open square*) (Wittmann and Heinzle 2001a), (iv) non-producing *C. glutamicum* LE4 and lysine producing *C. glutamicum* MH 20–22B in continuous culture on glucose (*closed circles*) (Marx et al. 1997), (v) lysine-producing *C. glutamicum* MH 20–22B in continuous culture on glucose (*closed square*) (Marx et al. 1996), lysine-producing *C. glutamicum* ATCC 21526 in batch culture on glucose, fructose (*closed diamond*) (Kiefer et al. 2004) and on sucrose (*closed triangle*) (Wittmann et al. 2004a), and lysine-producing *C. glutamicum lysC*^{fbr} and *C. glutamicum lysC*^{fbr} P_{EFTU}fbp during batch culture on glucose (*open triangle*) (Becker et al. 2005)

ing batch culture or production on fructose allowing only a small PPP flux (Kiefer et al. 2004).

The apparent imbalance between supply and demand for NADPH indicates that in *C. glutamicum* so far unassigned metabolic reactions are active in vivo, which either consume or supply NADPH. Possible candidates for NADPH consumption comprise superoxide-generating NADPH oxidase in the respiratory chain (Matsushita et al. 2001) or a metabolic cycle around the pyruvate node involving malic enzyme as hypothesized previously (de Graaf 2000; Petersen et al. 2000). Clear experimental evidence supporting the role of either one of these reactions has not been obtained to date. However, regarding the transhydrogenase-like cycle consisting of pyruvate carboxylase, malate dehydrogenase, and PEP carboxykinase, it has been shown that deletion of *pckA* improves lysine production and *pckA* overexpression reduces lysine production (Riedel et al. 2001) due to altered fluxes within this transhydrogenase-like cycle (Petersen et al. 2001). Concerning NADPH supply, the contribution of malic enzyme has been demonstrated for growth on fructose (Dominguez et al. 1998). Under these conditions cells carry an extremely low PPP flux due to an unfavorable entry point of the substrate into the central metabolism which probably activates malic enzyme (Kiefer et al. 2004). However, overexpression of the malic enzyme gene *malE* in a genetically defined lysine producing strain did neither improve lysine product yields on glucose nor on fructose or sucrose (Georgi et al. 2005).

Since the unassigned NADPH consumption flux becomes minimal with increasing lysine yield (dashed line in Fig. 7A) or NADPH even shows an apparent deficiency, a NADPH limitation of lysine production appears likely. This and the simple fact that four NADPH are required for synthesis of one lysine has stimulated metabolic engineering of the NADPH supply in *C. glutamicum* (Marx et al. 1999, 2003). As a prerequisite for successful modification of the NADPH metabolism, the key pathways supplying NADPH for lysine production in *C. glutamicum* were identified by different approaches. Stoichiometric investigation of the lysine network in the early 1990s already predicted that an increased lysine yield is linked to an increased flux through the PPP and a decreased flux through the TCA cycle (Kiss and Stephanopoulos 1992). The importance of the PPP for efficient lysine production was later shown by metabolic flux analysis (Marx et al. 1996) and by genetic experiments (Georgi et al. 2005; Marx et al. 2003; Ohnishi et al. 2005). As a combined outcome of various flux studies with different strains and under different conditions a close correlation of lysine production with the flux through the PPP could be observed (Fig. 7B). The importance of the PPP for efficient lysine production becomes also obvious from the theoretical flux distribution corresponding to optimal production (Fig. 5). In contrast to the PPP, the contribution of isocitrate dehydrogenase decreases with increasing lysine production (Wittmann and Heinzle 2002). In summary, it turns out that (i) lysine production is likely limited by NADPH availability and that (ii) the

PPP is the major pathway for supply of NADPH. In light of this, different approaches have aimed at redirecting the flux through the PPP in order to increase lysine production by *C. glutamicum*. As example, deletion of the phosphoglucose isomerase gene *pgi*, forcing the cell to metabolize the substrate glucose completely via the PPP, resulted indeed in increased lysine production on glucose (Marx et al. 2003). This strategy, however, appears useful only for glucose-based processes, since sucrose-based processes require an active phosphoglucose isomerase for recycling carbon into the PPP and full NADPH supply (Becker et al. 2005; Wittmann et al. 2004a). Amplified expression of the fructose 1,6-bisphosphatase gene *fbp* in a genetically defined strain of *C. glutamicum*, only carrying a point mutation in the *lysC* gene, was shown to increase lysine yield on glucose, fructose, and sucrose up to about 40% (Becker et al. 2005). Hereby, flux analysis of the mutant revealed that the overexpression of fructose 1,6-bisphosphatase gene indeed resulted in a 10% enhanced PPP flux. On the other hand, it was shown that *fbp* overexpression in a genetically defined strain carrying point mutations in *lysC*, *hom*, *pyc*, and *zwf* improved lysine yields due to reduced intracellular concentrations of fructose 1,6-bisphosphate, an inhibitor of glucose 6-phosphate DH and 6-phosphogluconate DH, in the engineered strain only on sucrose (Georgi et al. 2005). These differences are very likely to be caused by the different background of the strains, as the strain used by Georgi et al. for amplified expression of *fbp* already had an engineered PPP (Georgi et al. 2005). Further successful examples comprise the introduction of point mutations into PPP genes, previously identified by comparative sequencing of the *C. glutamicum* wild type and a classically derived production strain. The substitution A243T in the *zwf* gene encoding for glucose 6-phosphate dehydrogenase (Georgi et al. 2005; Zelder et al. 2005) and the substitution T1083C in the *gnd* gene encoding for 6-phosphogluconate dehydrogenase (Ohnishi et al. 2005) both led to a significantly increased lysine titre. The modification of 6-phosphogluconate dehydrogenase caused an 8% increase in the PPP flux (Ohnishi et al. 2005) which probably resulted from positively changed kinetics of the enzyme.

4.4 Metabolic Engineering of Precursor Supply

Oxaloacetate is a direct precursor of aspartate-derived amino acids, including lysine. In *C. glutamicum*, the anaplerotic enzymes phosphoenolpyruvate carboxylase (Eikmanns et al. 1989; O'Regan et al. 1989; Ozaki and Shiio 1969) and pyruvate carboxylase (Park et al. 1997; Peters-Wendisch et al. 1996, 1998) are involved in supplying oxaloacetate. The importance of these enzymes for lysine production becomes obvious from the correlation between the lumped anaplerotic net carboxylation flux and the flux into the lysine biosynthetic pathway under various conditions determined by ^{13}C metabolic flux analysis (Fig. 8). Pyruvate carboxylase is today regarded as the major anaplerotic enzyme in *C. glutamicum* (Petersen et al. 2000) and overexpression of its gene has been shown to improve lysine production (Peters-Wendisch et al. 2001). This makes sense taking into account that, due to the presence of the phosphotransferase system for sugar uptake, high amounts of pyruvate are generally formed during growth of *C. glutamicum*. However, deletion of the pyruvate carboxylase gene *pyc* leads to the inability to grow on lactate as a sole carbon source (Peters-Wendisch et al. 1998), whereas overexpression of *pyc* in a lysine-producing strain strongly increased production (Peters-Wendisch et al. 2001). Knowing the importance of pyruvate carboxylase for lysine production, the point mutation C1372T, identified in a classically derived producer, was introduced into the *pyc* gene and resulted in a strong increase of lysine production (Ohnishi et al. 2002). It should be noted that, however, also overexpression of the phosphoenolpyruvate carboxylase gene *ppc* is beneficial for the formation of amino acids of the aspartate family (Sano et al. 1987).

The metabolism around the pyruvate node in *C. glutamicum*, was further unravelled by genetic, biochemical, and ^{13}C metabolic flux analyses,

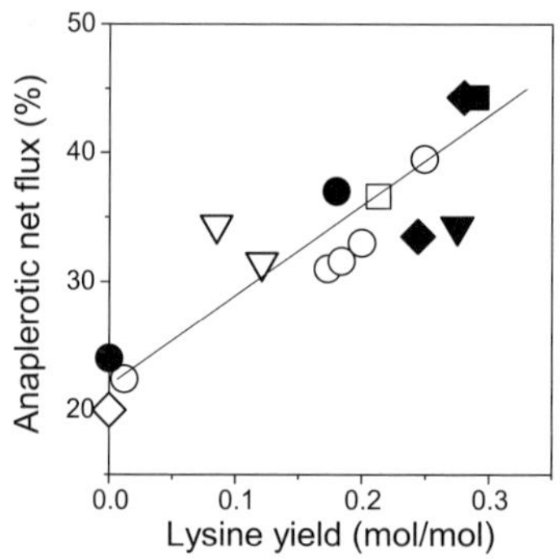

Fig. 8 Correlation between lysine yield and anaplerotic net flux in different *C. glutamicum* strains under various cultivation conditions as revealed by ^{13}C flux analysis. Studies involved flux analysis of (i) non-producing *C. glutamicum* ATCC 13032 in batch culture on glucose (*open diamond*) (Sonntag et al. 1995), (ii) non-producing *C. glutamicum* ATCC 13032 and lysine-producing *C. glutamicum* ATCC 13287, ATCC 21253, ATCC 21526, and ATCC 21543 in batch culture on glucose (*open circle*) (Wittmann and Heinzle 2002), (iii) lysine-producing *C. glutamicum* ATCC 21253 during maximum production phase in batch culture on glucose (*open square*) (Wittmann and Heinzle 2001a), (iv) non-producing *C. glutamicum* LE4 and lysine-producing *C. glutamicum* MH 20–22B in continuous culture on glucose (*closed circles*) (Marx et al. 1997), (v) lysine-producing *C. glutamicum* MH 20–22B in continuous culture on glucose (*closed square*) (Marx et al. 1996), lysine-producing *C. glutamicum* ATCC 21526 in batch culture on glucose, fructose (*closed diamond*) (Kiefer et al. 2004) and on sucrose (*closed triangle*) (Wittmann et al. 2004a), and lysine-producing *C. glutamicum* $lysC^{fbr}$ and *C. glutamicum* $lysC^{fbr}$ P_{EFTU}fbp during batch culture on glucose (*open triangle*) (Becker et al. 2005)

showing the presence of decarboxylating enzymes converting C_4 metabolites of the TCA cycle into C_3 metabolites of glycolysis. These enzymes, phosphoenolpyruvate carboxykinase and malic enzyme, operate in addition to the carboxylation enzymes and establish a highly flexible metabolic reaction cycle around the pyruvate node, which is also present in various other microorganisms (Sauer and Eikmanns 2005). It might function in wasting excess ATP under certain conditions (Sauer et al. 1997; Wittmann and Heinzle 2001b) or equilibrating the intracellular pool sizes of metabolites around the pyruvate node (Sauer and Eikmanns 2005). Additionally, a contribution to the NADPH metabolism has been hypothesized (Cocaign-Bousquet and Lindley 1995; de Graaf 2000). In light of these findings, deletion of the genes coding for the decarboxylating enzymes displayed a promising strategy to enhance the anaplerotic net flux. Indeed deletion of the phosphoenolpyruvate carboxykinase gene *pckA* resulted in a significant improvement of lysine production (Riedel et al. 2001). Neither the deletion nor the overexpression of the malic enzyme gene, however, influenced the metabolism of *C. glutamicum* on sugars markedly (Gourdon et al. 2000; Netzer et al. 2004). Moreover, overexpression of the malic enzyme gene did not change lysine production by *C. glutamicum* (Georgi et al. 2005). This might be due to the fact that typically phosphoenolpyruvate carboxykinase catalyzes the major decarboxylation flux in *C. glutamicum*, whereas malic enzyme obviously only plays a minor role (Gourdon et al. 2000; Petersen et al. 2000). It can, however, not be excluded that the situation might be different under certain physiological conditions.

4.5
Global Strain Engineering through Systems Biotechnology Approaches

The experience of the past clearly shows that detailed quantitative knowledge of metabolic physiology is required for rational design of superior production strains. Especially for the optimization of amino acid production by *C. glutamicum*, characterized by a close connection between central metabolism and product biosynthetic pathways, understanding of global metabolic regulation has turned out to be crucial. The powerful experimental and computational tools available today enable a detailed quantitative investigation of the metabolism of the industrial lysine producer *C. glutamicum*. A milestone in this research was the sequencing of the genome of *C. glutamicum* and the investigation of its genetic repertoire (Bathe et al. 1996; Haberhauer et al. 2001; Ikeda and Nakagawa 2003; Kalinowski et al. 2003; Tauch et al. 2002). Transcriptome analysis in *C. glutamicum* through DNA microarrays provided valuable insights into gene expression under various conditions (Wendisch 2003; Wendisch et al. 2006), such as growth on different carbon sources like glucose or acetate (Gerstmeir et al. 2003; Hayashi et al. 2002; Muffler et al. 2002) or production of lysine (Hayashi et al. 2006b; Krömer et al. 2004). Similarly, also the analysis of the proteome, based on 2-D gel electrophoresis (Bendt et al. 2003; Hermann et al. 2001; Schaffer and Burkovski 2005; Schluesener et al. 2005) has proven valuable to understand important metabolic processes including, for example, nitrogen starvation (Schmid et al. 2000). The enormous contribution of the quantification of metabolic fluxes (fluxome) to our current understanding of the *C. glutamicum* metabolism has already been pointed out above. For this purpose comprehensive approaches combining ^{13}C tracer experiments, metabolite balancing, and isotopomer modelling have been developed and applied to *C. glutamicum* (de Graaf 2000; Wittmann and de Graaf 2005). Important studies to be mentioned are the comparative analysis of fluxes during growth, glutamate, and lysine production (Marx et al. 1997; Sonntag et al. 1995), in different mutants of a lysine-producing strain genealogy (Wittmann and Heinzle 2002), during co-utilization of acetate and glucose (Wendisch et al. 2000), and during lysine production on different industrially relevant carbon sources (Kiefer et al. 2004; Wittmann et al. 2004a). Further achievements comprise the miniaturization of flux analysis to the μL scale for screening purposes (Sauer 2004; Wittmann et al. 2004b), novel approaches towards the analysis of large-scale production processes (Drysch et al. 2003, 2004; El Massaoudi et al. 2003; Yang et al. 2003, 2006a,b) and current developments aiming at flux analysis under dynamic conditions (Nöh et al. 2006; Nöh and Wiechert 2006; Wiechert and Nöh 2005).

These tools have contributed significantly to our current understanding of the *C. glutamicum* metabolism. To fully describe the physiological state of a biological system, however, not one but all of its functional components (genome, transcriptome, proteome, metabolome, and fluxome) have to be analyzed. First examples of such systems-oriented studies already reveal a great potential (Krömer et al. 2004; Lange et al. 2003; Silberbach et al. 2005) and have stimulated the development of systems biotechnology approaches for future characterization and engineering of *C. glutamicum* (Wendisch et al. 2006). Such approaches are especially promising for the targeted multidimensional alteration of complex regulatory networks towards better tolerance of production strains to high temperature or salt levels, or extreme pH values (Kelle et al. 2005).

5 Industrial Production Processes

5.1 Large-Scale Manufacturing

Today large plants are in use for industrial lysine production and the feed-grade amino acid market is developing towards a few major suppliers (Ajinomoto, ADM, BASF, Cheil Jedang, Degussa, Global Biochem, Kyowa Hakko).

Fig. 9 Lysine production plant of the BASF AG located in Gunsan, South Korea with an annual capacity of about 100 000 tons. Copyright BASF AG—The chemical company (2003). Reproduced with permission

As an example Fig. 9 shows the lysine production facility of BASF AG in Gunsan, located at the west coast of South Korea. This plant has an annual production capacity of about 100 000 tons, which accounts for about 15% of the total world market. Industrial large-scale manufacturing of lysine can be separated into different steps involved in upstream processing, the fermentation process itself, and the downstream processing. Upstream processing comprises raw material testing, delivery and storage, the preparation of media from the raw materials, and the preparation of the inoculum for the production. The major industrial carbon sources for lysine production are cane molasses, beet molasses, sucrose and dextrose, whereby the latter is obtained from hydrolysis of starch (Ikeda 2003). Because of batch-to-batch variation of these complex nutrient sources, extensive media testing is carried out to ensure the suitability of a certain raw material with respect to product yield or titre. The carbon source is the major cost factor in industrial lysine production (Kelle et al. 2005). Related to this, sugar suppliers and lysine producers are subject to close alliances or companies cover even both, sugar supply and lysine production. Because of the strong impact of the sugar costs the conversion yield is of major importance for the economy of the production process. The preparation of the inoculum typically involves successive cultivation of the production strain in increasing culture volumes, since *C. glutamicum* exhibits elevated lag phases when inoculated at a biomass concentration below 0.1 g L^{-1} (Kelle et al. 2005). Today, lysine-producing plants use large-scale fermenter vessels with 500 m^3 volume or even more to benefit from the economy of scale. The purification and formulation of the product in the downstream processing is a further important cost factor (Hermann 2003). One applied route comprises cell separation by vacuum filtration, evaporation, and spray drying for product formulation as shown in the flow chart in Fig. 9. Additionally, alternative strategies are used, mainly depending on the lysine preparation finally obtained (Kelle et al. 2005). During the past, lysine was mainly purified from the broth by ion exchange with separation of the biomass, followed by addition of HCl, evaporation, and drying (Hermann 2003). The crystalline lysine·HCl formed is much less hygroscopic than the corresponding sulfate salt (Kelle et al. 2005) and displayed the major product form through the past. In such processes the biomass of *C. glutamicum*, classified as GRAS organism and thus suitable for animal feed, is utilized separately for feed purposes. Today different developments allow a more economical downstream processing and have led to a number of different lysine preparations, such as liquid lysine (50% purity), granulated lysine sulfate (40–50% purity), or liquid lysine sulfate (20–30% purity), which are today well established on the market (Kelle et al. 2005). The granulated product contains the entire fermentation broth without separation of biomass reducing costs and adding additional nutritional value to the product.

5.2 Process Optimization

The optimization of the lysine production process can significantly contribute to reduction of the production costs. The improvement of the downstream processing leading to different lysine preparations as shown above is one illustrative example of the efficiency of such approaches. Other efforts aim at intensification of the process through a change of the operation mode. Repeated batch or fed-batch runs without preparation of the reactor or the inoculum display an interesting approach. After the production is finished a certain fraction of the broth is left in the reactor and is mixed with fresh medium, which significantly decreases the down-time and thus increases the volumetric productivity, i.e. the capacity of a given plant. Problems arise in cases where the production strain is genetically not stable, as in the case of lysine-producing mutants auxotrophic for other amino acids (Hermann 2003). Further improvement could be made by running the lysine production as a continuous process, but that has not been realized yet at a large scale (Ikeda 2003). Studies at a smaller scale, such as the optimization of lysine production in continuous culture under simultaneous phosphate and carbon source limitation, however, show the potential of such approaches (Hirao et al. 1989). Instability of production strains is still a major problem in converting current batch or fed-batch processes into processes with extended runtime, but might be overcome in the near future through new insights into the underlying metabolic processes related to genetic instability or application of, for example, flow cytometry to quantify inoculum viability.

Process optimization or process design can be additionally supported by process modelling. Process models do not only consider process stoichiometry via the mass streams into the process and operational parameters of the different unit operations, but also take environmental aspects into consideration by weighing the environmental burden of utilized and released components (Heinzle et al. 2006). In this regard, process modelling allows a detailed insight into the process, the estimation of process efficiency, and the identification of conditions for optimal yield, product titre, selectivity, or minimized environmental burden. Amongst various examples, this has been recently demonstrated for the case study of lysine production (Knoll and Büchs 2006). In this example, a process model of lysine production, related to the flow chart shown in Fig. 10, was established and implemented

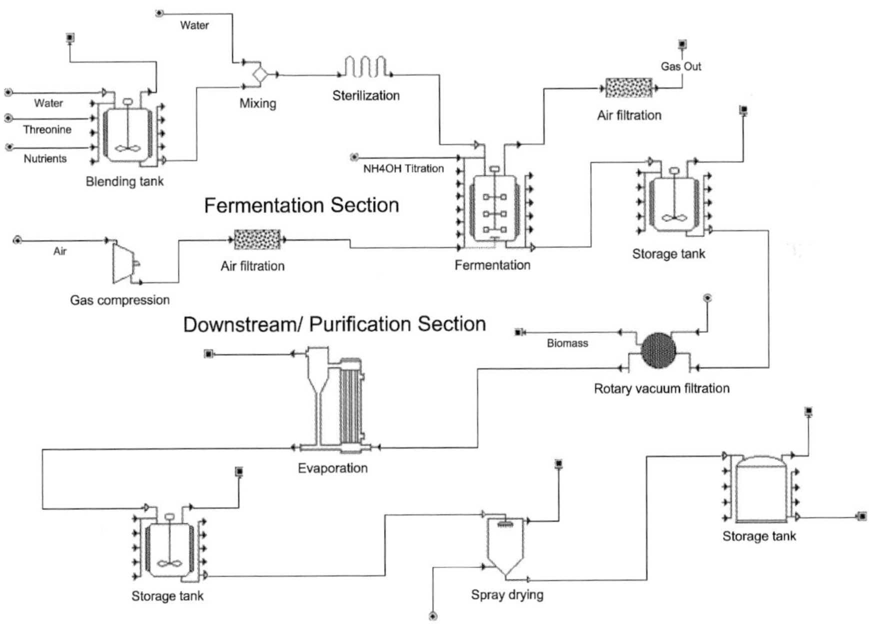

Fig. 10 Process flow diagram of a lysine production plant. Picture taken from: Knoll A, Büchs J (2006) L-lysine—coupling of bioreaction and process model. In: Heinzle, Biwer, Cooney, (eds) Development of Sustainable Bioprocesses—Modelling and Assessment. Copyright: Wiley. Reproduced with permission

into a process modelling software. For the given process setup, coupled to a simple biological model of lysine production, the authors could nicely show that the operational conditions allowing minimum production costs are different from those providing maximum space-time yield. Such modelling approaches are not only interesting for existing processes, but can also support optimal development of new plants. Through variation of the process setup, for example the comparison of alternative down-stream processing routes, optimal process configurations can be identified in an early phase of development, when the degree of freedom is still high.

References

Bartlett ATM, White PJ (1985) Species of *Bacillus* that make a vegetative peptidoglycan containing lysine lack diaminopimelate epimerase but have diaminopimelate dehydrogenase. J Gen Microb 131:2145–2152

Bathe B, Kalinowski J, Pühler A (1996) A physical and genetic map of the *Corynebacterium glutamicum* ATCC 13032 chromosome. Mol Gen Genet 252:255–265

Becker J, Klopprogge C, Zelder O, Heinzle E, Wittmann C (2005) Amplified expression of fructose 1,6-bisphosphatase in *Corynebacterium glutamicum* increases *in vivo* flux through the pentose phosphate pathway and lysine production on different carbon sources. Appl Environ Microbiol 71:8587–8596

Bellmann A, Vrljic M, Patek M, Sahm H, Krämer R, Eggeling L (2001) Expression control and specificity of the basic amino acid exporter LysE of *Corynebacterium glutamicum*. Microbiology 147:1765–1774

Bendt AK, Burkovski A, Schaffer S, Bott M, Farwick M, Hermann T (2003) Towards a phosphoproteome map of *Corynebacterium glutamicum*. Proteomics 3:1637–1646

Bonnassie S, Oreglia J, Sicard AM (1990) Nucleotide sequence of the *dapA* gene from *Corynebacterium glutamicum*. Nucleic Acids Res 18:6421

Born TL, Blanchard JS (1999) Structure/function studies on enzymes in the diaminopimelate pathway of bacterial cell wall biosynthesis. Curr Opin Chem Biol 3:607–613

Chatterjee SP, White PJ (1982) Activities and regulation of the enzymes of lysine biosynthesis in a lysine-excreting strain of *Bacillus megaterium*. J Gen Microb 128:1073–1081

Chen R, Yang H (2000) A highly specific monomeric isocitrate dehydrogenase from *Corynebacterium glutamicum*. Arch Biochem Biophys 383:238–245

Cocaign-Bousquet M, Lindley ND (1995) Pyruvate overflow and carbon flux within the central metabolic pathways of *Corynebacterium glutamicum* during growth on lactate. Enzyme Microb Technol 17:260–267

Cremer J, Eggeling L, Sahm H (1991) Control of the lysine biosynthesis sequence in *Corynebacterium glutamicum* as analyzed by overexpression of the individual corresponding genes. Appl Environ Microbiol 57(6):1746–1752

Cremer J, Treptow C, Eggeling L, Sahm H (1988) Regulation of enzymes of lysine biosynthesis in *Corynebacterium glutamicum*. J Gen Microbiol 134(Pt12):3221–3229

de Graaf AA (2000) Metabolic flux analysis of *Corynebacterium glutamicum*. In: Schügerl K, Bellgard KH (eds) Bioreaction engineering. Springer, Berlin Heidelberg New York, pp 506–555

de Graaf AA, Eggeling L, Sahm H (2001) Metabolic engineering for L-lysine production by *Corynebacterium glutamicum*. Adv Biochem Eng Biotechnol 73:9–29

Dominguez H, Lindley ND (1996) Complete sucrose metabolism requires fructose phosphotransferase activity in *Corynebacterium glutamicum* to ensure phosphorylation of liberated fructose. Appl Environ Microbiol 62(10):3878–3880

Dominguez H, Rollin C, Guyonvarch A, Guerquin-Kern JL, Cocaign-Bousquet M, Lindley ND (1998) Carbon-flux distribution in the central metabolic pathways of *Corynebacterium glutamicum* during growth on fructose. Eur J Biochem 254:96–102

Drysch A, El Massaoudi M, Mack C, Takors R, de Graaf AA, Sahm H (2003) Production process monitoring by serial mapping of microbial carbon flux distributions using a novel sensor reactor approach: ^{13}C-labeling-based metabolic flux analysis and L-lysine production. Metab Eng 5:96–107

Drysch A, El Massaoudi M, Wiechert W, de Graaf AA, Takors R (2004) Serial flux mapping of *Corynebacterium glutamicum* during fed-batch L-lysine production using the sensor reactor approach. Biotechnol Bioeng 85:497–505

Eggeling L, Bott M (2005) Handbook of *Corynebacterium glutamicum*. CRC Press, Boca Raton, Fl

Eggeling L, Oberle S, Sahm H (1998) Improved L-lysine yield with *Corynebacterium glutamicum*: Use of *dapA* resulting in increased flux combined with growth limitation. Appl Microbiol Biotechnol 49:24–30

Eggeling L, Sahm H (1999) L-glutamate and L-lysine: Traditional products with impetuous developments. Appl Microbiol Biotechnol 52:146–153

Eikmanns BJ (2005) Central metabolism: Tricarboxylic acid cycle and anaplerotic reactions. In: Eggeling L, Bott M (eds) Handbook of *Corynebacterium glutamicum*. CRC Press, Boca Raton, Fl, pp 241–276

Eikmanns BJ, Follettie MT, Griot MU, Sinskey AJ (1989) The phosphoenolpyruvate carboxylase gene of *Corynebacterium glutamicum*: Molecular cloning, nucleotide sequence, and expression. Mol Gen Genet 218:330–339

Eikmanns BJ, Rittmann D, Sahm H (1995) Cloning, sequence analysis, expression, and inactivation of the *Corynebacterium glutamicum icd* gene encoding isocitrate dehydrogenase and biochemical characterization of the enzyme. J Bacteriol 177:774–782

El Massaoudi M, Spelthahn J, Drysch A, de Graaf A, Takors R (2003) Production process monitoring by serial mapping of microbial carbon flux distributions using a novel sensor reactor approach: I—sensor reactor system. Metab Eng 5:86–95

Follettie MT, Peoples OP, Agoropoulou C, Sinskey AJ (1993) Gene structure and expression of the *Corynebacterium flavum* N13 *ask-asd* operon. J Bacteriol 175:4096–4103

Georgi T, Rittmann D, Wendisch VF (2005) Lysine and glutamate production by *Corynebacterium glutamicum* on glucose, fructose and sucrose: Roles of malic enzyme and fructose-1,6-bisphosphatase. Metab Eng 7:291–301

Gerstmeir R, Wendisch VF, Schnicke S, Ruan H, Farwick M, Reinscheid D, Eikmanns BJ (2003) Acetate metabolism and its regulation in *Corynebacterium glutamicum*. J Biotechnol 104:99–122

Goodfellow M, Collins MD, Minnikin DE (1976) Thin-layer chromatographic analysis of mycolic acid and other long-chain components in whole-organism methanolysates of coryneform and related taxa. J Gen Microbiol 96:351–358

Gourdon P, Baucher MF, Lindley ND, Guyonvarch A (2000) Cloning of the malic enzyme gene from *Corynebacterium glutamicum* and role of the enzyme in lactate metabolism. Appl Environ Microbiol 66:2981–2987

Gunji Y, Yasueda H (2006) Enhancement of L-lysine production in methylotroph methylophilus methylotrophus by introducing a mutant LysE exporter. J Biotechnol 127:1–13

Haberhauer G, Schröder H, Pompejus M, Zelder O, Kröger B (2001) *Corynebacterium glutamicum* genes encoding proteins involved in membrane synthesis and membrane transport. Patent WO 01/00805

Hartmann M, Tauch A, Eggeling L, Bathe B, Möckel B, Pühler A, Kalinowski J (2003) Identification and characterization of the last two unknown genes, *dapC* and *dapF*, in the succinylase branch of the L-lysine biosynthesis of *Corynebacterium glutamicum*. J Biotechnol 104:199–211

Hayashi M, Mizoguchi H, Ohnishi J, Mitsuhashi S, Yonetani Y, Hashimoto S, Ikeda M (2006a) A *leuC* mutation leading to increased L-lysine production and *rel*-independent global expression changes in *Corynebacterium glutamicum*. Appl Microbiol Biotechnol 72:783–789

Hayashi M, Mizoguchi H, Shiraishi N, Obayashi M, Nakagawa S, Imai J, Watanabe S, Ota T, Ikeda M (2002) Transcriptome analysis of acetate metabolism in *Corynebacterium glutamicum* using a newly developed metabolic array. Biosci Biotechnol Biochem 66:1337–1344

Hayashi M, Ohnishi J, Mitsuhashi S, Yonetani Y, Hashimoto S, Ikeda M (2006b) Transcriptome analysis reveals global expression changes in an industrial L-lysine producer of *Corynebacterium glutamicum*. Biosci Biotechnol Biochem 70:546–550

Heinzle E, Biwer A, Cooney CL (2006) Development of sustainable bioprocesses – modeling and assessment. Wiley, Chichester

Hermann T (2003) Industrial production of amino acids by coryneform bacteria. J Biotechnol 104:155–172

Hermann T, Pfefferle W, Baumann C, Busker E, Schaffer S, Bott M, Sahm H, Dusch N, Kalinowski J, Pühler A, Bendt AK, Krämer R, Burkovski A (2001) Proteome analysis of *Corynebacterium glutamicum*. Electrophoresis 22:1712–1723

Hirao T, Nakano T, Azuma T, Sugimoto M, Nakanishi T (1989) L-lysine production in continuous culture of an L-lysine hyperproducing mutant of *Corynebacterium glutamicum*. Appl Microbiol Biotechnol 32:269–273

Hollander JA (1994) Potential metabolic limitations in lysine production by *Corynebacterium glutamicum* as revealed by metabolic network analysis. Appl Microbiol Biotechnol 42:508–515

Ihnen ED, Demain AL (1969) Glucose-6-phosphate dehydrogenase and its deficiency in mutants of *Corynebacterium glutamicum*. J Bacteriol 98:1151–1158

Ikeda M (2003) Amino acid production processes. Adv Biochem Eng Biotechnol 79: 1–35

Ikeda M, Nakagawa S (2003) The *Corynebacterium glutamicum* genome: Features and impacts on biotechnological processes. Appl Microbiol Biotechnol 62:99–109

Imaizumi A, Kojima H, Matsui K (2006) The effect of intracellular ppGpp levels on glutamate and lysine overproduction in *Escherichia coli*. J Biotechnol 125:328–337

Imaizumi A, Takikawa R, Koseki C, Usuda Y, Yasueda H, Kojima H, Matsui K, Sugimoto S (2005) Improved production of L-lysine by disruption of stationary phase-specific *rmf* gene in *Escherichia coli*. J Biotechnol 117:111–118

Jetten M, Sinskey AJ (1993) Characterization of phosphoenolpyruvate carboxykinase from *Corynebacterium glutamicum*. FEMS Microbiol Lett 111:183–188

Jetten MS, Follettie MT, Sinskey AJ (1995) Effect of different levels of aspartokinase on the lysine production by *Corynebacterium lactofermentum*. Appl Microbiol Biotechnol 43:76–82

Jetten MS, Pitoc GA, Follettie MT, Sinskey AJ (1994) Regulation of phospho(enol)-pyruvate- and oxaloacetate-converting enzymes in *Corynebacterium glutamicum*. Appl Microbiol Biotechnol 41:47–52

Jetten MS, Sinskey AJ (1995) Purification and properties of oxaloacetate decarboxylase from *Corynebacterium glutamicum*. Antonie Van Leeuwenhoek 67:221–227

Kalinowski J, Bachmann B, Thierbach G, Pühler A (1990) Aspartokinase genes *lysC* alpha and *lysC* beta overlap and are adjacent to the aspartate beta-semialdehyde dehydrogenase gene *asd* in *Corynebacterium glutamicum*. Mol Gen Genet 224:317–324

Kalinowski J, Bathe B, Bartels D, Bischoff N, Bott M, Burkovski A, Dusch N, Eggeling L, Eikmanns BJ, Gaigalat L, Goesmann A, Hartmann M, Huthmacher K, Krämer R, Linke B, McHardy AC, Meyer F, Möckel B, Pfefferle W, Pühler A, Rey DA, Rückert C, Rupp O, Sahm H, Wendisch VF, Wiegrabe I, Tauch A (2003) The complete *Corynebacterium glutamicum* ATCC 13032 genome sequence and its impact on the production of L-aspartate-derived amino acids and vitamins. J Biotechnol 104:5–25

Kalinowski J, Cremer J, Bachmann B, Eggeling L, Sahm H, Pühler A (1991) Genetic and biochemical analysis of the aspartokinase from *Corynebacterium glutamicum*. Mol Microbiol 5:1197–1204

Kawahara Y, Yoshihara Y, Ikeda S, Yoshii H, Hirose Y (1990) Stimulatory effect of glycine betaine on L-lysine fermentation. Appl Microbiol Biotechnol 34:87–90

Kelle R, Hermann T, Bathe B (2005) L-lysine production. In: Eggeling L, Bott M (eds) Handbook of *Corynebacterium glutamicum*. CRC Press, Boca Raton, Fl, pp 465–488

Kiefer P, Heinzle E, Wittmann C (2002) Influence of glucose, fructose and sucrose as carbon sources on kinetics and stoichiometry of lysine production by *Corynebacterium glutamicum*. J Ind Microbiol Biotechnol 28:338–343

Kiefer P, Heinzle E, Zelder O, Wittmann C (2004) Comparative metabolic flux analysis of lysine-producing *Corynebacterium glutamicum* cultured on glucose or fructose. Appl Environ Microbiol 70:229–239

Kim HM, Heinzle E, Wittmann C (2006) Deregulation of aspartokinase by single nucleotide exchange leads to global flux rearrangement in the central metabolism of *Corynebacterium glutamicum*. J Microbiol Biotechnol 8:1174–1179

Kinoshita S, Nakayama K, Kitada S (1961) Method of producing L-lysine by fermentation. In: Office USP (ed) US Patent 2 979 439

Kinoshita S, Shigezo U, Shimono M (1957) Studies on the amino acid fermentation, Part I. Production of L-glutamic acid by various microorganisms. J Gen Appl Microbiol 3:193–205

Kiss RD, Stephanopoulos G (1992) Metabolic characterization of a L-lysine-producing strain by continuous culture. Biotechnol Bioeng 39:565–574

Knoll A, Büchs J (2006) L-lysine – coupling of bioreaction and process model. In: Heinzle E, Biwer A, Cooney CL (eds) Development of sustainable bioprocesses – modeling and assessment. Wiley, Chichester, pp 155–168

Krömer JO, Sorgenfrei O, Klopprogge K, Heinzle E, Wittmann C (2004) In-depth profiling of lysine-producing *Corynebacterium glutamicum* by combined analysis of the transcriptome, metabolome, and fluxome. J Bacteriol 186:1769–1784

Krömer JO, Wittmann C, Schröder H, Heinzle E (2006) Metabolic pathway analysis for rational design of L-methionine production by *Escherichia coli* and *Corynebacterium glutamicum*. Metab Eng 8:353–369

Lange C, Rittmann D, Wendisch VF, Bott M, Sahm H (2003) Global expression profiling and physiological characterization of *Corynebacterium glutamicum* grown in the presence of L-valine. Appl Environ Microbiol 69:2521–2532

Leuchtenberger W (1996) Amino acids—technical production and use. In: Rehm HJ, Reed G, Pühler A, Stadler P (eds) Biotechnology, Vol 6. VCH, Weinheim, Germany, pp 465–502

Liebl W (2005) *Corynebacterium* taxonomy. In: Eggeling L, Bott M (eds) Handbook of *Corynebacterium glutamicum*. CRC, Boca Raton, Fl, pp 9–34

Liebl W, Ehrmann M, Ludwig W, Schleifer KH (1991) Transfer of *Brevibacterium divaricatum* DSM 20297t, *Brevibacterium flavum* DSM 20411, *Brevibacterium lactofermentum* DSM 20412 and DSM 1412, and *Corynebacterium glutamicum* and their distinction by rRNA gene restriction patterns. Int J Syst Bacteriol 41:255–260

Malin GM, Bourd GI (1991) Phosphotransferase-dependent glucose transport in *Corynebacterium glutamicum*. J Appl Bacteriol 71:517–523

Malumbres M, Martin JF (1996) Molecular control mechanisms of lysine and threonine biosynthesis in amino acid-producing *Corynebacteria*: Redirecting carbon flow. FEMS Microbiol Lett 143:103–114

Marx A, de Graaf A, Wiechert W, Eggeling L, Sahm H (1996) Determination of the fluxes in the central metabolism of *Corynebacterium glutamicum* by nuclear magnetic resonance spectroscopy combined with metabolite balancing. Biotechnol Bioeng 49(2):111–129

Marx A, Eikmanns BJ, Sahm H, de Graaf AA, Eggeling L (1999) Response of the central metabolism in *Corynebacterium glutamicum* to the use of an NADH-dependent glutamate dehydrogenase. Metab Eng 1:35–48

Marx A, Hans S, Möckel B, Bathe B, de Graaf AA (2003) Metabolic phenotype of phosphoglucose isomerase mutants of *Corynebacterium glutamicum*. J Biotechnol 104:185–197

Marx A, Striegel K, de Graaf A, Sahm H, Eggeling L (1997) Response of the central metabolism of *Corynebacterium glutamicum* to different flux burdens. Biotechnol Bioeng 56(2):168–180

Matsushita K, Otofuji A, Iwahashi M, Toyama H, Adachi O (2001) NADH dehydrogenase of *Corynebacterium glutamicum*. Purification of an NADH dehydrogenase II homolog able to oxidize NADPH. FEMS Microbiol Lett 204:271–276

McCoy AJ, Adams NE, Hudson AO, Gilvarg C, Leustek T, Maurelli AT (2006) L,L-diaminopimelate aminotransferase, a trans-kingdom enzyme shared by *Chlamydia* and plants for synthesis of diaminopimelate/lysine. Proc Natl Acad Sci USA 103:17909–17914

Michal G (1999) Biochemical pathways. Wiley, Chichester

Molenaar D, van der Rest ME, Drysch A, Yucel R (2000) Functions of the membrane-associated and cytoplasmic malate dehydrogenases in the citric acid cycle of *Corynebacterium glutamicum*. J Bacteriol 182:6884–6891

Moon MW, Kim HJ, Oh TK, Shin CS, Lee JS, Kim SJ, Lee JK (2005) Analyses of enzyme II gene mutants for sugar transport and heterologous expression of fructosekinase gene in *Corynebacterium glutamicum* ATCC 13032. FEMS Microbiol Lett 244:259–266

Moritz B, Striegel K, De Graaf AA, Sahm H (2000) Kinetic properties of the glucose-6-phosphate and 6-phosphogluconate dehydrogenases from *Corynebacterium glutamicum* and their application for predicting pentose phosphate pathway flux *in vivo*. Eur J Biochem 267:3442–3452

Muffler A, Bettermann S, Haushalter M, Horlein A, Neveling U, Schramm M, Sorgenfrei O (2002) Genome-wide transcription profiling of *Corynebacterium glutamicum* after heat shock and during growth on acetate and glucose. J Biotechnol 98:255–268

Nakayama K, Araki K (1973) Process for producing L-lysine. US Patent 3 708 395

Nakayama K, Araki K, Kase H (1978) Microbial production of essential amino acid with *Corynebacterium glutamicum* mutants. Adv Exp Med Biol 105:649–661

Netzer R, Krause M, Rittmann D, Peters-Wendisch P, Eggeling L, Wendisch VF, Sahm H (2004) Roles of pyruvate kinase and malic enzyme in *Corynebacterium glutamicum* for growth on carbon sources requiring gluconeogenesis. Arch Microbiol 182:354–363

Nöh K, Wahl A, Wiechert W (2006) Computational tools for isotopically instationary ^{13}C labeling experiments under metabolic steady state conditions. Metab Eng 8:554–577

Nöh K, Wiechert W (2006) Experimental design principles for isotopically instationary ^{13}C labeling experiments. Biotechnol Bioeng 94:234–251

O'Regan M, Thierbach G, Bachmann B, Villeval D, Lepage P, Viret JF, Lemoine Y (1989) Cloning and nucleotide sequence of the phosphoenolpyruvate carboxylase-coding gene of *Corynebacterium glutamicum* ATCC 13032. Gene 77:237–251

Ohnishi J, Katahira R, Mitsuhashi S, Kakita S, Ikeda M (2005) A novel *gnd* mutation leading to increased L-lysine production in *Corynebacterium glutamicum*. FEMS Microbiol Lett 242:265–274

Ohnishi J, Mitsuhashi S, Hayashi M, Ando S, Yokoi H, Ochiai K, Ikeda M (2002) A novel methodology employing *Corynebacterium glutamicum* genome information to generate a new L-lysine-producing mutant. Appl Microbiol Biotechnol 58:217–223

Ozaki H, Shiio I (1969) Regulation of the TCA and glyoxylate cycles in *Brevibacterium flavum*. II. Regulation of phosphoenolpyruvate carboxylase and pyruvate kinase. J Biochem (Tokyo) 66:297–311

Park SM, Shaw-Reid C, Sinskey AJ, Stephanopoulos G (1997) Elucidation of anaplerotic pathways in *Corynebacterium glutamicum* via ^{13}C-NMR spectroscopy and GC-MS. Appl Microbiol Biotechnol 47:430–440

Patek M, Bilic M, Krumbach K, Eikmanns B, Sahm H, Eggeling L (1997) Identification and transcriptional analysis of the *dapB*-orf2-*dapA*-orf4 operon of *Corynebacterium glutamicum*, encoding two enzymes involved in L-lysine synthesis. Biotechnol Letters 19(11):1113–1117

Peters-Wendisch PG, Kreutzer C, Kalinowski J, Patek M, Sahm H, Eikmanns BJ (1998) Pyruvate carboxylase from *Corynebacterium glutamicum*: Characterization, expression and inactivation of the *pyc* gene. Microbiology 144(4):915–927

Peters-Wendisch PG, Schiel B, Wendisch VF, Katsoulidis E, Möckel B, Sahm H, Eikmanns BJ (2001) Pyruvate carboxylase is a major bottleneck for glutamate and lysine production by *Corynebacterium glutamicum*. J Mol Microbiol Biotechnol 3:295–300

Peters-Wendisch PG, Wendisch VF, de Graaf AA, Eikmanns BJ, Sahm H (1996) C_3-carboxylation as an *anaplerotic* reaction in phosphoenolpyruvate carboxylase-deficient *Corynebacterium glutamicum*. Arch Microbiol 165:387–396

Petersen S, de Graaf AA, Eggeling L, Mollney M, Wiechert W, Sahm H (2000) *In vivo* quantification of parallel and bidirectional fluxes in the anaplerosis of *Corynebacterium glutamicum*. J Biol Chem 275:35932–35941

Petersen S, Mack C, de Graaf AA, Riedel C, Eikmanns BJ, Sahm H (2001) Metabolic consequences of altered phosphoenolpyruvate carboxykinase activity in *Corynebacterium glutamicum* reveal anaplerotic regulation mechanisms *in vivo*. Metab Eng 3:344–361

Pfefferle W, Möckel B, Bathe B, Marx A (2003) Biotechnological manufacture of lysine. Adv Biochem Eng Biotechnol 79:59–112

Pisabarro A, Malumbres M, Mateos LM, Oguiza JA, Martin JF (1993) A cluster of three genes (*dapA*, orf2, and *dapB*) of *Brevibacterium lactofermentum* encodes dihydrodipicolinate synthase, dihydrodipicolinate reductase, and a third polypeptide of unknown function. J Bacteriol 175:2743–2749

Riedel C, Rittmann D, Dangel P, Möckel B, Petersen S, Sahm H, Eikmanns BJ (2001) Characterization of the phosphoenolpyruvate carboxykinase gene from *Corynebacterium glutamicum* and significance of the enzyme for growth and amino acid production. J Mol Microbiol Biotechnol 3:573–583

Sahm H, Eggeling L, de Graaf AA (2000) Pathway analysis and metabolic engineering in *Corynebacterium glutamicum*. Biol Chem 381:899–910

Sano K, Ito K, Miwa K, Nakamori S (1987) Amplification of the phosphoenolpyruvate carboxylase gene of *Brevibacterium lactofermentum* to improve amino acid production. Agric Biol Chem 51(2):597–599

Sauer U (2004) High-throughput phenomics: Experimental methods for mapping fluxomes. Curr Opin Biotechnol 15:58–63

Sauer U, Eikmanns BJ (2005) The PEP-pyruvate-oxaloacetate node as the switch point for carbon flux distribution in bacteria. FEMS Microbiol Rev 29:765–794

Sauer U, Hatzimanikatis V, Bailey JE, Hochuli M, Szyperski T, Wüthrich K (1997) Metabolic fluxes in riboflavin-producing *Bacillus subtilis*. Nat Biotechnol 15:448–452

Schaffer S, Burkovski A (2005) Proteomics. In: Eggeling L, Bott M (eds) Handbook of *Corynebacterium glutamicum*. CRC, Boca Raton, Fl, pp 99–118

Schilling CH, Schuster S, Palsson BO, Heinrich R (1999) Metabolic pathway analysis: Basic concepts and scientific applications in the post-genomic era. Biotechnol Prog 15:296–303

Schluesener D, Fischer F, Kruip J, Rogner M, Poetsch A (2005) Mapping the membrane proteome of *Corynebacterium glutamicum*. Proteomics 5:1317–1330

Schmid R, Uhlemann EM, Nolden L, Wersch G, Hecker R, Hermann T, Marx A, Burkovski A (2000) Response to nitrogen starvation in *Corynebacterium glutamicum*. FEMS Microbiol Lett 187:83–88

Schrumpf B, Eggeling L, Sahm H (1992) Isolation and prominent characteristics of an L-lysine hyperproducing strain of *Corynebacterium glutamicum*. Appl Microbiol Biotechnol 37:566–571

Schrumpf B, Schwarzer A, Kalinowski J, Pühler A, Eggeling L, Sahm H (1991) A functionally split pathway for lysine synthesis in *Corynebacterium glutamicium*. J Bacteriol 173:4510–4516

Schuster S, Hilgetag C, Woods JH, Fell DA (2002) Reaction routes in biochemical reaction systems: Algebraic properties, validated calculation procedure and example from nucleotide metabolism. J Math Biol 45:153–181

Silberbach M, Schäfer M, Huser AT, Kalinowski J, Pühler A, Krämer R, Burkovski A (2005) Adaptation of *Corynebacterium glutamicum* to ammonium limitation: A global analysis using transcriptome and proteome techniques. Appl Environ Microbiol 71:2391–2402

Sonntag K, Eggeling L, De Graaf AA, Sahm H (1993) Flux partitioning in the split pathway of lysine synthesis in *Corynebacterium glutamicum*. Quantification by ^{13}C- and ^{1}H-NMR spectroscopy. Eur J Biochem 213:1325–1331

Sonntag K, Schwinde J, de Graaf A, Marx A, Eikmanns B, Wiechert W, Sahm H (1995) ^{13}C NMR studies of the fluxes in the central metabolism of *Corynebacterium glutamicum* during growth and overproduction of amino acids in batch cultures. Appl Microbiol Biotechnol 44:489–495

Sugimoto M, Tanaka A, Suzuki T, Matsui H, Nakamori S, Takagi H (1997) Sequence analysis of functional regions of homoserine dehydrogenase genes from L-lysine and L-threonine-producing mutants of *Brevibacterium lactofermentum*. Biosci Biotechnol Biochem 61:1760–1762

Tauch A, Homann I, Mormann S, Ruberg S, Billault A, Bathe B, Brand S, Brockmann-Gretza O, Rückert C, Schischka N, Wrenger C, Hoheisel J, Möckel B, Huthmacher K, Pfefferle W, Pühler A, Kalinowski J (2002) Strategy to sequence the genome of *Corynebacterium glutamicum* ATCC 13032: Use of a cosmid and a bacterial artificial chromosome library. J Biotechnol 95:25–38

Tauch A, Kaiser O, Hain T, Goesmann A, Weisshaar B, Albersmeier A, Bekel T, Bischoff N, Brune I, Chakraborty T, Kalinowski J, Meyer F, Rupp O, Schneiker S, Viehoever P, Pühler A (2005) Complete genome sequence and analysis of the multiresistant nosocomial pathogen *Corynebacterium jeikeium* K411, a lipid-requiring bacterium of the human skin flora. J Bacteriol 187:4671–4682

Thierbach G, Kalinowski J, Bachmann B, Pühler A (1990) Cloning of a DNA fragment from *Corynebacterium glutamicum* conferring aminoethyl cysteine resistance and feedback resistance to aspartokinase. Appl Microbiol Biotechnol 32:443–448

Udaka S (1960) Screening method for microorganisms accumulating metabolites and its use in the isolation of *Micrococcus glutamicus*. J Bacteriol 79:754–755

Vasicova P, Patek M, Nesvera J, Sahm H, Eikmanns B (1999) Analysis of the *Corynebacterium glutamicum dapA* promoter. J Bacteriol 181:6188–6191

Velasco AM, Leguina JI, Lazcano A (2002) Molecular Evolution of the Lysine Biosynthetic Pathways. J Mol Evol 55:445–459

Vrljic M, Garg J, Bellmann A, Wachi S, Freudl R, Malecki MJ, Sahm H, Kozina VJ, Eggeling L, Saier MH Jr, Eggeling L, Saier MH Jr (1999) The LysE superfamily: Topology of the lysine exporter LysE of *Corynebacterium glutamicum*, a paradyme for a novel superfamily of transmembrane solute translocators. J Mol Microbiol Biotechnol 1:327–336

Vrljic M, Sahm H, Eggeling L (1996) A new type of transporter with a new type of cellular function: L-lysine export from *Corynebacterium glutamicum*. Mol Microbiol 22:815–826

Wehrmann A, Phillipp B, Sahm H, Eggeling L (1998) Different modes of diaminopimelate synthesis and their role in cell wall integrity: A study with *Corynebacterium glutamicum*. J Bacteriol 180:3159–3165

Weinberger S, Gilvarg C (1970) Bacterial distribution of the use of succinyl and acetyl blocking groups in diaminopimelic acid biosynthesis. J Bacteriol 101:323–324

Wendisch VF (2003) Genome-wide expression analysis in *Corynebacterium glutamicum* using DNA microarrays. J Biotechnol 104:273–285

Wendisch VF, Bott M, Kalinowski J, Oldiges M, Wiechert W (2006) Emerging *Corynebacterium glutamicum* systems biology. J Biotechnol 124:74–92

Wendisch VF, de Graaf AA, Sahm H, Eikmanns BJ (2000) Quantitative determination of metabolic fluxes during coutilization of two carbon sources: Comparative analyses with *Corynebacterium glutamicum* during growth on acetate and/or glucose. J Bacteriol 182:3088–3096

White PJ (1983) The essential role of diaminopimelate dehydrogenase in the biosynthesis of lysine by *Bacillus spaericus*. J Gen Microb 129:739–749

Wiechert W, Nöh K (2005) From stationary to instationary metabolic flux analysis. Adv Biochem Eng Biotechnol 92:145–172

Wittmann C, de Graaf A (2005) Metabolic flux analysis in *Corynebacterium glutamicum*. In: Eggeling L, Bott M (eds) Handbook of *Corynebacterium glutamicum*. CRC, Boca Raton, Fl, pp 277–304

Wittmann C, Heinzle E (2001a) Application of MALDI-TOF MS to lysine-producing *Corynebacterium glutamicum*: A novel approach for metabolic flux analysis. Eur J Biochem 268:2441–2455

Wittmann C, Heinzle E (2001b) MALDI-TOF MS for quantification of substrates and products in cultivations of *Corynebacterium glutamicum*. Biotechnol Bioeng 72:642–647

Wittmann C, Heinzle E (2002) Genealogy profiling through strain improvement by using metabolic network analysis: Metabolic flux genealogy of several generations of lysine-producing corynebacteria. Appl Environ Microbiol 68:5843–5859

Wittmann C, Kiefer P, Zelder O (2004a) Metabolic fluxes in *Corynebacterium glutamicum* during lysine production with sucrose as carbon source. Appl Environ Microbiol 70:7277–7287

Wittmann C, Kim HM, Heinzle E (2004b) Metabolic network analysis of lysine producing *Corynebacterium glutamicum* at a miniaturized scale. Biotechnol Bioeng 87:1–6

Yang TH, Wittmann C, Heinzle E (2003) Dynamic calibration and dissolved gas analysis using membrane inlet mass spectrometry for the quantification of cell respiration. Rapid Commun Mass Spectrom 17:2721–2731

Yang TH, Wittmann C, Heinzle E (2006) Respirometric ^{13}C flux analysis, part I: Design, construction and validation of a novel multiple reactor system using on-line membrane inlet mass spectrometry. Metab Eng 8:417–431

Yang TH, Wittmann C, Heinzle E (2006) Respirometric ^{13}C flux analysis, part II. *In vivo* flux estimation of lysine-producing *Corynebacterium glutamicum*. Metab Eng 8:432–446

Yokota A, Lindley ND (2005) Central metabolism: Sugar uptake and conversion. In: Eggeling L, Bott M (eds) Handbook of *Corynebacterium glutamicum*. CRC, Boca Raton, Fl, pp 215–240

Zelder O, Pompejus M, Schröder H, Kröger B, Klopprogge C, Haberhauer G (2005) Genes coding for glucose-6-phosphate-dehydrogenase proteins. US Patent 20050014235 A1

Microbiol Monogr (5)
V. F. Wendisch: Amino Acid Biosynthesis
DOI 10.1007/7171_2006_065/Published online: 9 December 2006

L-Threonine

Mechthild Rieping[1] · Thomas Hermann[2] (✉)

[1]Feed Additives, Research and Development, Degussa AG, Kantstr. 2, 33790 Halle/Westfalen, Germany

[2]Feed Additives, Agroferm ZRt., Degussa AG, Nádudvari útfél, 4183 Kaba, Hungary
thomas.hermann@degussa.com

Abstract This review focuses on the principles and recent progress in production of the essential amino acid L-threonine. Behind glutamic acid, methionine and lysine, threonine is one of the most important amino acids almost exclusively used in the feed industry. Basic principles of threonine producers, like amplified genes coding for enzymes involved in the biosynthesis, are explained. Possible modifications of parts of the metabolism that are not directly related and a section about recent findings on global regulation round up the review of strain improvement or strain breeding. In a second section important necessities of the bioprocess industry, such as reduction of gradient formation and other scale-up related topics, are discussed. Strategies for avoiding such problems by improved reactor design or process modifications are presented and discussed in relation to recent advances in strain improvements. Finally, future steps are presented and discussed at the end of the review.

1 Introduction

Among the "white biotechnology" industries, production of amino acids is one of the most important in terms of volumes and annual turnover. A couple of recent reviews summarize the status quo in this branch of biotechnology (Leuchtenberger et al. 2005; Hermann 2003). The current total annual worldwide consumption of amino acids is estimated to be significantly above two million tons. The biggest contributors to this are amino acids like monosodium glutamate, used as flavour enhancer, or L-lysine, D,L-methionine and L-threonine that are mainly used as feed additives. Their market volume is estimated to be significantly higher than one million tons for each group: food and feed applications.

The essential amino acid L-threonine, which is the focus of this review, belongs to the aspartate family of amino acids. Almost exclusively used as a feed additive, L-threonine is primarily added to pig and poultry diets. While, for example, corn germ meal contains similar amounts of L-threonine (0.38%) and L-lysine (0.42%), soybean meal contains almost twice as much L-lysine (2.80%) as L-threonine (1.81%) (Degussa 2006a). The increase of the L-threonine concentration from 0.55 to 0.75% in a corn–sorghum–peanut meal based diet for young broilers increases the breast meat deposition by more than 15% (Degussa 2006b; Kidd et al. 1999).

In 2005 the L-threonine world market had a volume of about 70 000 tons. Thus, L-threonine ranks third in production volume among the biotechnologically produced amino acids behind L-lysine and L-glutamic acid. However, the growth of market size decelerates and the volatility of prices is high with a decreasing tendency (Fig. 1). This underlines the need for process improvements and a strong and sustainable strain development. Major producers of L-threonine are Ajinomoto, Archer Daniels Midland and Degussa.

Even if there are possibilities to manufacture L-threonine by chemical synthesis or separation from other amino acids after hydrolysis of protein sources, the usage of L-threonine-producing bacteria nowadays is the exclusively applied production method. The focus of this review is the description of recent progress in the production of L-threonine through usage of bac-

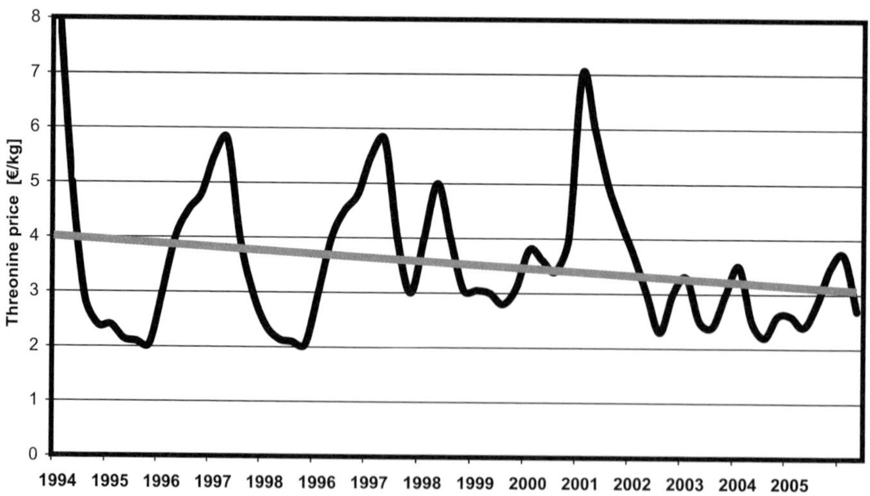

Fig. 1 Historical L-threonine price development 1995–2005

teria, especially *Escherichia coli* and subsequent purification of the product. In the frame of this, we deal with the applied bacterial strains, their physiology and networks of biosynthesis pathways, the production process itself and, finally, the drawbacks of physiology for the production process and vice versa. Although other processes are described and these are even in industrial use, nowadays L-threonine is usually produced by strains of *E. coli*. Hence, this review will focus on bioprocesses using *E. coli* strains to produce L-threonine. Due to the limited available information on this topic—only a few recent publications deal with the production of L-threonine–we also took previously unpublished information into consideration and introduced publications without direct connection to threonine production but with an indirect contribution.

2
How Does L-Threonine Production Work?

The synthesis of threonine is normally tightly controlled to produce only the amount of threonine required to support cellular activity. Bacteria such as *E. coli* do not naturally synthesize excess threonine or export threonine into the medium. If threonine begins to accumulate, the cell uses several mechanisms like feedback regulation at the enzyme level or at the transcriptional

Table 1 Genes and enzymes involved in threonine biosynthesis in *E. coli*

Gene	Enzyme	EC number	Inhibitory ligand(s) [a]	Transcription unit	Inhibitor of expression
thrA	Aspartate kinase + homoserine dehydrogenase	2.7.2.4 1.1.1.3	Threonine	*thrLABC*	Threonine isoleucine
metL	Aspartate kinase + homoserine dehydrogenase	2.7.2.4 1.1.1.3		*metBL*	Methionine
lysC	Aspartate kinase	2.7.2.4	Lysine isoleucine	*lysC*	Lysine
asd	Aspartate semialdehyde dehydrogenase	1.2.1.11		*asd*	Lysine threonine methionine
thrB	Homoserine kinase	2.7.1.39	Threonine homoserine lysine	*thrLABC*	Threonine isoleucine
thrC	Threonine synthase	4.2.3.1		*thrLABC*	Threonine isoleucine

[a] Only the most relevant effectors are shown

level to reduce its synthesis (Camajova et al. 2002b; Table 1). The threonine pathway is embedded into a large metabolic network, and exchange of information with the rest of the metabolic system becomes crucial for improvement of the performance of microbial strains in fermentation processes.

2.1
Strategies for Basic Engineering of the Threonine Pathway

In order to construct strains which are able to synthesize excess threonine and, preferably, transport it into the medium, several basic modifications are required. The threonine biosynthetic pathway starting from aspartate involves five enzymatic steps (Table 1): the first two steps, aspartate kinase and aspartate semialdehyde dehydrogenase, are common to lysine and methionine biosyntheses; a homoserine dehydrogenase activity leads to homoserine biosynthesis; and threonine synthesis is achieved through two further enzymatic steps catalysed by homoserine kinase and threonine synthase. The enzymes are co-ordinately controlled by feedback mechanisms at transcription and enzyme levels (summarized in Table 1) which have to be removed, such as described by Ogawa-Miyata et al. (2001), Park et al. (2002), Lee et al. (2003) and in Research Disclosure (2006). Selection of mutants resistant to the inhibitory action of amino acid analogues and amino acids is a routine approach to gain feedback-resistant mutants, and the ability to produce threonine is further enhanced by introduction of plasmids containing the *thrABC* operon (Debabov 2003; Jin-Ho et al. 1992). As construction of threonine production strains should be designed to achieve good genetic stability and feasible growth rate, metabolic burden effects resulting from multiple plasmid copies and the complex interaction between plasmids and hosts have to be considered (Thiry and Cingolani 2002). In this respect, low-copy plasmids may be more useful than high-copy plasmids (Jones et al. 2000), or threonine production might be stabilized by tightly regulated expression of *thrABC* genes (Camajova et al. 2002a). An alternative approach is the construction of plasmid-free strains which contain chromosomally integrated *thr* operons under the control of a strong non-native promoter (Wang et al. 1999).

Blocking pathways that metabolize threonine, increasing export rates and reducing re-uptake are further basic requirements to achieve high threonine production. Reduction or prevention of threonine degradation is done by attenuation or inactivation of *ilvA* and *tdh* genes (Debabov 2003; Fig. 2). Threonine transport processes play an important role in the production process (Okamoto et al. 1997; Burkovski and Krämer 2002), but complete understanding and elucidation of threonine transport systems is complicated by the presence of multiple transporters with overlapping functions in *E. coli*. SstT (TC: 2.A.23.4.1), TdcC (TC: 2.A.42.2.2) and LivJ (TC: 3.A.1.4.1) are known carriers for threonine uptake (Camajova et al. 2002b; Debabov 2003; Fig. 2), but the activity of at least one additional serine/threonine transporter

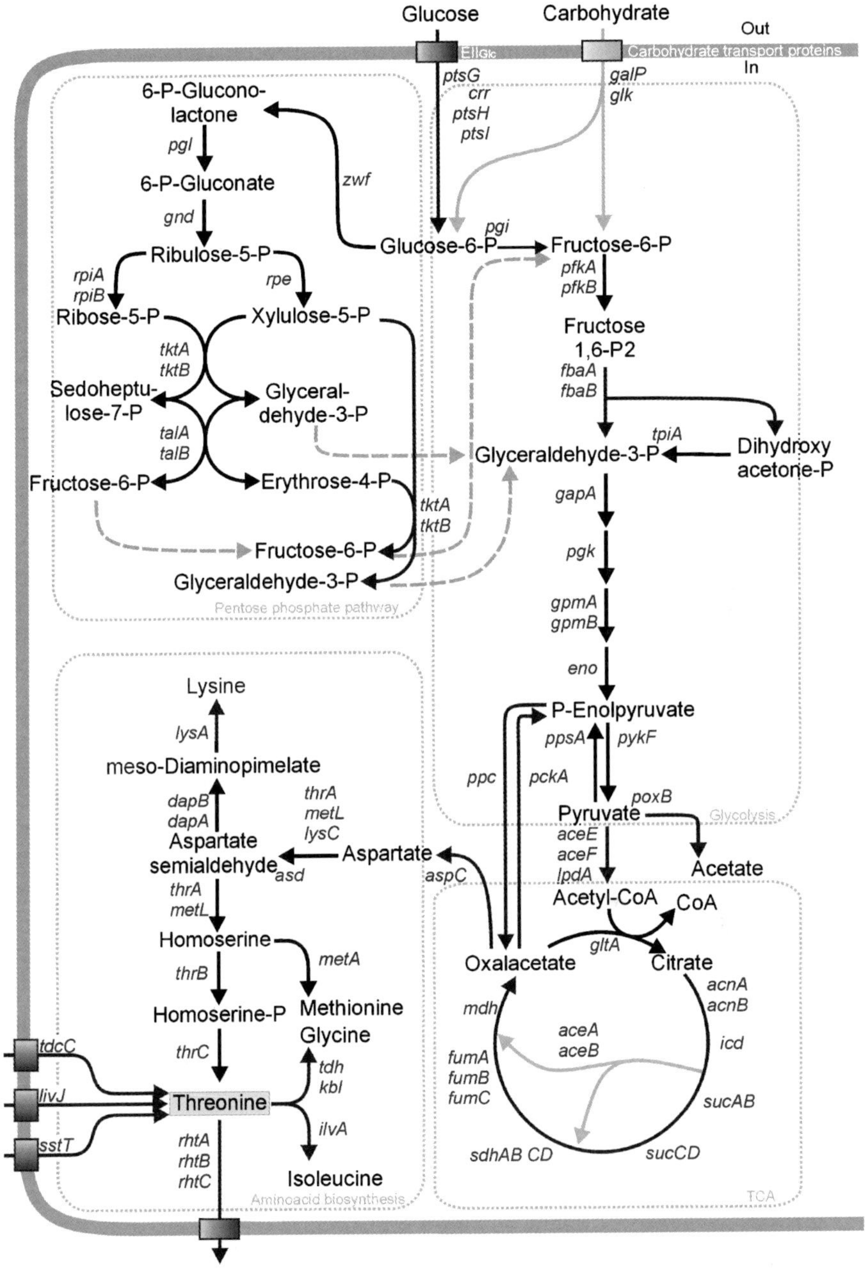

Fig. 2 Overview of the threonine pathway in *E. coli* and connection to central carbon metabolic pathways (glycolysis; pentose phosphate pathway; tricarboxylic acid (TCA) cycle)

was demonstrated in *E. coli* (Kruse et al. 2001). Three *E. coli* genes, *rhtA*, *rhtB* and *rhtC*, coding for membrane proteins involved in amino acid efflux (TC: 2.A.7.3.6 (RhtA), TC: 2.A.76. (RhtB), TC: 2.A.76.1.2 (RhtC)) were described (Fig. 2; Zakataeva et al. 1999; Livshits et al. 2003). It was shown that the respective gene products are involved in threonine excretion, but as in the case of the uptake systems the presence of an additional threonine-excreting carrier was implied by deletion experiments (Kruse et al. 2002).

2.2 Modification of the Central Carbohydrate Metabolism and Associated Pathways

Most studies focus primarily on proteins/enzymes directly involved in the threonine process itself, and less is known about regulation of other associated pathways and events that may be important in engineering improved threonine production strains.

We performed systematic studies on the physiology of our strain lines of threonine producers. The response of increase/decrease of expression levels of several genes involved in the pathway from sugar uptake to threonine export out of the cell (Fig. 2), but also in further associated pathways (nitrogen/sulphur metabolism; energy metabolism), was analysed (Table 2). To

Table 2 Functional classification of genes analysed in threonine production strains after specific modification of expression levels [a]

	Genes	Functional group
Carbohydrate metabolism	*aceEF, aldH, crr, eno, fba, gapA, glpX, gpmA, lpdA, pfkB, pgm, ptsG, pykF, tpiA*	Glycolysis/gluconeogenesis
	gltA, icd, pckA, sdhB, sucAB, sucCD	Citrate cycle (TCA)
	eda, phnN, rpiB, talB, zwf	Pentose phosphate pathway
	aceK	Fructose and mannose metabolism
	aldA, aldB, mqo, pflB, poxB	Pyruvate metabolism
	aceA, aceB	Glyoxylate and dicarboxylate metabolism
Energy metabolism	*aspA*	Nitrogen metabolism
	cysC, cysDN, cysE, cysH, cysJI, cysK, cysM	Sulphur metabolism
Lipid metabolism	*dhaK, dhaL, gldA, glpK*	Glycerolipid metabolism
	glpABCD, glpQ, ugpQ	Glycerophospholipid metabolism

Table 2 (continued)

	Genes	Functional group
Nucleotide metabolism	*adk*	Purine metabolism
	udp	Pyrimidine metabolism
Amino acid biosynthesis	*betB*	Glycine, serine and threonine metabolism, glycine betaine biosynthesis
Metabolism of cofactors and vitamins	*cysG, yggW*	Porphyrin metabolism
Transcription	*rpoE, rpoS*	RNA polymerase
	fadR, fruR, glpR, lrp, malT, phnF	HTH family of transcriptional regulators
	dhaR	Other transcriptional regulators
Folding, sorting, degradation	*ahpCF, clpB, dnaJ, dnaK, htpG, mopA, mopB*	Protein folding and associated processing
Replication and repair	*dps*	Replication, recombination and repair
Membrane transport	*cysA, cysP, cysU, cysW, malE, mglB, phnC, phnD, phnE, phnK, phnL, pstA, pstB, pstC, pstS, sbp, ugpA, ugpB, ugpC, ugpE, yodA, ytfQ*	ABC transporters
	galP, glpT, yaaU	Major facilitator superfamily (MFS)
	ybaL, yjcG	Ion-coupled transporter
	glpF, lamB, ompF, yohJ	Pores, ion channels
	dhaM, ptsH, ptsI	Phosphotransferase system
Signal transduction	*csrA, phoB, phoR*	Two-component system
Unassigned	*fsaAB, glpE, pepB, sodA, yibD*	Fructose 6-phosphate aldolase, thiosulphate sulphurtransferase, peptidase B, superoxide dismutase, putative glycosyl transferase
	bfr, cysB, glpG, hdeA, hdeB, hns, iclR, mlc, phnG, phnJ, phnM, phnO, phnP, phoE, rseA, rseB, rseC, yfiD, ygaC, yjfA, yjgF, yjgP, ytfP	Bacterioferritin, Cys regulon activator, GlpG, HdeA, HdeB, DNA binding protein (histone-like), acetate operon repressor, Mlc, PhnG, PhnJ, PhnM, PhnO, PhnP, outer membrane pore protein E precursor, sigmaE factor regulatory protein (RseA, RseB, RseC), putative formate acetyltransferase (YfiD), hypothetical protein (YgaC, YjfA, YjgF, YjgP, YtfP)

[a] Up-regulation of threonine biosynthesis genes and down-regulation of side reactions of the aspartate family of amino acids and of degradation of threonine is obvious and not included in this list

generate an optimized expression level of genes encoding proteins, which influence distribution of carbon and nitrogen fluxes, is one prerequisite to achieve threonine flux maximization. Blocking of overflow metabolites or side reactions should be combined with increasing conversion of the upstream metabolites to the product. Varying availability of carbon and nitrogen and the maximum attainable oxygen uptake commonly found in large-scale bioreactors (Sect. 3) constitute environmental signals which elicit transcriptional responses, ultimately determining the adaptation of the metabolic networks to (stressful) physiological conditions. Our more systematic approach of analysing parallel developments with different strains and conditions allowed the identification of specific metabolic targets, and subsequently led to significant improvements in product yield and productivity (Sect. 4, Fig. 4).

Detailed analyses were also performed based on gene expression levels obtained by DNA microarray technologies at different growth/production phases. One DNA microarray analysis was carried out to compare the overall gene expression pattern of two threonine production strains representing two basic production levels of our strain development. Figure 3 shows the relative expression values of the subset of genes involved in the central carbohydrate metabolism. RNA was isolated when the maximum threonine concentration was reached. Interestingly, in the strain which produces about 50% more threonine with a more than doubled volumetric productivity in the bioprocess, several genes of the glycolytic and pentose phosphate pathway and of the tricarboxylic acid (TCA) cycle were significantly down-regulated. Obviously, this strain is able to produce high amounts of threonine although a decreased transcriptional activity of central carbon metabolic genes is observed. At that level of strain development we focused on further steps, such as modification of stress responses and of overall maintenance, i.e. supply of building blocks and reducing power. According to changed energetic requirements, more biosynthetic capabilities to produce threonine are left in the more productive strain. Lee et al. (2003) and Kim et al. (2004b) described a global analysis of transcriptomes and proteomes of *E. coli* W3110 and a threonine overproducing mutant thereof. Genes involved in the glyoxylate shunt, the TCA cycle and amino acid biosynthesis were significantly up-regulated, while genes coding for membrane components and genes coding for ribosomal proteins were down-regulated in the mutant strain. Increased synthesis of glycolytic enzymes was also observed at the intracellular protein level. These various findings imply that conclusions about expression profiling analyses of different threonine production strains should be considered with caution and cannot be generalized.

An approach combining transcriptomic, proteomic and ^{13}C flux data was used to solve problems obtained during fermentation process development (Hermann and Rieping 2003). Process intensification by usage of repeated fed-batch technology with high-end threonine mutant strains showed that

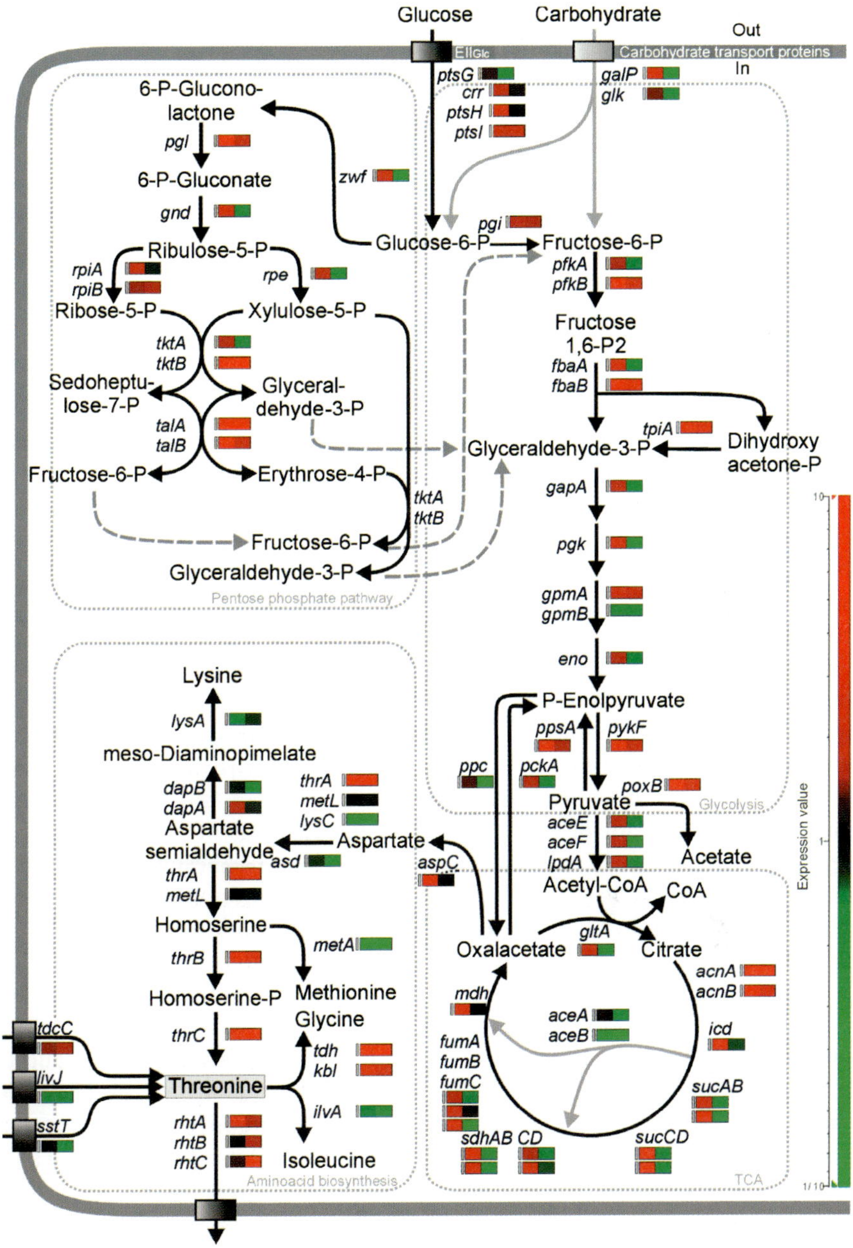

Fig. 3 Expression value projection. The *colour-coded* expression values for two threonine production strains are projected on the *right side* or *below* the respective genes. Relative expression levels of the precursor strain are shown immediately adjacent to the *grey box*; values of the more productive strain are projected on the second position *right of the grey box*. The *legend* for the *colour codes* and corresponding expression value ratios is shown on the *right side*

the performance of the strains decreased from batch to batch. Flux analysis experiments revealed that the carbon flux through PEP carboxykinase was significantly higher than that described for wild-type strains. Expression studies showed that the coding *pck* gene and a number of further genes were induced during the production process compared to other genes coding for anaplerotic reactions. Deletion of this and other targets led to an increase of the volumetric threonine productivity of more than 40% in model strains (Rieping et al. 2003).

2.3
Global Regulation

E. coli strain development clearly benefits from the huge amount of available data. Regulons, operons, regulatory proteins, structural genes, effectors and metabolic pathways related to *E. coli* can easily be researched through (among others) RegulonDB, EcoCyc and KEGG on the Web (Salgado et al. 2004, 2006; Keseler et al. 2005; Kanehisa et al. 2006). A regulatory analysis of the biosynthesis pathway of the aspartate family of amino acids which takes into account environmental conditions, such as regulation and effectors, was performed through an extensive database search by Jin et al. (2004). As the threonine pathway is embedded into a large metabolic network modification of signal integrating regulators, such as Crp, NtrBC, CysB, Fnr, Arc and Fur, controlling exchange of information of metabolic subsystems should occur. Technical advances in simultaneously manipulating multiple steps in metabolic pathways also include use of modified expression levels of transcriptional regulators (Table 2). A lot of studies on knockout mutants of global regulators were performed with *E. coli* (e.g. Gosset et al. 2004; Gyaneshwar et al. 2005; Levanon et al. 2005; Perrenoud and Sauer 2005; Salmon et al. 2003; Zhang et al. 2005), but industrial production strain optimization does not always benefit from these findings and such experiments should be repeated with threonine producers. One example is the defined modification of *rpoS* by introducing a stop codon mutation which is partially suppressed by corresponding tRNA mutations. RpoS controls a large regulatory network induced in response to various environmental stresses (Hengge-Aronis 2002) and it cannot be excluded that the RpoS regulator is required during fermentation, especially for long-term culture conditions. Recent global analyses of *rpoS* gene knockout effects on the metabolism of *E. coli* support this view (Rahman et al. 2006; Weber et al. 2005). In contrast to a complete knock out, it was found that reduction of translation of wild-type *rpoS* mRNA or expression of a modified protein is more beneficial for threonine production (Rieping and Siebelt 2003). The development of screening systems for the isolation of regulators with novel specificities, such as recently described by Galvao and de Lorenzo (2006), is one critical item for further progress in metabolic engineering.

3 What Is Required by the Fermentation Industry?

Bioprocesses for the production of amino acids have been developed since the end of the 1950s. In the beginning, the technologies served only a small market, but over the years the demand grew and new applications were found. With the growing market the size of the plants and the bioreactors increased stepwise. Nowadays, bioreactor sizes from 50 to 500 m^3 are standard in amino acid production (depending on the product). Hence, the fermentation industry requires first of all strains that work perfectly under conditions obtained in reactor volumes up to 500 m^3.

3.1 Scale-up Related Challenges

Like in all bigger bioreactors, mixing is a critical issue especially under consideration of bioreactor volumes up to 500 m^3. A recent overview about this topic and the effects on bioprocess performance can be found in Schmidt (2005). Previously, a European Community funded study about scale-up of an *E. coli* process was performed with several groups (Enfors et al. 2001). In a 22-m^3-scale bioreactor equipped with (commonly used) Rushton turbines, organic acids like formate accumulated, indicating oxygen-limited zones, although the dissolved oxygen signal did not indicate any oxygen limitation. A reduced biomass yield at this scale was suggested to be due to repeated formation and re-consumption of organic acids originating from overflow metabolism or mixed acid fermentation (Xu et al. 1999a). Later, a dynamic model of glucose overflow metabolism in batch and fed-batch cultivations of *E. coli* under fully aerobic conditions was developed, allowing prediction of physiological reactions under certain process conditions (Xu et al. 1999b). Overflow metabolism or mixed acid fermentation induced stress responses of the *E. coli* cells, which were determined by analysing mRNA levels. Stress responses were relaxed when the cells returned to the substrate-limited and oxygen-sufficient compartment of the reactor. Corresponding analyses in the large reactor showed that the concentration of mRNA of four stress-induced genes was lowest at the sampling port most distant from the feed zone. It was assumed that repeated induction/relaxation of stress responses in a large bioreactor may contribute to altered physiological properties of the cells grown in large-scale bioreactors (Schweder et al. 1999).

In a similar study focusing on transcriptional and metabolic effects, Lara et al. (2005) examined the response of *E. coli* on oxygen gradients. In a two-step scale-down system, green fluorescent protein (GFP) producing cells were cultivated in an anaerobic bioreactor with a residence time of 17 s before transfer into a reactor operating under aerobic conditions, where cells were cultivated with a residence time of 33 s. Transcription levels of mixed acid fermentation genes (*ldhA*, *poxB*, *frdD*, *ackA*, *adhE*, *pflD* and *fdhF*) increased from 1.5-fold to over sixfold under conditions of oscillatory oxygen supply compared to entirely aerobic cultures. In addition, the transcript level of *fumB* increased whereas those of *sucA* and *sucB* decreased, suggesting that the TCA cycle was functioning as two open branches. Gene transcription levels revealed that cytochrome bd, which has higher affinity to oxygen but lower energy efficiency, was preferred over cytochrome bO_3 in oscillatory cultures. Simulated oxygen gradients also affected the transcription of genes of the glyoxylate shunt (*aceA*), of global regulators of aerobic and anaerobic metabolism (*fnr*, *arcA* and *arcB*), and other relevant genes (*luxS*, *sodA*, *fumA* and *sdhB*). Transcriptional changes explained the observed alterations in overall stoichiometric and kinetic parameters, and production of ethanol and organic acids. Differences in transcription levels between aerobic and anaerobic compartments were also observed by Lara et al. (2005), indicating that *E. coli* can respond very fast to intermittent oxygen concentrations.

Production of organic acids like acetate or lactate lowers the product yield in bioprocesses additionally by wasting energy through a transport futile cycle, if they are exported under consumption of energy. Once exported, they could be able to re-enter the cell by passive diffusion in their protonated and electroneutral form. This process was examined by Axe and Bailey (1996) for the production of lactate and acetate by *E. coli*.

As mentioned above, a wide field for putative necessities in strain development exists. However, not every new strain can and should be tested on the production scale. This gives room for the development of testing and simulation tools on the lab scale, which allow meaningful scale-up studies. Transfer of a lab process into a production process running in bioreactors with volumes up to 500 m^3 or more needs efficient know-how of process scale-up (for overview, see Thiry and Cingolani 2002). In the past, scale-up was often performed by a stepwise increase of bioreactor volumes, i.e. because of a plant or production extension. However, some groups worked on the development of efficient tools for simulating process scale-up. Problems in scale-up involve the increased (hydrostatic) pressure, higher pCO_2, different sterilization conditions, inhomogeneous substrate or ammonia ion distribution, shear forces at the stirrer tip, power input and gradient formation in bioreactors of an industrial scale. The latter includes gradients of pH, dissolved oxygen, ammonia or carbon sources caused by higher mixing times. A model based approach studying metabolic responses of *E. coli* to gradients and mixing time effects in the bioreactor was described by Delvigne et al. (2005). However, this group concluded that the metabolic response of *E. coli* is not easy to interpret because of the possible simultaneous developments of overflow metabolism and mixed acid fermentation induced by the strong carbon source concentration in the reactor. This shows both the still existing limitations of modelling approaches and the general difficulty to understand

and improve culture conditions for considered-as-easy organisms like *E. coli*. More sophisticated modelling approaches like neural network based setups often show better results than mechanistic unstructured models (Zelic et al. 2006).

For studying gradient formation on the lab scale, at least two principal types of bioreactors were developed: (1) stirred tank reactors with a second external volume (i.e. vessel) used to decrease mixing efficiency and (2) reactors with internal installations inhibiting sufficient mixing. Several examples have been published for the first type of reactor. Xu et al. (1999a) describe a system in which a certain part of the process liquid from a stirred tank reactor is bypassed through a plug-flow reactor before re-entering the main reactor again. Due to the lack of aeration and mixing in the plug-flow reactor the part of the process liquid which is in this reactor is usually oxygen-limited. By adapting the residence time in these tubes, large-scale reactors with badly mixed zones of different volumes could be simulated. This effect was intensified by adding the carbon source to the process liquid just before it entered the plug-flow reactor. This system was used to study physiological responses of the cell to partial oxygen limitation (Schweder et al. 1999). By taking samples from different levels of the plug-flow reactor, the responses of the microorganisms could be quantified. Schilling et al. (1999) describe a bioreactor system with internal installations. In this type of reactor, mixing is inhibited by installation of five horizontal discs separating the vessel into chambers. Combined with Rushton turbines, axial mixing was reduced and the apparent mixing time θ_{90} was prolonged from 10 to 130 s. Although the system is described for *Corynebacterium glutamicum*, it is helpful for threonine processes with *E. coli* as well. Cellular sensing of gradients and the physiological response of the organism to it is of increasing interest. Furthermore, knowledge of genotype, phenotype and physiology of the used microorganism is very valuable for process scale-up. Besides this, the development and use of mixing systems with thorough axial mixing brought improvements in process performance. For scale-up of Rushton turbine-agitated tanks and details on stirrer design see, e.g., Ståhl-Wernersson and Trägårdh (1999) and Campolo et al. (2002).

3.2 Reactor and Process Designs: from Batch to Continuous Processes

The development of new reactor designs contributes to the above-mentioned problems like oxygen limitation or gradient formation. Besides the classical design of a stirred tank reactor (STR), other principles for mixing and dispergation like rotor stator concepts (aeration turbine) are applied more frequently. Lettner and Hermann (2004, unpublished results) evaluated such a system, which is also applied in oxygen-demanding processes like acetate production, in a threonine production process with model strain *E. coli*

B-3996 (Debabov 2003). In a 10-L model reactor, the rotor stator principle (turbine aerator; Ebner 1995) and the classical STR were compared. Remarkably and unexpectedly, it was found that the STR reached ca. 10–20% higher $k_L a_{O\_2}$ than the rotor stator more or less independent of the applied power uptake of the stirrer. Even more important were differences in mixing times: the mixing time ($\tau_{95\%}$, same ion strength, aeration 0.66 vvm) was about 50% higher in the case of the aeration turbine than in the case of the STR, which was again observed independently from the power uptake of the stirrer. Higher mixing times combined with lower oxygen transfer led to lower yields, productivities and biomass formation (31% in the STR experiments compared to 27% in the aeration turbine).

Nowadays, amino acids are produced in bioprocesses operated mainly in fed-batch, repeated fed-batch or continuous production modes. Even if batch technology seems to be rather old-fashioned, it is still used in plants all over the world. Batch technology is applied as a very simple process in terms of process control (no nutrient supply necessary), technology (no vessel for feed solution necessary) or sterility (fewer feed lines). The major boundary for a possible intensification of a batch process is the osmolarity of the initial medium. Especially, high initial concentrations of the carbon source interfere with the growth and productivity of bacteria (Wood et al. 2001). The principal disadvantage of most batch processes is the relatively low productivity due to increased lag phases.

The fed-batch process is still the standard process for manufacturing products from bioprocesses. This is also valid for production of L-threonine. Two of the advantages of fed-batch processes can be summarized as follows:

1. The control of the nutrient supply allows running the process under conditions of optimum oxygen transfer into the bioreactor. Besides this, the control of the nutrient concentration reduces influences of the nutrient on the productivity or yield of the process. If the nutrient concentration exceeds a certain level, unwanted effects like substrate inhibition or overflow metabolism might occur. Typical overflow metabolites also observed in threonine production are lactate or acetate (Enfors et al. 2001). Johannson et al. (2005) describe a shikimate production process with an *E. coli* W3110 derivative under phosphate- and carbon-limited conditions. In comparison to carbon-limited conditions, phosphate limitation resulted in a higher yield of shikimic acid and a lower yield of by-products from the shikimate pathway. For both the production strain and W3110, it was found that more acetate was produced under phosphate limitation than under carbon limitation. Svensson et al. (2005) describe the advantages of temperature-limited cultures in comparison to carbon source-limited cultures. The temperature-limited fed-batch process is a technique where the oxygen consumption rate is controlled by a gradually declining temperature profile rather than a growth-limiting carbon

source feeding profile. At growth rates below $0.1\,h^{-1}$, it was shown that certain strains show reduced lipopolysaccharide formation. Strains producing β-lactamase showed higher performance in temperature-controlled processes than in carbon source-limited processes.

2. Due to the feed of one or more nutrient flows into the bioreactor, osmotic stress for the cell is reduced. In some processes 20% of the final mass or more of a batch has its origin in sugar fed into the process (Ikeda 2003). By reduction of the initial nutrient concentration and further control of the nutrient concentration, the lag phase of the bacteria can be reduced. Raman et al. (2005) examined physiological changes in *E. coli* grown under dextrose-limited fed-batch conditions by applying proteome analysis. Proteins up-regulated during the exponential phase include ribonucleotide biosynthesis enzymes and ribosomal recycling factor. Proteins up-regulated during the fed-batch phase include those involved in high-affinity glucose uptake, transport and degradation of alternate carbon sources and in the TCA cycle, suggesting an enhanced role of the TCA cycle under glucose- and energy-limited conditions. They also report the up-regulation of several putative proteins (*ytfQ*, *ygiS*, *ynaF*, *yggX*, *yfeX*) (Raman et al. 2005). Feedback control of glucose feeding in *E. coli* bioprocesses to overcome the mentioned effects was described recently (Akesson et al. 2001). Other feeding strategies like pH-based control concepts are described, e.g. by Kim et al. (2004a), as alternatives.

A repeated fed-batch process with several L-threonine producers of the species *E. coli* was described by us (Hermann and Rieping 2003). By applying the model strain *E. coli* B-3996 a volumetric L-threonine productivity of $1.77\,g\,L^{-1}h^{-1}$ was achieved in 36 h cultivation time. Considering a preparation time of 10 h, the process productivity would only be $1.39\,g\,L^{-1}h^{-1}$. In the repeated fed-batch process operated with six repeats, the process productivity was increased by more than 20% reaching $1.69\,g\,L^{-1}h^{-1}$. Thus, the advantages of the repeated fed-batch process are often higher process productivities due to decreased lag times of the bacteria on the one hand, and higher plant productivities due to reduced preparation times between the batches on the other. In comparison to the classical fed-batch process, additional advantages of the repeated fed-batch process include a decreased number of seed bioprocesses, reducing variable costs up to 10%, or a reduction of investment costs in the case of newly constructed plants. Disadvantages include a higher risk of contamination, which, however, might be covered by more sophisticated contamination detection methods.

From repeated fed-batch processes only a small step is necessary to establish continuous processes. In this type of process the cells are cultivated until a particular biomass density; the product concentration or quality is achieved before the bioprocess is switched to the continuous mode. Hence, fresh nutrients are fed to the process while the same or similar amount of process liquid is withdrawn from the vessel keeping the process volume constant. Carbon- or phosphate-limited continuous cultures as described by Johannson et al. (2005) are favourable because they allow a tight control of the process and reduce the risk of contamination. Recent developments of repeated fed-batch and continuous processes in threonine production indicate the tremendous potential in process intensification of fermentative amino acid production (Hermann et al. 2005; Kruse et al. 2005a–c).

Independent of the kind of process applied, the media composition and process parameters have to be optimized. For example, it was found by Okamoto and Ikeda (2000) that processes with threonine producing *E. coli* showed fluctuating performance in dependency on corn steep liquor (CSL) batches. CSL is one of the usually applied complex raw materials containing a wide variety of nutrients, such as phosphate, amino acids as nitrogen sources, vitamins or trace elements. Okamoto and Ikeda (2000) showed that by addition of inorganic iron the process became independent from the fluctuating iron concentrations in the CSL batches. Optimization of vitamin concentrations in the media was also shown to improve production significantly. In *E. coli* K-12 derived strains, the optimization of biotin usage led to a more than threefold increase in L-threonine production (Lee et al. 2006). Even if biotin improved threonine production, the molecular cause remains obscure.

4
Where Is It Going To?

In a very competitive threonine market, improvement in both strain and bioprocess development is an inevitable precondition for economic success. Figure 4 shows the development of bioprocess performance in L-threonine production processes achieved at Degussa production sites. This tremendous success story was possible due to intense and fruitful cooperation between strain development, bioprocess development and downstream technology development on the one hand, and tight cooperation between production and research on the other.

Applications of amino acids, either in food and pharmaceuticals or in animal feed nutrition, are expected to grow further during the next few years. While the bigger products will continue to grow at a lower level, smaller products will develop with increasing speed and will thereby be supported through the development of the bigger products and vice versa. New technologies will be developed in all parts of the L-amino acid manufacturing process. Process intensifications like the mentioned repeated fed-batch or continuous bioprocess technology will be implemented into the production sites. Conversely, improvements in one part of the production plant often lead to bottlenecks in other parts. Therefore, efforts in up-

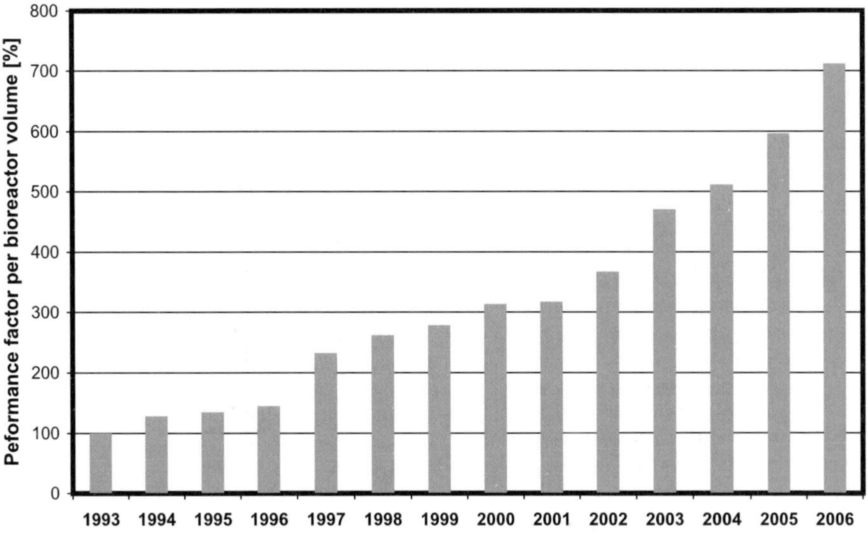

Fig. 4 Development of performance factor in L-threonine production processes achieved at Degussa production sites. Performance factor consists of development of volumetric productivity and fermentation yield. 1993 = 100%. *Not shown* is the parallel development of factory yield or reduction of by-product formation

stream or downstream technology, such as faster or more efficient sterilization technologies or biomass separation units, will accompany these developments.

The biotechnology industry and academic research institutions continuously explore new ways to improve the performance of microbial strains in bioprocesses (Wendisch et al. 2006). Integration of parallel experimental datasets into dynamic simulation tools forms one of the remaining challenges for the elucidation of biological networks and holds promise for biotechnological applications. A purely metabolic prediction cannot be equal to a prediction in which regulatory effects are taken into account. So, differences between models and experimental results do exist. This lack of understanding is the subject of research which is more focused on the behaviour of a system rather than on the behaviour of its single components. There is a general consensus that functional genomics has an enormous potential in metabolic engineering and in biotechnology, but how to use these technologies most efficiently is an area of active research (Hermann 2004). Other techniques like the reduction of genome size in production strains also promise a high potential in improvement of production strains, and will be a typical tool of tomorrow's strain development (Pósfai et al. 2006).

Rapid screening for microorganisms exhibiting specific patterns of gene expression and protein production is critical for progress in microbiology and biotechnology. Many metabolic and genetic experiments can now be designed and performed in bacteria, but only a small fraction can be tested under appropriate conditions.

Determination of experimental data on by-product formation, threonine production and transcriptional response to dynamic fermentation conditions in many individual cases is important to derive an understanding of the molecular response of *E. coli* to the heterogeneous environments commonly found in large-scale bioreactors. For complex transcriptional networks, more sophisticated tools are required to reveal the contribution of each regulator. This is important not only for improving bioreactor design and scale-up procedures, but also for constructing robust strains that can contend with typical industrial conditions.

Even if the speed of improvements slow down a little bit, we will see threonine production processes with a performance coming close to the physiologically feasible maximum.

Acknowledgements We would like to thank Klaus Huthmacher for continuous support and discussion. Furthermore, we are grateful to Patrick Lettner for providing previously unpublished data. We thank Nils Niedner, Reiner Beste and Ralf Kelle for many helpful comments and Stephan Hans for providing drawings.

References

Akesson M, Hagander P, Axelsson JP (2001) Avoiding acetate accumulation in *Escherichia coli* cultures using feedback control of glucose feeding. Biotechnol Bioeng 73:223–230

Axe DD, Bailey JE (1995) Transport of lactate and acetate through the energized cytoplasmic membrane of *Escherichia coli*. Biotechnol Bioeng 47:8–19

Burkovski A, Krämer R (2002) Bacterial amino acid transport proteins: occurrence, functions, and significance for biotechnological applications. Appl Microbiol Biotechnol 58:265–274

Camajova J, Camaj P, Minarik P, Timko J, Turna J (2002a) Stabilization of threonine production in an *Escherichia coli* overproducing strain using tightly regulated T7 promoter system. Biotechnol Lett 24:1893–1897

Camajova J, Camaj P, Timko J (2002b) Biosynthesis and transport of threonine in *Escherichia coli*. Biol Bratislava 57:695–705

Campolo M, Paglianti A, Soldati A (2002) Fluid dynamic efficiency and scale-up of a retreated blade impeller CSTR. Ind Eng Chem Res 41:164–172

Debabov VG (2003) The threonine story. In: Scheper T (ed) Advances in biochemical engineering, vol 79. Springer, Berlin Heidelberg New York, pp 113–136

Degussa (2006a) AminoDat 3.0. www.aminoacidsandmore.com

Degussa (2006b) Facts and figures poultry No.12. www.aminoacidsandmore.com

Delvigne F, Destain F, Thonart P (2005) Bioreactor hydrodynamic effect on *Escherichia coli* physiology: experimental results and stochastic simulations. Bioprocess Biosyst Eng 28:131–137

Ebner H (1995) Process and system for increasing the gas uptake by a liquid being aerated. US Patent 5,332,534

Enfors SO, Jahic M, Rozkov A, Xu B, Hecker M, Jurgen B, Kruger E, Schweder T, Hamer G, O'Beirne D, Noisommit-Rizzi N, Reuss M, Boone L, Hewitt C, McFarlane C, Nienow A,

Kovacs T, Tragardh C, Fuchs L, Revstedt J, Friberg PC, Hjertager B, Blomsten G, Skogman H, Hjort S, Hoeks F, Lin HY, Neubauer P, van der Lans R, Luyben K, Vrabel P, Manelius A (2001) Physiological responses to mixing in large-scale bioreactors. J Biotechnol 85:175–185

Galvao TC, de Lorenzo V (2006) Transcriptional regulators à la carte: engineering new effector specificities in bacterial regulatory proteins. Curr Opin Biotechnol 17:34–42

Gosset G, Zhang Z, Nayyar S, Cuevas WA, Saier MH Jr (2004) Transcriptome analysis of Crp-dependent catabolite control of gene expression in *Escherichia coli*. J Bacteriol 186:3516–3524

Gyaneshwar P, Paliy O, McAuliffe J, Jones A, Jordan MI, Kustu S (2005) Lessons from *Escherichia coli* genes similarly regulated in response to nitrogen and sulfur limitation. Proc Natl Acad Sci USA 102:3453–3458

Hengge-Aronis R (2002) Recent insights into the general stress response regulatory network in *Escherichia coli*. J Mol Microbiol Biotechnol 4:341–346

Hermann T (2003) Industrial production of amino acids by coryneform bacteria. J Biotechnol 104:155–172

Hermann T (2004) Using functional genomics to improve productivity in the manufacture of industrial biochemicals. Curr Opin Biotechnol 15:444–448

Hermann T, Rieping M (2003) Fermentation process for the preparation of L-threonine. US Patent 6,562,601

Hermann T, Kruse D, Rieping M, Thierbach G (2005) Verfahren zur Herstellung von L-Threonin. German Patent Application DE 2004/028859

Ikeda M (2003) Amino acid production processes. In: Scheper T (ed) Advances in biochemical engineering, vol 79. Springer, Berlin Heidelberg New York, pp 1–36

Jin JH, Choi KK, Jung US, In YH, Lee SY, Lee J (2004) Regulatory analysis of amino acid synthesis pathway in *Escherichia coli*: aspartate family. Enzyme Microb Technol 35:694–706

Jin-Ho L, Oh JW, Noh KS, Lee HH, Lee JH (1992) Construction of L-threonine overproducing *Escherichia coli* by cloning of the threonine operon. J Microbiol Biotechnol 2:243–247

Johansson L, Lindskog A, Silfversparre G, Cimander C, Nielsen KF, Lidén G (2005) Shikimic acid production by a modified strain of *E. coli* (W3110shik1) under phosphate-limited and carbon-limited conditions. Biotechnol Bioeng 92:384–392

Jones KL, Kim SW, Keasling JD (2000) Low-copy plasmids can perform as well as or better than high-copy plasmids for metabolic engineering of bacteria. Metab Eng 2:328–338

Kanehisa M, Goto S, Hattori M, Aoki-Kinoshita KF, Itoh M, Kawashima S, Katayama T, Araki M, Hirakawa M (2006) From genomics to chemical genomics: new developments in KEGG. Nucleic Acids Res 34:D354–D357

Keseler IM, Collado-Vides J, Gama-Castro S, Ingraham J, Paley S, Paulsen IT, Peralta-Gil M, Karp PD (2005) EcoCyc: a comprehensive database resource for *Escherichia coli*. Nucleic Acids Res 33:D334–D337

Kidd MT, Lerner SP, Allard JP, Rao SK, Halley JT (1999) Threonine needs for finishing broilers: growth, carcass, and economic responses. J Appl Poultry Res 8:160–169

Kim BS, Lee SC, Lee YS, Chang YC, Chang HN (2004a) High cell density fed-batch cultivation of *Escherichia coli* using exponential feeding combined with pH-stat. Bioprocess Biosyst Eng 26:147–150

Kim YH, Park JS, Cho JY, Cho KM, Park YH, Lee J (2004b) Proteomic response analysis of a threonine-overproducing mutant of *Escherichia coli*. Biochem J 381:823–829

Kruse D, Six S, Krämer R, Burkovski A (2001) Analysis of threonine uptake in *Escherichia coli* threonine production strains. Biotechnol Lett 23:401–404

Kruse D, Krämer R, Eggeling L, Rieping M, Pfefferle W, Tchieu JH, Chung YJ, Saier MH Jr, Burkovski A (2002) Influence of threonine exporters on threonine production in *Escherichia coli*. Appl Microbiol Biotechnol 59:205–210

Kruse D, Hermann T, Rieping M, Thierbach G (2005a) Verfahren zur Herstellung von L-Threonin. German Patent Application DE 2004/029340

Kruse D, Hermann T, Thierbach G, Rieping M (2005b) Verfahren zur Herstellung von L-Threonin. German Patent Application DE 2004/029639

Kruse D, Hermann T, Rieping M (2005c) Verfahren zur Herstellung von L-Threonin. German Patent Application DE 2004/030417

Lara AR, Leal L, Flores N, Gosset G, Bolívar G, Ramírez OT (2005) Transcriptional and metabolic response of recombinant *Escherichia coli* to spatial dissolved oxygen tension gradients simulated in a scale-down system. Biotechnol Bioeng 93:6–13

Lee JH, Lee DE, Lee BU, Kim HS (2003) Global analysis of transcriptomes and proteomes of a parent strain and an L-threonine-overproducing mutant strain. J Bacteriol 185:5442–5451

Lee MH, Lee HW, Park JH, Ahn JO, Jung JK, Hwang YI (2006) Improved L-threonine production of *Escherichia coli* mutant by optimization of culture conditions. J Biosci Bioeng 101:127–130

Leuchtenberger W, Huthmacher K, Drauz K (2005) Biotechnological production of amino acids and derivatives: current status and prospects. Appl Microbiol Biotechnol 69:1–8

Levanon SS, San KY, Bennett GN (2005) Effect of oxygen on the *Escherichia coli* ArcA and FNR regulation systems and metabolic responses. Biotechnol Bioeng 89:556–564

Livshits VA, Zakataeva NP, Aleshin VV, Vitushkina MV (2003) Identification and characterization of the new gene *rhtA* involved in threonine and homoserine efflux in *Escherichia coli*. Res Microbiol 154:123–135

Ogawa-Miyata Y, Kojima H, Sano K (2001) Mutation analysis of the feedback inhibition site of aspartokinase III of *Escherichia coli* K-12 and its use in L-threonine production. Biosci Biotechnol Biochem 65:1149–1154

Okamoto K, Kino K, Ikeda M (1997) Hyperproduction of L-threonine by an *Escherichia coli* mutant with impaired L-threonine uptake. Biosci Biotechnol Biochem 61:1877–1882

Okamoto K, Ikeda M (2000) Development of an industrial stable process for L-threonine fermentation by an L-methionine-auxotrophic mutant of *Escherichia coli*. J Biosci Bioeng 89:87–89

Park EY, Kim JH, Hyun HH, Lee HH (2002) Removal of attenuator region of *thr* operon increases the production of threonine in *Escherichia coli*. Biotechnol Lett 24:1815–1819

Perrenoud A, Sauer U (2005) Impact of global transcriptional regulation by ArcA, ArcB, Cra, Crp, Cya, Fnr, and Mlc on glucose catabolism in *Escherichia coli*. J Bacteriol 187:3171–3179

Pósfai G, Plunkett G III, Fehér T, Frisch D, Keil GM, Umenhoffer K, Kolisnychenko V, Stahl B, Sharma SS, de Arruda M, Burland V, Harcum SW, Blattner FR (2006) Emergent properties of reduced-genome *Escherichia coli*. Science 312:1044–1046

Rahman M, Hasan MR, Oba T, Shimizu K (2006) Effect of *rpoS* gene knockout on the metabolism of *Escherichia coli* during exponential growth phase and early stationary phase based on gene expressions, enzyme activities and intracellular metabolite concentrations. Biotechnol Bioeng 94:585–595

Raman B, Nandakumar MP, Muthuvijayan V, Marten MR (2005) Proteome analysis to assess physiological changes in *Escherichia coli* grown under glucose-limited fed-batch conditions. Biotechnol Bioeng 92:384–392

Research Disclosure Journal (2006) RD 505054

Rieping M, Siebelt N (2003) Amino acid-producing bacteria and a process for preparing L-amino acids. WO Patent Application 03/74719

Rieping M, Bastuck C, Hermann T, Thierbach G (2003) Fermentation process for the preparation of L-amino acids using strains of the family *Enterobacteriaceae*. US Patent 6,916,637

Salgado H, Gama-Castro S, Martinez-Antonio A, Diaz-Peredo E, Sanchez-Solano F, Peralta-Gil M, Garcia-Alonso D, Jimenez-Jacinto V, Santos-Zavaleta A, Bonavides-Martinez C, Collado-Vides J (2004) RegulonDB (version 4.0): transcriptional regulation, operon organization and growth conditions in *Escherichia coli* K-12. Nucleic Acids Res 32:D303–D306

Salgado H, Santos-Zavaleta A, Gama-Castro S, Peralta-Gil M, Penaloza-Spinola IM, Martinez-Antonio A, Karp PD, Collado-Vides J (2006) The comprehensive updated regulatory network of *Escherichia coli* K-12. BMC Bioinformatics 7:5

Salmon K, Hung S, Mekjian K, Baldi P, Hatfield GW, Gunsalus RP (2003) Global gene expression profiling in *Escherichia coli*. J Biol Chem 278:29837–29855

Schilling BM, Pfefferle W, Bachmann B, Leuchtenberger W, Deckwer W-D (1999) A special reactor design for investigations of mixing time effects in a scaled-down industrial L-lysine fed-batch fermentation process. Biotechnol Bioeng 64:599–606

Schmidt FR (2005) Optimization and scale-up of industrial fermentation processes. Appl Microbiol Biotechnol 68:425–435

Schweder T, Kruger E, Xu B, Jurgen B, Blomsten G, Enfors SO, Hecker M (1999) Monitoring of genes that respond to process-related stress in large-scale bioprocesses. Biotechnol Bioeng 65:151–159

Ståhl-Wernersson E, Trägårdh C (1999) Scale-up of Rushton turbine-agitated tanks. Chem Eng Sci 54:4245–4256

Svensson M, Svensson I, Enfors SO (2005) Osmotic stability of the cell membrane of *Escherichia coli* from a temperature-limited fed-batch process. Appl Microbiol Biotechnol 67:345–350

Thiry M, Cingolani D (2002) Optimizing scale-up fermentation processes. Trends Biotechnol 20:103–105

Wang MD, Bradshaw JS, Swisher SL, Liaw HJ, Hanke PD, Binder TP (1999) Strains of *Escherichia coli*, methods of preparing the same and use thereof in fermentation processes for L-threonine production. US Patent 5,939,307

Weber H, Polen T, Heuveling J, Wendisch VF, Hengge R (2005) Genome-wide analysis of the general stress response network in *Escherichia coli*: σ^S-dependent genes, promoters, and sigma factor selectivity. J Bacteriol 187:1591–1603

Wendisch VF, Bott M, Eikmanns BJ (2006) Metabolic engineering of *Escherichia coli* and *Corynebacterium glutamicum* for biotechnological production of organic acids and amino acids. Curr Opin Microbiol 9:268–274

Wood JM, Bremer E, Csonka LN, Kraemer R, Poolman B, van der Heide T, Smith LT (2001) Osmosensing and osmoregulatory compatible solute accumulation by bacteria. Comp Biochem Physiol A Mol Integr Physiol 130:437–60

Xu B, Jahic M, Blomsten G, Enfors SO (1999a) Glucose overflow metabolism and mixed-acid fermentation in aerobic large-scale fed-batch processes with *Escherichia coli*. Appl Microbiol Biotechnol 51:564–571

Xu B, Jahic M, Enfors SO (1999b) Modeling of overflow metabolism in batch and fed-batch cultures of *Escherichia coli*. Biotechnol Prog 15:81–90

Zakataeva NP, Aleshin VV, Tokmakova IL, Troshin PV, Livshits VA (1999) The novel transmembrane *Escherichia coli* proteins involved in the amino acid efflux. FEBS Lett 452:228–232

Zelić B, Bolf N, Vasić-Rački Đ (2006) Modeling of the pyruvate production with *Escherichia coli*: comparison of mechanistic and neural networks-based models. Bioprocess Biosyst Eng 29:39–47

Zhang Z, Gosset G, Barabote R, Gonzalez CS, Cuevas WA, Saier MH Jr (2005) Functional interactions between the carbon and iron utilization regulators, Crp and Fur, in *Escherichia coli*. J Bacteriol 187:980–990

Microbiol Monogr (5)
V. F. Wendisch: Amino Acid Biosynthesis
DOI 10.1007/7171_2006_067/Published online: 9 December 2006

Aromatic Amino Acids

Georg A. Sprenger

Institute of Microbiology, Universität Stuttgart, Allmandring 31, 70569 Stuttgart, Germany
Georg.sprenger@imb.uni-stuttgart.de

Abstract Biosynthesis of the three aromatic amino acids (L-phenylalanine, L-tryptophan, L-tyrosine) and its regulation in *Escherichia coli* and corynebacteria are reviewed. The common aromatic biosynthetic pathway (shikimate pathway) starts with the condensation of phosphoenolpyruvate and erythrose 4-phosphate. Through six biosynthetic steps the pathway proceeds via shikimate to chorismate, from which the terminal pathways to tryptophan, phenylalanine and tyrosine branch. The first step in the common pathway is performed in *E. coli* by three isoenzymes, which are specifically feedback-inhibited by the three terminal products.

The pathway to tryptophan starts with anthranilate formation and includes reactions with L-serine and 5-phosphoribosyl-pyrophosphate. Phenylalanine and tyrosine biosyntheses proceed via prephenate, and each include a decarboxylation and transamination step. The first committed steps of each terminal pathway are strictly regulated by feedback inhibition, repression and partly through attenuation (in *E. coli*).

L-Tryptophan and L-phenylalanine are essential amino acids for man and most livestock. Main microbial producer strains are *E. coli* and *Corynebacterium glutamicum*. Strain development includes alleviation of the various regulatory levels (feedback inhibition resistance, derepression), both in the common aromatic pathway and in the terminal pathways. In recent years, metabolic engineering has also taken into account the fact that precursor supply may become limiting once the other impediments for carbon flux are gone. Strains with improved phosphoenolpyruvate and/or erythrose 4-phosphate supply have successfully been developed. Applications for L-tryptophan are the feed and pharmaceutical markets, while L-phenylalanine is mainly used as building block for the artificial sweetener, aspartame®. A possible application for L-tyrosine is as a building block for the synthesis of L-DOPA.

1 Introduction

The three proteinogenic aromatic amino acids (AAA) are L-tryptophan (Trp), L-phenylalanine (Phe) and L-tyrosine (Tyr). They have in common a side group with an aromatic ring structure. AAA absorb UV light with absorption maxima around 260 nm (Phe) or 280 nm (Tyr, Trp). This is the basis of an easy spectrophotometric measurement of protein concentrations at wavelength 280 nm (A_{280}) (Layne 1957). On average, the three AAA together constitute less than 10% of proteins, L-tryptophan being the largest and the rarest of the 20 different amino acids in proteins (SwissProt 2005).

Our knowledge of AAA biosynthesis is based mainly on the studies with *Escherichia coli* and closely related bacteria in the 1950s through 1970s, especially to the pioneering work performed in the groups of B. Davis, C. Yanofsky, D. Sprinson, F. Gibson, A.J. Pittard and K.M. Herrmann (referred to in Pittard 1996; Herrmann and Weaver, 1999; Pittard and Yang, 2005). Biosynthesis of AAA in most microorganisms (e.g. yeast or bacteria), in unicellular organisms (as in the apicomplexan malaria parasite *Plasmodium falciparum*, Roberts et al. 1998), in fungi, or in plants is via the common (or general) aromatic (amino acid) biosynthetic pathway. This is often also called the shikimate pathway for the intermediate compound shikimic acid (or shikimate) (Bentley 1990; Herrmann 1995; Dewick 1998; Knaggs 2001, 2003). Shikimate was found to fulfill, as sole compound, all growth requirements of bacterial strains auxotrophic for Phe, Tyr and Trp (Davis 1951). The last common compound in this pathway, however, is chorismic acid (or chorismate; Greek "*khoris*" for "separate"; Gibson et al. 1962; Gibson 1964), which is thus the precursor for all aromatic amino acids. Although it could be reasoned that the pathway should therefore be appropriately renamed the "chorismate pathway", it should be noted that chorismate fails to act as growth factor in triple auxotrophic strains, presumably as it is unable to enter the cell (Pittard 1996). Table 1 lists the enzymes, genes and substrates of the common AAA pathway, depicted in Fig. 1, and of the three terminal pathways leading to Trp, Phe and Tyr, shown in Fig. 2.

The shikimate pathway occurs in microorganisms, fungi and plants, but not in animals (Herrmann and Weaver, 1999). As a consequence, L-tryptophan and L-phenylalanine are termed essential amino acids for man and most livestock (with the exception of the protein self-sufficient ruminants), and both amino acids must be present in the diet (Leuchtenberger et al. 2005). L-Tyrosine is not an essential amino acid as it can be formed in mammals from L-phenylalanine via an enzymatic hydroxylation step. Of course, a source of L-phenylalanine must be present.

The general AAA pathway uses two molecules of PEP and one molecule of erythrose 4-phosphate (plus reducing equivalents and ATP) to form, in seven steps, chorismate as the precursor of all three aromatic amino acids. In bac-

Table 1 Enzymes, substrate(s) and genes of the AAA pathway

Step	Enzyme name/EC number	Substrate(s)	Product(s)	Gene(s)
I	3-Deoxy-D-arabino-heptulosonate 7-phosphate synthase, EC 25154	Phospho*enol*pyruvate (PEP) Erythrose 4-phosphate	3-Deoxy-D-arabino-heptulosonate 7-phosphate (DAHP)	*aroF, aroG, aroH*(*E. coli*) *aro, aroII* (DAHP) (*C. glutamicum*)
II	3-Dehydroquinate synthase, EC 4234	DAHP	3-Dehydroquinate	*aroB*
III	3-Dehydroquinate dehydratase (Dehydroquinase) EC 42110	3-Dehydroquinate	3-Dehydroshikimate	*aroD*
IV	Shikimate dehydrogenase EC 11125	3-Dehydroshikimate, NAD(P)	Shikimate	*aroE, ydiB*
V	Shikimate kinase EC 27171	Shikimate, ATP	Shikimate 3-phosphate (Shi3P)	*aroK, aroL* *aroK* only in *C.glut*
VI	5-Enolpyruvoylshikimate 3-phosphate synthase EC 25119	Shi3P, PEP	5-Enolpyruvoylshikimate 3-phosphate (EPSP)	*aroA*
VII	Chorismate synthase EC 4235	EPSP	Chorismate	*aroC*
VIII	Chorismate mutase EC 54995	Chorismate	Prephenate	*tyrA**, *pheA**, *aroM* *csm* in *C.glut*
IX	Prephenate dehydrogenase EC 13112	Prephenate	*p*-Hydroxyphenylpyruvate	*tyrA**
(IXa	Prephenate aminotransferase EC 26178 or 26179)	Prephenate	Arogenate (pretyrosine)	gene unknown in *C.glut*
X	Prephenate dehydratase EC 42151	Prephenate	Phenylpyruvate	*pheA**

Gene designations and functions of gene products are the same for both *E. coli* and *C. glutamicum* unless indicated otherwise
* Bifunctional gene product

Table 1 (continued)

Step	Enzyme name/EC number	Substrate(s)	Product(s)	Gene(s)
XI	Tyrosine aminotransferase EC 2615 or 26157	*p*-Hydroxyphenylpyruvate or phenylpyruvate	Tyrosine or phenylalanine	*tyrB, (ilvE, aspC)* *tyrB*
(XIa	Arogenate dehydrogenase EC 13178)	Arogenate (pretyrosine)	Tyrosine	*tyrA* in *C.glut.*
(XIb	Phenylpyruvate transaminase EC 2615 or 26157)	Phenylpyruvate	Phenylalanine	*pat* in *C.glut.*
XII	Anthranilate synthase EC 41327	Chorismate, glutamine	Anthranilate, glutamate Pyruvate	*trpE, trpD** *trpE,G* in *C.glut.*
XIII	Anthranilate phosphoribosyl transferase EC 24218	Anthranilate, PRPP	Phosphoribosyl anthranilate (PRA) Pyrophosphate	*trpD** *trpD* in *C.glut.*
XIV	Phosphoribosyl anthranilate isomerase EC 53124	PRA	1-(*o*-Carboxyphenylamino)-1-deoxy-ribulose 5-phosphate (CPADRP)	*trpC** *trpC(F)* in *C.glut.*
XV	Indoleglycerol phosphate synthetase EC 41148	CPADRP	Indoleglycerol 3-phosphate (InGP) CO_2	*trpC** *trpC(F)* in *C.glut.*
XVI	Tryptophan synthase EC 42120	Indole-glycerol 3-phosphate L-Serine	tryptophan, Glyceraldehyde 3-phosphate	*trpA, trpB*

* = bifunctional gene product; Gene designations and functions of gene products are the same for both *E. coli* and *C. glutamicum* unless indicated otherwise.

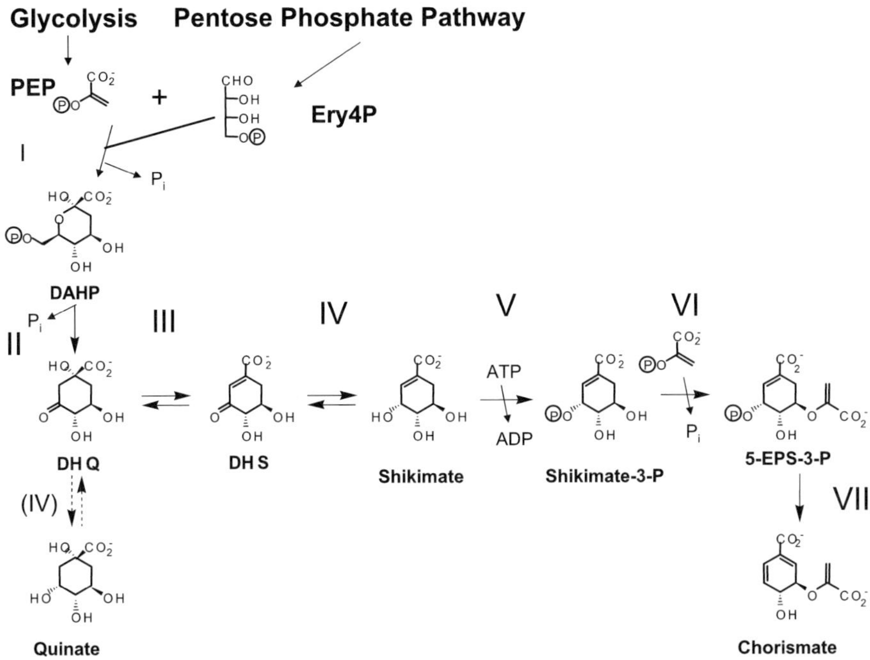

Fig. 1 General aromatic biosynthesis pathway up to chorismate. The precursors, phosphoenolpyruvate (*PEP*) and erythrose 4-phosphate (*Ery4P*) for the general aromatic biosynthesis pathway are provided by the central metabolic pathways, glycolysis and the pentose phosphate pathway. Enzymatic steps of the pathways are numbered from I to VII and are explained in the text and in Table 1. *DAHP* 3-deoxy-D-arabinoheptulosonate 7-phosphate, *DHQ* dehydroquinate, *DHS* dehydroshikimate, *5-EPS-3-P* 5-enolpyruvoylshikimate 3-phosphate, P_i inorganic phosphate

teria, chorismate is not only the precursor of aromatic amino acids but also of isochorismate (biosynthesis of menaquinone and siderophores like enterobactin), *p*-aminobenzoate (folic acid biosynthesis) and *p*-hydroxybenzoate (ubiquinone biosynthesis) (Dosselaere and Vanderleyden 2001). All enzymes of the pathway, their corresponding genes and metabolic intermediates are known and well studied, especially from *Escherichia coli* and the related *Salmonella enterica* serovar Typhimurium. Regulation also has been studied in detail (Pittard 1996). Apart from fermentation, AAA can be produced by extraction from protein sources, through chemical and chemo-enzymatic methods, or through biotransformations (e.g. tryptophan from indole and L-serine). In biotechnology, metabolic engineering of the AAA pathway has become powerful in recent years and will be described together with current uses of AAA (Sects. 2.4 and 2.5).

This chapter presents a brief outline of the biosynthesis of AAA, including regulation and genetics in microorganisms. However, due to space constraints, not all original articles can be quoted. Thus, the reader is referred to

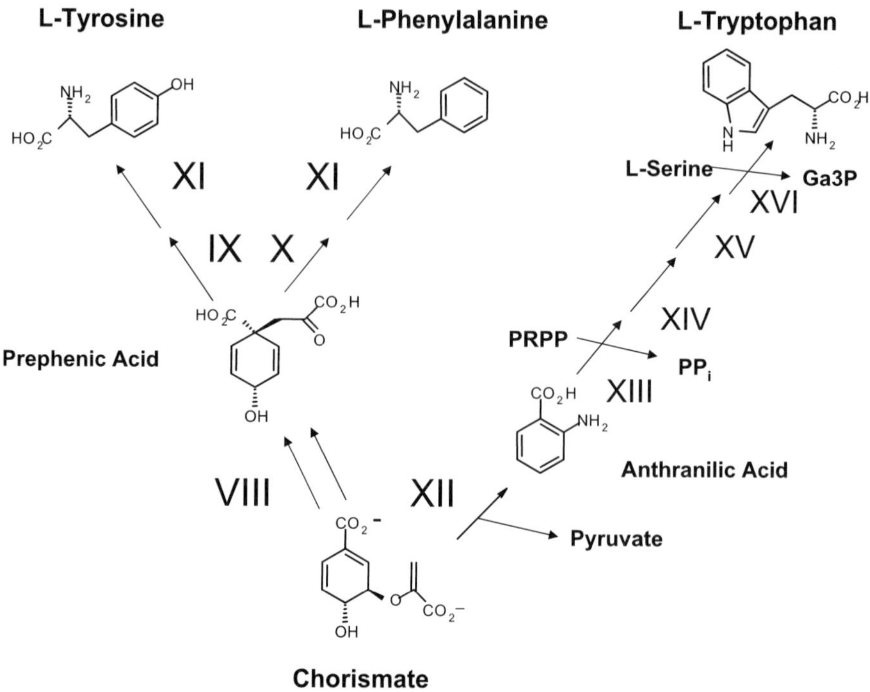

Fig. 2 Terminal pathway to L-tyrosine, L-phenylalanine and L-tryptophan. From chorismate, the three terminal pathways of aromatic amino acid biosynthesis diverge. Enzymatic steps of the pathways are numbered from VIII to XI (phenylalanine and tyrosine pathways) and from XII to XVI (tryptophan pathway) and are explained in the text and in Table 1. *PRPP* 5-phosphoribosyl pyrophosphate, *PP$_i$* pyrophosphate, *Ga3P* D-glyceraldehyde 3-phosphate

several excellent reviews and more recent original articles (usually from 2000 and later), which contain further references (for recent reviews see: Haslam 1993; Herrmann 1995; Pittard 1996; Dewick 1998; Herrmann and Weaver 1999; Knaggs 2001, 2003; Pittard and Yang 2005).

2 The General Aromatic Amino Acid Pathway (Shikimate Pathway)

The initial step in the biosynthesis of aromatic amino acids is the condensation of phospho*enol*pyruvate (PEP; from the glycolytic pathway) and of erythrose 4-phosphate (Ery4P; from the pentose phosphate pathway) to deliver DAHP (3-deoxy-D-*arabino*-heptulosonate 7-phosphate) and inorganic phosphate (see Fig. 1). This aldol condensation is catalysed by a metal-dependent enzyme, DAHP-synthase (EC 2.5.1.54; Srinivasan and Sprinson 1959; Herrmann and Weaver 1999). The shikimate pathway proceeds further from DAHP via 3-dehydroquinate and 3-dehydroshikimate to shikimate. In an ATP-dependent phosphorylation, shikimate 3-phosphate is formed, which then accepts a second molecule of PEP to form 5-enolpyruvyl-shikimate 3-phosphate. In the last step of the common pathway, chorismate is formed (see Fig. 1 and Table 1 for a description of the genes and enzymes involved).

DAHP-synthase, as the first committed step of AAA synthesis, controls carbon flow into the pathway (Ogino et al. 1982). In *E. coli*, three isoenzymes of DAHP synthase occur, encoded by the genes *aroF* (tyrosine-sensitive DAHP synthase AroF), *aroG* (phenylalanine-sensitive enzyme AroG) and *aroH* (tryptophan-sensitive enzyme AroH) (Umbarger 1978; Herrmann 1983; Pittard 1996). Both the Phe- and Tyr-sensitive isoenzymes can be completely inhibited by about 0.1 mM of the corresponding amino acid (K_i values of 13 μM and 82 μM, respectively; McCandliss et al. 1978). The three isoenzymes contribute to the overall DAHP synthase activity in *E. coli* wild-type cells grown on minimal medium (Phe-enzyme 80% or more, Tyr-enzyme about 15%, very little activity of Trp-enzyme) (Herrmann 1983; Pittard 1996). Amino acid residues within the allosteric sites of the isoenzymes have been identified from structural analysis of feedback-insensitive mutants. Some of these sites overlap with the active sites (Herrmann and Weaver, 1999). Mutations conferring feedback resistance have been described, some of them leading to single amino acid exchanges (Edwards et al. 1987; Ray et al. 1988; Weaver and Herrmann 1990; Ger et al. 1994; Pittard 1996; Kikuchi et al. 1997; Jossek et al. 2001).

In *Corynebacterium glutamicum*, two types of DAHP synthases are found, one is Tyr-sensitive (gene *aro*), and the other is sensitive to both L-tyrosine and L-phenylalanine (gene *aroII*). Moreover, AroII forms a polypeptide complex with chorismate mutase, which converts chorismate to prephenate (Ikeda, 2006). Mutants in allosterically regulated DAHP synthases have been described (Liao et al. 2001). In *Bacillus subtilis*, chorismate inhibits DAHP synthase (Gollnick et al. 2002). Recently, it has been proposed that initiation of AAA biosynthesis in some archaea is completely different and requires 6-deoxy-5-ketofructose 1-phosphate and L-aspartate semialdehyde as precursors (White 2004).

The second step of the pathway is from DAHP to dehydroquinate (DHQ) and is catalysed by DHQ synthase (EC 4.2.3.4). This enzymatic reaction comprises a cascade of reactions: oxidation, β-elimination of inorganic phosphate, ring opening and finally an intramolecular aldol condensation. The *E. coli* DHQ synthase requires divalent cations for activity and, although the catalysed reaction is redox neutral, catalytic amounts of NAD in addition (Bender et al. 1989; Herrmann and Weaver 1999).

In strains with feedback-resistant DAHP synthase, DHQ synthase activity may become limiting and DAHP (or its dephosphorylated derivative, DAH) accumulates (Ogino et al. 1982; Rüffer et al. 2004). *E. coli* mutants with lesions in *aroB* readily excrete DAHP or DAH into the culture supernatant (Frost and Knowles 1984). This feature was used by several groups to study the effects of varying precursor supplies for the DAHP synthase reaction by quantitative measurements with whole cells (Frost and Knowles 1984; Draths and Frost 1990; Patnaik and Liao 1994; Patnaik et al. 1995; Liao 1996; Lu and Liao 1997; Sprenger et al. 1998a,b; Krämer et al. 1999; Baez et al. 2001; Oldiges et al. 2004)

The subsequent step of water elimination from DHQ is catalysed by 3-DHQ dehydratase (dehydroquinase, EC 4.2.1.10; Fig. 1) leading to 3-dehydroshikimate. This introduces the first double bond into the ring. 3-Dehydroshikimate is reduced in the following step to shikimate (Fig. 1) by shikimate dehydrogenase (EC 1.1.1.25). *E. coli* appears to have two isoenzymes encoded by *aroE* (NADPH-dependent; Anton and Coggins 1988) and *ydiB* (NADH/NADPH-dependent enzyme; EC 1.1.1.282; Michel et al. 2003; Benach et al. 2003), which was recently discovered from a structural genomics effort. Shikimate kinase (EC 2.7.1.71) catalyses the phosphorylation of shikimate to yield shikimate 3-phosphate (Shi3P) (Fig. 1, Table 1). *E. coli* has two genes for shikimate kinase (*aroK, aroL*). The *aroL* gene apparently encodes the major enzyme (isoenzyme II) with the highest affinity to its substrates (Whipp and Pittard, 1995). Mutants blocked in *aroL* or in both genes excrete shikimic acid. As a by-product of shikimate-producing *E. coli* strains, 3-dehydroshikimate and quinate appear (Frost and Draths 1995; Draths et al. 1999; Knop et al. 2001; Chandran et al. 2003; Johansson et al. 2005, Fig. 1). Quinate is formed reversibly through a side activity of the shikimate dehydrogenases on 3-DHQ, a phenomenon called "hydroaromatic equilibrium" (Knop et al. 2001; Johansson et al. 2005). Shikimate produced by these *E. coli* mutants is used as building block for the synthesis of tamiflu®, an anti-influenza drug from Hoffmann-La Roche (Krämer et al. 2003; Johansson et al. 2005).

A second PEP molecule enters the AAA pathway at the sixth step, in which PEP is condensed with Shi3P by EPSP synthase (EC 2.5.1.19; Fig. 1, Table 1) to yield 5-enolpyruvyl-shikimate 3-phosphate (EPSP). The three-carbon fragment introduced from PEP is destined to become the side-chain of Phe and Tyr afterwards, and is replaced in Trp biosynthesis. Whereas plant EPSP synthases are the target of the herbicide glyphosate (Roundup®) which is a competitive inhibitor of PEP at the enzyme, bacterial enzymes are either insensitive or inhibited to a lesser degree (Herrmann and Weaver 1999). The seventh and last step is catalysed by chorismate synthase (EC 4.2.3.5, Fig. 1, Table 1) which performs a *trans*-1,4 elimination of phosphate from EPSP to yield chorismate. This introduces the second double bond to form a cyclohexadiene ring system.

Beyond chorismate, the pathways divide into the terminal pathways, which are specific for each aromatic amino acid. The initial reaction of the L-tryptophan path yields anthranilate from chorismate. The L-phenylalanine and L-tyrosine paths diverge at prephenate, which is derived from chorismate. As chorismate is the common substrate of several enzymes, it is interesting to note that the K_m values for chorismate range from 1.2 μM (for purified anthranilate synthase, Trp pathway), via 45 μM (chorismate mutase-prephenate dehydratase, Phe pathway), to 92 μM (chorismate mutase-prephenate dehydrogenase, Tyr pathway) (Herrmann and Weaver 1999).

For simplicity, the following steps are only considered for the main producing organisms, Corynebacteria and *E. coli*. In other bacteria, in plants and fungi, slightly deviating pathways exist (Herrmann 1983, 1995; Jung et al. 1986; Niederberger et al. 1992; Pittard 1996; Herrmann and Weaver 1999; Knaggs 2001, 2003; Heimstaedt et al. 2005). The vast plentitude of secondary plant or fungal compounds derived from phenylalanine or tyrosine (lignins, pigments, flavonoids, tocopherols) or from tryptophan (alkaloids) cannot be discussed here (Dewick 1998; Knaggs 2001, 2003).

2.1 Phenylalanine and Tyrosine Pathways

The first reaction of both the Phe and Tyr pathways involves the conversion of chorismate to prephenate and is catalysed by chorismate mutase (EC 5.4.99.5). In *E. coli*, chorismate mutases are moieties of bifunctional enzymes of the Tyr and Phe pathways and are associated with the enzyme activities that carry out the second reaction, an oxidation with concomitant decarboxylation: chorismate mutase-prephenate dehydrogenase (T protein; EC 5.4.99.5/EC 1.3.1.12, NAD-dependent and decarboxylating; delivering *p*-hydroxyphenylpyruvate; gene *tyrA*; Chen et al. 2003) in Tyr, and chorismate mutase-prephenate dehydratase (P protein; EC 5.4.99.5/EC 4.2.1.51; delivering phenylpyruvate; gene *pheA*; Gething et al. 1976; Pohnert et al. 1999) for Phe, respectively (see Fig. 2 and Table 1). Prephenate as a dissociable intermediate is unstable and spontaneously converts into phenylpyruvate at acidic pH. Due to this reaction, some phenylalanine auxotrophs with lesions in *pheA* were reported to slowly recover growth in the absence of added phenylalanine (Pittard 1996).

In *E. coli*, the activity of chorismate mutase-prephenate dehydrogenase is feedback-inhibited by tyrosine (up to 95% inhibition of the prephenate dehydrogenase and 45% of the chorismate mutase activity), and the chorismate mutase-prephenate dehydratase is inhibited by phenylalanine (up to 90% of the prephenate dehydratase and 55% of the mutase activity) (Pittard 1996; Pittard and Yang 2005). Feedback-resistant mutant forms are known and are used in biotechnology for the production of these two amino acids (Hagino and Nakayama 1974; Ito et al. 1990a; Nelms et al. 1992; Zhang et al. 1998; Pohnert et al. 1999; Chen et al. 2003). For applications see Sects. 2.4 and 2.5.

The last step in formation of Phe and Tyr is a transamination reaction onto the respective α-keto acids, phenylpyruvate and *p*-hydroxyphenylpyruvate, with glutamate as amino donor. At least two aminotransferases, TyrB ("aromatic aminotransferase", EC 2.6.1.5 or 2.6.1.57) and AspC ("aspartic aminotransferase", EC 2.6.1.1), both with broad substrate specificities, are able to form Phe or Tyr (Pittard 1996) (Table 1). The *tyrB* and *aspC* double mutants of *E. coli* require Tyr but not Phe. An additional mutation in the *ilvE* gene for the branched-chain aminotransferase renders these mutants also auxotrophic for Phe (Gelfand and Steinberg 1977; Pittard 1996).

In corynebacteria (*C. glutamicum* including those *C. glutamicum* strains previously misclassified as *Brevibacterium flavum* and *B. lactofermentum*; Liebl et al. 1991) a slightly different pathway for both amino acids is used and the pathways diverge only at prephenate. Chorismate mutase (EC 5.4.99.5, gene *csm*, Table 1) is a monofunctional enzyme and forms prephenate (Sugimoto and Shiio 1980). A monofunctional prephenate dehydratase (EC 4.2.1.51; gene *pheA*) forms phenylpyruvate from prephenate by oxidative decarboxylation. Phenylpyruvate eventually is the substrate for a phenylpyruvate aminotransferase (gene *pat*), which forms Phe (McHardy et al. 2003). Prephenate dehydratase is feedback inhibited by Phe; feedback resistant mutants are known (Chan and Hsu 1996; Hsu et al. 2004). In the Tyr path, a prephenate aminotransferase (EC 2.6.1.78 or 2.6.1.79; gene unknown so far) forms arogenate (pretyrosine) from prephenate, as is also known from cyanobacteria and plants (Fazel et al. 1980). In the last step, arogenate is decarboxylated and aromatized to Tyr by an arogenate dehydrogenase (EC 1.3.1.78; NADP-dependent; gene *tyrA*) (Song et al. 2005).

2.2 The Tryptophan Pathway

The first step in the terminal pathway of tryptophan synthesis is the reaction of anthranilate synthase (EC 4.1.3.27), which aminates chorismate and then aromatizes it to anthranilate with the concomitant loss of the enol-pyruvyl group as pyruvate through β-elimination (Pittard 1996; Nichols 1996). Ammonia or glutamine (via an amidotransferase reaction) can be utilized as amino group donors for anthranilate synthase, which is an aggregate of two dissimilar subunits (genes *trpE* and *trpD*, Table 1). The *trpE* gene encodes the larger subunit (component I, chorismate-binding), which converts chorismate and ammonia to anthranilate. *trpD* encodes the smaller subunit (component II, glutamine-binding) of anthranilate synthase. This subunit confers the ability to utilize glutamine as amino donor and thus acts as a glutamidotransferase (Nichols 1996; Pittard 1996; Herrmann and Weaver 1999). This activity channels nitrogen (presumably in the form of ammonia) from glutamine to the active site of anthranilate production (Pittard 1996). The enzyme-bound, aminodeoxyisochorismate (ADIC), is then cleaved into pyruvate and the aromatic anthranilate. Free ADIC is only detectable in bacteria with certain mutant forms of component I (Morollo and Bauerle 1993). Component II is a bifunctional enzyme that also performs the consecutive step, i.e. conversion of anthranilate and 5-phosphoribosyl-pyrophosphate

(PRPP) to anthranilate 5-phosphoribosyl pyrophosphate (Fig. 1, Table 1), and as such the enzyme is termed anthranilate phosphoribosyl (PRA) transferase (EC 2.4.2.18). In *E. coli* and corynebacteria, anthranilate synthase and PRA transferase activities are both feedback inhibited by the end product, tryptophan (Shiio et al. 1972; Somerville 1983; Pittard 1996). Feedback resistant mutant forms are known (Hagino and Nakayama 1975; Matsui et al. 1987a; Caligiuri and Bauerle 1991; Pittard 1996; Ikeda 2003, 2006).

The steps from PRA via 1-(*o*-carboxyphenylamino)-1-deoxy-ribulose 5-phosphate (CPADRP) to indole-glycerol-3-phosphate (InGP, see Table 1) are catalysed by a single, bifunctional enzyme (PRA isomerase/ Indole-glycerol-3-phosphate synthase EC 5.3.1.24/EC 4.1.1.48) encoded by the *trpC* gene. In non-enteric bacteria, these two functions may be encoded by separate genes (Pittard 1996). Finally, tryptophan synthase (EC 4.2.1.20; Yanofsky and Crawford 1972) forms, in an overall reaction from InGP and L-serine, glyceraldehyde 3-phosphate and L-tryptophan. Tryptophan synthase is a heterotetramer ($\alpha_2\beta_2$) of two protein components (α encoded by *trpA*, and β by *trpB* gene, respectively). The pyridoxal phosphate-dependent enzyme actually carries out three reactions: InGP to indole and D-glyceraldehyde 3-phosphate; indole plus L-serine to L-tryptophan (plus H_2O), and InGP plus L-serine to L-tryptophan plus D-glyceraldehyde 3-phosphate (and H_2O). The purified α_2 component is able to convert InGP to indole and glyceraldehyde 3-phosphate; the β_2 subunit converts indole plus L-serine in the presence of pyridoxal phosphate to tryptophan. In the complex, a tunnel is prominent, which apparently allows the rapid diffusion of indole from the active site of α subunit to the active site of β subunit, preventing the release of indole during catalysis (Hyde et al. 1988).

In other bacteria, features of the pathway (e.g. the number of separate or fused enzyme units) or of gene order can be slightly different, but the overall biochemical pathway is conserved (Jensen 1996, for a recent review see Xie et al. 2003).

2.3
Regulation and Genetics of the Pathways

As a pacemaker enzyme, the intracellular level of DAHP synthase(s) is controlled by transcriptional repression (Brown and Somerville 1971) and by feedback inhibition. The three *E. coli* genes encoding DAHP synthases are controlled at the transcriptional level by repression through the repressors TyrR and TrpR, which bind the aromatic amino acids (Pittard 1996; Herrmann and Weaver 1999). Transcription of *aroH* is controlled by both repressors (Herrmann and Weaver 1999). Deletion of the *tyrR* and *trpR* genes, or inactivation of the respective repressor proteins alleviates the transcriptional control (Berry 1996). *Cis*-acting repressor binding sites (operators) are present in the regulatory regions of *aroH* and *aroF* (Herrmann and Weaver

1999). All six enzymes past DAHP synthase appear to be synthesized constitutively; there is some transcriptional control on *aroA* (Man et al. 1997), and expression of *aroL* gene is regulated by both the TyrR and TrpR repressors (Lawley and Pittard 1994). Enzyme activities are not subject to feedback inhibition in *E. coli* (Pittard 1996) with the exception of a linear mixed-type inhibition of AroE (K_i of 0.16 mM) by shikimate (Dell and Frost 1993).

The pioneering work of Charles Yanofsky on the biochemistry and regulation of tryptophan biosynthesis in *E. coli* has led to the attenuator concept, which is now textbook knowledge, as well as to the understanding of the TrpR repressor system (Khodursky et al. 2000; Yanofsky 2001, 2003). Attenuation is one regulatory element in the biosynthesis of tryptophan (leader region *trpL* upstream of the *trpEDCBA* genes) and phenylalanine (gene *pheL* encoding a leader region is found upstream of *pheA* and of the *pheST* operon encoding phenylalanyl-tRNA synthetases). Availability of tryptophan- or phenylalanine-charged tRNA is crucial for this type of regulation. Due to space constraints, attenuation cannot be described here in detail and the reader is referred to some recent reviews (Henkin and Yanofsky 2002; Gollnick and Babitzke 2002; Vitreschak et al. 2004).

In many bacteria, end-products of the AAA pathway are negative effectors of the DAHP synthase(s) and in vivo, feedback inhibition is quantitatively the major regulatory mechanism (Ogino et al. 1982). The immediate response to the intracellular changes in the concentration of aromatic amino acids is by allosteric control of the first enzyme of the common pathway (DAHP synthase) by feedback inhibition. An overview of genes and feedback regulations in the general AAA pathway of *E. coli* is given in Fig. 3. After the divergence into the terminal pathways, again the first committed step is regulated by the end products (e.g. PheA, TyrA, TrpED proteins) in *E. coli* (Pittard and Yang 2005).

Genes of the common AAA pathway and for Phe and Tyr biosynthesis are scattered all over the chromosome of *E. coli*, few of them are clustered with genes of the common aromatic pathway (e.g. *aroK-aroB*, *aroF-tyrA-pheA*). The genes of the Trp pathway in *E. coli* form an operon (*trpEDCBA*) under control of the *trp* promoter, operator and leader region (*trpL*). The repressor genes, *tyrR* and *trpR* are located elsewhere (Pittard 1996 and refs. therein). TrpR is the repressor (with Trp as its co-repressor molecule) of the TrpR-regulon. This regulon consists of the *trpEDCBA* operon, the *aroH* gene, *trpR* gene itself, *aroL* and *mtr* (encoding a tryptophan permease, 2.A.42.1.2; Pittard 1996). The TyrR protein can act either as repressor or as activator of transcription of genes of the TyrR regulon. TyrR binds to so-called TyrR boxes, which lie upstream or downstream of the promoter regions of its various transcription units. TyrR has as effectors the three aromatic amino acids. These can act differently with TyrR, thereby exerting a range of regulatory effects (Pittard et al. 2005). The TyrR regulon consists of the genes *aroF*, *aroG*, *aroL*, *tyrR* itself, *mtr*, plus *tyrP* (encoding a tyrosine transporter, 2.A.42.1.1),

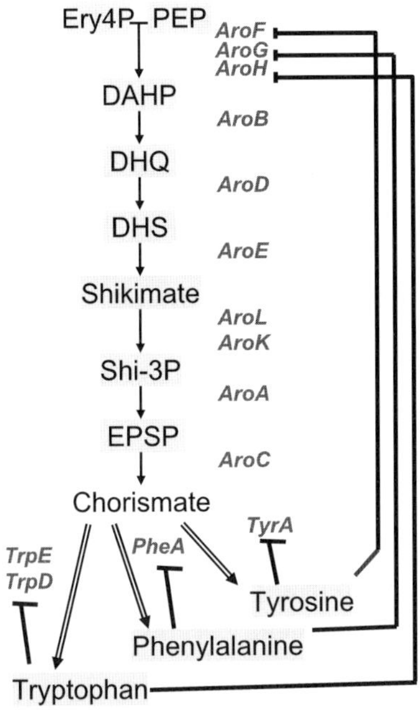

Fig. 3 Genes and feedback regulation of the AAA pathway in *E. coli*. The AAA pathway in *E. coli* is regulated at the levels of transcription, translation and enzyme activity. The end products inhibit the first committed steps of the general pathway (DAHP synthase isoenzymes AroF, AroG, AroH) individually (indicated by *straight lines*) and the first steps of the terminal pathways (e.g. PheA is inhibited by phenylalanine)

aroP (general aromatic transporter, 2.A.3.1.3) and *tyrB* (aminotransferase) (Pittard et al. 2005). Some genes, e.g. *aroL* or *mtr*, are regulated by both TrpR and TyrR (Lawley and Pittard 1994; Pittard 1996).

Aromatic amino acid biosynthesis and regulation of tryptophan biosynthesis in other proteobacteria (Panina et al. 2001) or in Gram-positive bacteria such as *Bacillus subtilis* uses novel regulatory mechanisms including a tryptophan-activated RNA-binding attenuation protein, which is different from the attenuation control found in proteobacteria such as *E. coli* (Gollnick et al. 2002, 2005).

2.4 Applications and Sources of Aromatic Amino Acids

In general, aromatic amino acids are in use mainly as food or feed additives, in infusion fluids, or as building blocks for chemical syntheses of pharmaceutically active compounds. In the feed market, L-tryptophan is of increasing importance, as it is the fourth limiting amino acid after L-lysine, D/L-methionine and L-threonine in plant-based animal feed (Leuchtenberger et al.

2005). However, its current price of 22–24 $/kg may be prohibitive for a wider use. The market size is estimated at 3000 tons per year (Ajinomoto 2006). A huge market has evolved for L-phenylalanine as building block for the low-calorie artificial sweetener, aspartame®. Table 2 lists applications, main producers and market volumes for the L-enantiomers of aromatic amino acids.

A general method for obtaining pure amino acids is the hydrolytic cleavage of abundant proteins and subsequent chromatographic purification to isolate the single amino acids. These processes are still in use for small but profitable niches aiming at bio-oriented consumers. Chemical synthesis of amino acids delivers racemic mixtures as in the DSM/Tosoh process of D,L-phenylalanine production for aspartame (Schmid et al. 2001; Bongaerts et al. 2001). Enzymatic methods have been described for L-phenylalanine (Hummel et al. 1987; Nakamichi et al. 1989) and L-tryptophan (Zeman et al. 1990). Biotransformations have been described, too (Bang et al. 1983; Lloyd-George and Chang 1995; Chao et al. 1999; Faurie and Fries 1999). The majority of production processes, however, are now based on microbial fermentations with the two main producers, *C. glutamicum* and *E. coli* (Leuchtenberger et al. 2005; Ikeda 2006).

2.4.1 Tyrosine, Phenylalanine and Aspartame

As L-tyrosine is not an essential amino acid, the demand for this amino acid is rather low (< 200 tons per year; Table 2, Ikeda 2006). Microbial production strains (e.g. *Erwinia herbicola*), which convert ammonia, pyruvate and phenol into L-tyrosine in biotransformations via the enzyme tyrosine phenol-lyase (EC 4.1.99.2) have been described (Lloyd-George and Chang 1995). A potential use of tyrosine is for the production of L-3,4-dihydroxyphenylalanine (L-DOPA), which is used to treating Parkinson's disease. *Corynebacterium* strains have been developed for fermentative L-tyrosine formation (Ito et al. 1990a; Ikeda and Katsumata 1992; Maiti et al. 1995; Hermann 2003; Ikeda 2003, 2006).

L-Phenylalanine can be obtained by chemical, enzymatic or microbial processes. Enzymatic steps are either amination of *trans*-cinnamic acid or transamination or reductive amination of phenylpyruvate (Hummel et al. 1987; Nakamichi et al. 1989; Chao et al. 1999). Microbial production involves either biotransformations of phenylpyruvate and aspartate with recombinant *E. coli* cells with elevated levels of aminotransferases and PEP carboxykinase (Chao et al. 1999) or microbial fermentations. Especially, recombinant strains of *C. glutamicum* and *E. coli* are in industrial use for fermentative production of phenylalanine (Choi and Tribe 1982; Ito et al. 1990b,c; Choi et al. 1992; Weikert et al. 1998; Bongaerts et al. 2001; Ikeda 2003, 2006 and refs therein).

Table 2 Applications, main producers and market volumes of aromatic amino acids*

	Use	Methods of production	Main producers	Market size#
L-Tryptophan	Feed additive Sleep aid (precursor of 5-hydroxytryptophan, serotonin), nutritional therapy, intravenous solutions, dietetic foods, antidepressant	Microbial fermentations Biotransformation Enzymatic methods Protein hydrolysates	Ajinomoto (70–80% feed-market share) Mitsui Chem. Tanabe Seiyaku Kyowa Hakko Kogyo Archer Daniels Midland Amino Rexim / Degussa	Feed-use: 2000–2200 tons per year
L-Phenylalanine	Aspartame precursor	Microbial fermentations	Nutrasweet Kelco	Aspartame: ~ 18 000 (Aspartame 40–45% share)
	Flavour enhancer, infusion fluids	Chemical synthesis	Ajinomoto	
	Dietetical and nutraceutical building block for pharma (rennin inhibitors, HIV protease inhibitor, anti-inflammatory drugs)	Enzymatic syntheses	Miwon HSC (aspartame process) Kyowa Hakko Several Chinese companies	
L-Tyrosine	Raw material for L-DOPA production, treatment of Basedow's disease, dietary supplement	Microbial fermentations Protein hydrolysates	Ajinomoto Kyowa Hakko Daebong LS Tanabe Seiyaku	100 ~ 200

* Data from Bongaerts et al. 2001; Hermann 2003; Ikeda 2003, 2006; Ajinomoto 2006, and additional data from Internet research August 2006
# Estimations for 2006

Apart from uses in feed and medical applications (infusion solutions) and as building block for pharmaceuticals (Table 2), the demand for free L-phenylalanine has constantly increased in recent years. This is primarily to satisfy the needs for its main use in synthesis of the artificial sweetener, aspartame (*N*-L-*α*-aspartyl-L-phenylalanine 1-methyl ester) (Schmid et al. 2001). Aspartame is a low-caloric sweetener which is about 200× as sweet as sucrose (Leuchtenberger et al. 2005). It is increasingly used in diet drinks or food (Bongaerts et al. 2001) with an estimated total market volume of 18 000 tons in 2006 (Table 2; Ajinomoto 2006). Up to 20 000 tons more may be produced in China.

2.4.2
L-Tryptophan

Industrial production of L-tryptophan nowadays is mostly for feed and pharmaceutical purposes (Table 2). For 2005, a world requirement of 3000 tons was projected (Ajinomoto 2006). Due to its low abundance in proteins and its lability during acidic hydrolytic processes, L-tryptophan is not prepared from acidic protein hydrolysates. In biotransformations with *E. coli* cells, which overproduce tryptophanase (gene *tnaA*), L-tryptophan is produced from indole, pyruvate and ammonia by the reverse reaction of tryptophanase (Zeman et al. 1990). Another biotransformation uses *E. coli* cells, which overproduce tryptophan synthase, and indole and L-serine as substrates (Bang et al. 1983, Faurie and Fries 1999). The majority of tryptophan production, however, is by microbial fermentations mainly with corynebacteria and *E. coli* (Ikeda and Katsumata 1999; Bongaerts et al. 2001; Ikeda 2003, 2006).

In the late 1980s, tryptophan-containing over-the-counter (OTC) formulations had become increasingly popular for self-medication (e.g. to treat insomnia), especially in North America, until the so-called *Showa Denko* disaster occurred. In late 1989, reports of severe sickness (so-called EMS, *eosinophilia myalgia syndrome*) in consumers of tryptophan appeared. Eventually, these cases of EMS led to 37 casualties and more than 1000 severely afflicted victims. Apparently, only tryptophan lots that originated from the *Showa Denko* process (using recombinant *Bacillus amyloliquefaciens* strains and a different downstream processing protocol), elicited this syndrome. Various impurities were detected in these lots, including adducts of tryptophan dimers with acetaldehyde (the notorious "peak E" of HPLC diagrams). While not all aspects of this tragic event are clear yet (for discussions see: Mayeno et al. 1994; Müller et al. 1999; Simat et al. 1999; Stahl et al. 2001), regulatory agencies such as the Food and Drug Administration (FDA) of the USA have banned the OTC purchase of tryptophan-containing medications over several years and have set up strict regulations for the use of tryptophan in food applications. Use of biotechnologically produced tryptophan for human consumption is still restricted in the USA. However, due to improvements in production, downstream processing and analytical methods, the impurity problems could be solved throughout. Not surprisingly, prices for pharma-grade L-tryptophan are still rather high ($\sim$ 100 $/kg). As tryptophan is a limiting compound in feed of livestock (Leuchtenberger et al. 2005; Ajinomoto 2006) it should become interesting again to enlarge production capacities.

2.5
Biotechnological Production of AAA

Microbial strains, mainly mutants of *E. coli* and *C. glutamicum*, have been isolated that allow overproduction of the three aromatic amino acids. To achieve overproduction, metabolic controls had to be removed to enlarge the carbon flux into the common aromatic pathway and into the terminal pathways. After decades of classical strain improvements by mutagenesis and selection, product titers were not high enough to be economic. Only after the introduction of genetic engineering did production strains become available that could meet the demands of economic production. Further improvements are necessary as the current production yields towards sugar (weight %) are still lagging behind those of L-lysine or L-threonine (Ikeda 2006).

2.5.1
Classical Strain Improvements

Microbial fermentation strains allow the production of various amino acids from cheap and renewable carbon sources such as molasses, sucrose or glucose (Leuchtenberger 1996; Eggeling and Sahm 1999; Eggeling and Bott 2005) and, therefore, are usually more favourable than biotransformation processes. Through classical strain improvement and recently also through genetic engineering, producer strains have been developed, which allow production of each of the three aromatic amino acids (refs. in Bongaerts et al. 2001; Ikeda 2003, 2006; Leuchtenberger et al. 2005).

Of course, a fundamental knowledge of the underlying metabolic pathways and especially of the genes, enzymes and regulatory circuits of the aromatic amino acid biosynthesis pathways is essential for microbial strain development and improvement (Berry 1996; Eggeling and Sahm 1999; Bongaerts et al. 2001; Ikeda 2003, 2006).

As with other amino acid producers, classical genetic breeding of microbial strains was usually performed by rounds of mutagenesis and selection or screening for useful mutants (Ikeda 2003). These were either auxotrophic or regulatory mutants. Auxotrophic mutants in the case of AAA are mostly those that require another AAA (e.g. L-Tyr in the case of L-Phe, or L-Phe in the case of L-Tyr). This auxotrophy is due to a biosynthetic block of a competing pathway, thus eliminating unwanted by-products (which may be difficult to remove during DSP) or drain of carbon source and energy (which limits the productivity). However, auxotrophy also means that mutants require a source of the respective amino acid during the growth phase (for examples see: Hwang et al. 1985; Backman et al. 1990; Takagi et al. 1996; Gerigk et al. 2002a,b).

For classical strain improvement, random mutagenesis and appropriate screening or selection methods were used. Antimetabolites are helpful tools as they allow the selection of feedback inhibition resistant mutants. These compounds (e.g. fluoro- or chlorophenylalanine, 5-methyltryptophan) exert the same feedback inhibition features as the natural amino acids on, e.g. DAHP synthase or the first committed steps of the terminal AAA pathways (Hagino and Nakayama 1974, 1975; De Boer and Dijkhuizen 1990). Thus, these antimetabolites induce starvation for the respective AAA, and microbial cells are unable to grow in the absence of this amino acid. Mutants that are resistant against these antimetabolites are able to grow and can easily be selected on suitable agar plates (De Boer and Dijkhuizen 1990). In many cases the resistance relies on altered allosteric binding sites of the pacemaker enzymes (see Sects. 2.1 and 2.2) and leads to deregulation of a pathway and product accumulation in the culture supernatant. Mutations in the AAA uptake systems may also lead to resistance against antimetabolites; these mutants, however, rarely overproduce AAA. The approach of feedback deregulation (Hagino and Nakayama 1974; Sugimoto and Shiio 1982; De Boer and Dijkhuizen 1990) and knockout of competing metabolic pathways was successful (Hwang et al. 1985; Tsuchida et al. 1987), but the productivity of these strains was not sufficient for industrial productions (Ikeda 2006).

2.5.2 Genetic and Metabolic Engineering

With the advent of genetic engineering in the late 1970s and early 1980s, microbial strains were developed that carried extra copies of biosynthesis genes for AAA. Cloning of genes for enzymes of the general AAA pathway (e.g. DAHP-synthase, shikimate kinase) or from the terminal pathways, and their expression in homologous or heterologous host strains helped to increase the productivity of these strains (del Real et al. 1985; Ozaki et al. 1985; Sano and Matsui 1987; Ito et al. 1990a,b; Ikeda et al. 1993). An example is an L-tyrosine producer of *C. glutamicum* (*B. lactofermentum*), which after expression of the cloned *aroL* gene showed a tenfold increased shikimate kinase activity in cell-free extracts and accumulated 21.6 g/L of Tyr instead of 17.4 g/L (Table 3, Ito et al. 1990a). The gene-dosage effect alone was already useful to (partially) overcome regulations and limitations, and to improve yields. Use of alternative promoter systems (e.g. of the bacteriophage lambda), allowed controllable expression of recombinant genes and overproduction of phenylalanine in *E. coli* (Sugimoto et al. 1987). Competing pathways could be knocked out

Table 3 Yields of biotechnological production of aromatic amino acids

Product	Microorganism	C or R*	Yields (g/L from glucose)	Refs.
L-Tryptophan	*Escherichia coli* (recombinant)	R	6.2 (in 27 h)	Aiba et al. 1982
		R	9.2	Chan et al. 1993
		R	20 (in 91 h)	Azuma et al.1993
		R	> 50 (+ anthranilate feed)	
		R	45	Berry 1996
	Corynebacterium glutamicum	R	~ 50	Ikeda et al. 1994
		R	43	Katsumata and Ikeda 1993
		R	58 (from sucrose)	Ikeda and Katsumata 1999
	Bacillus amyloliquefaciens	R	14.2 (+ anthranilate feed)	Yajima et al. 1990
L-Phenylalanine	*Escherichia coli*	R	16.8	Sugimoto et al. 1987
		R	38 (on maltose)	Grinter 1998
		R	~ 50	Miller et al. 1987
		R	50	Backman et al. 1990
		R	~ 50	Konstantinov et al. 1991
		R	~ 50	Konstantinov and Yoshida 1992
		R	50.8	Choi et al. 1992
		R	~ 50	Takagi et al. 1996
		R	38	Rüffer et al. 2004
			51 (+ISPR)	
	Corynebacterium	R	28 (from sucrose)	Ikeda and Katsumata 1992
		R	23.2	Shu and Liao 2002
	(*B.lactofermentum*)	C	48	Igarashi et al. 1993
L-Tyrosine	*Corynebacterium glutamicum*		21.6	Ito et al. 1990a
			26	Ikeda and Katsumata 1992

Data compiled from Eggeling and Sahm 1999, Bongaerts et al. 2001, Ikeda 2003, 2006
* C/R Classical strain development with random multagenesis and resistance against antimetabolites (C) or recombinant development (sometimes including classical steps)
ISPR In situ product recovery

by gene disruptions or deletions introducing auxotrophies (Miller et al. 1987; Mascarenhas et al. 1991; Konstantinov et al. 1991; Ikeda and Katsumata 1992, 1993; Liu et al. 2004). Moreover, genes encoding feedback-resistant pacemaker enzymes have been cloned and overexpressed (Matsui et al. 1987a,b; Weaver and Herrmann 1990; Itoh et al. 1990b,c; Chan and Hsu 1996; Tonouchi et al. 1997). Combinations of genes on expression plasmids allowed limitations to be overcome in pathways (Edwards et al. 1987; Backman et al. 1990).

Metabolic engineering (Bailey 1991) for the production of AAA and other products of the shikimate pathway has been reviewed recently (Bongaerts et al. 2001; Ikeda 2003, 2006; Krämer et al. 2003). For the AAA, engineering of regulatory circuits, amplification and overexpression of genes of the general and terminal pathways, and improvements in the precursor supply were most successful. Metabolic engineering of AAA pathways requires first alleviation of all control levels (repression, attenuation and feedback inhibition) (see Sect. 2.5.2.1), to identify and remove rate-limiting steps by the appropriate overproduction of enzymes of the general AAA pathway (Dell and Frost 1993; Ikeda 2003; Rüffer et al. 2004), and then to reduce competing pathways and to improve and balance precursor supply both in the common pathway as well as in the specific branch (see Sect. 2.5.2.2) (reviewed in Bongaerts et al. 2001; Ikeda 2003, 2006). If degrading enzymes exist (e.g. tryptophanase in *E. coli*), these have to be removed as well (Aiba et al. 1982). Transport and re-uptake of products can cause unwanted futile cycles (Krämer 1994). This problem can be solved by deletion of appropriate transporter genes as shown for Trp production in *C. glutamicum* (Ikeda and Katsumata 1994, 1995; Ikeda 2003).

Nowadays, strain improvements in *E. coli* and *C. glutamicum* are possible by precise "surgical molecular genetics" where genes of interest can be purposefully cloned, amplified, deleted, rearranged or altered in their base sequence by site-directed mutagenesis (for a recent review see Vertes et al. 2005). To avoid genetic instability due to difficulties in plasmid maintenance, examples of synthetic modification of the *E. coli* chromosome for the shikimate pathway intermediates have been given by Frost (Dell and Frost 1993; Snell et al. 1996; Krämer et al. 2003). This allows one to work without antibiotics and with no (or fewer) plasmid-borne genes, and the strategy can also be applied for AAA production strains.

2.5.2.1 Engineering Regulatory Circuits

To overcome regulatory circuits, knockout of repressor genes (for *E. coli*: *trpR*, *tyrR*; Berry 1996), removal or alteration of operator and/or attenuator sequences (*trpL*, *pheL*, Heery and Dunican 1993), and the use of genes encoding feedback-resistant variants of pacemaker enzymes have been very useful (see above). Tribe and Pittard (1979) were the first to apply recombinant DNA technology to strain improvement in *E. coli* AAA producers. A cloned, amplified and deregulated *trpE* gene encoding anthranilate synthase component I (see Sect. 2.2) led to a final titer of about 1 g/L (in those days termed "hyperproduction", Tribe and Pittard 1979). Aiba and coworkers used *E. coli* mutants with lesions in *trpR* and *tnaA* (encoding tryptophanase) as host strains to express a multicopy plasmid, which carried the *trp* operon-containing mutant *trpE* and *trpD* genes for deregulated enzyme versions. By cultivation in a glucose medium with continuous feeding of anthranilate (a semi-biotransformation approach) 6.2 g/L of Trp were obtained after 27 h (Aiba et al. 1982). Based on this strain (with additional resistances against the antimetabolites 6-fluoro-Trp and 8-azaguanine), Azuma et al. (1993) managed to operate a Trp process with accumulation of more than 50 g/L of Trp after 91 h. About 20 g/L of Trp was derived from glucose and about 30 g/L from anthranilate in the feed. Trp crystallized during the process after addition of non-ionic detergents (Azuma et al. 1993). Table 3 gives an overview of the evolution of product yields for the three aromatic amino acids. While product titers of Trp and Phe are now up to 50 g/L, current production yields towards sugar (weight %) are between 20 and 25% (Ikeda 2003, 2006).

Interestingly, the use of wild-type pacemaker enzymes (e.g. DAHP synthase encoded by *aroF* of *E. coli*) may be advantageous and can help to overcome problems of instability in feedback-resistant enzymes (Jossek et al. 2001), provided that the feedback inhibiting effector (L-tyrosine) is strictly controlled during fermentations with Phe producers (Rüffer et al. 2004). Further improvements in reaction engineering include in situ product recovery from fermenters, including filtration through hollow-fibre membranes or centrifugations (Maass et al. 2002; Gerigk et al. 2002a,b; Rüffer et al. 2004).

2.5.2.2
Strategies to Increase Precursor Supply

The precursors for the general AAA pathway, PEP and Ery4P, derive from glycolysis and from the pentose phosphate pathway, respectively (Frost and Draths 1995; Sprenger 1995; Fraenkel 1996). Each may become limiting for the production of AAA once there are no more limits in the specific pathways or in the common AAA pathway. Therefore, attempts to improve supply of either PEP or Ery4P or of both have been reported (for recent reviews see Bongaerts et al. 2001; Ikeda 2003, 2006). Early approaches were successful in increasing PEP supply by knocking out the genes of the competing PEP carboxylase or pyruvate kinases in *E. coli* Phe producers (Miller et al. 1987; Backman 1992; Gosset et al. 1996) or to combine PEP carboxylase gene mutations with overproduction of the gluconeogenic PEP carboxykinase for improvement of carbon flux into the shikimate pathway (Chao and Liao 1993). On the other hand, Ery4P supply was improved by knocking out the phosphoglucose isomerase gene. This led to a stronger flux through the pentose phosphate pathway and increased Trp production in *E. coli* (Mascarenhas et al. 1991). Cloning and overexpression of the transketolase gene also increased intracellular Ery4P and led to a raised production of DAHP (Draths and Frost 1990; Draths et al. 1992). Formation of DAH or DAHP in culture supernatants of *aroB* mutants of *E. coli* (Sect. 2.1) was taken as measure for the flux into the common AAA pathway. Detailed analyses on factors that influence this flux have been made and led to hypotheses about balanced precursor supplies for the AAA pathway. Overproduction of enzymes from the pentose phosphate pathway, such as transketolase or transaldolase, were found to improve Ery4P supply (Draths et al. 1992; Frost and Draths 1995; Liao et al. 1996; Lu and Liao 1997; Sprenger et al. 1998b) and, if correctly dosed, also Trp formation (Ikeda et al. 1999). Enzymes from glycolysis, gluconeogenesis or the glyoxylate cycle (Koehn et al. 1994) were studied for their contribution to the supply of PEP. For example, PEP carboxylase (Miller et al. 1987; Backman 1992) or pyruvate kinase activities were reduced by gene mutations (Gosset et al. 1996; Berry 1996; Grinter 1998). Pyruvate formed from glucose was recycled to PEP using PEP synthetase (Liao 1996) or PEP carboxykinase (Gulevich et al. 2004). Enzyme activities improving supply of PEP and Ery4P were also combined and analysed for synergistic effects on the flux into the shikimate pathway (Chao and Liao 1993; Patnaik and Liao 1994; Liao et al. 1994, 1996; Frost and Draths 1995; Patnaik et al. 1995; Flores et al. 1996; Gosset et al. 1996; Baez et al. 2001; Oldiges et al. 2004). These ideas and results were in the same line as pathway engineering approaches to improve formation of other shikimate pathway intermediates (Dell and Frost 1993; Frost and Draths 1995; Snell et al. 1996; Draths et al. 1999; Yi et al. 2002; Chandran et al. 2003; Oldiges et al. 2004).

Nevertheless, the most limiting factor in PEP supply is the PEP-dependent sugar phosphotransferase system (PTS), which is responsible for glucose uptake in both *E. coli* and *C. glutamicum* wild-type strains. About 50% of PEP generated in glycolysis during growth on glucose has to be spent for uptake of glucose and is concomitantly converted into pyruvate (Postma et al. 1993). Coupled to uptake by the PTS, the carbohydrate (glucose, fructose, sucrose or others) is phosphorylated and can enter the central metabolic pathways. For wild-type *E. coli* cells growing in minimal medium, only roughly 3% of the PEP enters the AAA pathway (Valle et al. 1996). Theoretically, with an intact PTS, less than 50% of the available PEP from carbon sources such as glucose can be funnelled into the AAA pathway (Foerberg et al. 1988; Patnaik and Liao 1994; Sauer and Eikmanns 2005). Thus, an exchange of the glucose-PTS for other transport systems that do not depend on PEP, was studied in *E. coli* mutant strains deficient for the glucose-PTS (Flores et al. 1996; Berry 1996; Gosset et al. 1996; Chen et al. 1997). Indeed, several groups showed that a PEP saving sugar uptake is possible. Some bacteria such as the ethanologenic *Zymomonas mobilis* harbour glucose transport systems (glucose facilitator Glf) that neither require PEP nor ATP for the uptake by facilitated diffusion. Inside the cell, an ATP-dependent sugar kinase (e.g. glucokinase) activates the glucose to enter glycolysis. The genes *glf* (or *glf* and *glk*) have been used to construct *E. coli* strains with altered glucose uptake systems (Snoep et al. 1994; Weisser et al. 1995; Parker et al. 1995; Sprenger et al. 1998a,b; Krämer et al. 1999). The galactose permease system of *E. coli* also transports glucose in a PEP-independent way into the cell by a proton symport mechanism. GalP must be activated through mutation to allow grow on glucose, however. By doing so, glucose-positive, PTS-negative strains were developed that showed improved Phe formation (Gosset et al. 1996; Berry 1996). Finally, pathway engineering by combinations of improved Ery4P and/or PEP supplies has already succeeded in improving Trp or Phe formation in both *E. coli* and *C. glutamicum* strains (Flores et al. 1996; Berry 1996; Gosset et al. 1996; Grinter 1998; Ikeda and Katsumata 1999; Rüffer et al. 2004; Baez-Vileros et al. 2004).

In the case of Trp, intracellular supply of L-serine and phosphoribosyl 5-pyrophosphate (PRPP) are additional bottlenecks for production (for a review see Ikeda 2006). Models to improve Trp production in *E. coli* have taken these prerogatives into account (Marin-Sanguino and Torres 2000; Schmid et al. 2004). If L-serine becomes limiting, *C. glutamicum* Trp producers show an overflow of indole (Ikeda et al. 1994). Additional genetic modification of the serine biosynthetic pathway helped to increase Trp production in *C. glutamicum* (Ikeda et al. 1994). Amplification of the gene *prs* encoding phosphoribosyl 5-pyrophosphate synthetase enhanced tryptophan production in *Bacillus* strains (Yajima et al. 1990).

2.5.3
Outlook

Future prospects for the biotechnological use of the AAA pathway are in further improving precursor supply (see above) and in carbon flux analysis (for example see Wahl et al. 2004). It should be noted, however, that drastic alterations of precursor supply could also lead to unwanted by-product formation (e.g. acetate, pyruvate) or growth impairment, as recognized by several groups (Miller et al. 1987; Mascarenhas et al. 1991; Kim et al. 2000, Ikeda 2006). Thus a balanced supply of both PEP and Ery4P is to be achieved (Ikeda and Katsumata 1999). Models that take into account the various levels of regulations or the different contributions of precursor supply have already been developed (Schmid et al. 2004). Efflux of AAA from producer cells is generally assumed to proceed via simple diffusion as all three compounds are hydrophobic enough to pass the membrane. However, an involvement of excretion mechanisms cannot be ruled out yet. More detailed studies are needed. Data from transcriptional analyses of *E. coli* cells in the presence of high concentrations (5 g/L) of L-phenylalanine have been obtained and point to an involvement of the *tyrR* gene product (Polen et al. 2005). Transcriptomics and genomics (Ikeda and Nakagawa 2003; Kalinowski et al. 2003) should help to better understand the behaviour of producer strains and in the long run can improve microbial production of aromatic amino acids.

Besides the proteinogenic AAA, other L- or D-amino acids derived from the AAA pathway have recently become of interest as building blocks for pharmaceutical compounds as antibiotics or protease inhibitors. For example, D-phenylalanine was prepared from an *E. coli* strain that lacks all three transaminase genes for L-phenylalanine formation from phenylpyruvate. A D-specific transaminase (using D-alanine as amino donor) and an alanine racemase were expressed to obtain D-phenylalanine (Fotheringham et al. 1998; Taylor et al. 2000). D-Phenylglycine (for semi-synthetic ampicillin or cephalexin) can be obtained from recombinant *E. coli* strains via an artificial pathway including a deregulated AAA biosynthesis up to phenylpyruvate. Additional genes from *Amycolatopsis* allow the formation of phenylglyoxalate and eventually D-phenylglycine through a stereoinverting aminotransferase (Müller et al. 2006).

As an example for the production of commodity compounds, a complete pathway to the denim dye indigo was established in recombinant *E. coli* tryptophan producers by adding a gene encoding a naphthalene dioxygenase from *Pseudomonas putida* (Ensley et al. 1983; Murdock et al. 1993; Berry et al. 2002). This dioxygenase distracts indole (provided either through tryptophanase from tryptophan or by *trpB* mutants with an inactivated terminal step in tryptophan biosynthesis) and delivers indoxyl which is further oxidized chemically by air to indigo. Although this process is not yet industrially relevant, it shows that the AAA pathway can be successfully engineered to deliver commodity fine chemicals. It can be envisioned that this may be of great interest in future.

References

Aiba S, Tsunekawa H, Imanaka T (1982) New approach to tryptophan production by *Escherichia coli*: genetic manipulation of composite plasmids in vitro. Appl Environ Microbiol 43:289–297

Ajinomoto (2006) Fact sheet: Feed-use amino acids business. http://www.ajinomoto.com/mx_03/ar/i_r/pdf/presentation/FY2005data.pdf

Anton JA, Coggins JR (1988) Sequencing and overexpression of the *Escherichia aroE* gene encoding shikimate dehydrogenase. Biochem J 249:319–326

Azuma S, Tsunekawa H, Okabe M, Okamoto R, Aiba S (1993) Hyper-production of L-tryptophan via fermentation with crystallization. Appl Microbiol Biotechnol 39:471–476

Backman KC (1992) Method of biosynthesis of phenylalanine. US Patent 5 169 768

Backman K, O'Connor MJ, Maruya A, Rudd E, McKay D, Balakrishnan R, Radjai M, DiPasquantonio V, Shoda D, Hatch R, Venkatasubramanian K (1990) Genetic engineering of metabolic pathways applied to the production of phenylalanine. Ann NY Acad Sci 589:16–24

Baez JL, Bolivar F, Gosset G (2001) Determination of 3-deoxy-D-arabino-heptulosonate 7-phosphate productivity and yield from glucose in *Escherichia coli* devoid of glucose phosphotransferase system. Biotechnol Bioeng 73:530–535

Baez-Viveros JL, Osuna J, Hernandez-Chavez G, Soberon X, Bolivar F, Gosset G (2004) Metabolic engineering and protein directed evolution increase the yield of L-phenylalanine synthesized from glucose in *Escherichia coli*. Biotechnol Bioeng 87:516–524

Bailey JE (1991) Toward a science of metabolic engineering. Science 252:1668–1675

Bang W-G, Lang S, Sahm H, Wagner F (1983) Production of L-tryptophan by *Escherichia coli* cells. Biotechnol Bioeng 25:999–1011

Benach J, Lee I, Edstrom W, Kuzin AP, Chiang Y, Acton TB, Montelione GT, Hunt JF (2003) The 23-X crystal structure of the shikimate 5-dehydrogenase orthologue YdiB from *Escherichia coli* suggests a novel catalytic environment for an NAD-dependent dehydrogenase. J Biol Chem 278:19176–19182

Bender SL, Mehdi S, Knowles JR (1989) Dehydroquinate synthase: the role of divalent metal cations and of nicotinamide adenine dinucleotide in catalysis. Biochemistry 28:7555–7560

Bentley R (1990) The shikimate pathway- a metabolic tree with many branches. CRC Crit Rev Biochem Mol Biol 2:307-384

Berry A (1996) Improving production of aromatic compounds in *Escherichia coli* by metabolic engineering. Trends Biotechnol 14:250–256

Berry A. Dodge T, Pepsin M, Weyler W (2002) Application of metabolic engineering to improve both the production and use of biotech indigo. J Ind Microbiol Biotechnol 28:127–133

Bongaerts J, Krämer M, Müller U, Raeven L, Wubbolts M (2001) Metabolic engineering for microbial production of aromatic amino acids and derived compounds. Metabol Engin 3:289–300

Brown RL, Somerville KD (1971) Repression of aromatic amino acid biosynthesis in *Escherichia coli*. J Bacteriol 108:386–399

Caligiuri MG, Bauerle R (1991) Identification of amino acid residues involved in feedback regulation of the anthranilate synthase complex from *Salmonella typhimurium*. Evidence for an amino-terminal regulatory site. J Biol Chem 266:8328–8335

Chan E-C, Tsai H-L, Chen S-L, Mou D-G (1993) Amplification of the tryptophan operon gene in *Escherichia coli* chromosome to increase L-tryptophan biosynthesis. Appl Microbiol Biotechnol 40:301–305

Chan MS, Hsu WH (1996) Cloning of m-fluorophenylalanine-resistant gene and mutational analysis of feedback-resistant prephenate dehydratase from *Corynebacterium glutamicum*. Biochem Biophys Res Commun 219:537–542

Chandran SS, Yi J, Draths KM, von Daeniken R, Weber W, Frost JW (2003) Phosphoenolpyruvate availability and the biosynthesis of shikimic acid. Biotechnol Prog 19:808–814

Chao YP, Liao JC (1993) Alteration of growth yield by overexpression of phosphoenolpyruvate carboxylase and phosphoenolpyruvate carboxykinase in *Escherichia coli*. Appl Environ Microbiol 59:4261–4265

Chao Y-P, Lai ZJ, Chen P, Chern J-T (1999) Enhanced conversion rate of L-phenylalanine by coupling reactions of aminotransferases and phosphoenolpyruvate carboxykinase in *Escherichia coli* K-12. Biotechnol Prog 15:453–458

Chen R, Hatzimanikatis V, Yap WM, Postma PW, Bailey JE (1997) Metabolic consequences of phosphotransferase (PTS) mutation in a phenylalanine-producing recombinant *Escherichia coli*. Biotechnol Prog 13:768–775

Chen S, Vincent S, Wilson DB, Ganem B (2003) Mapping of chorismate mutase and prephenate dehydrogenase domains in the *Escherichia coli* T-protein. Eur J Biochem 270:757–763

Choi YJ, Tribe DE (1982) Continuous production of phenylalanine using an *Escherichia coli* regulatory mutant. Biotechnol Lett 4:223–228

Choi HK, Kim HY, Lee DJ, Lim BL, Rhym H, Won CH, Choi HG (1992) Method for production of L-phenylalanine by recombinant *E. coli*. US Patent 5 304 475

Davis BD (1951) Aromatic biosynthesis, I. The role of shikimic acid. J Biol Chem 191:315–325

De Boer L, Dijkhuizen L (1990) Microbial and enzymatic processes for L-phenylalanine production. Adv Biochem Eng Biotechnol 41:1–27

Dell KA, Frost JW (1993) Identification and removal of impediments to biocatalytic synthesis of aromatics from D-glucose: rate-limiting enzymes in the common pathway of aromatic amino acid biosynthesis. J Am Chem Soc 115:11581–11589

del Real G, Aguilar A, Martín JF (1985) Cloning and expression of tryptophan genes from *Brevibacterium lactofermentum* in *Escherichia coli*. Biochem Biophys Res Commun 133:1013–1019

Dewick PM (1998) The biosynthesis of shikimate metabolites. Nat Prod Rep 15:17–58

Dosselaere F, Vanderleyden J (2001) A metabolic node in action: chorismate-utilizing enzymes in microorganisms. Crit Rev Microbiol 27:75–131

Draths KM, Frost JW (1990) Synthesis using plasmid-based biocatalysis: Plasmid assembly and 3-deoxy-D-arabino-heptulosonate production. J Am Chem Soc 112:1657–1659

Draths KM, Ward TL, Frost JW (1992) Biocatalysis and nineteenth century organic chemistry: conversion of D-glucose into quinoid organics. J Am Chem Soc 114:9725–9726

Draths KM, Pompliano DL, Conley DL, Frost JW, Berry A, Disbrow GL, Staversky RJ, Lievense JC (1992) Biocatalytic synthesis of aromatics from D-glucose: the role of transketolase. J Am Chem Soc 114:3956–3962

Draths KM, Knop DR, Frost JW (1999) Shikimic acid and quinic acid: Replacing isolation from plant sources with recombinant microbial biocatalysis. J Am Chem Soc 121:1603–1604

Edwards RM, Taylor PP, Hunter MG, Fotheringham IG (1987) Composite plasmids for amino acid synthesis. WO 87/00,202

Eggeling L, Sahm H (1999) Amino acid production: principles of metabolic engineering. In: Lee SY, Papoutsakis ET (eds) Metabolic engineering. Marcel Dekker, New York, pp 153–176

Eggeling L, Bott M (2005) Handbook of *Corynebacterium glutamicum*. CRC, Boca Raton, FL

Ensley BD, Ratzkin BJ, Osslund TD, Simon MJ, Wackett LP, Gibson DT (1983) Expression of naphthalene oxidation genes in *Escherichia coli* results in the biosynthesis of indigo. Science 222:167–169

Faurie R, Fries G (1999) From sugar beet molasses to Lyphan®-integrated quality management from the raw material to the drug. In: Tryptophan, serotonin and melatonin: Basic aspects and applications. Adv Exp Med Biol 467:443–452

Fazel AM, Bowen JR, Jensen RA (1980) Arogenate (pretyrosine) is an obligatory intermediate of L-tyrosine biosynthesis: confirmation in a microbial mutant. Proc Natl Acad Sci USA 77:1270–1273

Flores N, Xiao J, Berry A, Bolivar F, Valle F (1996) Pathway engineering for the production of aromatic compounds in *Escherichia coli*. Nat Biotechnol 14:620–623

Foerberg C, Eliaeson T, Haeggstroem L (1988) Correlation of theoretical and experimental yields of phenylalanine from non-growing cells of a rec *Escherichia coli* strain. J Biotechnol 7:319–332

Fotheringham IG, Taylor PP, Ton JL (1998) Preparation of D-amino acids by direct fermentative means. US Patent 5728555

Fraenkel DG (1996) Glycolysis, pentose phosphate pathway, and Entner-Doudoroff pathway. In: Neidhardt FC (ed) *Escherichia coli* and *Salmonella*: cellular and molecular biology, 2 edn, vol 1. ASM, Washington DC, pp 189–198

Frost JW (1992) Enhanced production of common aromatic pathway compounds. US Patent 5168056

Frost JW, Knowles JR (1984) 3-Deoxy-D-arabino-heptulosonic acid 7-phosphate: chemical synthesis and isolation from *Escherichia coli* auxotrophs. Biochemistry 23:4465–4469

Frost JW, Draths KM (1995) Biocatalytic syntheses of aromatics from D-glucose: renewable microbial sources of aromatic compounds. Annu Rev Microbiol 49:557–579

Gelfand DH, Steinberg RA (1977) *Escherichia coli* mutants deficient in the aspartate and aromatic amino acid aminotransferases. J Bacteriol 130:429-440

Ger YM, Chen SL, Chiang HJ, Shiuan D (1994) A single Ser-180 mutation desensitizes feedback inhibition of the phenylalanine-sensitive 3-deoxy-D-arabino-heptulosonate 7-phosphate (DAHP) synthetase in *Escherichia coli.* J Biochem 116:986–990

Gerigk M, Bujnicki R, Ganpo-Nkwenkwa E, Bongaerts J, Sprenger G, Takors R (2002a) Process control for enhanced L-phenylalanine production using different recombinant *Escherichia coli* strains. Biotechnol Bioeng 80:746–754

Gerigk M, Maass D, Kreutzer A, Sprenger G, Bongaerts J, Wubbolts M, Takors R (2002b) Enhanced pilot-scale fed-batch L-phenylalanine production with recombinant *Escherichia coli* by fully integrated reactive extraction. Bioproc Biosyst Eng 25:43–52

Gething MJH, Davidson BE, Dopheide TAA (1976) Chorismate mutase/prephenate dehydratase from *Escherichia coli* K 12. Eur J Biochem 71:317–325

Gibson F (1964) Chorismic acid: purification and some chemical and physical studies. Biochem J 90:256–261

Gibson MI, Gibson F, Doy CH, Morgan P (1962) The branch point in the biosynthesis of the aromatic amino-acids. Nature (London) 195:1173–1175

Gil GH, Kom SR, Bae JC, Lee JH (1985) Pilot-scale production of L-phenylalanine from D-glucose. Enzyme Microb Technol 7:370–372

Gollnick P, Babitzke P (2002) Transcription attenuation. Biochim Biophys Acta 1577:240–250

Gollnick P, Babitzke P, Merino E, Yanofsky C (2002) Aromatic amino acid metabolism in *Bacillus subtilis*. In: Sonenshein AL, Hoch JA, Losick R (eds) Bacillus subtilis and its closest relatives: from genes to cells. American Society for Microbiology, Washington, DC, pp 233–244

Gollnick P, Babitzke P, Antson A, Yanofsky C (2005) Complexity in regulation of tryptophan biosynthesis in *Bacillus subtilis*. Annu Rev Genet 39:47–68

Gosset G, Yong-Xiao J, Berry A (1996) A direct comparison of approaches for increasing carbon flow to aromatic biosynthesis in *Escherichia coli.* J Ind Microbiol 17:47–52

Grinter N (1998) Developing an L-phenylalanine process. Chemtech 28:33–37

Gulevich AY, Biryukova IV, Zimenkov DV, Skorokhodova AY, Kivero AD, Belareva AV, Mashko SV (2004) Method for producing L-amino acid using bacterium having enhanced expression of *pckA* gene. WO 2004/090125

Hagino H, Nakayama K (1974) L-phenylalanine production by analog-resistant mutants of *Corynebacterium glutamicum*. Agric Biol Chem 38:157–161

Hagino H, Nakayama K (1975) Regulatory properties of anthranilate synthase from *Corynebacterium glutamicum*. Agric Biol Chem 39:323–330

Haslam E (1993) Shikimic acid metabolism and metabolites. Wiley, Chichester, UK

Heery DM, Dunican LK (1993) Cloning of the *trp* gene cluster from a tryptophan-hyperproducing strain of *Corynebacterium glutamicum*: identification of a mutation in the *trp* leader sequence. Appl Environ Microbiol 59:791–799

Heimstaedt K, Strittmatter A, Lipscomb WN, Braus GH (2005) Evolution of 3-deoxy-D-arabino-heptulosonate-7-phosphate synthase-encoding genes in the yeast *Saccharomyces cerevisiae*. Proc Natl Acad Sci USA 102:9784–9789

Henkin TM, Yanofsky C (2002) Regulation by transcription attenuation in bacteria: how RNA provides instructions for transcription termination/antitermination decisions. BioEssays 24:700–707

Hermann T (2003) Industrial production of amino acids by coryneform bacteria. J Biotechnol 104:155–172

Herrmann KM (1983) The common aromatic biosynthetic pathway. In: Herrmann KM, Somerville RL (eds) Amino acids: biosynthesis and genetic regulation. Addison-Wesley, Reading, MA, pp 301–322

Herrmann KM (1995) The shikimate pathway: early steps in the biosynthesis of aromatic compounds. Plant Cell 7:907–919

Herrmann KM, Weaver LM (1999) The shikimate pathway. Annu Rev Plant Physiol Plant Mol Biol 50:473–503

Hsu SK, Lin LL, Lo HH, Hsu WH (2004) Mutational analysis of feedback inhibition and catalytic sites of prephenate dehydratase from *Corynebacterium glutamicum*. Arch Microbiol 181:237–244

Hummel W, Schuette H, Schmidt E, Wandrey C, Kula MR (1987) Isolation of L-phenylalanine dehydrogenase from *Rhodococcus sp.* M4 and its application for the production of L-phenylalanine. Appl Microbiol Biotechnol 26:409–416

Hwang SO, Gil GH, Cho YJ, Kang KR, Lee JH, Bae JC (1985) The fermentation process for L-phenylalanine production using an auxotrophic regulatory mutant of *Escherichia coli*. Appl Microbiol Biotechnol 22:108–113

Hyde SC, Ahmed SA, Pallars EA, Miles EH, Davies DR (1988) Three dimensional structure of the tryptophan synthase $\alpha_2\beta_2$ multienzyme complex from *Salmonella typhimurium*. J Biol Chem 263:17857–17871

Igarashi K, Kishino M, Seki M, Takenouchi T, Kureyama M (1993) Production of L-phenylalanine. Japan Patent 5 304 971 A.

Ikeda M (2003) Amino acid production processes. In: Faurie R, Thommel J (eds) Microbial production of L-amino acids. Adv Biochem Eng Biotechnol 79:1–35

Ikeda M (2006) Towards bacterial strains overproducing L-tryptophan and other aromatics by metabolic engineering. Appl Microbiol Biotechnol 69:615–626

Ikeda M, Katsumata R (1992) Metabolic engineering to produce tyrosine or phenylalanine in a tryptophan-producing *Corynebacterium glutamicum* strain. Appl Environ Microbiol 58:781–785

Ikeda M, Katsumata R (1994) Transport of aromatic amino acids and its influence on overproduction of the amino acids in *Corynebacterium glutamicum*. J Ferment Bioeng 78:420–425

Ikeda M, Katsumata R (1995) Tryptophan production by transport mutants of *Corynebacterium glutamicum*. Biosci Biotech Biochem 59:1600–1602

Ikeda M, Katsumata R (1999) Hyperproduction of tryptophan by *Corynebacterium glutamicum* with the modified pentose phosphate pathway. Appl Environ Microbiol 65:2497–2502

Ikeda M, Nakagawa S (2003) The *Corynebacterium glutamicum* genome: features and impacts on biotechnological processes. Appl Microbiol Biotechnol 62:99–109

Ikeda M, Ozaki A, Katsumata R (1993) Phenylalanine production by metabolically engineered *Corynebacterium glutamicum* with the *pheA* gene of *Escherichia coli*. Appl Microbiol Biotechnol 39:318–323

Ikeda M, Okamoto K, Katsumata R (1999) Cloning of the transketolase gene and the effect of its dosage on aromatic amino acid production in *Corynebacterium glutamicum*. Appl Microbiol Biotechnol 51:201–206

Ikeda M, Nakanishi K, Kino K, Katsumata R (1994) Fermentative production of tryptophan by a stable recombinant strain of *Corynebacterium glutamicum* with a modified serine-biosynthetic pathway. Biosci Biotech Biochem 58:674–678

Ito H, Sato K, Enei H, Hirose Y (1990a) Improvement in microbial production of L-tyrosine by gene dosage effect of *aroL* gene encoding shikimate kinase. Agric Biol Chem 54:823–824

Ito H, Sato K, Matsui K, Sano K, Enei H, Hirose Y (1990b) Molecular breeding of a *Brevibacterium lactofermentum* L-phenylalanine producer using a cloned prephenate dehydratase gene. Appl Microbiol Biotechnol 33:190–195

Ito H, Sato K, Matsui K, Sano K, Nakamori S, Tanaka T, Enei H (1990c) Cloning and characterization of genes responsible for *m*-fluoro-D,L-phenylalanine resistance in *Brevibacterium lactofermentum*. Agric Biol Chem 54:707–713

Jensen RA (1996) Evolution of metabolic pathways in enteric bacteria. In: Neidhardt FC (ed) *Escherichia coli* and *Salmonella*: cellular and molecular biology, vol 2. ASM, Washington, DC, pp 2649–2662

Johansson L, Lindskog A, Silfversparre G, Cimander C, Nielsen KF, Liden G (2005) Shikimic acid production by a modified strain of *E. coli* (W3110.shik1) under phosphate-limited and carbon-limited conditions. Biotechnol Bioeng 92:541–552

Jossek R, Bongaerts J, Sprenger GA (2001) Characterization of a new feedback-resistant 3-deoxy-D-arabino-heptulosonate 7-phosphate synthase of *Escherichia coli*. FEMS Microbiol Lett 202:145–148

Jung E, Zamir LO, Jensen RA (1986) Chloroplasts of higher plants synthesize L-phenylalanine via L-arogenate. Proc Natl Acad Sci USA 83:7231–7235

Kalinowski J, Bathe B, Bartels D, Bischoff N, Bott M, Burkovski A, Dusch N, Eggeling L, Eikmanns B, Gaigalat L, Goesmann A, Hartmann M, Huthmacher K, Krämer R, Linke B, McHardy AC, Meyer F, Möckel B, Pfefferle W, Pühler A, Rey DA, Rückert C, Rupp O, Sahm H, Wendisch VF, Wiegräbe I, Tauch A (2003) The complete *Corynebacterium glutamicum* ATCC 13032 genome sequence and its impact on the production of L-aspartate-derived amino acids and vitamins. J Biotechnol 104:5–25

Katsumata R, Ikeda M (1993) Hyperproduction of tryptophan in *Corynebacterium glutamicum* by pathway engineering. Biotechnology 11:921–925

Khodursky AB, Peter BJ, Cozzarelli NR, Botstein D, Brown PO, Yanofsky C (2000) DNA microarray analysis of gene expression in response to physiological and genetic changes that affect tryptophan metabolism in *Escherichia coli*. Proc Natl Acad Sci USA 97:12170–12175

Kikuchi Y, Tsujimoto K, Kurahashi O (1997) Mutational analysis of the feedback sites of phenylalanine-sensitive 3-deoxy-D-arabino-heptulosonate-7-phosphate synthase of *Escherichia coli*. Appl Environ Microbiol 63:761–762

Kim TH, Namgoong S, Kwak JH, Lee SY, Lee HS (2000) Effects of *tktA*, $aroF^{fbr}$, and *aroL* expression in the tryptophan-producing *Escherichia coli*. J Microbiol Biotechnol 10:789–796

Knaggs AR (2001) The biosynthesis of shikimate metabolites. Nat Prod Rep 18:334–355

Knaggs AR (2003) The biosynthesis of shikimate metabolites. Nat Prod Rep 20:119–136

Knop DR, Draths KM, Chandran SS, Barker JL, Frost JW (2001) Hydroaromatic equilibrium during biosynthesis of shikimic acid. J Am Chem Soc 123:10173–10182

Koehn SJ, Evans TM, Nelson RA, Taylor PP (1994) Methods for increasing carbon conversion efficiency in microorganisms. WO 94/28154

Konstantinov KB, Yoshida T (1992) Knowledge-based control of fermentation processes. Biotech Bioeng 39:479–486

Konstantinov KB, Nishio N, Seki T, Yoshida T (1991) Physiologically motivated strategies for control of the fed-batch cultivation of recombinant *Escherichia coli* for phenylalanine production. J Ferment Bioeng 71:350–355

Krämer R (1994) Systems and mechanisms of amino acid uptake and excretion in prokaryotes. Arch Microbiol 162:1–13

Krämer M, Karutz M, Sprenger G, Sahm H (1999) Microbial preparation of substances from aromatic metabolism III. WO 99/55877

Krämer M, Bongaerts J, Bovenberg R, Kremer S, Müller U, Orf S, Wubbolts M, Raeven L (2003) Metabolic engineering for microbial production of shikimic acid. Metab Eng 5:277–283

Lawley B, Pittard AJ (1994) Regulation of *aroL* by TyrR protein and Trp repressor in *Escherichia coli* K-12. J Bacteriol 176:6921–6930

Layne E (1957) Spectrophotometric and turbidimetric methods for measuring proteins. Meth Enzymol 3:447–455

Leuchtenberger W (1996) Amino acids – technical production and use. In: Rehm HJ, Reed G, Pühler A, Stadler P, Roehr M (eds) Biotechnology, 2nd edn, vol 6. VCH, Weinheim, pp 465–502

Leuchtenberger W, Huthmacher K, Drauz K (2005) Biotechnological production of amino acids and derivatives: current status and prospects. Appl Microbiol Biotechnol 69:1–8

Liao JC (1996) Microorganisms and methods for overproduction of DAHP by cloned *pps* gene. WO 96/08,567

Liao JC, Chao YP, Patnaik R (1994) Alteration of the biochemical valves in the central metabolism of *Escherichia coli*. Ann NY Acad Sci 30:21–34

Liao JC, Hou SY, Chao YP (1996) Pathway analysis, engineering, and physiological considerations for redirecting central metabolism. Biotechnol Bioeng 52:129–140

Liao H-F, Lin L-L, Chien HR, Hsu W-H (2001) Serine 187 is a crucial residue for allosteric regulation of *Corynebacterium glutamicum* 3-deoxy-D-arabino-heptulosonate 7-phosphate synthase. FEMS Microbiol Lett 194:59–64

Liebl W, Ehrmann M, Ludwig W, Schleifer KH (1991) Transfer of *Brevibacterium divaricatum* DSM 20297T, "*Brevibacterium flavum*" DSM 20411, "*Brevibacterium lactofermentum*" DSM 20412 and DSM 1412, and *Corynebacterium glutamicum* and their distinction by rRNA gene restriction patterns. Int J Syst Bacteriol 41:255–260

Liu DX, Fan CS, Tao JH, Liang GX, Gao SE, Wang HJ, Li X, Song DX (2004) Integration of *E. coli aroG-pheA* tandem genes into *Corynebacterium glutamicum tyrA* locus and its effect on L-phenylalanine biosynthesis. World J Gastroenterol 10:3683–3687

Lloyd-George I, Chang TMS (1995) Characterization of free and alginate-polylysine-alginate microencapsulated *Erwinia herbicola* for the conversion of ammonia, pyruvate, and phenol into L-tyrosine. Biotechnol Bioeng 48:706–714

Lu JL, Liao JC (1997) Metabolic engineering and control analysis for production of aromatics: role of transaldolase. Biotechnol Bioeng 53:132–138

Maass D, Gerigk M, Kreutzer A, Weuster-Botz D, Wubbolts M, Takors R (2002) Membrane-based reactive extraction for integrated L-phenylalanine separation in an *Escherichia coli* fed-batch process: from lab to pilot scale. Bioproc Biosyst Engin 25:85–96

Maiti TK, Roy A, Mukherjee SK, Chatterjee SP (1995) Microbial production of L-tyrosine: a review. Hindustan Antibiot Bull 37:51–65

Man TK, Pease AJ, Winkler ME (1997) Maximization of transcription of the *serC* (*pdxF*)-*aroA* multifunctional operon by antagonistic effects of the cyclic AMP (cAMP) receptor protein-cAMP complex and Lrp global regulators of *Escherichia coli* K-12. J Bacteriol 179:3458–3469

Marin-Sanguino A, Torres NV (2000) Optimization of tryptophan production in bacteria. Design of a strategy for genetic manipulation of the tryptophan operon for tryptophan flux maximization. Biotechnol Prog 16:133–145

Mascarenhas D, Ashworth DJ, Chen CS (1991) Deletion of *pgi* alters tryptophan biosynthesis in a genetically engineered strain of *Escherichia coli*. Appl Environ Microbiol 57:2995–2999

Matsui K, Miwa K, Sano K (1987a) Two single base pair substitutions causing desensitization to tryptophan feedback inhibition of anthranilate synthase and enhanced expression of tryptophan genes of *Brevibacterium lactofermentum*. J Bacteriol 109:5330–5332

Matsui K, Miwa K, Sano K (1987b) Cloning of tryptophan genes of *Brevibacterium lactofermentum*, a glutamic acid-producing bacterium. Agric Biol Chem 51:823–828

Mayeno AN, Gleich GJ (1994) Eosinophilia-Myalgia syndrome and tryptophan production: a cautionary tale. Tibtech 12:346–352

McCandliss RJ, Poling MD, Herrmann KM (1978) 3-Deoxy-D-arabino-heptulosonate 7-phosphate synthase. Purification and molecular characterization of the phenylalanine-sensitive isoenzyme from *Escherichia coli*. J Biol Chem 253:4259–4265

McHardy AC, Tauch A, Rückert C, Pühler A, Kalinowski J (2003) Genome-based analysis of biosynthetic aminotransferase genes of *Corynebacterium glutamicum*. J Biotechnol 104:229–240

Michel G, Roszak AW, Sauve V, Maclean J, Matte A, Coggins JR, Cygler M, Lapthorn AJ (2003) Structures of shikimate dehydrogenase AroE and its paralog YdiB. J Biol Chem 278:19463–19472

Miller JE, Backman KC, O'Connor MJ, Hatch RT (1987) Production of phenylalanine and organic acids by phosphoenolpyruvate carboxylase-deficient mutants of *Escherichia coli*. J Ind Microbiol 2:143–149

Morollo AA, Bauerle R (1993) Characterization of composite aminodeoxy-isochorismate synthase and aminodeoxyisochorismate lyase activities of anthranilate synthase. Proc Natl Acad Sci USA 90:9983–9987

Müller B, Pacholski C, Simat T, Steinhart H (1999) Synthesis and formation of an EMS correlated contaminant in biotechnologically manufactured L-tryptophan. In: Tryptophan, serotonin and melatonin: Basic aspects and applications. Adv Exp Med Biology 467:481–486

Müller U, van Assema F, Gunsior M, Orf S, Kremer S, Schipper D, Wagemans AN, Townsend CA, Sonke T, Bovenberg R, Wubbolts M (2006) Metabolic engineering of the *E. coli* L-phenylalanine pathway for the production of D-phenylglycine (D-Phg). Metabol Engin 8:196–208

Murdock D, Ensley BD, Serdar C, Thalen M (1993) Construction of metabolic operons catalyzing the de novo biosynthesis of indigo in *Escherichia coli*. Bio-Technol 11:381-386

Nakamichi K, Nabe K, Nishida Y, Tosa T (1989) Production of L-phenylalanine from phenylpyruvate by *Paracoccus denitrificans* containing aminotransferase activity. Appl Microbiol Biotechnol 30:243–246

Nelms J, Gonzalez DH, Yoshida T, Fotheringham I (1992) Novel mutations in the *pheA* gene of *Escherichia coli* K-12 which result in highly feedback inhibition-resistant variants of chorismate mutase/prephenate dehydratase. Appl Environ Microbiol 58:2592–2598

Nichols BP (1996) Evolution of genes and enzymes of tryptophan biosynthesis. In: Neidhardt FC (ed) *Escherichia coli* and *Salmonella*: cellular and molecular biology, vol 2. ASM, Washington, DC, pp 2638–2648

Niederberger P, Prasad R, Miozzari G, Kacser H (1992) A strategy for increasing an in vivo flux by genetic manipulations. The tryptophan system of yeast. Biochem J 287:473–479

Ogino T, Garner C, Markley JL, Herrmann KM (1982) Biosynthesis of aromatic compounds: ^{13}C NMR spectroscopy of whole *Escherichia coli* cells. Proc Natl Acad Sci USA 79:5828–5832

Oldiges M, Kunze M, Degenring D, Sprenger GA, Takors R (2004) Stimulation, monitoring and analysis of pathway dynamics by *metabolic profiling* in the aromatic amino acid pathway. Biotechnol Prog 20:1623–1633

Ozaki A, Katsumata R, Oka T, Furuya A (1985) Cloning of the genes concerned in phenylalanine biosynthesis in *Corynebacterium glutamicum* and its application to breeding of a phenylalanine producing strain. Agric Biol Chem 49:2925–2930

Panina EM, Vitreschak AG, Mironov AA, Gelfand MS (2001) Regulation of aromatic amino acid biosynthesis in gamma-proteobacteria. J Mol Microbiol Biotechnol 3:529–543

Parker C, Barnell WO, Snoep JL, Ingram LO, Conway T (1995) Characterization of the *Zymomonas mobilis* glucose facilitator gene product (*glf*) in recombinant *Escherichia coli*: examination of transport mechanism, kinetics and the role of glucokinase in glucose transport. Mol Microbiol 15:795–802

Patnaik R, Liao JC (1994) Engineering of *Escherichia coli* central metabolism for aromatic metabolite production with near theoretical yield. Appl Environ Microbiol 60:3903–3908

Patnaik R, Spitzer R, Liao JC (1995) Pathway engineering for production of aromatics in *Escherichia coli*: confirmation of stoichiometric analysis by independent modulation of AroG, TktA, and Pps activities. Biotechnol Bioeng 46:361–370

Pittard J (1996) Biosynthesis of the aromatic amino acids. In: Neidhardt FC (ed) *Escherichia coli* and *Salmonella*: cellular and molecular biology, vol 1. ASM, Washington, DC, pp 458–484

Pittard J, Yang J (2005) Biosynthesis of phenylalanine and tyrosine. EcoSal Module 3618. In: Böck A, Curtiss R III, Kaper JB, Neidhardt FC, Nyström T, Rudd KE, Squires CL (eds) Ecosal *Escherichia coli* and *Salmonella*: cellular and molecular biology. www.ecosal.org, ASM, Washington, DC

Pittard J, Camakaris H, Yang J (2005) The TyrR regulon. Mol Microbiol 55:16–26

Pohnert G, Zhang S, Husain A, Wilson DB, Ganem B (1999) Regulation of phenylalanine biosynthesis. Studies on the mechanism of phenylalanine binding and feedback inhibition in the *Escherichia coli* P-protein. Biochemistry 38:12212–12217

Polen T, Krämer M, Bongaerts J, Wubbolts M, Wendisch VF (2005) The global gene expression response of *Escherichia coli* to L-phenylalanine. J Biotechnol 115:221–237

Postma PW, Lengeler JW, Jacobson GR (1993) Phosphoenolpyruvate: carbohydrate phosphotransferase systems of bacteria. Microbiol Rev 57:543–594

Ray JM, Yanofsky C, Bauerle (1988) Mutational analysis of the catalytic and feedback sites of the tryptophan-sensitive 3-deoxy-D-arabino-heptulosonate 7-phosphate synthase of *Escherichia coli.* J Bacteriol 170:5500–5506

Roberts F, Roberts CW, Johnson JJ, Kyle DE, Krell T, Coggins JR, Coombs GH, Milhous WK, Tzipori S, Ferguson DJ, Chakrabarti D, McLeod R (1998) Evidence for the shikimate pathway in apicomplexan parasites. Nature 393:801–805

Rüffer N, Heidersdorf U, Kretzers I, Sprenger GA, Raeven L, Takors R (2004) Fully integrated L-phenylalanine separation and concentration using reactive-extraction with liquid-liquid centrifuges in a fed-batch process with *E. coli*. Bioproc Biosyst Engin 26:239–248

Sano K, Matsui K (1987) Structure and function of the *trp* operon control regions of *Brevibacterium lactofermentum*, a glutamic-acid-producing bacterium. Gene 53:191–200

Sauer U, Eikmanns BJ (2005) The PEP pyruvate-oxaloacetate node as the switch point for carbon flux distribution in bacteria. FEMS Microbiol Rev 29:765–794

Schmid A, Dordick JS, Hauer B, Kiener A, Wubbolts M, Witholt B (2001) Industrial biocatalysis today and tomorrow. Nature 409:258–268

Schmid JW, Mauch K, Reuss M, Gilles ED, Kremling A (2004) Metabolic design based on a coupled gene expression–metabolic network model of tryptophan production in *Escherichia coli*. Metabol Engin 6:364–377

Shiio I, Miyajima R, Nakagawa M (1972) Regulation of aromatic amino acid biosynthesis in *Brevibacterium flavum*, I, Regulation of anthranilate synthetase. J Biochem 72:1447–1455

Shu CH, Liao CC (2002) Optimization of L-phenylalanine production of *Corynebacterium glutamicum* under product feedback inhibition by elevated oxygen transfer rate. Biotechnol Bioeng 77:131–141

Simat TJ, Kleeberg KK, Müller G, Sierts A (1999) Synthesis, formation, and occurrence of contaminants in biotechnologically manufactured L-tryptophan. In: Tryptophan, serotonin and melatonin: basic aspects and applications. Adv Exp Med Biol 467:469–480

Snell KD, Draths KM, Frost JW (1996) Synthetic modification of the *Escherichia coli* chromosome: enhancing the biocatalytic conversion of glucose into aromatic chemicals. J Am Chem Soc 116:5605–5614

Snoep JL, Arfman N, Yomano LP, Fliege RK, Conway T, Ingram LO (1994) Reconstruction of glucose uptake and phosphorylation in a glucose-negative mutant of *Escherichia coli* by using *Zymomonas mobilis* genes encoding the glucose facilitator protein and glucokinase. J Bacteriol 176:2133–2135

Somerville RL (1983) Tryptophan: biosynthesis, regulation, and large-scale production. In: Herrmann KM, Somerville RL (eds) Amino acids: biosynthesis and genetic regulation. Addison-Wesley, Reading, MA, pp 351–378

Song J, Bonner CA, Wolinsky M, Jensen RA (2005) The TyrA family of aromatic-pathway dehydrogenases in phylogenetic context. BMC Biology 3:13

Sprenger GA (1995) Genetics of pentose-phosphate pathway enzymes of *Escherichia coli* K-12. Arch Microbiol 164:324–330

Sprenger G, Siewe R, Sahm H, Karutz M, Sonke T (1998a) Microbial preparation of substances from aromatic metabolism/I. WO 98/18936

Sprenger G, Siewe R, Sahm H, Karutz M, Sonke T (1998b) Microbial preparation of substances from aromatic metabolism/II. WO 98/18937

Srinivasan PR, Sprinson DB (1959) 2-keto-3-deoxy-D-arabo-heptonic acid 7-phosphate synthetase. J Biol Chem 234:713–716

Stahl JL, Cook EB, Pariza MA, Cook ME, Graziano FM (2001) Effect of L-tryptophan supplementation on eosinophils and eotaxin in guinea pigs. Exp Biol Med 226:177–184

Sugimoto S, Shiio I (1980) Purification and properties of dissociable chorismate mutase from *Brevibacterium flavum*. J Biochem 88:167–176

Sugimoto S, Shiio I (1982) Tryptophan synthase and production of L-tryptophan in regulatory mutants. Agric Biol Chem 46:2711–2718

Sugimoto S, Yabuta M, Kato N, Seki T, Yoshida T, Taguchi H (1987) Hyperproduction of phenylalanine by *Escherichia coli*: application of a temperature-controllable expression vector carrying the repressor-promoter system of bacteriophage lambda. J Biotechnol 5:237–253

SwissProt (2005) UniProtKB/Swiss-Prot protein knowledgebase Swiss-Prot headline. Release 466 of 26-Apr-2005 http://us.expasy.org/sprot/relnotes/spwrnew.html

Takagi M, Nishio Y, Oh G, Yoshida T (1996) Control of L-phenylalanine production by dual feeding of glucose and L-tyrosine. Biotechnol Bioeng 52:653–660

Taylor PP, Grinter NJ, McCarthy SL; Pantaleone DP; Ton JL, Yoshida RK; Fotheringham IG (2000) D-Phenylalanine biosynthesis using *Escherichia coli*: Creation of a new metabolic pathway. In: Saha BC, Demirjian DC (eds) Applied Biocatalysis in Specialty Chemicals and Pharmaceuticals. ACS Symposium Series 776, pp 65–75

Tonouchi N, Kojima H, Matsui H (1997) Recombinant DNA sequences encoding feedback inhibition released enzymes, plasmids comprising the recombinant DNA sequences, transformed microorganisms useful in the production of aromatic amino acids, and a process for preparing aromatic amino acids by fermentation. EP 0,488,424 B1

Tribe DE, Pittard J (1979) Hyperproduction of tryptophan by *Escherichia coli*: genetic manipulation of the pathways leading to tryptophan formation. Appl Environ Microbiol 38:181–190

Tsuchida T, Kubota K, Morinaga Y, Matsui H, Enei H, Yoshinaga F (1987) Production of L-phenylalanine by a mutant of *Brevibacterium lactofermentum* 2256. Agric Biol Chem 51:2095–2101

Umbarger HE (1978) Amino acid biosynthesis and its regulation. Annu Rev Biochem 47:533–606

Valle F, Munoz E, Ponce E, Flores N, Bolivar F (1996) Basic and applied aspects of metabolic diversity: the phosphoenolpyruvate node. J Ind Microbiol Biotechnol 17:458–462

Vertes AA, Inui M, Yukawa H (2005) Manipulating corynebacteria, from individual genes to chromosomes. Appl Environ Microbiol 71:7633–7642

Vitreschak AG, Lyubetskaya EV, Shirshin MA, Gelfand MS, Lyubetsky VA (2004) Attenuation regulation of amino acid biosynthetic operons in proteobacteria: comparative genomics analysis. FEMS Microbiol Lett 234:357–370

Wahl A, El Massaoudi M, Schipper D, Wiechert W, Takors R (2004) Serial ^{13}C-based flux analysis of an L-phenylalanine-producing *E. coli* strain using the sensor reactor. Biotechnol Prog 20:706–714

Weaver LM, Herrmann KM (1990) Cloning of an *aroF* allele encoding a tyrosine-insensitive 3-deoxy-D-arabino-heptulosonate 7-phosphate synthase. J Bacteriol 172: 6581–6584

Weikert C, Sauer U, Bailey JE (1998) Increased phenylalanine production by growing and nongrowing *Escherichia coli* strain CWML2. Biotechnol Prog 14:420–424

Weisser P, Krämer R, Sahm H, Sprenger GA (1995) Functional expression of the glucose transporter of *Zymomonas mobilis* leads to restoration of glucose and fructose uptake in *Escherichia coli* mutants and provides evidence for its facilitator action. J Bacteriol 177:351–354

Whipp MJ, Pittard AJ (1995) A reassessment of the relationship between *aroK*- and *aroL*-encoded shikimate kinase enzymes of *Escherichia coli.* J Bacteriol 177:1627–1629

White RH (2004) L-Aspartate semialdehyde and a 6-deoxy-5-ketohexose 1-phosphate are the precursors to the aromatic amino acids in *Methanocaldococcus jannaschii.* Biochemistry 43:7618–7627

Xie G, Keyhani NO, Bonner CA, Jensen RA (2003) Ancient origin of the tryptophan operon and the dynamics of evolutionary change. Microbiol Mol Biol Rev 67:303–342

Yajima Y, Sakimoto K, Takahashi K, Miyao K, Kudome Y, Aichi K (1990) L-Tryptophan-producing microorganism and production of L-tryptophan. Japan Patent Appl 02,190, 182

Yanofsky C (2001) Advancing our knowledge in biochemistry, genetics, and microbiology through studies on tryptophan metabolism. Annu Rev Biochem 70:1–37

Yanofsky C (2003) Using studies on tryptophan metabolism to answer basic biological questions. J Biol Chem 278:10859–10878

Yanofsky C, Crawford IP (1972) Tryptophan synthetase. In: Boyer PD (ed) The Enzymes, vol 7, 3rd edn. Academic, New York, pp 1–31

Yi J, Li K, Draths KM, Frost JW (2002) Modulation of phosphoenolpyruvate synthase expression increases shikimate pathway product yields in *E. coli*. Biotechnol Prog 18:1141–1148

Zeman R, Plachy J, Bulantova H, Sikyta B, Pavlasova E, Stejskalova E (1990) Enzyme synthesis of L-tryptophan. Folia Microbiol (Praha) 35:200–204

Zhang S, Pohnert G, Kongsaere P, Wilson DB, Clardy J, Ganem B (1998) Chorismate mutase-prephenate dehydratase from *Escherichia coli*. Study of catalytic and regulatory domains using genetically engineered proteins. J Biol Chem 273:6248–6253

Microbiol Monogr (5)
V. F. Wendisch: Amino Acid Biosynthesis
DOI 10.1007/7171_2006_070/Published online: 1 February 2007

Branched-Chain Amino Acids

Miroslav Pátek

Institute of Microbiology, Academy of Sciences of the Czech Republic, Vídeňská 1083, 14220 Prague 4, Czech Republic
patek@biomed.cas.cz

Abstract The branched-chain amino acids (BCAAs) leucine, isoleucine, and valine are synthesized by bacteria, fungi, and plants, but are essential for vertebrates including humans, who must receive them from their diet. The interest to construct overproducing industrial strains therefore stems from the need to supplement the food or feed with these amino acids to use them in medical treatment and as precursors in biochemical synthesis. Regulation of the biosynthesis pathways of branched-chain amino acids has many features, such as homologous reactions catalyzed by a single enzyme, branching of the pathways and multivalent regulation of both gene expression and enzyme activity, which make their analysis both interesting and challenging. The structural similarity of these three amino acids and their precursors causes their alternative binding to the proteins as substrates, inhibitors, activators, and passengers in transporters with different affinities. Studies of threonine deaminase, the first enzyme specific for isoleucine biosynthesis, and of acetohydroxyacid synthase, the first common enzyme in the pathways of BCAA biosynthesis, promoted the discovery of feedback inhibition and may serve as paradigms for this regulatory mechanism. Regulation of the *Escherichia coli* operons *ilvGMEDA* and *ilvBN* provides examples of typical translation-mediated transcriptional termination (attenuation). Mechanisms of regulation by the seemingly similar structures found in *ilvBNC* operon and *leuA* gene of the industrial amino acid producer *Corynebacterium glutamicum* still have to be unveiled. A wide range of different specific and global regulatory mechanisms being gradually uncovered in various microorganisms will contribute to the knowledge of genetic control of BCAA biosynthesis.

1 Introduction

The branched-chain amino acids (BCAAs), leucine, isoleucine and valine, are the most hydrophobic amino acids due to their unsubstituted aliphatic chain with a branched alkyl group (Fig. 1). These BCAAs are often located in the core of the proteins and play a crucial role in determining the structures of globular proteins as well as in the interaction of the transmembrane domains of membrane proteins with phospholipid bilayers. In terms of protein secondary structure, valine and isoleucine exhibit a definite preference for the β-structure, whereas leucine has a higher preference for the α-helix. BCAAs, which are essential for human and other vertebrates, are synthesized de novo by most bacteria, fungi, and plants.

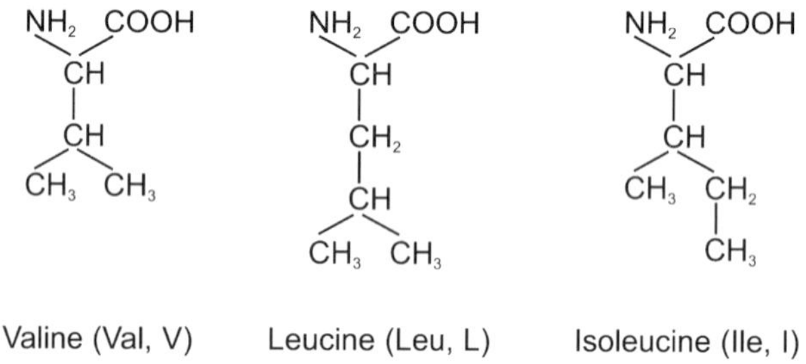

Fig. 1 Chemical structures of branched-chain amino acids

Biosynthesis pathways of isoleucine, valine and leucine (all in their L-forms) are closely tied, and run in most organisms as shown in Fig. 2. Valine and isoleucine are synthesized in parallel pathways in which homologous reaction steps are catalyzed by the same enzymes. Leucine is formed in a specific series of reactions branching off from the valine pathway, thus, sharing common precursors with valine. A complex regulatory network, which controls the metabolite flux through these pathways in bacteria, includes mul-

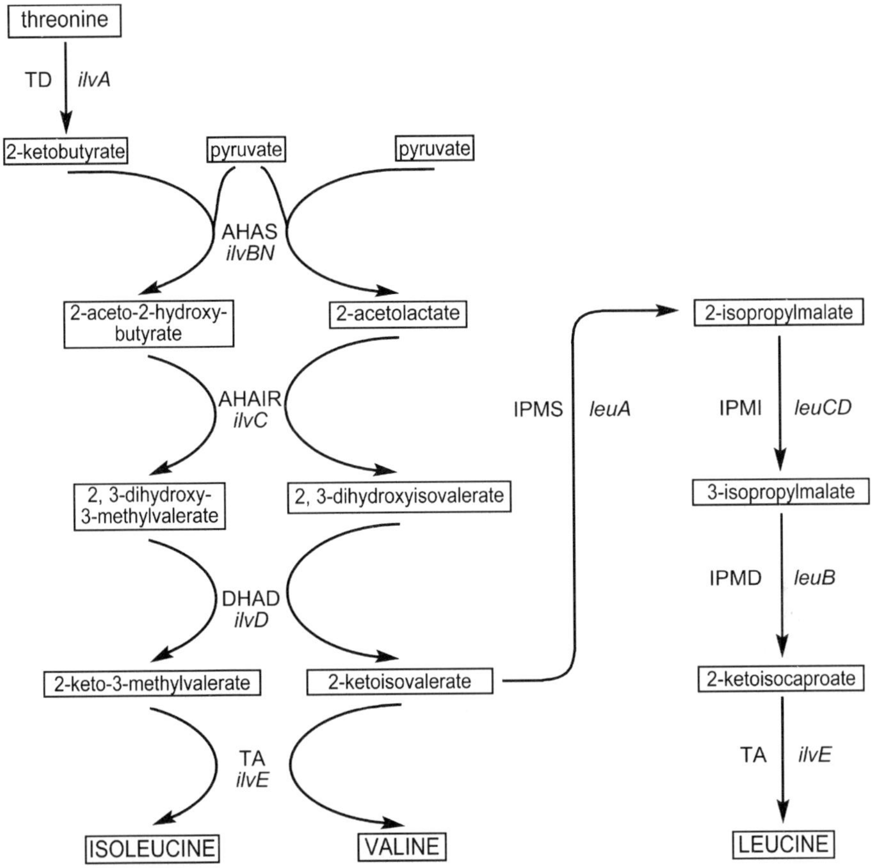

Fig. 2 Biosynthesis of isoleucine, valine, and leucine in bacteria. Enzymes and the corresponding genes: TD (*ilvA*), threonine deaminase; AHAS (*ilvBN*), acetohydroxyacid synthase; AHAIR (*ilvC*), acetohydroxyacid isomeroreductase; DHAD (*ilvD*), dihydroxy-acid dehydratase; TA (*ilvE*), transaminase, IPMS (*leuA*), isopropylmalate synthase; IPMI (*leuCD*) isopropylmalate dehydratase; IPMD (*leuB*) isopropylmalate dehydrogenase. The genes are designated as those in *C. glutamicum*. The other AHAS isoenzymes are encoded by *ilvIH* and *ilvGM* in *E. coli*. Other transaminases active in BCAA pathways are encoded by *aroT*, *avtA* and *pdxR* in *C. glutamicum* and by *avtA* and *tyrB* in *E. coli*

tivalent regulation of both gene expression and enzyme activity. Regulatory circuits are also complicated by the structural similarity of branched-chain amino acids which mimic each other in their abilities to bind to the catalytic or regulatory sites of the proteins.

Binding of these near-isomorphic amino acids varies only with the different affinities of the respective proteins for the individual amino acids and this blurred recognition has a consequence in metabolic antagonism of the three BCAAs (Umbarger 1996). This review is focused on the synthesis of BCAAs and its regulation in *Escherichia coli* (or *Salmonella enterica* serovar Typhimurium) and in *Corynebacterium glutamicum* as well as on their biotechnological production. *E. coli* is a model organism in which the regulation of the pathways has been most thoroughly described, whereas *C. glutamicum* is the industrial microorganism frequently used for manufacturing amino acids by fermentation processes.

2 Isoleucine and Valine

Since the structures of valine and isoleucine, as well as of the respective intermediates of their biosynthesis, differ only by a single methylene group, four reactions of their biosynthesis pathways are catalyzed by common enzymes.

2.1 Regulation of Enzyme Activity

The major roles in regulation of isoleucine and valine biosynthesis are played by threonine deaminase (converting threonine into 2-ketobutyrate), which is strongly inhibited by isoleucine and acetohydroxyacid synthase, which has higher affinity to 2-ketobutyrate than to pyruvate and thus favors the reaction within the isoleucine biosynthesis pathway. In this way, balancing the levels of valine and leucine on one side and isoleucine on the other, is ensured. The enzymes of the pathways have been studied particularly in plants and fungi as potential targets for herbicides and fungicides (McCourt and Duggleby 2006). Mutagenesis studies in yeast confirmed that the herbicides (which resemble neither the substrates nor feedback inhibitors) bind neither to the active site nor to the allosteric site (Duggleby et al. 2003). Structural and regulatory properties of the major enzymes of the pathways in gram-positive bacteria are similar to that from *E. coli* or *C. glutamicum* described here. The activity of the single acetohydroxyacid synthase in Gram-positive bacteria, such as in bacilli (Porat et al. 2004) or mycobacteria (Zohar et al. 2003), is essentially regulated as acetohydroxyacid synthase III from *E. coli*.

2.1.1 Threonine Deaminase (TD)

Escherichia Coli

The enzyme with the systematic name L-threonine ammonia-lyase (EC 4.3.1.19) is referred to as threonine deaminase or threonine dehydratase. This enzyme utilizes pyridoxal-5′-phosphate as a cofactor and catalyzes the conversion of threonine to 2-ketobutyrate. It is the only enzyme specific for isoleucine synthesis (Fig. 2). The end product of the pathway, isoleucine, acts as an allosteric effector and exerts inhibition of TD activity, while valine, the end product of the parallel pathway, acts as its activator in both *E. coli* and *S. enterica* serovar Typhimurium (Changeux 1963; Eisenstein 1991). These heterotropic effectors bind to the site that is distinct from the active site in the molecule of the enzyme. Threonine deaminase can therefore serve as a useful model for structure-based mechanism of allostery. The early studies of threonine deaminase (Changeux 1961) actually influenced formation of a model of allosteric regulation of enzyme activity. Threonine deaminase of *E. coli* forms tetramers with identical subunits and cooperative binding of effectors (Eisenstein et al. 1994). The model assumed that binding of a single inhibitor or substrate or activator to a subunit would favor binding of an identical ligand molecule to the other subunits. This notion proposes equilibrium between the active and inactive form of the enzyme in which all subunits of the tetramer bind the same effector. Mutagenesis studies showed that the amino-terminal domain of TD is involved in catalysis whereas the carboxy-terminal domain is important for its regulation (Fisher and Eisenstein 1993).

Corynebacterium Glutamicum

TD from *C. glutamicum*, formed by four identical subunits of 47 kDa, is (according to its sequence) highly similar (59 to 78% identity) to TDs found in related nocardioform bacteria (corynebacteria, mycobacteria, nocardiae and rhodococci), but much less similar to TDs of Gram-negative bacteria like *E. coli*. While the N-terminal part shows 35% identity with that part of *E. coli* TD, the C-terminal part is shorter than its *E. coli* counterpart and differs substantially in sequence, although it is also involved in the allosteric regulation of the enzyme activity (Möckel et al. 1992). As in *E. coli*, *C. glutamicum* threonine deaminase is inhibited by isoleucine and activated by valine. The $K_{0.5}$ for threonine is 21 mM and it increases to 78 mM or decreases to 12 mM in the presence of isoleucine or valine, respectively.

The site-directed mutagenesis of the *ilvA* gene coding for TD resulted in the abolishment of feedback inhibition (Möckel et al. 1994). These mutations resided in the C-terminal part of the polypeptide, which represents the regulatory domain of the enzyme, spatially separated from the catalytic domain according to the 3D model of the threonine deaminase structure. Other random mutations decreasing the inhibitory effect of isoleucine as well as the basic enzyme activity were also located in the C-terminal part (residues within the region 266–349). The location of the mutations thus supported the assignment of the regulatory domain to the C-terminal part of the protein. However, the enzyme is apparently not of a modular structure and most probably interactions of the catalytic and regulatory domains are required for its proper function (Möckel et al. 1994).

2.1.2
Acetohydroxyacid Synthase (AHAS)

Acetohydroxyacid synthase (EC 2.2.1.6), also called acetolactate synthase, is the first common enzyme in the pathways of BCAA biosynthesis (Fig. 2). It catalyzes decarboxylation of pyruvate and its condensation either with another molecule of pyruvate to produce acetolactate (a precursor of valine and leucine) or with 2-ketobutyrate to produce acetohydroxybutyrate (a precursor of isoleucine). AHAS has been studied in plants as a target for herbicides (imidazolines and sulfonylureas), which has encouraged determination of the 3D structure of the protein (McCourt and Duggleby 2006). The AHAS catalytic subunit from *Arabidopsis thaliana* was crystallized (McCourt et al. 2005) and its structure was described in the presence of all three cofactors and each of five sulfonylurea herbicides (McCourt and Duggleby 2006). The crystal structure of the AHAS catalytic subunit from *Saccharomyces cerevisiae*, which is considered a suitable model for plant AHAS, was also analyzed (Pang et al. 2002). The structure and role of AHAS in biosynthesis of BCAAs in fungi and plants has recently been reviewed (McCourt and Duggleby 2006).

Escherichia Coli

Three AHAS isoenzymes, differing in their genetic determination, biochemical properties, and regulation of both biosynthesis and activity (Table 1), were described for wild-type strains of *E. coli* and *S. enterica* serovar Typhimurium (Umbarger 1996). The isoenzymes AHAS I, II, and III are composed of large subunits and small subunits encoded by the genes *ilvBN*, *ilvGM*, and *ilvIH*, respectively. The large subunits (IlvB, IlvG, and IlvI) are responsible for catalytic activity of the three isoenzymes, while the small subunits (IlvN, IlvM and IlvH) mediate their inhibition. The differences between the AHAS isoenzymes are based particularly on the size, sequence, and properties of regulatory subunits, whereas the catalytic subunits are much more similar to each other. The catalytic subunits exhibit about 40% sequence identity between any two isoenzymes and are of similar size (about 60 kDa), in contrast to the

Table 1 Characteristics of *E. coli* AHAS isoenzymes

Isoenzyme	Genes	Inhibition	R^{a}	Repression
AHAS I	*ilvBN*	Val	2	Val, Leu
AHAS II	*ilvGM*	–	60	Val, Ile, Leu
AHAS III	*ilvIH*	Val, Ile, Leu	40	Leu

[a] A specific ratio of alternative products (acetohydroxybutyrate and acetolactate) formed in homologous reactions catalyzed by AHAS, defined in Eq. 1

regulatory subunits, which are far more diverse in sequence and size (11 kDa, 9.5 kDa, and 18 kDa, respectively) (Ibdah et al. 1996). Although the catalytic machinery is entirely contained within the large subunits, they have much lower activity (which is insensitive to inhibition), when separated from the small subunits (Weinstock et al. 1992). AHAS catalytic subunits thus require the presence of regulatory subunits for full activity and for regulation. Active holoenzymes may be reconstituted from the respective isolated subunits. Due to the specific interactions, the heterologous combinations of large and small subunits provide no detectable activity, despite their sequence similarities (Weinstock et al. 1992). Activity of AHAS depends on thiamine diphosphate (ThDP) and on flavin adenine dinucleotide (FAD). The large (catalytic) subunits show significant sequence similarities to pyruvate oxidase (EC 1.2.3.3) and other ThDP-dependent enzymes catalyzing reactions in which the first step is decarboxylation of a 2-ketoacid (Bar-Ilan et al. 2001).

The activities of AHAS I and AHAS III are inhibited by valine, while AHAS II is valine-resistant. A striking feature of AHAS II and AHAS III is their much higher affinity to 2-ketobutyrate than to pyruvate. In the presence of 2-ketobutyrate, synthesis of isoleucine is therefore preferred. To determine the rates of the formation of the alternative products (acetolactate and acetohydroxybutyrate), an analytical method for simultaneous assaying of the two separate reactions was developed (Gollop et al. 1987). A specific ratio of the two products was defined (Barak et al. 1987):

$$R = \frac{\text{(rate of acetohydroxybutyrate formation)/(rate of acetolactate formation)}}{\text{[2-ketobutyrate]/[pyruvate]}} . \qquad (1)$$

The parameter R is constant for a particular isoenzyme over a wide range of concentrations. However, it depends to some degree on the conditions of its experimental determination. The larger the value of R, the stronger the preference of the isoenzyme for 2-ketobutyrate (Table 2).

Many microorganisms possess only a single AHAS, which is predominantly similar to AHAS III (IlvIH) in its sequence. The respective orthologous genes should therefore be designated *ilvIH*, although they have been named *ilvBN* in some cases (e.g., in *C. glutamicum*). Orthologs of *ilvBN* encoding AHAS I are present only in enterobacteria. In the most commonly used strain of *E. coli*, K-12, the valine-insensitive isoenzyme (AHAS II) activity is missing due to a frameshift mutation in the *ilvG* gene (Lawther et al. 1981). This is the reason why the growth of K-12-derived strains is inhibited by higher concentrations of valine if isoleucine is missing in the medium.

AHAS I is markedly different from the other two AHAS isoenzymes in its sequence and, consequently, in its properties (Vinogradov et al. 2006). This isoenzyme shows only a two-fold preference for 2-ketobutyrate over pyruvate ($R \sim 2$), whereas AHAS II and III show 60-fold and 40-fold higher affinities to 2-ketobutyrate than to pyruvate, respectively (Gollop et al. 1989;

Vinogradov et al. 2006). AHAS I exhibited the highest activity of the three isoenzymes in vitro (ratio of specific activities AHAS I: AHAS II: AHAS III in $\mu mol\ min^{-1}\ mg^{-1}$ was 60 : 20 : 2.6) (Gollop et al. 1989; Vinogradov et al. 2006). However, AHAS I can be completely inhibited by valine even in the presence of high concentrations of pyruvate, whereas AHAS III is only partially inhibited by valine. Separated from the regulatory subunit IlvN, the catalytic subunit IlvB has 10 to 20% activity of the holoenzyme, in contrast to the large subunit IlvI, which has alone only 3 to 5% activity of the IlvI–IlvH holoenzyme (Vyazmensky et al. 1996). Synthesis of AHAS I is induced when the available carbon source is poor (e.g., acetate, succinate), since expression of the *ilvBN* operon is regulated by carbon catabolite repression (Sutton and Freundlich 1980; Tedin and Norel 2001). All these unique features suggest that AHAS I fulfils a special metabolic role in *E. coli* (Vinogradov et al. 2006). The AHAS I isoenzyme ensures the synthesis of valine and leucine under conditions, when the intracellular concentration of pyruvate is low (Dailey and Cronan 1986). Thus, AHAS I is important for growth on other substrates than glucose and provides *E. coli* and other enterobacteria with the ability to adapt to a wider range of carbon sources.

AHAS II is insensitive to inhibition by any of the BCAA (Lawther et al. 1987). This isoenzyme therefore sustains growth of the cells in the presence of excess valine, which inhibits activity of both AHAS I and AHAS III (McCourt and Duggleby 2006). AHAS isoenzymes have 26–30% sequence identity with pyruvate oxidases (EC 1.2.3.3) and an evolutionary relationship of these proteins may explain their peculiar requirement for FAD, which is not directly involved in catalysis. A 3D model of the catalytic subunit of AHAS II from *E. coli* could be created (Ibdah et al. 1996) by homology modelling on the basis of sequence similarity with pyruvate oxidase (Muller and Schulz 1993). The structural model suggests which parts of the protein dimer are involved in coenzyme binding and in substrate specificity. The replacement of Trp464 with other AA by site-directed mutagenesis confirmed its proposed role in preference of the isoenzyme for 2-ketobutyrate (Ibdah et al. 1996).

AHAS III, which has orthologs in those microorganisms possessing only a single AHAS, seems to represent an archetype of this enzyme. A structural model of its regulatory subunit (IlvH) is based on the mutagenesis of *ilvH* resulting in abolishment of the ability of IlvH to bind valine (Vyazmensky et al. 1996; Mendel et al. 2001). In addition to changes in valine binding, some mutants of the small subunit had altered affinity to the large subunit. The similarity search showed that the residues, whose mutation (e.g., Gly-14-Asp) abolished the valine-mediated inhibition of AHAS, are conserved in all regulatory subunits of AHAS in bacteria which possess only a single valine-sensitive AHAS. Mutation Asn-11-Ala (constructed on the basis of the IlvH model) resulted in a valine-resistant enzyme, as expected. IlvH of *E. coli* and a number of regulatory subunits from other microorganisms (e.g., from streptomycetes, corynebacteria, and yeast) belong, according to the sequence homology and folding, to the ACT sequence family, which includes regulatory domains of 3-phosphoglycerate dehydrogenase (EC 1.1.1.95) and aspartokinase (EC 2.7.2.2.4) (Mendel et al. 2001). It is noteworthy that IlvM, the regulatory subunit of the valine-insensitive AHAS II, is not a member of this family. Deletion studies of the regulatory subunit IlvH showed that its N-terminal portion (80 of 163 AAs) is sufficient for recognition and activation of the catalytic subunit. The C-terminal part is required for valine binding and inhibition (Mendel et al. 2003). The subunits IlvI and IlvH were isolated separately and the holoenzyme was reconstituted (Vyazmensky et al. 1996, 2000), which confirmed the stoichiometry of 1 : 1. The active sites in the holoenzyme (2 catalytic + 2 regulatory subunits) are located at the interface of a dimer of the catalytic subunits, which is stable only in the presence of the regulatory subunits (Mendel et al. 2001). Among the regulatory subunits of bacterial AHASs it is IlvH from *E. coli*, whose crystal structure was first described (Kaplun et al. 2006). The crystal structure of IlvH indicates that the positively charged C-terminal domain of the small subunit interacts with the negatively charged surface of the large subunit and that the valine-binding sites are located in two symmetrical positions in the interface between a pair of N-terminal ACT domains of the small subunits (Kaplun et al. 2006).

Enterobacteria living in a feast-or-famine style are equipped with three AHAS isoenzymes constituting a flexible set of enzymes with differences in (1) regulation of gene expression, (2) enzyme activity and its inhibition, (3) stability, (4) substrate affinity, and (5) response to stress and to carbon and nitrogen source availability. This set of isoenzymes allows the bacteria to cope with any growth conditions and produce the three amino acids prevalent in proteins in an appropriate level.

Corynebacterium Glutamicum

Only a single AHAS was found in *C. glutamicum* (Eggeling et al. 1987). As in other organisms, the active holoenzyme is composed of two large (catalytic) subunits and two small (regulatory) subunits. The subunits are encoded by the genes *ilvB* and *ilvN*, respectively, which form an operon together with *ilvC*, which codes for AHAIR (Keilhauer et al. 1993). The large subunit IlvB shows the highest degree of similarity (except for other corynebacteria) to the analogous proteins from nocardioform actinomycetes: *Nocardia farcinica*, *Rhodococcus* sp. and mycobacteria (69 to 65% identity). There is a 45% identity shared with the *E. coli* AHAS III large subunit (IlvI). The degree of identities shared with the large subunits of AHAS II (IlvG) and AHAS I (IlvB) from *E. coli* is 45 and 44%, respectively. As to the small (regulatory) subunit (172 AAs coded by *ilvN*) of *C. glutamicum* AHAS, the analogous proteins from nocardioforms show 51 to 48% identity. The AHAS III small subunit (IlvH) of *E. coli* exhibits 39% identity, while the much shorter AHAS I IlvN subunit (96

AAs) is identical at 30% of its residues. No obvious similarity could be recognized with the AHAS II IlvM subunit, which is only about half in size (86 AAs) and does not mediate sensitivity to valine.

Activity of *C. glutamicum* AHAS can be inhibited by any of the three BCAAs (Eggeling et al. 1987; Elišáková et al. 2005), with valine as the strongest inhibitor. The concentrations of the amino acids required for 50% inhibition (IC_{50}) were 0.9 mM for valine, 3.1 mM for isoleucine and 6.0 mM for leucine. In the presence of 5 mM amino acid, the inhibition was 56% (valine), 49% (isoleucine) and 48% (leucine), respectively. It should be noted that inhibition by any combination of two amino acids or by all three amino acids did not exceed 57% (Elišáková et al. 2005).

Similar to AHAS II and AHAS III from *E. coli* (Umbarger 1996), *C. glutamicum* AHAS has a much higher affinity to 2-ketobutyrate than to pyruvate (Eggeling et al. 1987). In the presence of 2-ketobutyrate, synthesis of isoleucine is therefore preferred, and synthesis of valine and leucine is reduced. Accordingly, high 2-ketobutyrate concentrations (100 mM) resulted in valine and leucine deficiency (Eggeling et al. 1987). The ratio R (Eq. 1), calculated on the basis of these results (Eggeling et al. 1987), was approximately 20 to 25, which is closest to the ratio for AHAS III from *E. coli* ($R = 40$).

Site-directed mutagenesis of the regulatory subunit was performed to remove the inhibition of AHAS in *C. glutamicum* (Elišáková et al. 2005). The bases within *ilvN* to be altered were selected according to the alignment of amino acid sequence of *C. glutamicum* IlvN with sequences of homologous subunits of *E. coli* and *Streptomyces cinnamonensis* mutants. Amino acid alterations in the mutants *E. coli ilvH** (Mendel et al. 2001) and *S. cinnamonensis ilvN** (Kopecký et al. 1999), which resulted in AHASs resistant to inhibition by valine, were found within the conserved N-terminus of the regulatory subunits. Alteration of three consecutive AAs in the N-terminus of *C. glutamicum* IlvN resulted in a mutant AHAS entirely resistant to inhibition by all three BCAAs (Elišáková et al. 2005). The results indicate that there is a single binding (allosteric) site for the three BCAAs at the enzyme molecule, which exhibits different affinities to the individual amino acids.

2.1.3
Acetohydroxyacid Isomeroreductase (AHAIR)

The active AHAIR (EC 1.1.1.86) of *E. coli* is a tetramer with identical 53-kDa subunits. It catalyzes the conversion of acetohydroxyacids into dihydroxyacids (Fig. 2). The reaction including alkyl isomeration and reduction requires Mg^{2+} and NADPH as the hydrogen donor. The enzyme is inhibited by valine and leucine in *C. glutamicum* ($IC_{50} = 7$ mM for both AAs) (Leyval et al. 2003). The protein encoded by the *ilvC* gene was found to display low ketopantoate reductase (EC 1.1.1.169) activity in *E. coli*, being able to substitute for the *panC* gene product (Elischewski et al. 1999) by converting

2-ketopantoate to pantoate. Moreover, IlvC is the only protein showing activity in this reaction of pantothenate synthesis pathway in *C. glutamicum* (Merkamm et al. 2003). Interestingly, the IlvC was identified in two variants differing by *pI* among the phosphorylated proteins of *C. glutamicum* in phosphoproteome analysis (Bendt et al. 2003). Similarly to AHAS, AHAIR is considered a promising target for the design of antituberculosis agents (Grandoni et al. 1998) and herbicides (Wang et al. 2005). New compounds that mimic the structure of intermediates metabolized by AHAIR are therefore designed and tested as competitive inhibitors (Wang et al. 2005).

2.1.4 Dihydroxyacid Dehydratase (DHAD)

The third step of valine and isoleucine biosyntheses, conversion of dihydroxyacids to 2-ketoacids, is catalyzed by dihydroxyacid dehydratase (EC 4.2.1.9) (Fig. 2). The *E. coli* enzyme encoded by the *ilvD* gene is a dimer of 66-kDa subunits. The apoprotein of *E. coli* DHAD, which contains a [4Fe – 4S] cluster essential for catalysis, was used after inactivation of the enzyme in vivo by hyperbaric oxygen to study Fe – S cluster biosynthesis involved in its reactivation (Flint et al. 1996). *C. glutamicum* DHAD is weakly inhibited by valine and leucine with IC_{50} values of 170 mM and 120 mM, respectively. For both DHAD and AHAIR, no cooperative inhibition was observed in the presence of two or all three BCAAs (Leyval et al. 2003).

2.1.5 Transaminases (TAs)

Transaminases typically catalyze the final step in biosynthesis of AAs. A number of these pyridoxal-5′-phosphate-dependent enzymes with both narrow and overlapping substrate specificities developed probably by the functional specialization of primeval enzymes with broad range of activities. In contrast to the other enzymes of BCAA biosynthesis, TAs also occur in animals where they participate in BCAA catabolism.

Transaminases B and C and aromatic transaminase (encoded by *ilvE*, *avtA* and *tyrB*, respectively) are active in BCAA synthesis in *E. coli*. Transaminase B (EC 2.6.1.6) is the main enzyme fulfilling this function for all three BCAAs in *E. coli*. According to the study of its crystal structure, the branched-chain amino acid TA is a trimer of dimers made up of six identical subunits of 32 kDa (Goto et al. 2003). Although the reactions catalyzed by transaminase B are reversible, the balance of the reaction greatly favors the direction to BCAA synthesis. The enzyme produces isoleucine most efficiently of the three BCAAs. Whereas transaminase B uses glutamate as an amino group donor, the alternative transaminase C (EC 2.6.1.66) uses alanine or 2-aminobutyrate (Umbarger 1996) in valine synthesis. This alanine-valine transaminase ensures sufficient synthesis of valine in *ilvE* mutants and even sufficient synthesis of isoleucine (and weak synthesis of leucine) when the *avtA* gene is present in high copy number (Berg et al. 1988). The third TA involved in BCAA synthesis, aromatic transaminase (EC 2.6.1.57), catalyzes efficient conversion of 2-ketoisocaproate to leucine (Powell and Morrison 1978; Vartak et al. 1991), in addition to its function in formation of the aromatic AAs.

Based on the complete *C. glutamicum* genome sequence, comprehensive knowledge of TAs in *C. glutamicum* has been gathered by the systematic analysis of all putative TA genes (McHardy et al. 2003) and of in vitro enzymatic activities of the respective isolated proteins (Marienhagen et al. 2005). Deletion studies (McHardy et al. 2003) confirmed that TA encoded by *ilvE* (Radmacher et al. 2002) is the main enzyme catalyzing the final step in biosynthesis of all three BCAAs (Leyval et al. 2003). Pairwise sequence comparisons of presumed protein family members with the motif-based approach resulted in identification of the *pdxR* gene product, which is active in valine biosynthesis (McHardy et al. 2003). In addition to the aminotransferase domain, the PdxR protein possesses a motif described in the MocR subfamily of GntR transcriptional regulators. According to its role in pyridoxine biosynthesis, the PdxR may function also as a transcriptional activator of the respective genes (McHardy et al. 2003). Assays of purified AvtA revealed that this protein provides a strong activity in valine synthesis in *C. glutamicum*, as in *E. coli*, and a low activity in isoleucine synthesis using alanine as the amino donor (Marienhagen et al. 2005). In another line of similarity to *E. coli*, corynebacterial aromatic TA encoded by *aroT* also showed detectable activity in leucine formation. In conclusion, only IlvE ensures sufficient synthesis of isoleucine and leucine to sustain significant growth, whereas both IlvE and AvtA are sufficient for substantial valine synthesis. The role of *pdxR* is not completely clear. The activities of other TAs involved in BCAA synthesis alone would not secure synthesis of any of the BCAAs in quantities sufficient for growth.

2.2 Regulation of Gene Expression

The enterobacteria possess three isoenzymes of AHAS, encoded by distinct and differently regulated operons (Umbarger 1996). Expression of the respective genes is subject to multivalent repression by branched-chain amino acids. In *E. coli*, expression of the operons *ilvGMEDA* and *ilvBN* is controlled by transcriptional attenuation, while transcription of *ilvIH* is regulated by the leucine-responsive protein. In most Gram-positive bacteria, the genes for the first two enzymes of the pathways (AHAS and AHAIR) belong to the same operon (*ilvBNC*), as in streptomycetes (De Rossi et al. 1995), corynebacteria or in mycobacteria (Gusberti et al. 1996), often together with the *leu* genes (*ilvBNC-leuACBD*), as in *Bacillus subtilis* (Zahler et al. 1991). A wide variety of molecular machineries regulating expression of *ilv* and *leu* operons occur in microorganisms. These regulatory mechanisms include classical attenuation in proteobacteria (Vitreschak et al. 2004a), a modified transcriptional control in actinobacteria like *C. glutamicum* (Seliverstov et al. 2005), T-box transcriptional antitermination (Grandoni et al. 1992; Marta et al. 1996), expression control by global transcriptional regulators of nitrogen metabolism (Tojo et al. 2004) and carbon catabolite repression (Tojo et al. 2005) in *Bacillus subtilis*.

2.2.1 Regulation of the *ilvGMEDA* Operon in *Escherichia Coli*

The *ilvGMEDA* operon provides a unique example of a transcriptional unit whose expression depends on multivalent control by all three BCAAs (Lawther et al. 1987). Transcriptional attenuation of the whole operon is the main control mechanism, although fine tuning of expression of individual genes in the operon is ensured by transcription from internal promoters, differential stability of specific mRNA segments, and binding of regulatory proteins. Premature transcriptional termination depends on the formation of alternative secondary RNA structures, which is in turn affected by availability of the regulatory amino acids (Fig. 3). The leader peptide of 32 AAs encoded in the *ilvGMEDA* operon includes 15 BCAAs (Table 2). The regulatory mechanism depends on a coupling of transcription and translation. Depending on the availability of regulatory amino acids and the respective

Table 2 Amino acid sequences of the leader peptides of attenuators

Operon	Amino acid sequence[a]	Refs.
ilvGMEDA, *E. coli*	MTALLRVISLVVISVVVIIIPPCGAALGRGKA	(Lawther et al. 1987)
ilvBN, *E. coli*	MTTSMLNAKLLPTAPSAAVVVVRVVVVVGNAP	(Friden et al. 1982)
ilvBNC, *C. glutamicum*	MTIIRLVVVTARRLP	(Keilhauer et al. 1993)
leuABCD, *E. coli*	MTHIVRFIGLLLLNASSLRGRRVSGIQH	(Wessler and Calvo 1981)
leuABCD, *S. enterica* serovar Typhimurium	MSHIVRFTGLLLLNAFIVRGRPVGGIQH	(Gemmill et al. 1979)
leuA, *C. glutamicum*	MTSRANLLLLRRGGSQRS	(Pátek et al. 1994)

[a] Tandem amino acid residues supposed to exert the regulatory functions are underlined

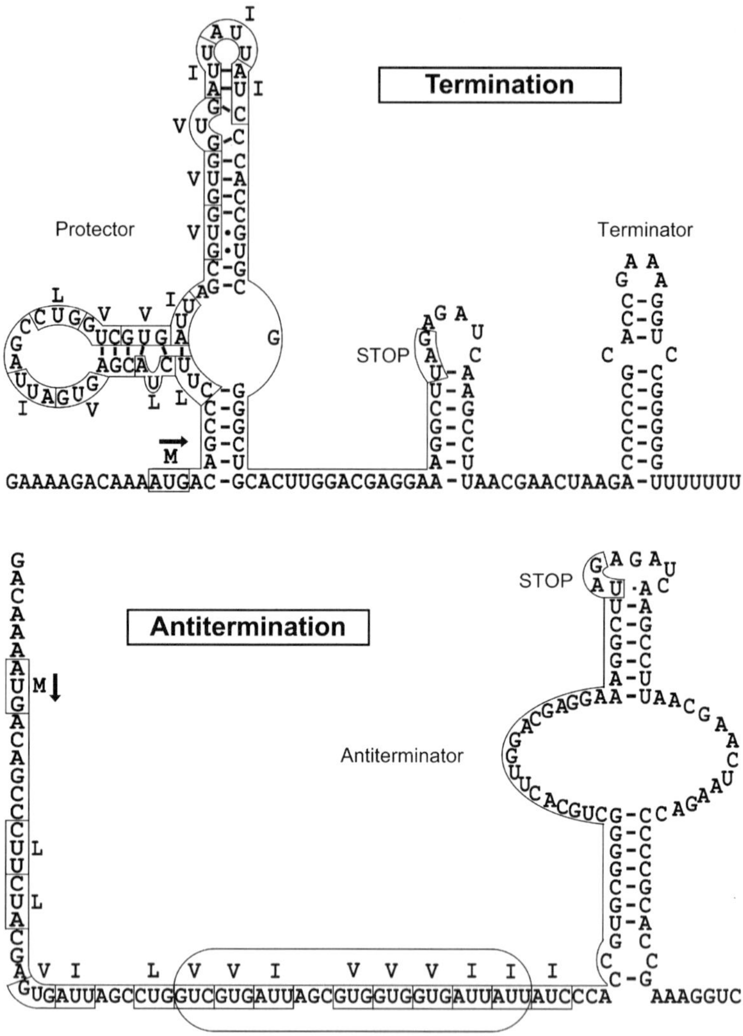

Fig. 3 Proposed alternative secondary RNA structures of the leader transcript of the *E. coli ilvGMEDA* operon. The *line* along the sequence marks the region encoding the leader peptide. The initial methionine codon (M), the tandem control codons (L, V and I) and the stop codon are *boxed*; the *oval* delimits the region masked by ribosome stalling at the central of three consecutive valine codons. Based on the results of Lawther et al. (1987) and Chen et al. (1991)

charged tRNAs, transcriptional terminator or antiterminator RNA structures are formed and transcription either is terminated or continues, accordingly. The concept of attenuation supposes the following mechanism (Chen et al. 1991): If any BCAA is limiting, the ribosome translating the leader transcript stalls at one of the control codons for BCAA, thus precluding formation of the protector RNA structure (Fig. 3). In this case, the bases of the downstream nascent transcript are not covered by the ribosome and form an antiterminator RNA structure, which includes the upper arm of the potential terminator. The formation of antiterminator thus excludes formation of the terminator and transcription continues further into the *ilvGMEDA* operon (readthrough). Under conditions of excess of all BCAAs, translation does not stall but proceeds to the stop codon. The downstream bases in the leader, which may form secondary structure, are covered by the translating ribosome and can not interfere with formation of the protector, which includes the lower part of the potential antiterminator. Consequently, the terminator (typical rho-independent terminator hairpin) is formed, which results in transcriptional termination (Fig. 3). This description of the mechanism of attenuation is quite simplified, since additional transient secondary structures influencing the repression mechanism may be formed within the leader transcript.

There are also several BCAA-independent factors, which affect expression of the operon. The leucine-responsive regulatory protein (Lrp), known to affect expression of many operons and genes in *E. coli* (Willins et al. 1991; Brinkman et al. 2003), binds two sites in the DNA region between the attenuator and the *ilvG* gene and represses transcription of the operon about three-fold (Rhee et al. 1996; Hung et al. 2002). Another global regulator, the integration host factor (IHF), binds to two sites within the promoter region (Pereira et al. 1988). Upstream of the *ilvG* gene, two transcriptional start sites were detected, pertinent to the promoters P1-*ilvG* and P2-*ilvG*. One of the IHF-binding sites overlaps the P1-*ilvG* promoter, which suppresses transcription from this promoter, but activates transcription from the promoter P2-*ilvG*. This seven-fold activation of transcription from P2-*ilvG* may be accounted for by the enhanced DNA bending in the region, which itself exhibits intrinsic bending. P2-*ilvG* is an example of a promoter with a discriminator region that is probably responsible for a global regulation (activation) by the stringent response alarmone ppGpp during growth in AA-limited medium. Conversely, in rich medium, the expression of the operon is reduced, due to the low level of ppGpp. In addition to P1-*ilvG* and P2-*ilvG*, the weaker internal promoters P-*ilvE*, P-*ilvD* and P-*ilvA* were identified upstream of the genes *ilvE*, *ilvD* and *ilvA*, respectively (Huang et al. 1992). Promoter P-*ilvA* is preferentially active in cells grown under anaerobic conditions. Activity of P-*ilvE* depends neither on AA levels in the cell nor on the growth rate of the culture (Harms and Umbarger 1991). Transcription initiation at these promoters, potential transcriptional termination within the operon and processing (RNA decay) of the transcripts result in relatively stable transcripts covering *ilvEDA* and *ilvD* and unstable transcripts covering *ilvGMEDA* and *ilvDA* (Huang et al. 1992). Transcripts originated from different promoters and displaying different stability provide additional means for the cell to meet the momentary requirements of the cell metabolism.

Expression of the *ilvGMEDA* operon represents a complex transcriptional pattern due to a set of independent regulatory mechanisms of various nature and level that have gradually been uncovered over decades of investigations. A large number of overlapping regulatory signals encoded by a short DNA region demonstrates the capacity of DNA to code for exceptionally dense information.

2.2.2
Regulation of the *ilvIH* Operon in *Escherichia Coli*

Activation of the *ilvIH* operon by Lrp protein is one of the most thoroughly studied examples of regulation of gene expression by this transcriptional factor. Expression of the *ilvIH* genes was found to be repressed by leucine (Ricca et al. 1989). However, there is no sequence reminding any section of the attenuator structure upstream of *ilvIH*. Instead, six binding sites for Lrp were found within a 250-bp sequence upstream of the coding regions (Wang and Calvo 1993). Lrp directly activates the *ilvIH* promoter by enhancing the intrinsic bending of the region upstream of the promoter and thus positively regulates expression of the operon. Interaction dependent on distance and phasing between Lrp molecules bound to two sites centered at 231 (relative to the position of transcriptional start) and to four sites centered at 95 is essential for efficient activation of the promoter (Jafri et al. 2002; Chen et al. 2005). Activity of the *ilvIH* promoter was reduced more than 30-fold in the strain lacking Lrp (Platko et al. 1990). In the presence of leucine, the amount of the bound Lrp protein is reduced in the wild-type strain and, consequently, the activity of the *ilvIH* promoter decreased 5- to 10-fold (Jafri et al. 2002). In addition to regulation by Lrp, *ilvIH* was found to be repressed by the stringent response factor ppGpp under the conditions of amino acid starvation (Baccigalupi et al. 1995), in contrast to *ilvGMEDA* and *ilvBN*, which are activated by ppGpp.

2.2.3
Regulation of the *ilvBN* Operon in *Escherichia Coli*

In accordance with the unusual properties of AHAS I, which occurs exclusively in enterobacteria, regulation of the *ilvBN* operon expression exhibits unique features. It is regulated specifically by transcriptional attenuation and at the global level by carbon catabolite repression and by the stringent response to amino acid starvation.

The attenuator of *ilvBN* exhibits features quite similar to the features of other typical attenuators of amino acid biosynthesis operons. The leader transcript coding for a 32-AAs peptide (Table 2) contains a tandem of leucine codons and nine valine codons (runs of four and five), but lacks isoleucine codons (Friden et al. 1982; Hauser and Hatfield 1983). The operon is therefore not regulated by isoleucine availability. The leader transcript may potentially form a number of mutually exclusive stem-loop RNA structures, however, due to the transient folding of RNA and different conceptions of their temporal appearance, the proposed secondary RNA structures differ in different models (Friden et al. 1982; Hauser and Hatfield 1983). On the basis of in vitro transcription, the termination site of the leader transcript (184 nt) was found downstream of the run of 5 U preceded by a G + C-rich region, which may fold into a typical terminator hairpin (Hauser and Hatfield 1983).

Another type of expression control, independent of attenuation, was found on the basis of the sequence of the *ilvBN* operon. A cAMP-receptor protein (CRP) binding site centered near position 65 upstream of the supposed transcriptional start point was recognized (Friden et al. 1982). The *ilvBN* operon is the only biosynthetic operon in *E. coli* induced by CRP, which is generally involved in carbon catabolite repression (Sutton and Freundlich 1980; Tedin and Norel 2001). Very little AHAS I activity was detected when the cAMP level in the cell was low. Addition of cAMP under these conditions increased *ilvBN* expression 24-fold. Activation of the *ilvBN* promoter through CRP thus probably permits efficient growth of *E. coli* on substrates different from glucose (e.g., acetate or oleate) (Dailey and Cronan 1986).

Similar to expression of *ilvGMEDA* and in contrast to expression of *ilvIH*, expression of *ilvBN* is stimulated by the alarmone of the stringent response, ppGpp (Freundlich 1977; Sutton and Freundlich 1980; Tedin and Norel 2001), although the mechanism of its action has not been described.

The recent finding that *ilvBN* is derepressed by the dephospho-form of enzyme IIA^{Ntr} (involved in nitrogen-metabolic phosphotransferase system), when the *E. coli* cells are transferred from BCAA-containing medium to a minimal medium (Lee et al. 2005) shows that not all factors influencing the expression of the long-studied genes have as yet been revealed.

2.2.4
Regulation of the *ilvBNC* Operon in *Corynebacterium Glutamicum*

The genes coding for the subunits of the single AHAS in *C. glutamicum* were named *ilvBN*, although they are orthologs of *ilvIH* of *E. coli*. The *ilvBNC* operon of *C. glutamicum* is transcribed into three different transcripts covering *ilvBNC*, *ilvNC* and the *ilvC* gene alone (Keilhauer et al. 1993). The leader region of *ilvB* (292 bp) is comprised of features typical for regulation by attenuation: (1) a short ORF with two Ile codons, three Val codons and two Leu codons, (2) a sequence with a potential to form alternative RNA stem-loop structures, (3) a sequence potentially forming a transcriptional terminator RNA structure (Morbach et al. 2000). Formation of a very short abundant RNA (about 200 bp) corresponding to the prematurely terminated *ilvBNC* transcript and increased formation of *ilvBNC* transcript under growth limitation by all BCAAs or by any one of them suggested that the expression is regulated by transcriptional attenuation. Moreover, the necessity of translation of the leader peptide (Table 2) for the regulatory response was confirmed, although the transcriptional start and the supposed translation initiation codon are only 1 bp apart and a ribosome-binding site is thus missing. However, some results failed to consistently support the attenuation model: (1) the regulation was abolished by mutational alteration of valine codons, but not by alteration of isoleucine codons, and (2) evidence for the function of the putative terminator structure could not be obtained (Morbach et al. 2000). In a similar attenuator-like structure in *Streptomyces coelicolor*, replacement of the supposed control codons of the *ilvBNC* operon did not eliminate the regulatory effect of BCAAs. This led to the conclusion that some other regulatory mechanism is involved, although typical features of an attenuator are present in *S. coelicolor* (Craster et al. 1999). Induction of both the complete *ilvBNC* transcript and the short leader transcript by 2-ketobutyrate in *C. glutamicum* indicates that yet another independent regulation operates in control of *ilvBNC* expression. The *ilvNC* transcript was also induced by 2-ketobutyrate (Keilhauer et al. 1993). The *ilvN* promoter, which was repressed by valine, was localized about 300 bp upstream of the gene, within the *ilvB* coding region, but no sequences reminiscent of attenuators or other control elements were recognized in the untranslated region of the transcript (Pátek, unpublished results). The *ilvNC* transcript in *C. glutamicum* may ensure an excess of the regulatory subunit coded by *ilvN*, which is necessary for full AHAS activity and/or for efficient inhibition. Regulation of the most abundant *ilvC* transcript has not as yet been described. The complex expression pattern of the *ilvBNC* operon and the respective control mechanisms thus must still be revealed.

3 Leucine

The isopropylmalate pathway of leucine synthesis described here occurs in nearly all bacteria, fungi, and plants studied so far. An alternative isovalerate carboxylation pathway was found in some bacteria such as the anaerobic ruminal cellulose digester *Bacteroides ruminicola* (Allison et al. 1984).

3.1 Enzymes

Isopropylmalate synthase (LeuA; EC 4.1.3.12) catalyzing the first reaction in the leucine-specific biosynthesis pathway (Fig. 2) forms a homotetramer with subunits of about 50 kDa in *E. coli* and *S. enterica* serovar Typhimurium (Leary and Kohlhaw 1970). Proteins of the same type as LeuA from *E. coli* were identified in most γ-proteobacteria (e.g., *Erwinia* and *Vibrio*) and in some α-proteobacteria (e.g., *Rhodobacter*), whereas enzymes of the LeuA2 type (sharing approximately 30% identity) were found particularly in fungi (e.g., *Saccharomyces*), in actinobacteria including *C. glutamicum* (Leary and Kohlhaw 1970; Pátek et al. 1994; Vitreschak et al. 2004a), in bacilli and in pseudomonads (Vitreschak et al. 2004a). Isopropylmalate synthase is sensitive to feedback inhibition by leucine in these bacteria (Ward and Zahler 1973) as well as in yeast (Hu and Kohlhaw 1995).

In both *S. enterica* serovar Typhimurium and *C. glutamicum*, inhibition of isopropylmalate synthase by leucine is the main regulatory mechanism on the level of enzyme activity in the leucine synthesis pathway (Leary and Kohlhaw 1970; Pátek et al. 1994). The apparent K_i value was 0.4 mM in *C. glutamicum* (Pátek et al. 1994). The inhibition is noncompetitive with respect to 2-ketoisovalerate and competitive with respect to acetyl-CoA. The inhibitor disaggregates the tetrameric form of the enzyme (Leary and Kohlhaw 1970).

Isopropylmalate dehydratase (LeuCD; EC 4.2.1.33) from *E. coli* consists of a heterodimer of 50-kDa and 23-kDa subunits encoded by *leuC* and *leuD*. The homologous genes of *C. glutamicum* encode proteins of 52 and 22 kDa, respectively. Isopropylmalate dehydratase from *Saccharomyces cerevisiae*, which was more deeply studied, is formed by a monomer of approximately 90 kDa (Kohlhaw 1988, 2003).

Isopropylmalate dehydrogenase (LeuB; EC 1.1.1.85) isolated from *S. enterica* serovar Typhimurium in its active form is a dimer of identical 35-kDa subunits (Parsons and Burns 1969). The *leuB* gene of *C. glutamicum* specifies a protein of 36 kDa (Pátek et al. 1998). It was cloned and overexpressed in the plasmid pEKEx2 (Eikmanns et al. 1991) providing 10-fold higher activity than the wild-type strain (Pátek et al. 1998). Knowledge of regulatory mechanisms, which govern the activity of Leu-enzymes is rather limited.

3.2 Regulation of Gene Expression

Leucine genes are clustered in operons in many organisms. In some γ-proteobacteria (e.g., *E. coli*, *Yersinia pestis* and *Vibrio cholerae*) the *leuABCD* operon is regulated by translation-mediated transcriptional attenuation (Wessler and Calvo 1981; Vitreschak et al. 2004a). In other groups of bacteria (e.g., *Agrobacterium tumefaciens*, *Rhodobacter sphaeroides* and *Caulobacter crescentus*), the leucine genes are located on different parts of the chromosome and only *leuA* is regulated by attenuation, although a typical terminator-like structure is missing within the attenuator sequence (Vitreschak et al. 2004a). In actinomycetes (e.g., corynebacteria, streptomycetes and mycobacteria) LEU elements were revealed upstream of the *leuA* gene, within the sequence with some features of an attenuator (Seliverstov et al. 2005). In *Bacillus subtilis*, the genes of leucine synthesis form together with the *ilv* genes a single operon, *ilvBNC-leuACBD* (Sect. 2.2).

3.2.1 Regulation of the Leucine Operon in *Escherichia Coli* and *Salmonella Enterica* Serovar Typhimurium

The *leuABCD* operon in *E. coli* (and probably in many other γ-proteobacteria) is regulated mainly by leucine-mediated transcriptional attenuation (Wessler and Calvo 1981). Expression of the leucine operon varies over a 40- to 50-fold range in both *E. coli* and *S. enterica* serovar Typhimurium depending on the intracellular concentration of leucine. There are four consecutive Leu codons in their leader transcripts (Table 2) and their crucial role in transcriptional attenuation in *S. enterica* serovar Typhimurium was proven by mutagenesis. Changing the three rare CTA codons to the more frequent CTG codons reduced the basal level of expression of *leuABCD* operon as well as the sensitivity of the response to leucine starvation (Carter et al. 1986). After converting the four Leu codons into four Thr codons by mutation, 90% regulation by leucine was abolished (Carter et al. 1985). Whether an additional type of regulation (e.g., Lrp control) is responsible for the remaining threefold derepression of the operon needs to be solved. Surprisingly, the mutant with Thr codons responded significantly less to threonine starvation than the wild-type in which the regulation by threonine was unexpectedly found. Explanation of these observations may lay in modification of some tRNAs by threonine and in the presence of a single Thr codon upstream of the four Leu codons in the WT-leader (Carter et al. 1985). More generally, the lesson from mutagenesis studies of attenuators is that the results may differ from those anticipated from the models, and this is probably due to the involvement of both machineries of transcription and translation in the regulatory mechanism and due to the complex structure of the leader transcript in which codon-specific response and potential secondary structures cannot be altered independently.

3.2.2 Regulation of Leucine Genes in *Corynebacterium Glutamicum*

In *C. glutamicum*, the leucine biosynthesis genes (*leuA*, *leuB* and *leuCD*) are scattered over the chromosome. Upstream of the *leuA* gene, features of a typical attenuator are present. However, caution is needed in inferring a regulatory mechanism even if the DNA and the deduced AA sequences strongly suggest the presence of a standard attenuator, according to a study of a very similar structure upstream of *leuA* in the related actinomycete *S. coelicolor* (Craster et al. 1999). Replacement of three supposed leucine control codons within the leader transcript with threonine codons did not clearly abolish the regulation of *S. coelicolor leuA* by leucine. In *C. glutamicum*, the 18-AA leader peptide (Table 2) of *leuA* would contain four consecutive leucine residues as the *E. coli leu*-leader. In comparison to cells grown in medium with leucine, the specific activity of isopropylmalate synthase encoded by *leuA* was about two-fold higher in cells grown in minimal medium without leucine (Pátek et al. 1994) and expression of a fusion of the *leuA* promoter region to the promoter-less chloramphenicol acetyltransferase gene was six-fold higher when the cells were starved for leucine (Pátek et al., unpublished). The regulatory effect of leucine was abolished when the AUG start codon of the supposed leader peptide was altered to AAG or when four consecutive leucine codons were exchanged for phenylalanine codons (Pátek et al., unpublished). These results indicate that (1) translation into the leader peptide is essential for the regulation, and (2) leucine codons play a crucial role in the regulatory mechanism. These lines of evidence suggest that it is indeed an attenuation mechanism as that in *E. coli*, which controls *leuA* expression in *C. glutamicum*.

However, according to the comparative analysis of RNA regulatory elements based on the genome sequences, *C. glutamicum* and a number of other actinomycete species (corynebacteria, mycobacteria, and streptomycetes) contain regulatory structures designated LEU elements in the leader regions of their *leuA* genes (Seliverstov et al. 2005). Although the LEU element resembles an attenuator, the transcript is supposed to fold into a pseudoknot with an additional stem at its base formed by pairing of the leucine control codons with the Shine-Dalgarno sequence of *leuA*. To accomplish the regulatory role, the same region might form an alternative hairpin with the same base stem. There is a similarity of the secondary structure to riboswitches (Vitreschak et al. 2004b), but binding of a small molecule to the LEU elements seems unlikely (Seliverstov et al. 2005). No experimental evidence that the element is a binding site of a protein has been provided. The role of ribosome stalling at the control Leu codons should also be taken into consideration. The actual nature and function of the *leuA* regulatory element thus remain to be elucidated.

Data on the regulation of the other *leu* genes in *C. glutamicum* are scarce. The *leuB* gene is transcribed into a monocistronic transcript and is repressed by leucine (Pátek et al. 1998). Upstream of *leuB* and *leuCD* no prominent sequences indicating the type of control are apparent.

The specific activity of isopropylmalate dehydratase (LeuCD) is about half in minimal medium with leucine as compared to medium without leucine and ten-fold lower in complex medium (Pátek et al. 1994). This suggests that expression of *leuCD* gene cluster is also subject to a complex regulation.

4
Use of Branched-Chain Amino Acids and their Biotechnological Production

BCAAs are used as additives to food and feedstuff for infusions and for synthesis of pharmaceuticals (Sahm et al. 1999). Sustained dynamic exercise stimulates in human amino acid oxidation, chiefly of the BCAAs (Rennie and Tipton 2000). During endurance sport events, BCAAs (particularly leucine) attenuate the breakdown of muscle proteins, promote protein synthesis in skeletal muscle, and prevent premature muscle fatigue (Rennie et al. 2006; Shimomura et al. 2006). BCAAs in special dietary supplements for sport nutrition are therefore expected to promote maintaining the skeletal muscle and its regeneration. Since BCAAs are essential for immune cell functions and may improve immunity when administered to postsurgical or septic patients (Calder 2006), they are applied in nutritional or medical treatment.

Valine was found among the five most limiting amino acids in reduced protein corn-soybean meal diets for livestock (Mavromichalis et al. 1998). Isoleucine is more deficient than valine for pigs during both early and late finishing stages (Liu et al. 1999). It is expected that the decreasing price of the biotechnologically produced BCAAs, would permit wider application of BCAA in human and animal nutrition (Kelle et al. 1996).

BCAAs are mainly produced by fermentation, enzymatic catalysis and by extraction of protein hydrolyzates. The alternative to the production of pure AAs for nutrition purposes is to construct transgenic plants in which deficiency in some BCAAs (particularly isoleucine) is overcome by gene manipulations (Weaver et al. 2005). Isoleucine, having the most common use of BCAAs, is produced in quantities about 400 tons per year, with *C. glutamicum* and *E. coli* as the best described producers (Hermann 2003). A few hundred tons of L-valine are produced each year using enzyme membrane reactor (Leuchtenberger et al. 2005).

The productions of amino acid increases by about 10% annually (Hermann 2003) and BCAAs are particularly strong in the growing market. In response to the growing demand for BCAAs, the largest fermentation company in terms of volume, Ajinomoto, has opened a new plant in Brazil with initial production of four amino acid products, including valine, leucine and isoleucine for use in food and pharmaceutical industry.

4.1 Isoleucine Overproducing Strains

Isoleucine may be produced with microorganisms by conversion of precursors (e.g., 2-ketobutyrate) or by fermentation using mutant or recombinant strains (Eggeling et al. 1997). Biosynthesis of isoleucine, in which aspartate and pyruvate are the starting metabolites, consists of a number of reaction steps (Fig. 4). Along this pathway, several steps are rate-limiting and must be engineered to overcome the metabolic barriers. The strains were first derived particularly as mutants resistant to analogs of threonine and isoleucine-like hydroxynorvaline and methylthreonine, respectively. An isoleucine-producing strain of *Serratia marcescens* derived by transduction

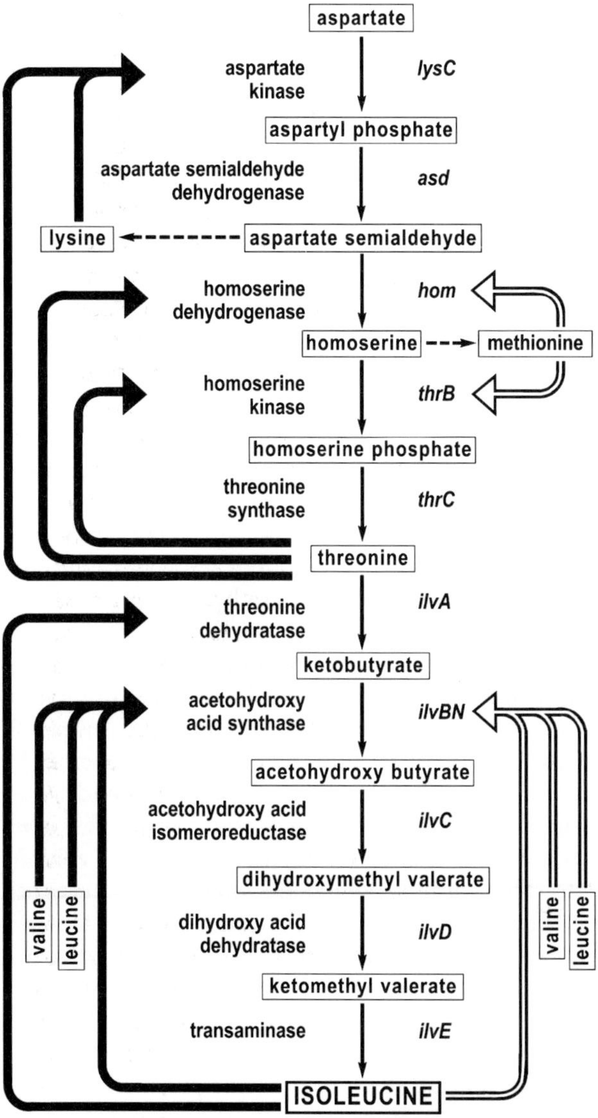

Fig. 4 Biosynthesis of isoleucine and its regulation by repression and inhibition in *C. glutamicum*. Feedback regulation by various amino acids is indicated by *thick arrows* for inhibition and by *empty arrows* for repression and/or attenuation. The genes encoding the respective enzymes are shown on the *right*

had six regulatory mutations for threonine and isoleucine biosyntheses and exhibited high specific activities of threonine deaminase and acetohydroxy-acid synthase, both of which were insensitive to feedback inhibition (Komatsubara et al. 1980). Methods of gene cloning permitted constructing recombinant isoleucine-producing strains of *E. coli* and *C. glutamicum*. The *E. coli* strain TVD5 (*ilvA*, *thrC*, thr^R, *tdh*::Tn5) contained two plasmids, one with the *thrA**BC* operon and the other with the *ilvGMEDA** operon (Hashiguchi et al. 1999b). Owing to the presence of the ThrA* protein, which is active both as aspartate kinase and as homoserine dehydrogenase and is resistant to inhibition by threonine and of TD (*ilvA*), which is resistant to inhibition by isoleucine, the strain produced isoleucine and valine. Since common enzymes synthesize both isoleucine and valine, the valine pathway cannot be shut down by metabolic block to achieve an isoleucine producer. When the *E. coli* strain with the *lysC** gene encoding the lysine-resistant aspartate kinase III was used as the host for the two plasmids, the resulting producer accumulated only isoleucine due to the increased 2-ketobutyrate supply (Hashiguchi et al. 1999a).

In *C. glutamicum*, at least five enzymes within the pathway from aspartate to isoleucine are inhibited by various AAs (Fig. 4): the end product (isoleucine), an intermediate (threonine), the product of a pathway which branches off the isoleucine pathway (lysine) and products of partially parallel pathways (valine and leucine). Repression of the genes *hom*, *thrB* and *ilvBN* involved in the pathway (Fig. 4) has also to be overcome in the construction of an isoleucine producing strain.

In a *C. glutamicum* strain with a feedback-resistant (FBR) aspartate kinase, genes coding for homoserine kinase (*thrB*), for FBR homoserine dehydrogenase (*hom*) and for FBR threonine deaminase (*ilvA*) were cloned in two plasmids, each with approximately 20 copies per chromosome (Morbach et al. 1996). Although high expression of *hom*(FBR) was found to be unstable in threonine-producing strains (Reinscheid et al. 1994), the *hom*(FBR) allele on a multicopy plasmid was stable in the presence of a highly expressed *ilvA*(FBR) allele. This finding documents the general necessity to first remove potential limitations within the downstream part of a biosynthesis pathway to avoid potentially deleterious accumulation of high intracellular concentrations of intermediates (Eggeling et al. 1997). Similar to threonine overproduction (Reinscheid et al. 1994), higher intracellular (110 mM) than extracellular (60 mM) isoleucine concentrations indicated that amino acid efflux has become the limiting step of isoleucine production in the engineered strain (Morbach et al. 1996). However, overexpression of *brnFE* encoding the exporter of BCAAs (Kennerknecht et al. 2002) did not increase isoleucine production (Eggeling 2005). Although in tryptophan and threonine producers the inactivation of the respective AA uptake systems improved their final yield (Ikeda and Katsumata 1995; Okamoto et al. 1997), deletion of *brnQ* coding for the isoleucine uptake carrier failed to prove beneficial (Eggeling 2005). Using another strategy of metabolic engineering, the *E. coli tdcB* gene coding for catabolic threonine deaminase was expressed together with *C. glutamicum thrB* and hom^{dr} (coding deregulated homoserine dehydrogenase) from a single plasmid vector harbored by *C. glutamicum* (Guillouet et al. 2001). Since

TdcB from *E. coli* converting threonine to 2-ketobutyrate is an isoleucine insensitive enzyme, its presence in *C. glutamicum* contributed substantially to the final yield of isoleucine.

The study of correlation of enzyme activity and mRNA abundance for *hom*, *thrB* and *ilvA* expression in *C. glutamicum* showed that *hom* mRNA and the level of homoserine dehydrogenase activity changed in profile and scale in a similar way, whereas in the case of *ilvA* and TD, increase in enzyme activity strongly exceeded the moderate increase in mRNA abundance. A third pattern emerged for *thrB* and homoserine kinase (ThrB): while the level of the transcript initially increased considerably and decreased later, HK activity showed a small but steady increase over the same time course (Glanemann et al. 2003). Since the relationship between the level of gene expression and product yield depends on a number of factors, including the rate of transcription initiation, the stability of the mRNA, the efficiency of translation and protein modification and stability, systems-level approaches (Wendisch et al. 2006), including monitoring mRNA abundances, enzyme activities, and metabolite concentrations over time are expected to provide deeper insights into the physiology of genetically engineered bacterial producers.

4.2
Valine Overproducing Strains

During the era of random mutagenesis of microorganisms and selection of AA producers, valine-producing strains of *C. glutamicum* (Tsuchida and Momose 1975) and *Serratia marcescens* (Kisumi et al. 1971) were isolated as mutants resistant to BCAA analogs such as norvaline or 2-aminobutyric acid. In these strains, deregulation of AHAS was probably the reason for valine overproduction (Tsuchida and Momose 1975). By the methods of directed metabolic engineering the valine-excreting strains *C. glutamicum ΔilvA ΔpanBC* (isoleucine and pantothenate auxotroph) carrying a plasmid with the cloned genes *ilvBNCD* or *ilvBNCE* were constructed (Sahm and Eggeling 1999; Radmacher et al. 2002). Deletion of *ilvA* coding for TD, the key enzyme of the isoleucine synthesis, seemed decisive for valine production. AHAS, which is active in both the isoleucine and valine pathways, is then available only for conversion of pyruvate to acetolactate (Fig. 2). When limiting amounts of isoleucine were used for supplementation of this isoleucine auxotrophic strain, the AHAS activity was not inhibited by isoleucine and the imposed growth limitation further improved valine production by this strain, probably by *ilvBNC* derepression under isoleucine starvation (Sahm and Eggeling 1999). Growth limitation of *C. glutamicum ilvA* strains, even in presence of sufficient levels of isoleucine in the medium, may also be induced by high concentrations of valine (Lange et al. 2003). Valine (50–200 mM) in the medium provokes unexpected effects: deficiency of isoleucine in the cell, increased AHAS activity, and higher valine excretion. Genome-wide expression profiling using DNA microarrays in combination with proteomics and genetic experiments proved that these effects are caused by competition of valine and isoleucine for uptake by the carrier BrnQ, which transports all BCAAs (Lange et al. 2003). These consequences may also appear in fermentations by isoleucine auxotrophs at stages when a high level of valine is accumulated in the medium and the concentration of the supplemented isoleucine dropped to a critical value.

An AHAS protein resistant to inhibition by all end products of the pathways (i.e., all three BCAAs) was constructed by site-directed mutagenesis of regulatory AHAS subunit coded by *ilvN*, designed according to the respective DNA sequences of deregulated mutants of *E. coli* and *S. cinnamonensis* (Elišáková et al. 2005). The respective allele (*ilvN*M13) encoding three altered AAs in IlvN was introduced into the chromosome of *C. glutamicum* *ΔilvA ΔpanBC* harboring the plasmid with the *ilvBNC* operon. The resulting strain showed high AHAS activity and steadily produced valine in parallel with growth of the culture and during the stationary phase (Elišáková et al. 2005). As an alternative to *ilvA* deletion, which causes isoleucine auxotrophy and consequently a requirement for isoleucine supplementation, down-mutations in promoters of *ilvA* or *leuA* (P-*ilvA*, P-*leuA*) were constructed, which severely decreased expression of the respective genes and caused bradytrophy for isoleucine or leucine, respectively (Pátek at al. unpublished). Growth of the isoleucine or leucine bradytrophic strains was limited by the respective AA and provided the same valine yields as isoleucine auxotrophs or even higher. The leucine bradytroph displayed high excretion of valine but no production of isoleucine, although the unimpaired isoleucine pathway was expected to be favored over the valine pathway due to the much higher affinity of AHAS to 2-ketobutyrate than to pyruvate. This may be explained by the strict inhibition of wild-type TD activity by isoleucine or by the change of the role of valine at high concentrations from an activator of TD to its inhibitor, as observed in *E. coli* (Umbarger 1996). The concept of tuning the individual reaction rates in the biosynthesis by promoter strength modulation was followed also in enhancing the activity of promoters P-*ilvD* (100-fold) and P-*ilvE* (20-fold) by specific mutations, which further increased the carbon flux through the pathway. A plasmid-free *C. glutamicum* valine producer was constructed by introducing these mutations (*ΔpanB*, *ilvN*M13, P-*ilvA*M1, P-*ilvD*M7, P-*ilvE*M6) into the chromosome (Pátek et al. unpublished). The mutations present in this valine producer might also prove beneficial for isoleucine production when combined with deregulation of *ilvA* as well as for leucine production if the leucine-specific pathway is deregulated. Techniques of metabolic engineering within the chromosome (Vertes et al. 2005), which do not introduce plasmids, markers or heterologous DNA sequences into the cell, may provide more stable producer strains, which do not exhibit theoretical hazards of horizontal transfer of genes with a potentially negative effect to the environment.

4.3
Leucine Overproducing Strains

Breeding of *C. glutamicum* leucine producers began as it did for other BCAAs by chemical mutagenesis and isolation of analog-resistant mutants. Mutants resistant to 2-azaleucine and 3-hydroxyleucine possessed isopropylmalate synthase and AHAS resistant to inhibition by leucine and by all three BCAAs, respectively (Tsuchida and Momose 1975; Tsuchida and Momose 1986). The repression of genes coding for AHAS could be overcome by a limiting supply of isoleucine. Leucine producers may be improved by genetic manipulation of *leu* genes (particularly *leuA*) in valine producing strains. In another approach, 2-ketoisocaproic acid was used as a precursor in continuous cultivations (Wichmann et al. 1990).

Acknowledgements The work was supported by Grant 525/04/0548 from the Czech Science Foundation and by Institutional Research Concept no.AV0Z5020903. The author thanks David Chipman, Lothar Eggeling, and Jan Nešvera for critical reading of the manuscript and helpful comments, the laboratory colleagues Jiří Holátko and Veronika Elišáková for providing their unpublished results and Lubomíra Kopecká for drawing the figures.

References

Allison MJ, Baetz AL, Wiegel J (1984) Alternative pathways for biosynthesis of leucine and other amino acids in *Bacteroides ruminicola* and *Bacteroides fragilis*. Appl Environ Microbiol 48:1111–1117

Baccigalupi L, Marasco R, Ricca E, De Felice M, Sacco M (1995) Control of ilvIH transcription during amino acid downshift in stringent and relaxed strains of *Escherichia coli*. FEMS Microbiol Lett 131:95–98

Barak Z, Chipman DM, Gollop N (1987) Physiological implications of the specificity of acetohydroxy acid synthase isozymes of enteric bacteria. J Bacteriol 169:3750–3756

Bar-Ilan A, Balan V, Tittmann K, Golbik R, Vyazmensky M, Hubner G, Barak Z, Chipman DM (2001) Binding and activation of thiamin diphosphate in acetohydroxyacid synthase. Biochemistry 40:11946–11954

Bendt AK, Burkovski A, Schaffer S, Bott M, Farwick M, Hermann T (2003) Towards a phosphoproteome map of *Corynebacterium glutamicum*. Proteomics 3:1637–1646

Berg CM, Wang MD, Vartak NB, Liu L (1988) Acquisition of new metabolic capabilities: multicopy suppression by cloned transaminase genes in *Escherichia coli* K-12. Gene 65:195–202

Brinkman AB, Ettema TJ, de Vos WM, van der Oost J (2003) The Lrp family of transcriptional regulators. Mol Microbiol 48:287–294

Calder PC (2006) Branched-chain amino acids and immunity. J Nutr 136:288S–293S

Carter PW, Bartkus JM, Calvo JM (1986) Transcription attenuation in *Salmonella typhimurium*: the significance of rare leucine codons in the leu leader. Proc Natl Acad Sci USA 83:8127–8131

Carter PW, Weiss DL, Weith HL, Calvo JM (1985) Mutations that convert the four leucine codons of the *Salmonella typhimurium* leu leader to four threonine codons. J Bacteriol 162:943–949

Changeux JP (1961) The feedback control mechanisms of biosynthetic L-threonine deaminase by L-isoleucine. Cold Spring Harb Symp. Quant Biol 26:313–318

Changeux JP (1963) Allosteric interaction on biosynthetic L-threonine deaminase from *E. coli* K-12. Cold Spring Harbor Symp Quant Biol 28:497–504

Chen JW, Bennett DC, Umbarger HE (1991) Specificity of attenuation control in the ilvGMEDA operon of *Escherichia coli* K-12. J Bacteriol 173:2328–2340

Chen S, Iannolo M, Calvo JM (2005) Cooperative binding of the leucine-responsive regulatory protein (Lrp) to DNA. J Mol Biol 345:251–264

Craster HL, Potter CA, Baumberg S (1999) End-product control of expression of branched-chain amino acid biosynthesis genes in *Streptomyces coelicolor* A3(2): paradoxical relationships between DNA sequence and regulatory phenotype. Microbiology 145:2375–2384

Dailey FE, Cronan JE Jr (1986) Acetohydroxy acid synthase I, a required enzyme for isoleucine and valine biosynthesis in *Escherichia coli* K-12 during growth on acetate as the sole carbon source. J Bacteriol 165:453–460

De Rossi E, Leva R, Gusberti L, Manachini PL, Riccardi G (1995) Cloning, sequencing and expression of the *ilvBNC* gene cluster from *Streptomyces avermitilis*. Gene 166:127–132

Duggleby RG, Pang SS, Yu H, Guddat LW (2003) Systematic characterization of mutations in yeast acetohydroxyacid synthase. Interpretation of herbicide-resistance data. Eur J Biochem 270:2895–2904

Eggeling I, Cordes C, Eggeling L, Sahm H (1987) Regulation of acetohydroxy acid synthase in *Corynebacterium glutamicum* during fermentation of alfa-ketobutyrate to L-isoleucine. Appl Microbiol Biotechnol 25:346–351

Eggeling L (2005) Export of amino acids and other solutes. In: Handbook of *Corynebacterium glutamicum*. CRC Press, Boca Raton, pp 187–214

Eggeling L, Morbach S, Sahm H (1997) The fruits of molecular physiology: engineering the L-isoleucine biosynthesis pathway in *Corynebacterium glutamicum*. J Biotechnol 56:167–182

Eikmanns BJ, Kleinertz E, Liebl W, Sahm H (1991) A family of *Corynebacterium glutamicum/Escherichia coli* shuttle vectors for cloning, controlled gene expression, and promoter probing. Gene 102:93–98

Eisenstein E (1991) Cloning, expression, purification, and characterization of biosynthetic threonine deaminase from *Escherichia coli*. J Biol Chem 266:5801–5807

Eisenstein E, Yu HD, Schwarz FP (1994) Cooperative binding of the feedback modifiers isoleucine and valine to biosynthetic threonine deaminase from *Escherichia coli*. J Biol Chem 269:29423–29429

Elischewski F, Puhler A, Kalinowski J (1999) Pantothenate production in *Escherichia coli* K12 by enhanced expression of the panE gene encoding ketopantoate reductase. J Biotechnol 75:135–146

Elišáková V, Pátek M, Holátko J, Nešvera J, Leyval D, Goergen JL, Delaunay S (2005) Feedback-resistant acetohydroxy acid synthase increases valine production in *Corynebacterium glutamicum*. Appl Environ Microbiol 71:207–213

Fisher KE, Eisenstein E (1993) An efficient approach to identify ilvA mutations reveals an amino-terminal catalytic domain in biosynthetic threonine deaminase from *Escherichia coli.* J Bacteriol 175:6605–6613

Flint DH, Tuminello JF, Miller TJ (1996) Studies on the synthesis of the Fe-S cluster of dihydroxy-acid dehydratase in *Escherichia coli* crude extract. Isolation of O-acetylserine sulfhydrylases A and B and beta-cystathionase based on their ability to mobilize sulfur from cysteine and to participate in Fe-S cluster synthesis. J Biol Chem 271:16053–16067

Freundlich M (1977) Cyclic AMP can replace the relA-dependent requirement for derepression of acetohydroxy acid synthase in *E. coli* K-12. Cell 12:1121–1126

Friden P, Newman T, Freundlich M (1982) Nucleotide sequence of the ilvB promoter-regulatory region: a biosynthetic operon controlled by attenuation and cyclic AMP. Proc Natl Acad Sci USA 79:6156–6160

Gemmill RM, Wessler SR, Keller EB, Calvo JM (1979) Leu operon of *Salmonella typhimurium* is controlled by an attenuation mechanism. Proc Natl Acad Sci USA 76:4941–4945

Glanemann C, Loos A, Gorret N, Willis LB, O'Brien XM, Lessard PA, Sinskey AJ (2003) Disparity between changes in mRNA abundance and enzyme activity in *Corynebacterium glutamicum*: implications for DNA microarray analysis. Appl Microbiol Biotechnol 61:61–68

Gollop N, Barak Z, Chipman DM (1987) A method for simultaneous determination of the two possible products of acetohydroxy acid synthase. Anal Biochem 160:323–331

Gollop N, Damri B, Barak Z, Chipman DM (1989) Kinetics and mechanism of acetohydroxy acid synthase isozyme III from *Escherichia coli.* Biochemistry 28:6310–6317

Goto M, Miyahara I, Hayashi H, Kagamiyama H, Hirotsu K (2003) Crystal structures of branched-chain amino acid aminotransferase complexed with glutamate and glutarate: true reaction intermediate and double substrate recognition of the enzyme. Biochemistry 42:3725–3733

Grandoni JA, Marta PT, Schloss JV (1998) Inhibitors of branched-chain amino acid biosynthesis as potential antituberculosis agents. J Antimicrob Chemother 42:475–482

Grandoni JA, Zahler SA, Calvo JM (1992) Transcriptional regulation of the ilv-leu operon of *Bacillus subtilis.* J Bacteriol 174:3212–3219

Guillouet S, Rodal AA, An GH, Gorret N, Lessard PA, Sinskey AJ (2001) Metabolic redirection of carbon flow toward isoleucine by expressing a catabolic threonine dehydratase in a threonine-overproducing *Corynebacterium glutamicum.* Appl Microbiol Biotechnol 57:667–673

Gusberti L, Cantoni R, De Rossi E, Branzoni M, Riccardi G (1996) Cloning and sequencing of the *ilvBNC* gene cluster from *Mycobacterium avium.* Gene 177:83–85

Harms EH, Umbarger HE (1991) The absence of branched-chain amino acid and growth rate control at the internal ilvEp promoter of the ilvGMEDA operon. J Bacteriol 173:6446–6452

Hashiguchi K, Matsui H, Kurahashi O (1999a) Effects of a feedback-resistant aspartokinase III gene on L-isoleucine production in *Escherichia coli* K-12. Biosci Biotechnol Biochem 63:2023–2024

Hashiguchi K, Takesada H, Suzuki E, Matsui H (1999b) Construction of an L-isoleucine overproducing strain of *Escherichia coli* K-12. Biosci Biotechnol Biochem 63:672–679

Hauser CA, Hatfield GW (1983) Nucleotide sequence of the ilvB multivalent attenuator region of *Escherichia coli* K12. Nucleic Acids Res 11:127–139

Hu Y, Kohlhaw GB (1995) Additive activation of yeast LEU4 transcription by multiple cis elements. J Biol Chem 270:5270–5275

Huang F, Coppola G, Calhoun DH (1992) Multiple transcripts encoded by the ilvGMEDA gene cluster of *Escherichia coli* K-12. J Bacteriol 174:4871–4877

Hung SP, Baldi P, Hatfield GW (2002) Global gene expression profiling in *Escherichia coli* K12. The effects of leucine-responsive regulatory protein. J Biol Chem 277:40309–40323

Ibdah M, Bar-Ilan A, Livnah O, Schloss JV, Barak Z, Chipman DM (1996) Homology modeling of the structure of bacterial acetohydroxy acid synthase and examination of the active site by site-directed mutagenesis. Biochemistry 35:16282–16291

Ikeda M, Katsumata M (1995) Tryptophan production by transport. mutants of *Corynebacterium glutamicum*. Biosci Biotechnol Biochem 59:1600–1602

Jafri S, Chen S, Calvo JM (2002) ilvIH operon expression in *Escherichia coli* requires Lrp binding to two distinct regions of DNA. J Bacteriol 184:5293–5300

Kaplun A, Vyazmensky M, Zherdev Y, Belenky I, Slutzker A, Mendel S, Barak Z, Chipman DM, Shaanan B (2006) Structure of the regulatory subunit of acetohydroxyacid synthase isozyme III from *Escherichia coli*. J Mol Biol 357:951–963

Keilhauer C, Eggeling L, Sahm H (1993) Isoleucine synthesis in *Corynebacterium glutamicum*: molecular analysis of the ilvB-ilvN-ilvC operon. J Bacteriol 175:5595–5603

Kennerknecht N, Sahm H, Yen MR, Pátek M, Saier MH Jr, Eggeling L (2002) Export of L-isoleucine from *Corynebacterium glutamicum*: a two-gene-encoded member of a new translocator family. J Bacteriol 184:3947–3956

Kisumi M, Komatsubara S, Chibata I (1971) Valine accumulation by alpha-aminobutyric acid-resistant mutants of *Serratia marcescens*. J Bacteriol 106:493–499

Kohlhaw GB (1988) Isopropylmalate dehydratase from yeast. Methods Enzymol 166:423–429

Kohlhaw GB (2003) Leucine biosynthesis in fungi: entering metabolism through the back door. Microbiol Mol Biol Rev 67:1–15

Komatsubara S, Kisumi M, Chibata I (1980) Transductional construction of an isoleucine-producing strain of *Serratia marcescens*. J Gen Microbiol 119:51–61

Kopecký J, Janata J, Pospíšil S, Felsberg J, Spížek J (1999) Mutations in two distinct regions of acetolactate synthase regulatory subunit from *Streptomyces cinnamonensis* result in the lack of sensitivity to end-product inhibition. Biochem Biophys Res Commun 266:162–166

Lange C, Rittmann D, Wendisch VF, Bott M, Sahm H (2003) Global expression profiling and physiological characterization of *Corynebacterium glutamicum* grown in the presence of L-valine. Appl Environ Microbiol 69:2521–2532

Leuchtenberger W, Huthmacher K, Drauz K (2005) Biotechnological production of amino acids and derivatives: current status and prospects. Appl Microbiol Biotechnol 69:1–8

Lawther RP, Calhoun DH, Adams CW, Hauser CA, Gray J, Hatfield GW (1981) Molecular basis of valine resistance in *Escherichia coli* K-12. Proc Natl Acad Sci USA 78:922–925

Lawther RP, Wek RC, Lopes JM, Pereira R, Taillon BE, Hatfield GW (1987) The complete nucleotide sequence of the ilvGMEDA operon of *Escherichia coli* K-12. Nucleic Acids Res 15:2137–2155

Leary TR, Kohlhaw G (1970) Dissociation of alpha-isopropylmalate synthase from *Salmonella typhimurium* by its feedback inhibitor leucine. Biochem Biophys Res Commun 39:494–501

Lee CR, Koo BM, Cho SH, Kim YJ, Yoon MJ, Peterkofsky A, Seok YJ (2005) Requirement of the dephospho-form of enzyme IIANtr for derepression of *Escherichia coli* K-12 ilvBN expression. Mol Microbiol 58:334–344

Leyval D, Uy D, Delaunay S, Goergen JL, Engasser JM (2003) Characterisation of the enzyme activities involved in the valine biosynthetic pathway in a valine-producing strain of *Corynebacterium glutamicum*. J Biotechnol 104:241–252

Liu H, Allee GL, Berkemeyer JJ, Touchette KJ, Spencer JD, Kim IB (1999) Effect of reducing protein level and adding amino acids on growth performance and carcass characteristics of finishing pigs. J Anim Sci 77(Suppl 1):69

Marienhagen J, Kennerknecht N, Sahm H, Eggeling L (2005) Functional analysis of all aminotransferase proteins inferred from the genome sequence of *Corynebacterium glutamicum*. J Bacteriol 187:7639–7646

Marta PT, Ladner RD, Grandoni JA (1996) A CUC triplet confers leucine-dependent regulation of the *Bacillus subtilis* ilv-leu operon. J Bacteriol 178:2150–2153

Mavromichalis I, Webel DM, Emmert JL, Moser RL, Baker DH (1998) Limiting order of amino acids in a low-protein corn-soybean meal-based diet for nursery pigs. J Anim Sci 76:2833–2837

McCourt JA, Duggleby RG (2006) Acetohydroxyacid synthase and its role in the biosynthetic pathway for branched-chain amino acids. Amino Acids 31:173–210

McCourt JA, Pang SS, Guddat LW, Duggleby RG (2005) Elucidating the specificity of binding of sulfonylurea herbicides to acetohydroxyacid synthase. Biochemistry 44:2330–2338

Mendel S, Elkayam T, Sella C, Vinogradov V, Vyazmensky M, Chipman DM, Barak Z (2001) Acetohydroxyacid synthase: a proposed structure for regulatory subunits supported by evidence from mutagenesis. J Mol Biol 307:465–477

Mendel S, Vinogradov M, Vyazmensky M, Chipman DM, Barak Z (2003) The N-terminal domain of the regulatory subunit is sufficient for complete activation of acetohydroxyacid synthase III from *Escherichia coli*. J Mol Biol 325:275–284

Merkamm M, Chassagnole C, Lindley ND, Guyonvarch A (2003) Ketopantoate reductase activity is only encoded by ilvC in *Corynebacterium glutamicum*. J Biotechnol 104:253–260

McHardy AC, Tauch A, Ruckert C, Puhler A, Kalinowski J (2003) Genome-based analysis of biosynthetic aminotransferase genes of *Corynebacterium glutamicum*. J Biotechnol 104:229–240

Möckel B, Eggeling L, Sahm H (1992) Functional and structural analyses of threonine dehydratase from *Corynebacterium glutamicum*. J Bacteriol 174:8065–8072

Möckel B, Eggeling L, Sahm H (1994) Threonine dehydratases of *Corynebacterium glutamicum* with altered allosteric control: their generation and biochemical and structural analysis. Mol Microbiol 13:833–842

Morbach S, Junger C, Sahm H, Eggeling L (2000) Attenuation control of *ilvBNC* in *Corynebacterium glutamicum*: evidence of leader peptide formation without the presence of a ribosome binding site. J Biosci Bioeng 90:501–507

Morbach S, Sahm H, Eggeling L (1996) L-isoleucine production with *Corynebacterium glutamicum*: further flux increase and limitation of export. Appl Environ Microbiol 62:4345–4351

Muller YA, Schulz GE (1993) Structure of the thiamine- and flavin-dependent enzyme pyruvate oxidase. Science 259:965–967

Okamoto K, Kino K, Ikeda M (1997) Hyperproduction of L-threonine by an *Escherichia coli* mutant with impaired L-threonine uptake. Biosci Biotechnol Biochem 61:1877–1882

Pang SS, Duggleby RG, Guddat LW (2002) Crystal structure of yeast acetohydroxyacid synthase: a target for herbicidal inhibitors. J Mol Biol 317:249–262

Parsons SJ, Burns RO (1969) Purification and properties of beta-isopropylmalate dehydrogenase. J Biol Chem 244:996–1003

Pátek M, Hochmannová J, Jelínková M, Nešvera J, Eggeling L (1998) Analysis of the *leuB* gene from *Corynebacterium glutamicum*. Appl Microbiol Biotechnol 50:42–47

Pátek M, Krumbach K, Eggeling L, Sahm H (1994) Leucine synthesis in *Corynebacterium glutamicum*: enzyme activities, structure of leuA, and effect of leuA inactivation on lysine synthesis. Appl Environ Microbiol 60:133–140

Pereira RF, Ortuno MJ, Lawther RP (1988) Binding of integration host factor (IHF) to the ilvGp1 promoter of the ilvGMEDA operon of *Escherichia coli* K12. Nucleic Acids Res 16:5973–5989

Platko JV, Willins DA, Calvo JM (1990) The ilvIH operon of *Escherichia coli* is positively regulated. J Bacteriol 172:4563–4570

Porat I, Vinogradov M, Vyazmensky M, Lu CD, Chipman DM, Abdelal AT, Barak Z (2004) Cloning and characterization of acetohydroxyacid synthase from *Bacillus stearothermophilus*. J Bacteriol 186:570–574

Powell JT, Morrison JF (1978) Role of the *Escherichia coli* aromatic amino acid aminotransferase in leucine biosynthesis. J Bacteriol 136:1–4

Radmacher E, Vaitsikova A, Burger U, Krumbach K, Sahm H, Eggeling L (2002) Linking central metabolism with increased pathway flux: L-valine accumulation by *Corynebacterium glutamicum*. Appl Environ Microbiol 68:2246–2250

Reinscheid DJ, Kronemeyer W, Eggeling L, Eikmanns BJ, Sahm H (1994) Stable expression of *hom-1-thrB* in *Corynebacterium glutamicum* and its effect on the carbon flux to threonine and related amino acids. Appl Environ Microbiol 60:126–132

Rennie MJ, Bohe J, Smith K, Wackerhage H, Greenhaff P (2006) Branched-chain amino acids as fuels and anabolic signals in human muscle. J Nutr 136:264S-268S

Rennie MJ, Tipton KD (2000) Protein and amino acid metabolism during and after exercise and the effects of nutrition. Annu Rev Nutr 20:457–483

Rhee KY, Parekh BS, Hatfield GW (1996) Leucine-responsive regulatory protein-DNA interactions in the leader region of the ilvGMEDA operon of *Escherichia coli*. J Biol Chem 271:26499–26507

Ricca E, Aker DA, Calvo JM (1989) A protein that binds to the regulatory region of the *Escherichia coli* ilvIH operon. J Bacteriol 171:1658–1664

Sahm H, Eggeling L (1999) D-Pantothenate synthesis in *Corynebacterium glutamicum* and use of *panBC* and genes encoding L-valine synthesis for D-pantothenate overproduction. Appl Environ Microbiol 65:1973–1979

Sahm H, Eggeling L, Morbach S, Eikmanns B (1999) Construction of L-isoleucine overproducing strains of *Corynebacterium glutamicum*. Naturwissenschaften 86:33–38

Seliverstov AV, Putzer H, Gelfand MS, Lyubetsky VA (2005) Comparative analysis of RNA regulatory elements of amino acid metabolism genes in Actinobacteria. BMC Microbiol 5:54–67

Shimomura Y, Yamamoto Y, Bajotto G, Sato J, Murakami T, Shimomura N, Kobayashi H, Mawatari K (2006) Nutraceutical effects of branched-chain amino acids on skeletal muscle. J Nutr 136:529S-532S

Sutton A, Freundlich M (1980) Regulation of cyclic AMP of the ilvB-encoded biosynthetic acetohydroxy acid synthase in *Escherichia coli* K-12. Mol Gen Genet 178:179–183

Tedin K, Norel F (2001) Comparison of ΔrelA strains of *Escherichia coli* and *Salmonella enterica* serovar Typhimurium suggests a role for ppGpp in attenuation regulation of branched-chain amino acid biosynthesis. J Bacteriol 183:6184–6196

Tojo S, Satomura T, Morisaki K, Deutscher J, Hirooka K, Fujita Y (2005) Elaborate transcription regulation of the *Bacillus subtilis* ilv-leu operon involved in the biosynthesis of branched-chain amino acids through global regulators of CcpA, CodY and TnrA. Mol Microbiol 56:1560–1573

Tojo S, Satomura T, Morisaki K, Yoshida K, Hirooka K, Fujita Y (2004) Negative transcriptional regulation of the ilv-leu operon for biosynthesis of branched-chain amino acids through the *Bacillus subtilis* global regulator TnrA. J Bacteriol 186:7971–7979

Tsuchida T, Momose H (1975) Genetic changes of regulatory mechanisms occurred in leucine and valine producing mutants derived from *Brevibacterium lactofermentum*. Agric Biol Chem 39:2193–2198

Tsuchida T, Momose H (1986) Improvement of an L-leucine-producing mutant of Brevibacterium lactofermentum 2256 by genetically desensitizing it to alfa-acetohydroxy acid synthase. Appl Environ Microbiol 51:1024–1027

Umbarger HE (1996) Biosynthesis of the branched-chain amino acids. In: Neidhardt FC, Ingraham JL, Low BL, Magasanik B, Schaechter M, Umbarger HE (eds) *Escherichia coli* and Salmonella typhimurium. Cell Mol Biol. ASM Press, Washington, DC, pp 442–457

Vartak NB, Liu L, Wang BM, Berg CM (1991) A functional leuABCD operon is required for leucine synthesis by the tyrosine-repressible transaminase in *Escherichia coli* K-12. J Bacteriol 173:3864–3871

Vertes AA, Inui M, Yukawa H (2005) Manipulating corynebacteria, from individual genes to chromosomes. Appl Environ Microbiol 71:7633–7642

Vinogradov V, Vyazmensky M, Engel S, Belenky I, Kaplun A, Kryukov O, Barak Z, Chipman DM (2006) Acetohydroxyacid synthase isozyme I from *Escherichia coli* has unique catalytic and regulatory properties. Biochim Biophys Acta 1760:356–363

Vitreschak AG, Lyubetskaya EV, Shirshin MA, Gelfand MS, Lyubetsky VA (2004a) Attenuation regulation of amino acid biosynthetic operons in proteobacteria: comparative genomics analysis. FEMS Microbiol Lett 234:357–370

Vitreschak AG, Rodionov DA, Mironov AA, Gelfand MS (2004b) Riboswitches: the oldest mechanism for the regulation of gene expression? Trends Genet 20:44–50

Vyazmensky M, Elkayam T, Chipman DM, Barak Z (2000) Isolation of subunits of acetohydroxy acid synthase isozyme III and reconstitution of holoenzyme. Meth Enzymol 324:95–103

Vyazmensky M, Sella C, Barak Z, Chipman DM (1996) Isolation and characterization of subunits of acetohydroxy acid synthase isozyme III and reconstitution of the holoenzyme. Biochemistry 35:10339–10346

Wang JG, Li ZM, Ma N, Wang BL, Jiang L, Pang SS, Lee YT, Guddat LW, Duggleby RG (2005) Structure-activity relationships for a new family of sulfonylurea herbicides. J Comput Aided Mol Des 19:801–820

Wang Q, Calvo JM (1993) Lrp, a global regulatory protein of *Escherichia coli*, binds co-operatively to multiple sites and activates transcription of ilvIH. J Mol Biol 229:306–318

Ward JB Jr, Zahler SA (1973) Regulation of leucine biosynthesis in *Bacillus subtilis*. J Bacteriol 116:727–735

Weaver LM, Mitsky TA, Rapp WD, Gruys KJ, Liang J, Vaduva G (2005) Plants with increased levels of one or more amino acids, US Patent application 20 050 289 668

Weinstock O, Sella C, Chipman DM, Barak Z (1992) Properties of subcloned subunits of bacterial acetohydroxy acid synthases. J Bacteriol 174:5560–5566

Wendisch VF, Bott M, Kalinowski J, Oldiges M, Wiechert W (2006) Emerging *Corynebacterium glutamicum* systems biology. J Biotechnol 124:74–92

Wessler SR, Calvo JM (1981) Control of leu operon expression in *Escherichia coli* by a transcription attenuation mechanism. J Mol Biol 149:579–597

Wichmann R, Wandrey C, Große-Wiesmann J (1990) Continuous microbial production of L-leucine with cell retention. Appl Microbiol Biotechnol 32:373–379

Willins DA, Ryan CW, Platko JV, Calvo JM (1991) Characterization of Lrp, and *Escherichia coli* regulatory protein that mediates a global response to leucine. J Biol Chem 266:10768–10774

Zahler SA, Najimudin N, Kessler DS, Vaneyar MA (1990) α-Acetolactate synthesis by *Bacillus subtilis*. In: Barak Z, Chipman DM, Schloss JV (ed) Biosynthesis of branched chain amino acids. Wiley-VCH, Weinheim, pp 25–32

Zohar Y, Einav M, Chipman DM, Barak Z (2003) Acetohydroxyacid synthase from *Mycobacterium avium* and its inhibition by sulfonylureas and imidazolinones. Biochim Biophys Acta 1649:97–105

Microbiol Monogr (5)
V. F. Wendisch: Amino Acid Biosynthesis
DOI 10.1007/7171_2006_059/Published online: 9 December 2006

Methionine Biosynthesis in *Escherichia coli* and *Corynebacterium glutamicum*

Rainer M. Figge

Metabolic Explorer, Biopôle Clermont-Limagne, 63360 St. Beauzire, France
rfigge@metabolic-explorer.com

Abstract The sulphur-containing amino acid methionine and its derivatives play important roles in cellular metabolism. These include initiation of protein biosynthesis, methyl transfer and synthesis of polyamines. Methionine cannot be synthesized by humans and animals and must therefore be obtained from the diet. Since concentrations of methionine in plant-based diets are low, the amino acid is routinely added to animal feed. For this purpose methionine is produced in large scale by chemical synthesis.

Currently, no fermentative methionine production process exists, but for economic reasons its development has recently received increasing attention. Research has mainly focused on methionine biosynthesis in the bacteria *Escherichia coli* and *Corynebacterium glutamicum*, which are already used for the production of several other amino acids. Investigation of methionine biosynthesis and its regulation in *C. glutamicum* has revealed major differences with the corresponding *E. coli* metabolism. This review intends to summarize the current knowledge on methionine biosynthesis in these organisms and discusses approaches for the construction of methionine producer strains.

1 Introduction

Methionine is a sulphur-containing amino acid that is essential for human and animal nutrition. Apart from its function as a structural component of proteins, methionine also plays an important role in methyl transfer reactions and in the initiation of protein biosynthesis via its derivatives *S*-adenosylmethionine (SAM) and *N*-formyl-methionine-*t*RNA, respectively.

Since methionine concentrations in many plant-based diets are low (Tabe and Higgins 1998), the amino acid or its hydroxyl analogue 2-hydroxy-4-(methylthio) butanoic acid, which is converted to methionine inside the animal by oxidation of the hydroxyl group and subsequent transamination (Dibner and Knight 1984), are routinely added to animal feed. For this purpose methionine and its analogue are produced at a scale of at least 400 000 tons per year (Bray 2002). Methionine production relies on chemical synthesis from acrolein, methylmercaptan and hydrogen cyanide. In contrast to all other amino acids produced at large scale, no fermentative production process has been established for methionine. Increasing precursor prices for chemical methionine synthesis have recently sparked interest in its fermentative production.

Most fermentative amino acid production processes rely on modified *Corynebacterium glutamicum* or *Escherichia coli*. These strains were often obtained by screening for analogue resistance, an approach that did not prove to be successful for methionine producers. Recent efforts have focused on metabolic engineering strategies in the two organisms. These approaches require the precise understanding of pathways and regulatory properties that are to be modified. For this purpose research on the methionine metabolism of *C. glutamicum*, which in contrast to its *E. coli* counterpart had so far received little attention, has recently been intensified. These studies have revealed major differences in methionine biosynthesis and its regulation in the two organisms. In this review the current knowledge of methionine biosynthesis in *C. glutamicum* and *E. coli* will be summarized and approaches for the construction of producer strains discussed.

2 Methionine Biosynthesis

In both, *C. glutamicum* and *E. coli*, methionine is derived from the amino acid aspartate. In addition, methionine biosynthesis requires precursors from sulphur and one-carbon metabolism. The first three reactions of methionine production, which transform aspartate into homoserine, are shared with the biosynthetic pathways of several other metabolites. The subsequent reactions, which integrate reduced sulphur into homoserine and transfer the C5-methyl group on the resulting homocysteine, are specific to the production of methionine and its derivatives (Fig. 1).

2.1 Conversion of Aspartate to Homoserine

Aspartate is not only a precursor in the biosynthesis of methionine, but serves also as a building block for the production of threonine, lysine and branched chain amino acids. The reactions shared by these biosynthetic pathways are described in more detail in Wittmann, Hermann and Rieping, and Patek, this volume. Briefly, in *E. coli* three enzymes with aspartokinase activity encoded by *lysC*, *thr*A and *metL* catalyze the phosphorylation of aspartate yielding aspartyl-4-phosphate (Patte J-C 1996 and references therein, Fig. 1). In *Corynebacterium* this reaction is exclusively catalyzed by LysC (Kalinowski et al. 2003 and references therein). For the conversion of aspartyl-phosphate to aspartate-semialdehyde both organisms use aspartate-semialdehyde dehydrogenase, encoded by *asd*. The last common step in threonine and methionine biosynthesis, the reduction of aspartate-semialdehyde to homoserine is catalyzed by homoserine dehydrogenase. Unlike *E. coli* that harbours two homoserine dehydrogenases encoded by the *thrA* and *metL* genes, *C. glutamicum* has only one homoserine dehydrogenase encoded by the *hom* gene. The *hom* gene is repressed transcriptionally by methionine and the activity of the gene product is inhibited by threonine, thus balancing the production of these two amino acids (Lee 2005). In *E. coli* the genes *metL*, *lysC* and *thrA* are all regulated on the transcriptional level, but only the enzymes LysC and ThrA are feed-back controlled by the end-products of their corresponding pathways lysine and threonine.

2.2 Homoserine Activation by Homoserine Transacetylase and Transsuccinylase

In the first specific reaction of methionine biosynthesis the γ-hydroxyl group of homoserine is activated. *C. glutamicum* and *E. coli* use different acyl substrates, acetyl-CoA and succinyl-CoA, for this activation step. The activating

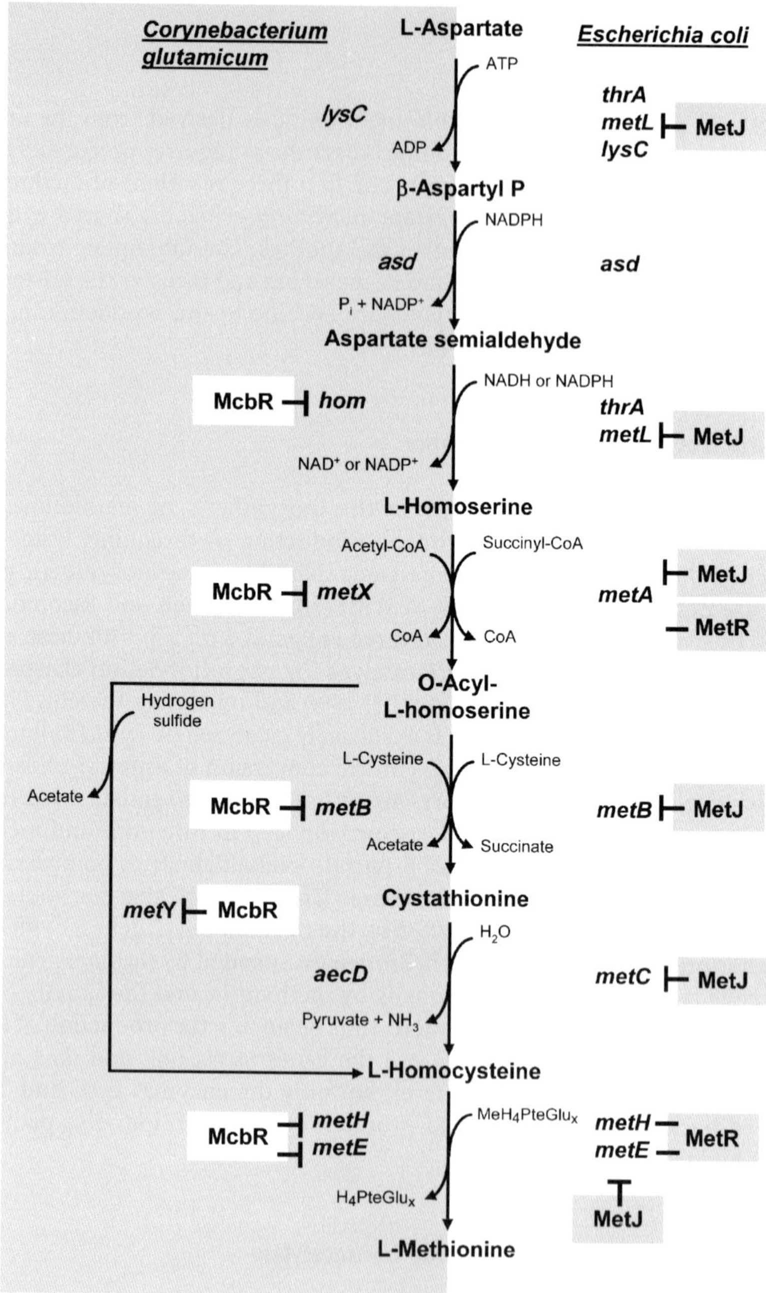

Fig. 1 Biosynthesis of methionine from aspartate in *Corynebacterium glutamicum* (*left side, shaded*) and *Escherichia coli* (*right side, white*). Transcriptional repression by the regulators McbR and MetJ and regulation by MetR are indicated. Abbreviations: $MeH_4PteGlu_x$: N^5-Methyl-tetrahydropteroyl with x bound glutamates

Table 1 *Escherichia coli* and *Corynebacterium glutamicum* genes involved in methionine biosynthesis

Protein function	*Escherichia coli* Gene	Length (aa)	b-number	Refs	*Corynebacterium glutamicum* Gene	Length (aa)	NCgl-number	Refs	Identity (%)
Homoserine dehydrogenase	*metL*	810	3940	(Zakin et al. 1983)	*hom*	445	1183	(Peoples et al. 1988)	– *
Homoserine transacylase	*metA*	309	4013	(Duclos et al. 1989)	*metX*	377	0652	(Park et al. 1998)	
Cystathionine-γ-synthase	*metB*	386	3939	(Duchange et al. 1983)	*metB*	386	2446	(Hwang et al. 1999)	41
O-acetylhomoserine sulfhydrylase	–	–	–	–	*metY*	437	0653	(Hwang et al. 2002)	–
Cystathionine-β-lyase	*metC*	395	3008	(Belfaiza et al. 1986)	*aecD*	368	2309	(Rossol & Pühler 1992)	41
B_{12}-dependent methionine synthase	*metH*	1,227	4019	(Banerjee et al. 1989)	*metH*	1,221	1507	(Rückert et al. 2003)	31
B_{12}-independent methionine synthase	*metE*	753	3829	(Maxon et al. 1989)	*metE*	745	1139	(Rückert et al. 2003)	45
5,10-Methylene tetra-hydrofolate reductase	*metF*	296	3941	(Saint-Girons et al. 1983)	*metF*	326	2171	(Rückert et al. 2003)	33
Transcriptional repressor	*metJ*	105	3938	(Saint-Girons et al. 1984)	*mcbR*	213	2941	(Rey et al. 2003)	–
S-adenosylmethionine synthetase	*metK*	384	2942	(Markham et al. 1984)	*metK*	407	1603	(Großman et al. 2000)	57
Transcriptional activator	*metR*	317	3828	(Maxon et al. 1990)	–	–	–	–	–

* not determined, since *metL* comprises homoserine dehydrogenase and aspartokinase domains

enzymes homoserine transacetylase and transsuccinylase share only 27% amino acid identity and belong to two different gene families (Gophna et al. 2005, Table 1).

In *E. coli* homoserine transsuccinylase (HTS; EC 2.3.1.46), encoded by the *metA* gene, catalyzes the synthesis of O-succinyl-homoserine (OSH) from succinyl-CoA and homoserine (Flavin and Slaughter 1965). HTS utilizes a ping pong mechanism in which the succinyl-group of succinyl-CoA is transferred to a nucleophilic residue in the active site, before its reaction with the γ-hydroxyl group of homoserine (Born and Blanchard 1999). Born and Blanchard (1999) suggested a cysteine residue as the nucleophile, but Rosen et al. (2004b) could demonstrate that the succinyl group is covalently bound to a lysine residue. K_m values for succinyl-CoA, homoserine and OSH were 0.17, 1.6 and 3.5 mM, respectively (Born and Blanchard 1999). HTS aggregates and loses activity at elevated temperature (Gur et al. 2002 and references therein). The enzyme is unstable and proteolyzed in an energy-dependent manner (Biran et al. 2000). As is known for other highly acidic proteins, phosphorylation may target HTS for degradation (Rosen et al. 2004a). MetA activity is feedback inhibited by its two end-products methionine and SAM in a synergistic manner (Lee et al. 1966). At concentrations of 1 mM methionine or 1 mM SAM, only 4% or 13%, respectively, of the initial activity was retained (Usuda and Kurahashi 2005).

In *Corynebacterium glutamicum* homoserine transacetylase (EC 2.3.1.31), encoded by the *metX* gene, catalyzes the synthesis of acetyl-homoserine from acetyl-CoA and homoserine (Park et al. 1998). The K_m values of the enzyme for acetyl-CoA and homoserine are 0.05 and 2.8 mM, respectively (Miyajima and Shiio 1973). As MetA of *E. coli,* MetX is subject to feedback control by methionine and SAM, with respective concentrations of 4.8 and 0.28 mM causing 50% inhibition of the enzyme activity (Miyajima and Shiio 1973). In contrast to *E. coli*, however, no synergistic action of the two inhibitors was detected and inhibition was only observed if the assay was performed in the presence of sulfhydryl compounds (Shiio and Ozaki 1981). Temperature sensitivity of the MetX enzyme has been suggested (Hwang et al. 2002).

2.3
Incorporation of Sulphur: Cystathionine-γ-synthase and Cystathionine-β-lyase versus Acetyl-homoserine Sulfhydrylase

Activation of homoserine is a prerequisite for the incorporation of sulphur into the methionine precursor. In transsulfuration, the nucleophile cysteine replaces the acyl moiety yielding γ-cystathionine, which is subsequently converted to homocysteine. In sulfhydrylation, homocysteine is directly produced from acyl-homoserine and hydrogen sulfide (Flavin and Slaughter 1967).

In *E. coli*, which relies almost exclusively on transsulfuration, cystathionine-γ-synthase (CGS, EC 4.2.99.9), a homotetrameric, pyridoxal 5′-phosphate (PLP) dependent enzyme, encoded by *metB*, catalyzes the γ-replacement reaction of O-succinyl-homoserine and cysteine yielding γ-cystathionine. The MetB enzyme has been purified (Holbrook et al. 1990) and its structure determined (Clausen et al. 1998). The reaction mechanism of MetB depends on the available cysteine concentration. At cysteine concentrations greater than the K_m^{cys} of the replacement reaction, the mechanism is ordered and cysteine binds to the enzyme-OSH complex prior to the release of succinate. If cysteine concentrations are below K_m^{cys}, the mechanism is ping-pong and succinate is eliminated prior to the binding of cysteine. In the absence of cysteine, the aldimine between the enzyme and OSH slowly decomposes to succinate, ammonia and α-ketobutyrate in a γ-elimination reaction (Aitken et al. 2003). In addition to cysteine, MetB can also use hydrogen sulfide or methylmercaptan with low affinity (Flavin and Slaughter 1967).

E. coli metC encodes cystathionine-β-lyase (CBL, EC 4.4.1.8), which catalyzes the reaction of γ-cystathionine to homocysteine, pyruvate and ammonia. MetC is organized in a hexameric structure with one pyridoxal phosphate bound per subunit (Belfaiza et al. 1986). The K_m value for γ-cystathionine is 0.04 mM (Dwivedi et al. 1982). MetC also catalyzes the β-elimination reactions of several other sulphur amino acids, such as cysteine, lanthionine and djenkolic acid (Dwivedi et al. 1982; Awano et al. 2003).

Unlike *E. coli*, *C. glutamicum* applies both transsulfuration and sulfhydrylation for the synthesis of homocysteine. PLP binding motifs are conserved in cystathionine-γ-synthase, encoded by *metB* (Lee 2005) and the enzyme has 41% identity with its *E. coli* counterpart (Table 1). In contrast to the *E. coli* enzyme, MetB from *C. glutamicum* preferentially accepts acetylated homoserine (Hwang et al. 1999, 2002). Enzyme activity is inhibited 60% in the presence of 5 mM SAM (Kase and Nakayama 1974).

The *C. glutamicum* CBL homolog, encoded by the *aecD* gene, was initially identified based on its ability to confer resistance to the lysine analog *S*-(2-aminoethyl)-L-cysteine (AEC) (Rossol and Pühler 1992). Subsequently, Kim et al. (2001) isolated *aecD* in a complementation screen of an *E. coli metC* mutant. As its *E. coli* counterpart, AecD from *C. glutamicum* can cleave several C – S bond containing substrates with a decreasing affinity from djenkolic acid, cysteine, DL-cystathionine, DL-lanthionine, cysteine to methionine (Rossol and Pühler 1992; Wada et al. 2002). The K_m value for the L-cystathionine-β-lyase reaction is 5 mM. The enzyme is inhibited by cysteine (83% at 10 mM) and glycine (53% at 10 mM) (Kase and Nakayama 1974).

Direct synthesis of homocysteine from hydrogen sulfide and O-acetyl-homoserine is catalyzed by acetyl-homoserine sulfhydrylase (EC 2.5.1.49), encoded by *metY* (Hwang et al. 2002). The *metY* gene is separated from *metX* by only 143 bp on the chromosome, but the two genes seem to be regulated independently (Hwang et al. 2002). MetY activity is inhibited by methionine (50% inhibition at 10 mM) and O-acetyl-serine (30% inhibition at 10 mM) (Ozaki and Shiio 1982). Even though a putative SAM binding site in MetY has been postulated (Bailey et al. 2004), MetY does not seem to be inhibited by this compound (Lee 2005). MetY from *C. glutamicum* may be phosphorylated, although not at a seryl residue (Bendt et al. 2003)

In *C. glutamicum*, both transsulfuration and sulfhydrylation seem to function equally well, since *metY* and *metB* single deletion mutants show comparable growth characteristics. In addition, strains with single deletions in genes *metB*, *metC* or *metY* are methionine prototroph and strains with deletions in both *metY* and *metB* or *metY* and *aecD* are methionine auxotroph (Hwang et al. 2002; Lee and Hwang 2003).

2.4
Methylation of Homocysteine to Methionine: Methionine Synthase

In both *E. coli* and *C. glutamicum*, the terminal step in methionine biosynthesis can be catalyzed by two apparently unrelated proteins, cobalamin-independent methionine synthase (MetE, EC 2.1.1.14) and cobalamin-dependent methionine synthase (MetH, EC 2.1.1.13). MetE and MetH from *C. glutamicum* and *E. coli* share 45% and 31% sequence identity, respectively (Table 1). Both enzymes catalyze the transfer of a methyl group to homocysteine. In contrast to *C. glutamicum*, where little is known about the mechanism of methionine synthase, in *E. coli* the two enzymes have been intensively studied (for recent reviews see Ludwig and Matthews 1997; Matthews 2001).

Whereas *E. coli* MetE exclusively uses derivatives of N^5-methyl-tetrahydropteroyl-glutamate with three or more glutamate molecules as the methyl donor, MetH can accept derivatives with fewer glutamates such as the monoglutamate N^5-methyl-tetrahydrofolate (CH_3-THF). MetE catalyzes a direct transfer of the methyl group to homocysteine, while MetH uses cobalamin as an intermediate methyl carrier. Mechanistically catalysis via cobalamin seems to be the better solution, since k_{cat} values for MetH are 50–100 times higher than for MetE. This may also explain the high abundance (up to 5% of the total protein) of the MetE enzyme in *E. coli* grown in minimal medium lacking vitamin B_{12}.

The MetH enzyme folds into four modules (Ludwig and Matthews 1997 and references therein). The N-terminal module binds and activates homocysteine, the second module activates CH_3-THF for methyl transfer and the third module binds the cobalamin cofactor (Banerjee et al. 1989; Drummond et al. 1993). The fourth C-terminal module binds SAM used for the reactivation of oxidized inactivated enzyme by reductive methylation (Dixon et al. 1996). Large domain motions of the cobalamin binding domain are required for the catalytic methyl transfer reactions (Evans et al. 2004).

MetE activity is dependent on inorganic phosphate (P_i) and stimulated by magnesium and dithiothreitol (Whitfield et al. 1970). The structure of the *Thermotoga maritima* MetE enzyme has been solved. MetE consists of two face-to-face $(\beta\alpha)_8$ barrels that are most likely the result of a gene duplication event (Pejchal and Ludwig 2005). MetE is a thermolabile protein that can be protected from aggregation at high temperature by the chaperone DnaK. Degradation of MetE upon long exposure to high temperature is Lon-dependent (Mogk et al. 1999). Previously, MetE was shown to be an in vivo substrate of GroEL (Houry et al. 1999). MetE is also inactivated by oxidative stress (Leichert and Jakob 2004), leading to cellular methionine limitation. Under these conditions, Cys-645 of MetE, which is localized at the entrance of the active site, most likely forms a mixed disulfide with glutathione. This may protect the active site in an easily reversible manner (Hondorp and Matthews 2004). Recently, it could be demonstrated that N^5-methyl-tetrahydropteroyl-triglutamate bound to MetE is activated by protonation before the methyl transfer reaction (Taurog and Matthews 2006).

The corynebacterial *metE* and *metH* genes have just recently been identified (Rückert et al. 2003). In contrast to *E. coli*, a corynebacterial *metE* deletion mutant was viable in liquid minimal medium lacking vitamin B_{12}. On solid medium, however, vitamin B_{12} was required for growth, which may be explained by differences in vitamin B_{12} biosynthesis under aerobic and micro-aerobic conditions (Rückert et al. 2003). The MetE protein had already been identified in a global *Corynebacterium* proteome map (Schaffer et al. 2001). MetE has also been found in the phospho-proteome, indicating that it is a phospho-protein presumably containing (a) phosphorylated serine residue(s) (Bendt et al. 2003). The methyl group providing substrates for the two methionine synthase reactions are currently not known.

2.5 Reduction of CH_2-THF to CH_3-THF by N^5,N^{10}-Methylenetetrahydrofolate Reductase

Reduction of N^5,N^{10}-methylenetetrahydrofolate (CH_2-THF) to CH_3-THF catalyzed by N^5,N^{10}-methylenetetrahydrofolate reductase, encoded by *metF*, commits methyl groups to the terminal reaction in methionine biosynthesis (Saint-Girons et al. 1983). The MetF protein has recently been purified together with its bound cofactor that was identified as FAD (Sheppard et al. 1999). The K_m for NADH was determined to be 13 μM and for CH_2-THF 0.8 μM. An X-ray structure of MetF revealed that MetF forms a $\beta_8\alpha_8$ barrel (Guenther et al. 1999). As MetE, the MetF protein has been shown to be a substrate of GroEL (Houry et al. 1999).

MetF has just recently been identified in *C. glutamicum* and its deletion leads to methionine auxotrophy (Rückert et al. 2003).

3
Methionine Transformation and Degradation

Apart from its role in protein biosynthesis, methionine is also the precursor of SAM, which is the major methyl donor in cellular metabolism and a precursor of several other important metabolites. In contrast to *E. coli*, in which methionine is only converted to 5-thiomethylribose, in *C. glutamicum*, methionine can be degraded to methylmercaptan, α-ketobutyrate and ammonia.

3.1
Biosynthesis of SAM

SAM is recognized as the major methyl donor for methylation reactions. In addition, it can be the source of methylene, ribosyl, and aminoalkyl groups, and it is the precursor of polyamines and 5′-deoxyadenosyl radicals (for a review see Fontecave et al. 2004). SAM synthesis is catalyzed by SAM synthetase (EC 2.5.1.6) encoded by the *metK* gene.

SAM synthetase of *E. coli* has been crystallized as the apo-enzyme, with ADP and P_i, with the bound products PP_i or P_i, and with methionine and the ATP analogue AMPPNP (Takusagawa et al. 1996a,b; Komoto et al. 2004). On the basis of NMR studies a binding site for SAM has been proposed (Schalk-Hihi and Markham 1999). Two subunits each contribute to an active site that requires binding of two divalent metal ions (e.g. Mg^{2+}) and a monovalent cation (K^+) for maximal activity. SAM is produced in an unusual two-step reaction from ATP and methionine, in which at first the entire triphosphate chain is cleaved from ATP while SAM is formed and subsequently triphosphate is hydrolyzed to PP_i and P_i prior to the release of SAM (Taylor and Markham 2000 and references therein). The *metK* gene is essential in *E. coli* and mutants in *metK* have been isolated by virtue of their resistance to ethionine (Greene et al. 1970), norleucine and γ-glutamyl ester (Kraus et al. 1979). Phenotypes of *metK* mutants include methionine overproduction (Greene et al. 1970; Usuda and Kurahashi 2005), vitamin B_{12} or methionine auxotrophy or complete lack of growth on defined media (Satishchandran et al. 1990). Morphologically, reduced *metK* expression leads to cell filamentation (Newman et al. 1998). Recently, the introduction of the SAM transporter (TC 2.A.7.3.7) of *Rickettsia prowazekii* into *E. coli* permitted the deletion of *metK* when SAM was added to the medium (Driskell et al. 2005).

The *C. glutamicum metK* gene product has 57% sequence identity to its homolog in *E. coli*. MetK transcripts were detected in exponential growth but not in stationary phase (Großmann et al. 2000).

3.2
Methionine as a Metabolic Intermediate

Apart from its role as a methyl donor, in *E. coli* SAM is also the precursor of autoinducer 2 (AI-2) involved in quorum sensing (Xavier and Bassler 2003) and provides an aminopropyl group for the biosynthesis of the polyamine spermidine (Tabor and Tabor 1985; Fig. 2).

Transfer of the methyl group from SAM to one of its numerous acceptor molecules yields *S*-adenosyl-homocysteine (SAH). In *E. coli*, SAH is recycled by cleavage to adenine and *S*-ribosylhomocysteine (SRH), a reaction catalyzed by *S*-adenosylhomocysteine nucleosidase, encoded by *pfs* (Cornell and Riscoe 1998). Subsequently, SRH conversion to homocysteine and 4.5-dihydroxy-2.3-pentadione is catalyzed by autoinducer 2 synthase encoded by *luxS* (Schauder et al. 2001). 4.5-Dihydroxy-2.3-pentadione is supposed to

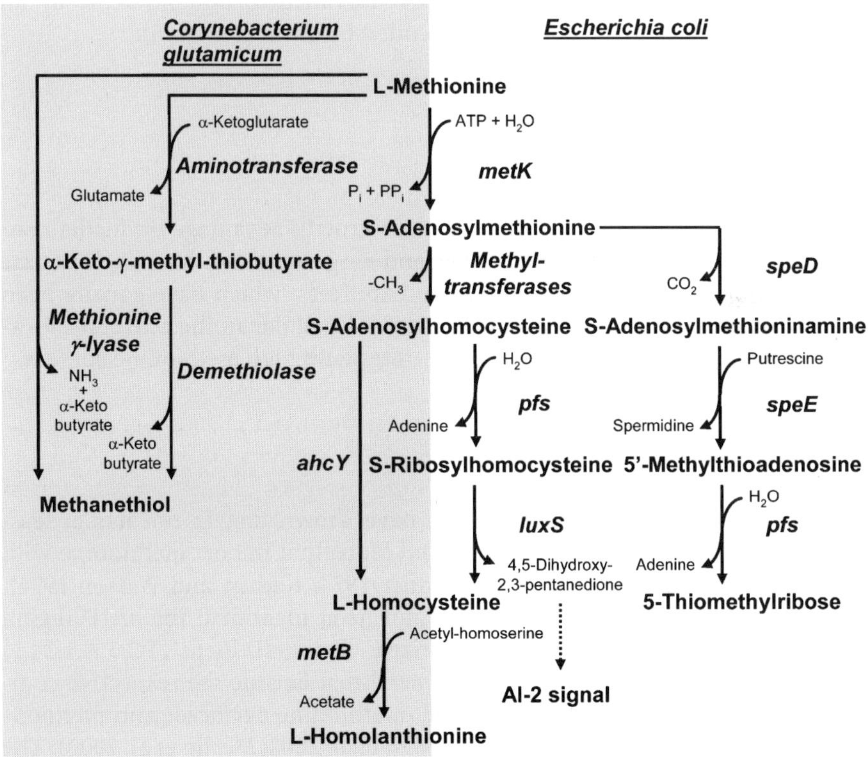

Fig. 2 Degradation of methionine in *Corynebacterium glutamicum* (*left, shaded*) and *Escherichia coli* (*right, white*). Aminotransferase and methionine-γ-lyase activities have been demonstrated, but the corresponding genes have not been identified. Demethiolase activity is speculative

cyclicize spontaneously to a mixture of AI-2 molecules that control various functions in *E. coli* (DeLisa et al. 2001).

Spermidine is required for optimal growth of *E. coli* (Chattopadhyay et al. 2003; Yoshida et al. 2004). Biosynthesis of spermidine starts with the decarboxylation of SAM to *S*-adenosylmethioninamine, catalyzed by *S*-adenosylmethionine decarboxylase, encoded by *speD* (Fig. 2). Subsequently, spermidine synthase, encoded by *speE* (Tabor et al. 1986), catalyzes the synthesis of spermidine and 5-methylthioadenosine. 5-Methylthioadenosine nucleosidase, which is identical to SAM nucleosidase (Pfs,; Greene 1996), produces adenine and 5-thiomethylribose from 5-methylthioadenosine. A recycling pathway for 5-thiomethylribose in *E. coli* is not known and the compound seems be excreted (Schroeder et al. 1973).

In *C. glutamicum*, methionine is slowly degraded to methylmercaptan, α-ketobutyrate and ammonia (Fig. 2). The reaction is most likely catalyzed either by a combination of methionine aminotransferase and α-keto-γ-methyl-thio-butyric acid demethiolase or by methionine-γ-lyase (Bonnarme et al. 2000). SAH may be directly degraded to homocysteine by an *S*-adenosylhomocysteine hydrolase, encoded by *ahcY* (Lee 2005).

4 Methionine Transport

Methionine importers give bacteria a clear growth advantage on media containing methionine. Therefore *E. coli* and *C. glutamicum* have at least one methionine import system. Methionine exporters, which have equally been found in both organisms, probably balance intracellular amino acid concentrations, when the organisms grow on substrates with high methionine content.

4.1 Methionine Import

Early biochemical and kinetic studies have shown that *E. coli* has at least two specific transport systems, MetD and MetP, that import methionine with high and low affinity, respectively (Kadner 1974; Kadner and Watson 1974). In contrast to the *metP* locus that has not been identified, the *metD* region has recently been cloned and characterized. The *metD* locus (TC 3.A.1.24.1) harbours three genes *metN*, *metI* and *metQ* that encode the respective components of an ABC transporter: ATPase, methionine permease and methionine binding protein (Paulsen et al. 1998; Gal et al. 2002; Merlin et al. 2002). The transporter encoded by the MetD locus imports both methionine isomers and their analogues (Zhang et al. 2003). The *C. glutamicum* genome harbours homologs to MetN, I and Q (Kalinowski et al. 2003), but the putative methionine importer has not been characterized.

4.2 Methionine Export

Amino acid export is an important factor in amino acid production (Aleshin et al. 1999; Eggeling & Sahm 2003; Eggeling 2005). In *E. coli* several putative methionine exporters have been found. Kutukova et al. (2005) showed that methionine production was increased in a methionine producer that overexpressed the *yeaS* (TC 2.A.76.1.5) gene. Further putative transporters have been reported in two recent patent applications. Maier et al. (2004) claims the overexpression of the gene for the inner membrane protein *yjeH* (TC 2.A.3.13.1) for methionine production. Tabolina et al. (2002) demonstrates increased methionine production upon the overexpression of the operon *ygaZH* (TC 2.A.78.1.3).

Recently, it could be shown that the corynebacterial amino acid transporter BrnFE (TC 2.A.78.1.2), which was originally shown to export branched chain amino acids (Kennerknecht et al. 2002), is the major methionine exporter (Trötschel et al. 2005). Interestingly, *ygaZ* is the closest homolog of *brnF* on the *E. coli* genome.

5 Regulation of Methionine Biosynthesis

Both *E. coli* and *C. glutamicum* regulate methionine biosynthesis on the transcriptional level. In *E. coli*, where transcriptional regulation is well understood, the repressor MetJ and the activator MetR have been shown to control methionine biosynthesis. In *C. glutamicum*, McbR has just recently been identified as the master regulator of methionine biosynthesis and other regulatory proteins are about to be characterized.

In both organisms methionine production is also regulated by feed-back inhibition of the aspartate amino acid branch and the first enzyme specific to methionine biosynthesis, homoserine acyltransferase. *C. glutamicum* controls methionine biosynthesis on the enzymatic level also by feed-back inhibiting cystathionine-γ-synthase, cystathionine-β-lyase and O-acetyl-homoserine sulfhydrylase (see above).

5.1 Transcriptional Regulation in *E. coli*

5.1.1 The Methionine Repressor MetJ

The structure and function of the methionine repressor MetJ have been intensively studied. MetJ together with its co-repressor SAM represses methionine biosynthesis by binding to variations of the consensus sequence 5′-AGACGTCT-3′, called the met-box, in the promoter regions of its target genes. Natural operators contain tandem sites of the palindrome. The three-dimensional structure of MetJ, which belongs to the ribbon-helix-helix family of DNA-binding proteins has been solved in the presence or absence of its co-repressor SAM and complexed with a perfect tandem met-box operator (Phillips 1991, 1994 and references therein). A pair of repressor dimers binds to a tandem repeat of two met-boxes. Each MetJ subunit has three alpha helices (A, B, C) and contributes one β-strand to an antiparallel β-ribbon, which is inserted into the major groove of the DNA. Formation of the repressor operator complex relies on (i) protein–DNA interactions between operator and repressor, (ii) protein–protein interactions between several bound repressor molecules and (iii) protein–small-molecule interactions between apo-repressor and co-repressor.

(i) Protein-DNA interactions involve direct and indirect read-out mechanisms. Direct hydrogen bonds are formed between Lys-23 and Thr-25 of each β-strand and four base pairs of one met-box. Indirect readout is achieved by a 23° kink at the centre of the met-box, which improves the interaction between the β-strands and the major groove. This distortion of the DNA seems to be dependent on the CG base pair in the centre of the met-box and mutation to TA entails a 48-fold loss in repressor binding affinity (Wild et al. 1996). Over-twisting of the TA base-pair step at the CTAG junction between two met-boxes allows additional contacts between the sugar phosphate backbone and the alpha helices of the repressor. Mutation of TA to AT resulted in a 76-fold reduced repressor binding affinity (Wild et al. 1996). Garvie and Phillips (2000) recently showed that the AT mutation leads to an unfavourable DNA conformation. The repressor makes sufficiently strong interactions with the sugar-phosphate backbone to accommodate some variations in the operator region. This plasticity can explain the functionality of wild-type promoters with several modifications in the consensus met-box (see below).

(ii) Protein–protein interactions between the repressor dimers seem to be essential for the stability of the protein-DNA complex and are effected via the A-helix of the repressor. In vitro, the minimal stable repression complex requires two MetJ dimers bound to two met-boxes (He et al. 2002). In contrast to wild-type MetJ, mutant MetJ Q44K can bind to an isolated met-box based on additional contacts made between the DNA phosphodiester backbone and the lysine residues. At the same time, protein-protein cooperativity is reduced in the mutant repressor. He et al. (2002) have suggested that the MetJ Q44K-DNA complex represents an intermediate on the pathway to operator saturation. MetJ may locate its operator through a sliding mechanism and achieve specificity by the formation of higher-order complexes requiring specific DNA duplex distortions (He et al. 2002; Lawrenson and Stockley 2004).

(iii) Binding of the co-repressor SAM to the apo-repressor does not significantly change its structure nor its DNA binding specificity, but increases affinity for operators by at least 100-fold (Cooper et al. 1994; Hyre and Spicer 1995). The two SAM molecules bind on the faces of the protein opposite to the one that contacts the DNA. SAH binds competitively to MetJ but is unable to act as a co-repressor. This indicates that the positive charge on the tertiary sulphur atom of SAM is important for the activation of the repressor (Phillips and Phillips 1994). SAM increases the affinity of MetJ to met-boxes probably by a long-range electrostatic mechanism, which has been called the "electric genetic switch" (Phillips and Phillips 1994; Parsons et al. 1995). Mutations decreasing the negative charge of the repressor surface have been investigated and confirm the sensitivity of the regulatory system to electrostatic effects (Lawrenson and Stockley 2004).

5.1.2 Repression of Methionine Biosynthesis through MetJ and its Co-Repressor SAM

Initially, met-boxes were identified in the regulatory regions of *metA*, *metBL*/*metJ*, *metC*, *metF*, *metE*/*metR* and *metK* (Old et al. 1991; Greene 1996 Fig. 3). Thanbichler et al. (1999) identified four met-boxes in the intergenic region of the genes *mmuP* and *yagD* that both encode enzymes involved in *S*-methylmethionine metabolism. A bioinformatic search for met-boxes identified additional binding sites in the regulatory regions of the genes *metN* encoding part of the major methionine importer (Gal et al. 2002), in genes encoding enzymes not directly related to methionine biosynthesis and in several hypothetical genes (Liu et al. 2001). On the basis of transcriptome analysis and electrophoretic mobility shift assays (EMSA), the MetJ regulon has recently been extended by the genes *folE*, *yaeS* and *cspA* (Marincs et al. 2006). For *folE* encoding GTP cyclohydrolase I, which catalyzes the first steps in THF biosynthesis, a link with methionine biosynthesis is evident. For the two other gene products, CspA, a cold shock protein, and YaeS, a subunit of undecaprenyl diphophosphate synthase, the relation with methionine biosynthesis is not clear.

Promoter regions contain from two (*metC*, *metNIQ*) up to five operators (*metBL*, *metF*) (Fig. 3). Naturally occurring operators differ from the consensus sequence to a greater extent as the number of met-boxes increases and in promoters with several met-boxes the identity increases towards the centre of the operator (Phillips and Stockley 1996). Surprisingly, only one of the *E. coli* operators (*metNIQ*) contains a perfect palindrome (Gal et al. 2002). This is in contrast to SELEX [Systematic Evolution of Ligands by EXponential enrichment (Tuerk and Gold 1990)] experiments in which 90% of the selected 20 base pair ligands contained two tandem repeats of perfect eight

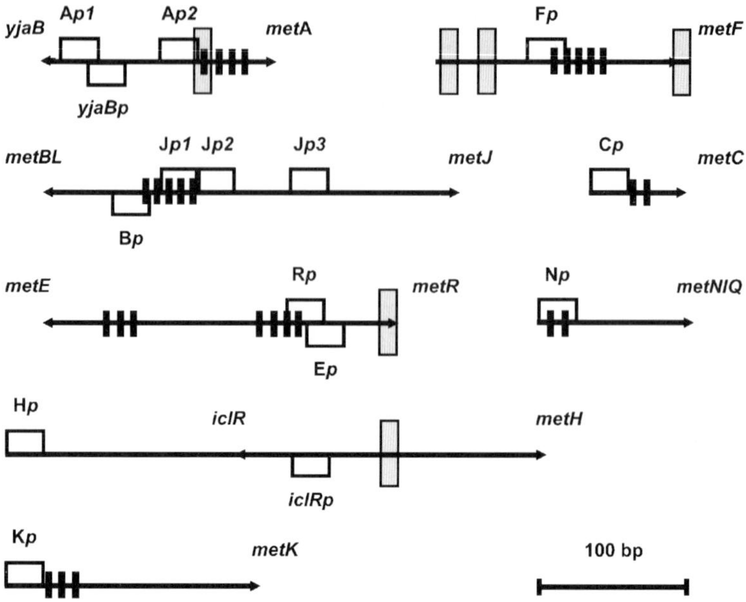

Fig. 3 Architecture of promoters regulating genes in methionine biosynthesis in *E. coli* (reproduced and adapted with permission from Marincs et al. 2006, © the Biochemical Society). The positions of promoters are indicated by open boxes. *Arrowheads* denote the positions of start codons and the direction of transcription. *Black and grey boxes* indicate positions of MetJ and MetR binding sites, respectively

base met-boxes (He et al. 1996). The quasi absence of consensus met-boxes on the *E. coli* genome may be explained by the fact that these sequences share sequence identity with Trp repressor operator sites, which would allow cross-talk between the two regulatory circuits (Phillips and Stockley 1996). Recent transcriptome analyses of *metJ* and *trp* deletion mutants confirm the absence of cross-talk between the two regulatory systems (Marincs et al. 2006).

5.1.3 Activation of Methionine Biosynthesis by MetR

The MetR protein together with its co-activator homocysteine activates the transcription of several methionine biosynthesis genes (Urbanowski et al. 1987; Maxon et al. 1989). MetR belongs to the LysR family of DNA binding proteins that generally contain a helix-turn-helix DNA binding and co-inducer recognition domain (Schell 1993). MetR has been shown to activate the transcription of *metA*, *metF*, *metE* and *metH* genes. The *metR* binding sequence consensus derived from the *E. coli* promoters is TGAannt/anntTCA (Greene 1996). Recently, it was shown that MetR contacts different regions on the α dimer subunit of RNA polymerase when activating the *metE* or *metH* gene (Fritsch et al. 2000).

5.1.4
Regulation of Specific Methionine Biosynthesis Genes

MetA

The transcription of the *metA* gene is activated by MetR and repressed by MetJ. The gene is transcribed from two promoters (Fig. 3), but only the one immediately upstream of the four *metJ* binding sites is subject to repression by MetJ (Michaeli et al. 1984). The promoter most distal to the *metA* coding sequence overlaps with the promoter of the *yjaB* gene encoding a putative acyltransferase that is transcribed in the opposite direction (Gery and Ron 1997). Derepression of *metA* transcription in a *metJ* deletion strain has been confirmed by transcriptome analysis (Marincs et al. 2006). Regulation by MetR has not been demonstrated in *E. coli*, but a MetR binding site is present. In *Salmonella typhimurium*, activation of *metA* expression by MetR has been established. The presence of homocysteine seems to reverse this effect by converting MetR to a repressor of *metA* (Mares et al. 1992). Transcriptional regulation of *metA* in response to heat shock has been reported (Biran et al. 1995).

MetBL/J

Apart from *metNIQ*, only the genes *metB* and *metL* are organized in an operon within the methionine regulon (Greene and Smith 1984). The *metBL* operon is divergently transcribed from the *metJ* gene (Kirby et al. 1986). The intergenic region of 276 bp between *metJ* and *metB* harbours three promoters of *metJ* and one promoter driving *metBL* expression (Kirby et al. 1986). MetJ has been shown to regulate the transcription of its own gene and the *metBL* operon in vivo and in vitro (Saint-Girons et al. 1984; Shoeman et al. 1985). The intergenic region comprises five met-boxes that overlap the *metB* promoter and one of the *metJ* promoters (Greene 1996). Only the *metJ* promoter closest to the coding sequence does not seem to be regulated by the methionine repressor (Kirby et al. 1986). Recent transcriptome analyses confirm that the expression of the genes *metB* and *metJ* is increased in a *metJ* deletion mutant. In contrast, *metL* expression is nearly unaffected by a *metJ* deletion, indicating that additional factors such as transcript stability may regulate *metL* expression (Marincs et al. 2006).

MetC

The *metC* operator harbours only two binding sites for the MetJ repressor and induction upon *metJ* deletion is slim (Marincs et al. 2006). In addition, *metC* expression seems to be regulated by other stimuli, e.g. it is increased in the presence of vitamin B_{12} in a *metJ* minus strain. Nevertheless, no potential MetR binding sites have been detected in the operator region (Greene 1996).

MetE/R and MetH

The genes *metE* and *metR* are divergently transcribed from overlapping promoters (Maxon et al. 1989). The 236 bp intergenic region contains a total of seven met-boxes and one MetR binding site that overlaps the coding region of *metR* (Cai et al. 1989a; Fig. 3). In agreement with its regulatory properties, the *metH* promoter region contains one MetR binding site, but no detectable met-boxes. Transcription of *metH* is initiated 324 bp upstream of the coding sequence within the *iclR* coding sequence that is divergently transcribed from the *metH* gene (Marconi et al. 1991).

Unlike the *metH* gene, both *metE* and *metR* are regulated by the methionine repressor (Kung et al. 1972; Maxon et al. 1989). Compared to other methionine biosynthesis genes *metE* and *metR* are only weakly derepressed in a *metJ* deletion strain (Marincs et al. 2006). Reducing intracellular SAM concentrations, however, increases *metE* and *metH* expression in a *metJ* deletion strain (Lamonte and Hughes 2006). Expression of both *metE* and *metH* is dependent on the presence of MetR (Urbanowski et al. 1987; Cai et al. 1989a), but differs in response to the presence of its co-activator homocysteine. Homocysteine is required for the full activation of *metE*, but represses *metH* expression (Cai et al. 1989b; Urbanowski and Stauffer 1989). Homocysteine levels and MetR are also instrumental in the response to vitamin B_{12}. Addition of vitamin B_{12} has been shown to repress the expression of the *metE* gene in the presence of functional proteins MetF and MetJ and CH_3-THF (Weissbach and Brot 1991 and references therein). The regulatory mechanism behind this phenomenon is controversial. In early work it was suggested that the MetH-B_{12} holoenzyme has a regulatory function in addition to its catalytic role (Weissbach and Brot 1991). Cai et al. (1992) argue that a 30% reduction of homocysteine levels upon the addition of vitamin B_{12} is unlikely to cause nearly complete repression of *metE*. Stauffer and co-workers have shown that in the presence of functional MetR the addition of homocysteine can override vitamin B_{12}-mediated repression of *metE*. Therefore, they argue that MetH-B_{12} may indirectly regulate *metE* expression by changing intracellular homocysteine concentrations, which results in a decrease of MetR-mediated activation (Wu et al. 1992).

MetE expression is affected by several other regulatory stimuli. Changing cultures from anaerobic to aerobic conditions results in a rapid increase in MetE abundance (Smith and Neidhardt 1983). High expression levels may compensate in part for the oxidative inactivation of MetE (see above). The transcription of the *metE* gene, unlike other methionine biosynthesis genes, decreases under *S*-limiting conditions (Gyaneshwar et al. 2005). Efficient *metE* transcription requires the presence of a functional *luxS* gene product (Wang et al. 2005).

The *metR* gene is mainly controlled by MetJ, but the *metR* gene product also represses its own transcription (Maxon et al. 1989; Cai et al. 1992). Recently, it was shown that high levels of cyclic diguanylic acid reduce *metR* expression (Mendez-Ortiz et al. 2006). Regulation of *metR* expression has also been related to biofilm formation (Herzberg et al. 2006). As shown for many other *met* genes, *metR* expression is increased upon reduction of intracellular SAM concentrations (Lamonte and Hughes 2006).

MetF

The *metF* gene is transcribed from a single promoter that is controlled by MetJ via five met-boxes (Greene 1996). In addition, the *metF* promoter region harbours three MetR binding sites, one of which overlaps the *metF* coding region (Fig. 3). Regulation of *metF* expression by MetR has not been studied in *E. coli*, but is well understood in *S. typhimurium*, which has a very similar promoter structure. MetR seems to antagonize repression exerted by MetJ (Cowan et al. 1993). In transcriptome analyses deletion of *metJ* increased *metF* expression by a factor of five (Marincs et al. 2006). MetF protein quantities increased in an *rpoS* minus mutant upon deletion of the *hfq* gene encoding an RNA chaperone (Vecerek et al. 2003).

MetK

The *metK* transcriptional start-site is located 140 bp upstream of the translational start site (Wei and Newman 2002). Three MetJ binding sites can be distinguished in the untranslated leader region (Greene 1996; Liu et al. 2001) and *metK* expression has been shown to be regulated by MetJ and SAM (Holloway et al. 1970; Lamonte and Hughes 2006; Marincs et al. 2006). In addition, the *metK* gene is also induced by the addition of leucine and repressed by the Lrp regulator (Tuan et al. 1990; Newman et al. 1998). The stability of the MetK protein seems to be regulated by ClpP-containing proteases (Weichart et al. 2003) and MetK accumulates when *E. coli* is subjected to cadmium stress (Ferianc et al. 1998).

5.2
Transcriptional Regulation in *C. glutamicum*

Detailed studies on the transcriptional regulation of methionine biosynthesis in *C. glutamicum* have just recently been initiated. Isolation of DNA-binding proteins with an affinity for the *metY* promoter region identified the regulator McbR. Proteome analysis of an *mcbR* deletion mutant showed that McbR plays a role in the regulation of sulphur assimilation and methionine biosynthesis

(Rey et al. 2003). McbR belongs to the family of TetR repressors, which generally harbour a conserved DNA binding domain at the N-terminus and a highly divergent substrate binding domain at the C-terminus (Ramos et al. 2005).

Transcriptome analyses of wildtype and *mcbR* deletion strains revealed that the transcription of 86 genes was increased and 51 genes decreased in the absence of McbR (Rey et al. 2005). Genes with reduced expression encode proteins mostly involved in transport, energy conversion and translation. Rey et al. (2005) assume that the down-regulation of these genes is most likely growth rate-dependent considering that the *mcbR* deletion mutant has a significantly reduced growth rate and that McbR is supposed to act as transcriptional repressor. Derepressed genes encode proteins that are mainly involved in transport and metabolism of sulphur-containing compounds. In addition, five genes encoding transcriptional regulators and several genes involved in oxidative stress are up-regulated. Bioinformatic analysis of the respective promoter regions identified the palindrome 5′-TAGAC-N6-GTCTA-3′ as the putative consensus binding sequence for McbR. It is present in promoter regions of 22 single genes and operons, bringing the total to 45 genes that are most likely directly regulated by McbR. These genes encode functions pertaining to five functional clusters: (i) sulphate reduction, (ii) synthesis of sulphur-containing amino acids, (iii) uptake and utilization of sulphur-containing compounds, (iv) ABC transport (two systems of unknown function) and (v) transcriptional regulation (two open reading frames). One of the transcriptional regulators, termed SsuR, has recently been shown to activate sulphonate utilization genes in *C. glutamicum* (Koch et al. 2005). EMSA confirmed the McbR binding site in the promoter regions of the *hom*, *cysI*, *cysK*, *metK* and *mcbR* genes. TetR-type repressors bind DNA in the absence of small effector molecules. Increasing the effector concentration leads to the dissociation of the repressor operator complex. On the basis of this assumption Rey et al. (2005) have assayed the affinity of McbR to its cognate binding sequence in the presence of various sulphur-containing compounds. Only SAH prevented binding of McbR to its binding sequence in EMSA and surface plasmon resonance studies. Surprisingly, the expression of *aecD* and *metF* was not affected by the *mcbR* deletion. The *mcbR* deletion strain does not overproduce methionine, but accumulates homolanthionine (Fig. 2). Homolanthionine synthesis from homocysteine and acetyl-homoserine is catalyzed by cystathionine-γ-synthase (Krömer et al. 2006a).

Recently, another open reading frame, NCgl2640, involved in the regulation of methionine biosynthesis has been identified by transposon mutagenesis followed by a screen for resistance to the methionine analogue ethionine. Deletion of NCgl2640 strongly derepresses the expression of the *cys* operon in the presence of methionine. Promoter activity of the *metY* gene is also affected in the deletion strain. The amino acid sequence of the open reading frame resembles proteins involved in glutathione synthesis or GTPase activation (Mampel et al. 2005).

Both, *E. coli* and *C. glutamicum*, regulate methionine biosynthesis with respect to the availability of SAM, but they differ in their approaches. In *E. coli*, the response to a decrease in SAM concentrations is direct and transmitted by reducing MetJ-mediated repression. In *C. glutamicum*, SAM consumption is measured by monitoring concentrations of the SAM derivative SAH. Repressor activity decreases with increasing concentrations of SAH. In *E. coli*, the in vivo role of SAH as a competitor for SAM binding to MetJ is not clearly established.

Regulatory networks for sulphur assimilation and methionine biosynthesis differ also in the two bacterial species. In *E. coli*, methionine biosynthesis is repressed by MetJ and sulphur assimilation and cysteine biosynthesis activated by CysB (Kredich 1996). In contrast to this, *C. glutamicum* seems to control its entire sulphur metabolism by one master regulator, McbR. Integrated control of sulphur assimilation and methionine biosynthesis in *C. glutamicum* is consistent with the presence of a functional sulfhydrylation pathway that directly incorporates reduced sulphur into methionine. Understanding the complete regulatory network of sulphur assimilation and methionine biosynthesis in the two organisms will require the characterization of other regulators and their interactions.

6
Construction of Methionine-Producing Strains

Until recently, the construction of methionine-producing strains largely relied on the selection of analogue resistant mutants (for a review see Kumar and Gomes 2005). The majority of mutations in these strains have been mapped to the *metA*/*metX*, *metJ* and *metK* genes and they affect MetJ-mediated repression and the feed-back regulation of MetA and MetX. Use of norleucine or ethionine as methionine analogues produces mutations in *metJ* or *metK* (Chattopadhyay et al. 1991). Feed-back resistant MetA alleles are obtained by selecting for α-methylmethionine resistance. In an early attempt, a *C. glutamicum* mutant resistant to the analogues ethionine, selenomethionine and methionine hydroxamate was isolated and produced 2 g/l methionine (Kase and Nakayama 1975). The expression of methionine biosynthesis encoding genes was derepressed in this strain. Mutant *E. coli* strains resistant to the threonine analogue α-amino-β-hydroxy valeric acid and ethionine or norleucine also produced up to 2 g/l methionine (Chattopadhyay et al. 1995). Mondal and colleagues have reported on *Brevibacterium heali* strains that produce up to 25.5 g/l methionine (Mondal et al. 1994). So far, none of the analogue resistant strains has permitted the implementation of a commercially viable methionine production process (Kumar and Gomes 2005).

With the advent of metabolic engineering, rational strain modification has become possible. In *E. coli*, these approaches have focused on removing the repression effected by MetJ and reducing feed-back inhibition of HTS. Nakamori et al. (1999) have produced 0.91 g/l methionine by replacing Ser-54 by Asn in the MetJ protein and thus derepressing the methionine regulon. Strains harbouring a combination of a *metJ* deletion with feed-back resistant HTS (triple mutant R27C, I296S, P298L) produced up to 0.24 g/l methionine, when the genes *thrBC* encoding homoserine kinase and threonine synthase were deleted and a mutant *metK* allele was added (Usuda and Kurahashi 2005). A strain harbouring a different feed-back resistant HTS (Q64E) and a *metJ* deletion produced 3.4 g/l methionine when *metL* was overexpressed (Bestel-Corre et al. 2005). Overexpression of the putative exporter *yjeH* gene in a *metJ* deletion strain with still another feed-back resistant homoserine transsuccinylase (Y294C) allowed the production of 4.8 g/l methionine (Maier et al. 2004). Möckel et al. (2002) showed that an AEC-resistant lysine producing *C. glutamicum*, in which the homologous *metX* and *metY* genes were overexpressed, produced 16 g/l methionine. In addition to high titers, process economy relies on methionine productivity and yield. A recent study that compares theoretical methionine yields indicates that the highest potential yield of 0.91 C – mol methionine/C-mol glucose can be obtained with *C. glutamicum* when methylmercaptan is used as the sulphur source (Krömer et al. 2006b). With this substrate no reduction of oxidized sulphur is required and e.g. 8 NADPH normally necessary for the reduction of sulfate to sulfite can be saved. At the same time the C5-methyl group of methionine is provided by the substrate, thus making methionine synthesis independent of sulphur reduction and one-carbon metabolism. Chateau et al. (2005) have isolated *E. coli* MetB alleles that have a 45 times reduced K_m for methylmercaptan and allow methionine production with methylmercaptan as the sulphur source. Fermentative production processes with reduced sulphur sources will rely on intelligent technical solutions that deal with the reactivity and toxicity of these compounds.

7
Conclusion

Since understanding methionine biosynthesis and its regulation is crucial for the construction of a methionine producer, research on this topic has recently received increased attention. Studies that have focused on the classic amino acid-producing bacteria *C. glutamicum* and *E. coli* revealed major differences in their methionine metabolism and corresponding regulation. Nevertheless, these advances have not yet permitted the construction of a methionine producer with properties that would make fermentative methionine production economically viable. The difficulty in constructing such a strain reflects the complexity of methionine biosynthesis that relies on sulphur assimilation, the production of aspartate and the availability of one-carbon metabolites. Integration of data from various disciplines, such as metabolomics, transcriptomics and proteomics are about to reveal the interactions between the different metabolic networks. Their understanding should permit the rational construction of methionine producers in the near future.

Acknowledgements The author thanks Tony Chevalier for help with preparing figures. Drs. Benjamin Gonzalez, Philippe Soucaille and Gwénaëlle Bestel-Corre are acknowledged for critical reading of the manuscript.

References

Aitken SM, Kim DH, Kirsch JF (2003) *Escherichia coli* cystathionine gamma-synthase does not obey ping-pong kinetics. Novel continuous assays for the elimination and substitution reactions. Biochemistry 42:11297–11306

Aleshin VV, Zakataeva NP, Livshits VA (1999) A new family of amino-acid-efflux proteins. Trends Biochem Sci 24:133–135

Awano N, Wada M, Kohdoh A, Oikawa T, Takagi H, Nakamori S (2003) Effect of cysteine desulfhydrase gene disruption on L-cysteine overproduction in *Escherichia coli*. Appl Microbiol Biotechnol 62:239–243

Bailey RB, Blomquist P, Doten R, Driggers EM, Madden KT, O'Leary J, O'Toole GA, Trueheart J, Walbridge MJ, Yorgey P (2004) Methods and compositions for amino acid production. Int Patent Application WO 2004/108894

Banerjee RV, Johnston NL, Sobeski JK, Datta P, Matthews RG (1989) Cloning and sequence analysis of the *Escherichia coli metH* gene encoding cobalamin-dependent methionine synthase and isolation of a tryptic fragment containing the cobalamin-binding domain. J Biol Chem 264:13888–13895

Belfaiza J, Parsot C, Martel A, de la Tour CB, Margarita D, Cohen GN, Saint-Girons I (1986) Evolution in biosynthetic pathways: two enzymes catalyzing consecutive steps in methionine biosynthesis originate from a common ancestor and possess a similar regulatory region. Proc Natl Acad Sci USA 83:867–871

Bendt AK, Burkovski A, Schaffer S, Bott M, Farwick M, Hermann T (2003) Towards a phosphoproteome map of *Corynebacterium glutamicum*. Proteomics 3:1637–1646

Bestel-Corre GAL, Chateau M, Figge RM, Raynaud C, Soucaille PNP (2005) Recombinant enzyme with altered feed-back sensitivity. Int Patent Application WO 2005111202

Biran D, Brot N, Weissbach H, Ron EZ (1995) Heat shock-dependent transcriptional activation of the *metA* gene of *Escherichia coli*. J Bacteriol 177:1374–1379

Biran D, Gur E, Gollan L, Ron EZ (2000) Control of methionine biosynthesis in *Escherichia coli* by proteolysis. Mol Microbiol 37:1436–1443

Bonnarme P, Psoni L, Spinnler HE (2000) Diversity of L-methionine catabolism pathways in cheese-ripening bacteria. Appl Environ Microbiol 66:5514–5517

Born TL, Blanchard JS (1999) Enzyme-catalyzed acylation of homoserine: mechanistic characterization of the *Escherichia coli metA*-encoded homoserine transsuccinylase. Biochemistry 38:14416–14423

Bray RG (2002) Amino Acids. SRI Consulting, Menlo Park, CA

Cai XY, Maxon ME, Redfield B, Glass R, Brot N, Weissbach H (1989a) Methionine synthesis in *Escherichia coli*: effect of the MetR protein on *metE* and *metH* expression. Proc Natl Acad Sci USA 86:4407–4411

Cai XY, Redfield B, Maxon M, Weissbach H, Brot N (1989b) The effect of homocysteine on MetR regulation of *metE*, *metR* and *metH* expression in vitro. Biochem Biophys Res Commun 163:79–83

Cai XY, Jakubowski H, Redfield B, Zaleski B, Brot N, Weissbach H (1992) Role of the *metF* and *metJ* genes on the vitamin B_{12} regulation of methionine gene expression: involvement of N^5-methyltetrahydrofolic acid. Biochem Biophys Res Commun 182:651–658

Chateau M, Gonzalez B, Meynial-Salles I, Soucaille PNP, Zink O (2005) Method for the production of evolved microorganisms which permit the generation or modification of metabolic pathways. European Patent Application EP 1597364

Chattopadhyay MK, Ghosh AK, Sengupta S (1991) Control of methionine biosynthesis in *Escherichia coli* K12: a closer study with analogue-resistant mutants. J Gen Microbiol 137:685–691

Chattopadhyay MK, Ghosh AK, Sengupta S, Sengupta D, Sengupta S (1995) Threonine analogue resistant mutants of *Escherichia coli* K-12. Biotechnol Lett 17:567–570

Chattopadhyay MK, Tabor CW, Tabor H (2003) Polyamines protect *Escherichia coli* cells from the toxic effect of oxygen. Proc Natl Acad Sci USA 100:2261–2265

Clausen T, Huber R, Prade L, Wahl MC, Messerschmidt A (1998) Crystal structure of *Escherichia coli* cystathionine gamma-synthase at 1.5 A resolution. EMBO J 17:6827–6838

Cooper A, McAlpine A, Stockley PG (1994) Calorimetric studies of the energetics of protein-DNA interactions in the *E. coli* methionine repressor (MetJ) system. FEBS Lett 348:41–45

Cornell KA, Riscoe MK (1998) Cloning and expression of *Escherichia coli* 5′-methylthioadenosine/*S*-adenosylhomocysteine nucleosidase: identification of the *pfs* gene product. Biochim Biophys Acta 1396:8–14

Cowan JM, Urbanowski ML, Talmi M, Stauffer GV (1993) Regulation of the *Salmonella typhimurium metF* gene by the MetR protein. J Bacteriol 175:5862–5866

DeLisa MP, Wu CF, Wang L, Valdes JJ, Bentley WE (2001) DNA microarray-based identification of genes controlled by autoinducer 2-stimulated quorum sensing in *Escherichia coli*. J Bacteriol 183:5239–5247

Dibner JJ, Knight CD (1984) Conversion of 2-hydroxy-4-(methylthio)butanoic acid to L-methionine in the chick: a stereospecific pathway. J Nutr 114:1716–1723

Dixon MM, Huang S, Matthews RG, Ludwig M (1996) The structure of the C-terminal domain of methionine synthase: presenting *S*-adenosylmethionine for reductive methylation of B_{12}. Structure 4:1263–1275

Driskell LO, Tucker AM, Winkler HH, Wood DO (2005) Rickettsial *metK*-encoded methionine adenosyltransferase expression in an *Escherichia coli metK* deletion strain. J Bacteriol 187:5719–5722

Drummond JT, Huang S, Blumenthal RM, Matthews RG (1993) Assignment of enzymatic function to specific protein regions of cobalamin-dependent methionine synthase from *Escherichia coli*. Biochemistry 32:9290–9295

Duchange N, Zakin MM, Ferrara P, Saint-Girons I, Park I, Tran SV, Py MC, Cohen GN (1983) Structure of the *metJBLF* cluster in *Escherichia coli* K12. Sequence of the *metB* structural gene and of the 5′- and 3′- flanking regions of the *metBL* operon. J Biol Chem 258:14868–14871

Duclos B, Cortay JC, Bleicher F, Ron EZ, Richaud C, Saint Girons I, Cozzone AJ (1989) Nucleotide sequence of the *metA* gene encoding homoserine trans-succinylase in *Escherichia coli*. Nucleic Acids Res 17:2856

Dwivedi CM, Ragin RC, Uren JR (1982) Cloning, purification, and characterization of beta-cystathionase from *Escherichia coli*. Biochemistry 21:3064–3069

Eggeling L (2005) Export of amino acids and other solutes. In: Eggeling L, Bott M (eds) Handbook of *Corynebacterium glutamicum*. CRC Press, Boca Raton, FL, pp 187–211

Eggeling L, Sahm H (2003) New ubiquitous translocators: amino acid export by *Corynebacterium glutamicum* and *Escherichia coli*. Arch Microbiol 180:155–160

Evans JC, Huddler DP, Hilgers MT, Romanchuk G, Matthews RG, Ludwig ML (2004) Structures of the N-terminal modules imply large domain motions during catalysis by methionine synthase. Proc Natl Acad Sci USA 101:3729–3736

Ferianc P, Farewell A, Nystrom T (1998) The cadmium-stress stimulon of *Escherichia coli* K-12. Microbiology 144:1045–1050

Flavin M, Slaughter C (1965) Synthesis of the succinic ester of homoserine, a new intermediate in the bacterial biosynthesis of methionine. Biochemistry 4:1370–1375

Flavin M, Slaughter C (1967) Enzymatic synthesis of homocysteine or methionine directly from O-succinyl-homoserine. Biochim Biophys Acta 132:400–405

Fontecave M, Atta M, Mulliez E (2004) *S*-adenosylmethionine: nothing goes to waste. Trends Biochem Sci 29:243–249

Fritsch PS, Urbanowski ML, Stauffer GV (2000) Role of the RNA polymerase alpha subunits in MetR-dependent activation of *metE* and *metH*: important residues in the C-terminal domain and orientation requirements within RNA polymerase. J Bacteriol 182:5539–5550

Gal J, Szvetnik A, Schnell R, Kalman M (2002) The *metD* D-methionine transporter locus of *Escherichia coli* is an ABC transporter gene cluster. J Bacteriol 184:4930–4932

Garvie CW, Phillips SE (2000) Direct and indirect readout in mutant Met repressor-operator complexes. Struct Fold Des 8:905–914

Gery S, Ron EZ (1997) An *Escherichia coli* gene divergently transcribed from a promoter overlapping the *metA* promoter. FEMS Microbiol Lett 154:219–222

Gophna U, Bapteste E, Doolittle WF, Biran D, Ron EZ (2005) Evolutionary plasticity of methionine biosynthesis. Gene 355:48–57

Greene RC (1996) Biosynthesis of Methionine. In: Neidhardt FC (ed) *Escherichia coli* and *Salmonella typhimurium*: Cellular and Molecular Biology, vol 1. ASM Press, Washington, DC, pp 542–560

Greene RC, Smith AA (1984) Insertion mutagenesis of the *metJBLF* gene cluster of *Escherichia coli* K-12: evidence for a *metBL* operon. J Bacteriol 159:767–769

Greene RC, Su CH, Holloway CT (1970) *S*-Adenosylmethionine synthetase deficient mutants of *Escherichia coli* K-12 with impaired control of methionine biosynthesis. Biochem Biophys Res Commun 38:1120–1126

Großmann K, Herbster K, Mack M (2000) Rapid cloning of *metK* encoding methionine adenosyltransferase from *Corynebacterium glutamicum* by screening a genomic library on a high density colony-array. FEMS Microbiol Lett 193:99–103

Guenther BD, Sheppard CA, Tran P, Rozen R, Matthews RG, Ludwig ML (1999) The structure and properties of methylenetetrahydrofolate reductase from *Escherichia coli* suggest how folate ameliorates human hyperhomocysteinemia. Nat Struct Biol 6:359–365

Gur E, Biran D, Gazit E, Ron EZ (2002) In vivo aggregation of a single enzyme limits growth of *Escherichia coli* at elevated temperatures. Mol Microbiol 46:1391–1397

Gyaneshwar P, Paliy O, McAuliffe J, Jones A, Jordan MI, Kustu S (2005) Lessons from *Escherichia coli* genes similarly regulated in response to nitrogen and sulfur limitation. Proc Natl Acad Sci USA 102:3453–3458

He YY, Stockley PG, Gold L (1996) In vitro evolution of the DNA binding sites of *Escherichia coli* methionine repressor MetJ. J Mol Biol 255:55–66

He YY, Garvie CW, Elworthy S, Manfield IW, McNally T, Lawrenson ID, Phillips SE, Stockley PG (2002) Structural and functional studies of an intermediate on the pathway to operator binding by *Escherichia coli* MetJ. J Mol Biol 320:39–53

Herzberg M, Kaye IK, Peti W, Wood TK (2006) YdgG (TqsA) controls biofilm formation in *Escherichia coli* K-12 through autoinducer 2 transport. J Bacteriol 188:587–598

Holbrook EL, Greene RC, Krueger JH (1990) Purification and properties of cystathionine gamma-synthase from overproducing strains of *Escherichia coli*. Biochemistry 29:435–442

Holloway CT, Greene RC, Su CH (1970) Regulation of *S*-adenosylmethionine synthetase in *Escherichia coli*. J Bacteriol 104:734–747

Hondorp ER, Matthews RG (2004) Oxidative stress inactivates cobalamin-independent methionine synthase (MetE) in *Escherichia coli*. PLoS Biol 2:e336

Houry WA, Frishman D, Eckerskorn C, Lottspeich F, Hartl FU (1999) Identification of in vivo substrates of the chaperonin GroEL. Nature 402:147–154

Hwang BJ, Kim Y, Kim HB, Hwang HJ, Kim JH, Lee HS (1999) Analysis of *Corynebacterium glutamicum* methionine biosynthetic pathway: isolation and analysis of *metB* encoding cystathionine gamma-synthase. Mol Cells 9:300–308

Hwang BJ, Yeom HJ, Kim Y, Lee HS (2002) *Corynebacterium glutamicum* utilizes both transsulfuration and direct sulfhydrylation pathways for methionine biosynthesis. J Bacteriol 184:1277–1286

Hyre DE, Spicer LD (1995) Thermodynamic evaluation of binding interactions in the methionine repressor system of *Escherichia coli* using isothermal titration calorimetry. Biochemistry 34:3212–3221

Kadner RJ (1974) Transport systems for L-methionine in *Escherichia coli*. J Bacteriol 117:232–241

Kadner RJ, Watson WJ (1974) Methionine transport in *Escherichia coli*: physiological and genetic evidence for two uptake systems. J Bacteriol 119:401–409

Kalinowski J, Bathe B, Bartels D, Bischoff N, Bott M, Burkovski A, Dusch N, Eggeling L, Eikmanns BJ, Gaigalat L, Goesmann A, Hartmann M, Huthmacher K, Krämer R, Linke B, McHardy AC, Meyer F, Möckel B, Pfefferle W, Pühler A, Rey DA, Rückert C, Rupp O, Sahm H, Wendisch VF, Wiegrabe I, Tauch A (2003) The complete *Corynebacterium glutamicum* ATCC 13032 genome sequence and its impact on the production of L-aspartate-derived amino acids and vitamins. J Biotechnol 104:5–25

Kase H, Nakayama K (1974) The regulation of L-methionine synthesis and the properties of cystathionine γ-synthase and β-cystathionase in *Corynebacterium glutamicum*. Agric Biol Chem 38:2235–2242

Kase H, Nakayama K (1975) L-methionine production by methionine analog-resistant mutants of *Corynebacterium glutamicum*. Agric Biol Chem 39:153–160

Kennerknecht N, Sahm H, Yen MR, Patek M, Saier MH Jr, Eggeling L (2002) Export of L-isoleucine from *Corynebacterium glutamicum*: a two-gene-encoded member of a new translocator family. J Bacteriol 184:3947–3956

Kim JW, Kim HJ, Kim Y, Lee MS, Lee HS (2001) Properties of the *Corynebacterium glutamicum metC* gene encoding cystathionine beta-lyase. Mol Cells 11:220–225

Kirby TW, Hindenach BR, Greene RC (1986) Regulation of in vivo transcription of the *Escherichia coli* K-12 *metJBLF* gene cluster. J Bacteriol 165:671–677

Koch DJ, Rückert C, Albersmeier A, Huser AT, Tauch A, Pühler A, Kalinowski J (2005) The transcriptional regulator SsuR activates expression of the *Corynebacterium glutamicum* sulphonate utilization genes in the absence of sulphate. Mol Microbiol 58:480–494

Komoto J, Yamada T, Takata Y, Markham GD, Takusagawa F (2004) Crystal structure of the *S*-adenosylmethionine synthetase ternary complex: a novel catalytic mechanism of *S*-adenosylmethionine synthesis from ATP and Met. Biochemistry 43:1821–1831

Kraus J, Soll D, Low KB (1979) Glutamyl-gamma-methyl ester acts as a methionine analogue in *Escherichia coli*: analogue resistant mutants map at the *metJ* and *metK* loci. Genet Res 33:49–55

Kredich NM (1996) Biosynthesis of Cysteine. In: Neidhardt FC (ed) *Escherichia coli* and *Salmonella typhimurium*: Cellular and Molecular Biology, vol 1. ASM Press, Washington, DC, pp 514–527

Krömer JO, Heinzle E, Schroder H, Wittmann C (2006a) Accumulation of homolanthionine and activation of a novel pathway for isoleucine biosynthesis in *Corynebacterium glutamicum* McbR deletion strains. J Bacteriol 188:609–618

Krömer JO, Wittmann C, Schroder H, Heinzle E (2006b) Metabolic pathway analysis for rational design of L-methionine production by *Escherichia coli* and *Corynebacterium glutamicum*. Metab Eng 8:353–369

Kumar D, Gomes J (2005) Methionine production by fermentation. Biotechnol Adv 23:41–61

Kung HF, Spears C, Greene RC, Weissbach H (1972) Regulation of the terminal reactions in methionine biosynthesis by vitamin B_{12} and methionine. Arch Biochem Biophys 150:23–31

Kutukova EA, Livshits VA, Altman IP, Ptitsyn LR, Zyiatdinov MH, Tokmakova IL, Zakataeva NP (2005) The *yeaS* (*leuE*) gene of *Escherichia coli* encodes an exporter of leucine, and the Lrp protein regulates its expression. FEBS Lett 579:4629–4634

Lamonte BL, Hughes JA (2006) In vivo hydrolysis of *S*-adenosylmethionine induces the *met* regulon of *Escherichia coli*. Microbiology 152:1451–1459

Lawrenson ID, Stockley PG (2004) Kinetic analysis of operator binding by the *E. coli* methionine repressor highlights the role(s) of electrostatic interactions. FEBS Lett 564:136–142

Lee H-S (2005) Sulfur Metabolism and its regulation. In: Eggeling L, Bott M (eds) Handbook of *Corynebacterium glutamicum*. CRC Press, Boca Raton, FL, pp 351–376

Lee H-S, Hwang BJ (2003) Methionine biosynthesis and its regulation in *Corynebacterium glutamicum*: parallel pathways of transsulfuration and direct sulfhydrylation. Appl Microbiol Biotechnol 62:459–467

Lee LW, Ravel JM, Shive W (1966) Multimetabolite control of a biosynthetic pathway by sequential metabolites. J Biol Chem 241:5479–5480

Leichert LI, Jakob U (2004) Protein thiol modifications visualized in vivo. PLoS Biol 2:e333

Liu R, Blackwell TW, States DJ (2001) Conformational model for binding site recognition by the *E.coli* MetJ transcription factor. Bioinformatics 17:622–633

Ludwig ML, Matthews RG (1997) Structure-based perspectives on B_{12}-dependent enzymes. Annu Rev Biochem 66:269–313

Maier T, Winterhalter C, Pfeiffer K (2004) Verfahren zur fermentativen Herstellung von L-Methionin. European Patent Application EP 1445310

Mampel J, Schroder H, Haefner S, Sauer U (2005) Single-gene knockout of a novel regulatory element confers ethionine resistance and elevates methionine production in *Corynebacterium glutamicum*. Appl Microbiol Biotechnol 68:228–236

Marconi R, Wigboldus J, Weissbach H, Brot N (1991) Transcriptional start and MetR binding sites on the *Escherichia coli metH* gene. Biochem Biophys Res Commun 175:1057–1063

Mares R, Urbanowski ML, Stauffer GV (1992) Regulation of the *Salmonella typhimurium metA* gene by the MetR protein and homocysteine. J Bacteriol 174:390–397

Marincs F, Manfield IW, Stead JA, McDowall KJ, Stockley PG (2006) Transcript analysis reveals an extended regulon and the importance of protein-protein co-operativity for the *Escherichia coli* methionine repressor. Biochem J 396:227–234

Markham GD, DeParasis J, Gatmaitan J (1984) The sequence of *metK*, the structural gene for *S*-adenosylmethionine synthetase in *Escherichia coli*. J Biol Chem 259:14505–14507

Matthews RG (2001) Cobalamin-dependent methyltransferases. Acc Chem Res 34:681–689

Maxon ME, Redfield B, Cai XY, Shoeman R, Fujita K, Fisher W, Stauffer G, Weissbach H, Brot N (1989) Regulation of methionine synthesis in *Escherichia coli*: effect of the MetR protein on the expression of the *metE* and *metR* genes. Proc Natl Acad Sci USA 86:85–89

Maxon ME, Wigboldus J, Brot N, Weissbach H (1990) Structure-function studies on *Escherichia coli* MetR protein, a putative prokaryotic leucine zipper protein. Proc Natl Acad Sci USA 87:7076–7079

Mendez-Ortiz MM, Hyodo M, Hayakawa Y, Membrillo-Hernandez J (2006) Genome-wide transcriptional profile of *Escherichia coli* in response to high levels of the second messenger 3′,5′-cyclic diguanylic acid. J Biol Chem 281:8090–8099

Merlin C, Gardiner G, Durand S, Masters M (2002) The *Escherichia coli metD* locus encodes an ABC transporter which includes Abc (MetN), YaeE (MetI), and YaeC (MetQ). J Bacteriol 184:5513–5517

Michaeli S, Mevarech M, Ron EZ (1984) Regulatory region of the *metA* gene of *Escherichia coli* K-12. J Bacteriol 160:1158–1162

Miyajima R, Shiio I (1973) Regulation of aspartate family amino acid biosynthesis in *Brevibacterium flavum*. VII. Properities of homoserine O-transacetylase. J Biochem (Tokyo) 73:1061–1068

Möckel B, Pfefferle W, Huthmacher K, Rückert C, Kalinowski J, Pühler A, Binder M, Greissinger D, Thierbach G (2002) Nucleotide sequences which code for the *metY* gene. International Patent Application WO 02/18613

Mogk A, Tomoyasu T, Goloubinoff P, Rudiger S, Roder D, Langen H, Bukau B (1999) Identification of thermolabile *Escherichia coli* proteins: prevention and reversion of aggregation by DnaK and ClpB. EMBO J 18:6934–6949

Mondal S, Das YB, Chatterjee SP (1994) L-Methionine production by double auxotrophic mutants of an ethionine resistant strain of *Brevibacterium heali*. Acta Biotechnol 14:61–66

Nakamori S, Kobayashi S, Nishimura T, Takagi H (1999) Mechanism of L-methionine overproduction by *Escherichia coli*: the replacement of Ser-54 by Asn in the MetJ protein causes the derepression of L-methionine biosynthetic enzymes. Appl Microbiol Biotechnol 52:179–185

Newman EB, Budman LI, Chan EC, Greene RC, Lin RT, Woldringh CL, D'Ari R (1998) Lack of *S*-adenosylmethionine results in a cell division defect in *Escherichia coli*. J Bacteriol 180:3614–3619

Old IG, Phillips SE, Stockley PG, Saint Girons I (1991) Regulation of methionine biosynthesis in the *Enterobacteriaceae*. Prog Biophys Mol Biol 56:145–185

Ozaki H, Shiio I (1982) Methionine biosynthesis in *Brevibacterium flavum*: properties and essential role of O-acetylhomoserine sulfhydrylase. J Biochem (Tokyo) 91:1163–1171

Park SD, Lee JY, Kim Y, Kim JH, Lee HS (1998) Isolation and analysis of *metA*, a methionine biosynthetic gene encoding homoserine acetyltransferase in *Corynebacterium glutamicum*. Mol Cells 8:286–294

Parsons ID, Persson B, Mekhalfia A, Blackburn GM, Stockley PG (1995) Probing the molecular mechanism of action of co-repressor in the *E. coli* methionine repressor-operator complex using surface plasmon resonance (SPR). Nucleic Acids Res 23:211–216

Patte J-C (1996) Biosynthesis of Threonine and Lysine. In: Neidhardt FC (ed) *Escherichia coli* and *Salmonella typhimurium*: Cellular and Molecular Biology, vol 1. ASM Press, Washington, DC, pp 528–541

Paulsen IT, Sliwinski MK, Saier MH Jr (1998) Microbial genome analyses: global comparisons of transport capabilities based on phylogenies, bioenergetics and substrate specificities. J Mol Biol 277:573–592

Pejchal R, Ludwig ML (2005) Cobalamin-Independent Methionine Synthase (MetE): A Face-to-Face Double Barrel That Evolved by Gene Duplication. PLoS Biol 3:e31

Peoples OP, Liebl W, Bodis M, Maeng PJ, Follettie MT, Archer JA, Sinskey AJ (1988) Nucleotide sequence and fine structural analysis of the *Corynebacterium glutamicum hom-thrB* operon. Mol Microbiol 2:63–72

Phillips K, Phillips SE (1994) Electrostatic activation of *Escherichia coli* methionine repressor. Structure 2:309–316

Phillips SE (1991) Specific beta-sheet interactions. Curr Opin Struct Biol 1:89–98

Phillips SE (1994) The beta-ribbon DNA recognition motif. Annu Rev Biophys Biomol Struct 23:671–701

Phillips SE, Stockley PG (1996) Structure and function of *Escherichia coli met* repressor: similarities and contrasts with *trp* repressor. Philos Trans R Soc Lond B Biol Sci 351:527–535

Ramos JL, Martinez-Bueno M, Molina-Henares AJ, Teran W, Watanabe K, Zhang X, Gallegos MT, Brennan R, Tobes R (2005) The TetR family of transcriptional repressors. Microbiol Mol Biol Rev 69:326–356

Rey DA, Nentwich SS, Koch DJ, Rückert C, Pühler A, Tauch A, Kalinowski J (2005) The McbR repressor modulated by the effector substance *S*-adenosylhomocysteine controls directly the transcription of a regulon involved in sulphur metabolism of *Corynebacterium glutamicum* ATCC 13032. Mol Microbiol 56:871–887

Rey DA, Pühler A, Kalinowski J (2003) The putative transcriptional repressor McbR, member of the TetR-family, is involved in the regulation of the metabolic network directing the synthesis of sulfur containing amino acids in *Corynebacterium glutamicum.* J Biotechnol 103:51–65

Rosen R, Becher D, Buttner K, Biran D, Hecker M, Ron EZ (2004a) Highly phosphorylated bacterial proteins. Proteomics 4:3068–3077

Rosen R, Becher D, Buttner K, Biran D, Hecker M, Ron EZ (2004b) Probing the active site of homoserine trans-succinylase. FEBS Lett 577:386–392

Rossol I, Pühler A (1992) The *Corynebacterium glutamicum aecD* gene encodes a C-S lyase with alpha, beta-elimination activity that degrades aminoethylcysteine. J Bacteriol 174:2968–2977

Rückert C, Pühler A, Kalinowski J (2003) Genome-wide analysis of the L-methionine biosynthetic pathway in *Corynebacterium glutamicum* by targeted gene deletion and homologous complementation. J Biotechnol 104:213–228

Saint-Girons I, Duchange N, Zakin MM, Park I, Margarita D, Ferrara P, Cohen GN (1983) Nucleotide sequence of *metF*, the *E. coli* structural gene for 5–10 methylene tetrahydrofolate reductase and of its control region. Nucleic Acids Res 11:6723–6732

Saint-Girons I, Duchange N, Cohen GN, Zakin MM (1984) Structure and autoregulation of the *metJ* regulatory gene in *Escherichia coli.* J Biol Chem 259:14282–14285

Satishchandran C, Taylor JC, Markham GD (1990) Novel *Escherichia coli* K-12 mutants impaired in *S*-adenosylmethionine synthesis. J Bacteriol 172:4489–4496

Schaffer S, Weil B, Nguyen VD, Dongmann G, Gunther K, Nickolaus M, Hermann T, Bott M (2001) A high-resolution reference map for cytoplasmic and membrane-associated proteins of *Corynebacterium glutamicum*. Electrophoresis 22:4404–4422

Schalk-Hihi C, Markham GD (1999) The conformations of a substrate and a product bound to the active site of *S*-adenosylmethionine synthetase. Biochemistry 38:2542–2550

Schauder S, Shokat K, Surette MG, Bassler BL (2001) The LuxS family of bacterial autoinducers: biosynthesis of a novel quorum-sensing signal molecule. Mol Microbiol 41:463–476

Schell MA (1993) Molecular biology of the LysR family of transcriptional regulators. Annu Rev Microbiol 47:597–626

Schroeder HR, Barnes CJ, Bohinski RC, Mallette MF (1973) Biological production of 5-methylthioribose. Can J Microbiol 19:1347–1354

Sheppard CA, Trimmer EE, Matthews RG (1999) Purification and properties of NADH-dependent 5, 10-methylenetetrahydrofolate reductase (MetF) from *Escherichia coli*. J Bacteriol 181:718–725

Shiio I, Ozaki H (1981) Feedback inhibition by methionine and *S*-adenosylmethionine, and desensitization of homoserine O-acetyltransferase in *Brevibacterium flavum*. J Biochem (Tokyo) 89:1493–1500

Shoeman R, Coleman T, Redfield B, Greene RC, Smith AA, Saint-Girons I, Brot N, Weissbach H (1985) Regulation of methionine synthesis in *Escherichia coli*: effect of *metJ* gene product and *S*-adenosylmethionine on the in vitro expression of the *metB*, *metL* and *metJ* genes. Biochem Biophys Res Commun 133:731–739

Smith MW, Neidhardt FC (1983) Proteins induced by aerobiosis in *Escherichia coli*. J Bacteriol 154:344–350

Tabe L, Higgins TJV (1998) Engineering plant protein composition for improved nutrition. Trends Plant Sci 3:282–286

Tabolina EA, Rybak KV, Khourges EM, Voroshilova EB, Gusyatiner MM (2002) Method for producing L-amino acid using bacteria belonging to the genus *Escherichia*. European Patent Application EP 1239041

Tabor CW, Tabor H (1985) Polyamines in microorganisms. Microbiol Rev 49:81–99

Tabor CW, Tabor H, Xie QW (1986) Spermidine synthase of *Escherichia coli*: localization of the *speE* gene. Proc Natl Acad Sci USA 83:6040–6044

Takusagawa F, Kamitori S, Markham GD (1996a) Structure and function of *S*-adenosylmethionine synthetase: crystal structures of *S*-adenosylmethionine synthetase with ADP, BrADP, and PP_i at 28 angstroms resolution. Biochemistry 35:2586–2596

Takusagawa F, Kamitori S, Misaki S, Markham GD (1996b) Crystal structure of *S*-adenosylmethionine synthetase. J Biol Chem 271:136–147

Taurog RE, Matthews RG (2006) Activation of methyltetrahydrofolate by cobalamin-independent methionine synthase. Biochemistry 45:5092–5102

Taylor JC, Markham GD (2000) The bifunctional active site of *S*-adenosylmethionine synthetase. Roles of the basic residues. J Biol Chem 275:4060–4065

Thanbichler M, Neuhierl B, Bock A (1999) *S*-methylmethionine metabolism in *Escherichia coli*. J Bacteriol 181:662–665

Trötschel C, Deutenberg D, Bathe B, Burkovski A, Krämer R (2005) Characterization of methionine export in *Corynebacterium glutamicum*. J Bacteriol 187:3786–3794

Tuan LR, D'Ari R, Newman EB (1990) The leucine regulon of *Escherichia coli* K-12: a mutation in *rblA* alters expression of L-leucine-dependent metabolic operons. J Bacteriol 172:4529–4535

Tuerk C, Gold L (1990) Systematic evolution of ligands by exponential enrichment: RNA ligands to bacteriophage T4 DNA polymerase. Science 249:505–510

Urbanowski ML, Stauffer GV (1989) Genetic and biochemical analysis of the MetR activator-binding site in the *metE metR* control region of *Salmonella typhimurium*. J Bacteriol 171:5620–5629

Urbanowski ML, Stauffer LT, Plamann LS, Stauffer GV (1987) A new methionine locus, *metR*, that encodes a trans-acting protein required for activation of *metE* and *metH* in *Escherichia coli* and *Salmonella typhimurium*. J Bacteriol 169:1391–1397

Usuda Y, Kurahashi O (2005) Effects of deregulation of methionine biosynthesis on methionine excretion in *Escherichia coli*. Appl Environ Microbiol 71:3228–3234

Vecerek B, Moll I, Afonyushkin T, Kaberdin V, Blasi U (2003) Interaction of the RNA chaperone Hfq with *m*RNAs: direct and indirect roles of Hfq in iron metabolism of *Escherichia coli*. Mol Microbiol 50:897–909

Wada M, Awano N, Haisa K, Takagi H, Nakamori S (2002) Purification, characterization and identification of cysteine desulfhydrase of *Corynebacterium glutamicum* and its relationship to cysteine production. FEMS Microbiol Lett 217:103–107

Wang L, Li J, March JC, Valdes JJ, Bentley WE (2005) *luxS*-dependent gene regulation in *Escherichia coli* K-12 revealed by genomic expression profiling. J Bacteriol 187:8350–8360

Wei Y, Newman EB (2002) Studies on the role of the *metK* gene product of *Escherichia coli* K-12. Mol Microbiol 43:1651–1656

Weichart D, Querfurth N, Dreger M, Hengge-Aronis R (2003) Global role for ClpP-containing proteases in stationary-phase adaptation of *Escherichia coli*. J Bacteriol 185:115–125

Weissbach H, Brot N (1991) Regulation of methionine synthesis in *Escherichia coli*. Mol Microbiol 5:1593–1597

Whitfield CD, Steers EJ Jr, Weisbach H (1970) Purification and properties of 5-methyltetrahydropteroyltriglutamate-homocysteine transmethylase. J Biol Chem 245:390–401

Wild CM, McNally T, Phillips SE, Stockley PG (1996) Effects of systematic variation of the minimal *Escherichia coli met* consensus operator site: *in vivo* and *in vitro met* repressor binding. Mol Microbiol 21:1125–1135

Wu WF, Urbanowski ML, Stauffer GV (1992) Role of the MetR regulatory system in vitamin B_{12}-mediated repression of the *Salmonella typhimurium metE* gene. J Bacteriol 174:4833–4837

Xavier KB, Bassler BL (2003) LuxS quorum sensing: more than just a numbers game. Curr Opin Microbiol 6:191–197

Yoshida M, Kashiwagi K, Shigemasa A, Taniguchi S, Yamamoto K, Makinoshima H, Ishihama A, Igarashi K (2004) A unifying model for the role of polyamines in bacterial cell growth, the polyamine modulon. J Biol Chem 279:46008–46013

Zakin MM, Duchange P, Ferrara P, Cohen GN (1983) Nucleotide sequence of the *metL* gene of *Escherichia coli*. Its product, the bifucntional aspartokinase II-homoserine dehydrogenase II and the bifunctional product of the *thrA* gene, aspartokinase L-homoserine dehydrogenase I derive from a common ancestor. J Biol Chem 258: 3028–3031

Zhang Z, Feige JN, Chang AB, Anderson IJ, Brodianski VM, Vitreschak AG, Gelfand MS, Saier MH Jr (2003) A transporter of *Escherichia coli* specific for L- and D-methionine is the prototype for a new family within the ABC superfamily. Arch Microbiol 180:88–100

Microbiol Monogr (5)
V. F. Wendisch: Amino Acid Biosynthesis
DOI 10.1007/7171_2006_060/Published online: 9 December 2006

Cysteine Metabolism and Its Regulation in Bacteria

Eric Guédon[1] · Isabelle Martin-Verstraete[2,3] (✉)

[1]Laboratoire de Génétique Microbienne, UR895, INRA, Domaine de Vilvert, 78352 Jouy en Josas, France

[2]Unité de Génétique des Génomes Bactériens, Institut Pasteur, URA CNRS 2171, 28, rue du Docteur Roux, 75724 Paris Cedex 15, France

[3]France and UFR de Biochimie, Université Paris 7-Denis Diderot, 2 place Jussieu, 75251 Paris, France

iverstra@pasteur.fr

Abstract Sulfur is necessary for the synthesis of cysteine. Microorganisms can use sulfate, thiosulfate or sulfonates as sole sulfur sources. These compounds are taken up by specific transporters followed by the conversion of sulfate or sulfonates into sulfide in 2 to 4 steps. The biosynthesis of cysteine from serine in bacteria is carried out by a two-step pathway beginning with the *O*-acetylation of serine, followed by *β*-replacement of the acetyl group by sulfide or thiosulfate. Some microorganisms can also use methionine or cysteine-derived compounds such as glutathione as sole sulfur source. Glutathione is degraded to liberate cysteine, whereas methionine is converted into cysteine via the reverse transsulfuration pathway or via methanethiol formation. Cysteine is also taken up directly from the environment by ABC transporters or symporters mainly as cystine, the disulfide-linked cysteine dimer. Several mechanisms are involved in the control of the intracellular concentration of cysteine, which is a highly reactive compound due to its –SH group. This amino acid is degraded mainly by cysteine desulfhydrases or is excreted by exporters. A large variety of molecular mechanisms participate in fine-tuning the regulation of cysteine metabolism: positive regulation by LysR-type regulators, negative control by repressors of the Rrf2 or TetR family and regulation by premature termination of transcription. In *Escherichia coli* and *Bacillus subtilis*, a global regulator, CysB and CymR, respectively, controls cysteine synthesis and transport in response to *O*-acetylserine or its derivative *N*-acetyl-serine availability. In *Lactococcus lactis* and *Corynebacterium glutamicum*, a unique regulator modulates the methionine and cysteine metabolisms. Cysteine or derivative compounds are biotechnically interesting. Fermentation processes with *E. coli* or *C. glutamicum* involving mutants insensitive to feedback inhibition by cysteine and also strains overproducing cysteine exporters or inactivated for cysteine degradative enzymes are currently being developed.

1 Introduction

Sulfur is a crucial atom in cysteine and methionine, as well as in several coenzymes and cofactors. Cysteine biosynthesis is the primary pathway for incorporating sulfur into cellular components. Cysteine is a precursor of methionine and also thiamine, biotin, lipoic acid, coenzyme A and coenzyme M. Cysteine also plays an important role in the biogenesis of [Fe – S] clusters, is found in the catalytic site of several enzymes and aids protein folding and assembly by forming disulfide bonds. Finally, cysteine-derived proteins such as thioredoxin or glutathione play a central role in protecting cells against oxidative stress. As a result of its crucial role in cellular physiology and the reactivity of the SH group, cysteine metabolism is tightly controlled in response to environmental variations. Cysteine is also biotechnologically interesting as a nutritional supplement, a pharmaceutical (antidote) or a precursor for drugs. Two major cysteine biosynthetic pathways are present in microorganisms: the thiolation pathway, which requires sulfide and the reverse transsulfuration pathway, which converts homocysteine to cysteine via a cystathionine intermediate. Homocysteine is synthesized from methionine, while sulfide is provided mostly from the reduction of sulfate. Finally, thiosulfate or glutathione can also be used as cysteine precursors.

2 Assimilation of Sulfate and Sulfonates

Several microorganisms can use inorganic sulfate, an abundant source of utilizable sulfur in the aerobic biosphere, or sulfonates, which are widespread in nature, for their growth. These compounds are taken up and converted into sulfide.

2.1 Transport of Sulfate

Three different sulfate-transport systems have been described. These belong either to the ATP-binding cassette (ABC) superfamily or to the major facilitator superfamily (MFS). The SulT family (sulfate uptake transporter; TC 3.A.1.6) of ABC-transporters has mostly been studied in *Escherichia coli* and *Salmonella typhimurium*. This sulfate transporter consists of the periplasmic sulfate binding protein, Sbp, the CysT and CysW permeases and an ATP-binding subunit, CysA (Kertesz 2001). This system can also take up other compounds such as sulfite and thiosulfate (Kredich 1996). A second periplasmic binding protein, CysP, which is more specific for thiosulfate has also been characterized. The *cysPTWA* genes form an operon, whereas the *sbp* gene is located elsewhere on the chromosome. Strains with single mutations in *sbp* or *cysP* can both grow with sulfate or thiosulfate as the sulfur source, and thus the substrate specificity of these two proteins overlaps (Kredich 1996). By contrast, *cysA*, *cysT* or the *sbp cysP* double mutant are cysteine auxotrophs indicating that CysATW with Sbp or CysP is the only sulfate transporter in *E. coli* (Sirko et al. 1995). The *sbp-cysPTWA* cluster is also found in a number of bacterial genomes (Kertesz 2001). In *Pseudomonas aeruginosa*, a similar system to that described in *E. coli* and *S. typhimurium* is derepressed during sulfate starvation (Kertesz 2001; Sect. 6.1). This system is the only sulfate transporter in *Mycobacterium tuberculosis* (Wooff et al. 2002). The second sulfate-transport system, CysP, which was characterized in *Bacillus subtilis* (Mansilla and de Mendoza 2000) is a symporter belonging to the Pit family (inorganic phosphate transporter; TC 2.A.20) of MFS. The *cysP* gene in *B. subtilis* that codes this sulfate permease forms an operon with genes of the sulfate assimilation pathway (Mansilla and de Mendoza 1997). The introduction of the *cysP* gene from *B. subtilis* into *E. coli* mutants deficient for sulfate transport restores sulfate uptake and growth in the presence of sulfate. The possible existence of other sulfate/thiosulfate transporters in *B. subtilis* has not yet been investigated. The third transporter belongs to the SulP family (sulfate permease; TC 2.A.53). This member of the MFS family consists of one subunit with 10 to 13 transmembrane domains. This type of permease, which acts as a proton/sulfate symporter, has been characterized in eukaryotes. Several bacterial genomes encode proteins similar to SulP (Kertesz 2001). However, in *E. coli* and *M. tuberculosis* at least, the ABC-transporters are the only sulfate uptake systems suggesting that SulP-like proteins have another function in these microorganisms (Sirko et al. 1995; Wooff et al. 2002).

2.2 Reduction of Sulfate to Sulfide

The pathway that assimilates sulfur from sulfate ions has been well studied in *E. coli* (Kredich 1996). ATP sulfurylase (sulfate adenylyltransferase, EC 2.7.7.4) catalyzes the first step in inorganic sulfate assimilation (Fig. 1). This enzyme activates sulfate via an ATP-dependent reaction leading to the formation of adenosine 5′-phosphosulfate (APS) and pyrophosphate. In *E. coli* and many other bacteria, this activity requires two proteins, the catalytic subunit encoded by the *cysD* gene and the GTP regulatory subunit encoded by the *cysN* gene. By contrast, in eukaryotes, archae and some bacteria such as *B. subtilis*, these two subunits are fused (Mansilla and de Mendoza 1997). An ATP-dependent phosphorylation of APS by APS kinase (EC 2.7.1.25), the *cysC* gene product, further leads to the formation of 3′-phosphoadenosine 5′-phosphosulfate (PAPS) (Fig. 1). In some organisms including *Mycobacterium* and *Pseudomonas*, the APS kinase CysC is fused to CysN (Pinto et al. 2004). The third step of the pathway is the release of sulfite and PAP (AMP-3′-P) from PAPS by PAPS sulfotransferase (PAPS reductase, EC 1.8.4.8), which is encoded by *cysH* (Fig. 1). *E. coli* CysH uses reduced thioredoxin or glutathione as an electron donor (Lillig et al. 1999). In plants, green algae and several bacteria, APS is directly reduced into sulfite by APS reductase (EC 1.8.4.9) (Mendoza-Cozatl et al. 2005) (Fig. 1). Enzymes having PAPS and APS reductase activity in vitro are found in *P. aeruginosa*, *M. tuberculosis* and *B. subtilis*. The corresponding genes are able to complement *E. coli cysH* and/or *cysC* mutants, indicating that these enzymes can bypass PAPS formation (Fig. 1) (Berndt et al. 2004; Bick et al. 2000; Williams et al. 2002). The final step is the reduction of sulfite into sulfide by NADPH sulfite reductase (EC 1.8.1.2) (Fig. 1). In *E. coli*, this complex enzyme is composed of two different polypeptides, α (CysJ) and β (CysI), and has an $\alpha_8\beta_4$ structure. The flavoprotein, α_8, contains four FAD and four FMN cofactors and has NADPH-cytochrome *c* reductase activity. CysI is a hemoprotein that contains a $[4Fe-4S]$ cluster and a siroheme. The flavoprotein accepts electrons from NADPH, and transfers them to CysI, which then reduces sulfite into sulfide (Kredich 1996). The sulfite reductase is present in *B. subtilis* (van der Ploeg et al. 2001a). Finally, the *cysG* gene encodes an uroporphyrinogen methyltransferase (EC 2.1.1.107), which catalyzes siroheme synthesis from uroporphyrinogen III (Kredich 1996). In *B. subtilis*, uroporphyrinogen methyltransferase activity is encoded by *ylnD* and *ylnF* corresponding to the 3′ and 5′ part of *cysG*, respectively (Johansson and Hederstedt 1999).

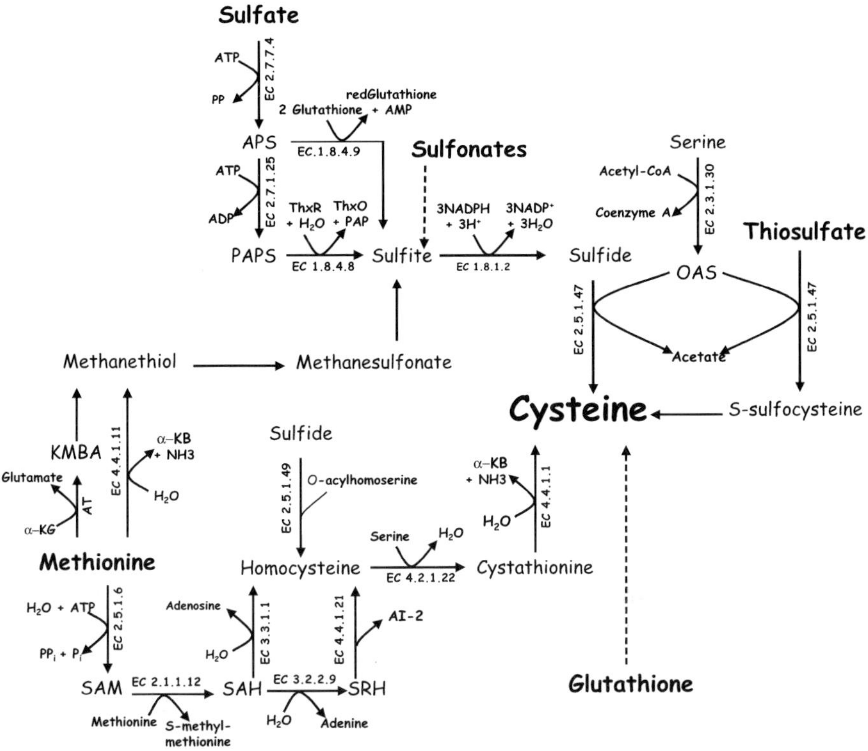

Fig. 1 Cysteine biosynthetic pathways. APS, Adenylyl sulfate; ATP, adenosine triphosphate; Pi, phosphate; PPi, diphosphate; PAPS, 3′-phosphoadenylyl sulfate; ThxR, reduced thioredoxin; ThxR, oxydated thioredoxin; PAP, adenosine 3′,5′-bisphosphate; AMP, adenosine monophosphate; redGlutathione, reduced Glutathione; OAS, *O*-acetylserine; SAM, S-adenosylmethionine; SAH, S-adenosylhomocysteine; SRH, S-ribosylhomocysteine; AI-2, Autoinducer 2; KMBA, α-keto-γ-methyl-thiobutyric acid; α-KB, α-ketobutyrate; AT, aminotransferase; α-KG, α-ketoglutarate. From sulfate : EC 2.7.7.4, sulfate adenylyltransferase; EC 2.7.1.25, adenylyl-sulfate kinase; EC 1.8.4.8, phosphoadenylyl-sulfate reductase; EC 1.8.4.9, APS reductase; EC 1.8.1.2, sulfite reductase; EC 2.3.1.30, serine *O*-acetyltransferase; EC 2.5.1.47, cysteine synthase. From methionine: EC 2.5.1.6, methionine adenosyltransferase; EC 2.1.1.12, methionine S-methyltransferase; EC 3.3.1.1, adenosylhomocysteinase; EC 3.2.2.9, adenosylhomocysteine nucleosidase; EC 4.4.1.21, S-ribosylhomocysteine lyase; EC 4.2.1.22, Cystathionine β-synthase; EC 4.4.1.1, cystathionine γ-lyase; EC 4.4.1.11, methionine γ-lyase; EC 2.5.1.49, *O*-acylhomoserine sulfhydrylase

In *E. coli*, genes coding for enzymes that reduce sulfate to sulfide are organized in three transcriptional units: *cysDNC* for the reduction of sulfate to PAPS and *cysJIH* and *cysG* for the reduction of PAPS to sulfide (Kredich 1996). A large *cysHPynlBCDEF* operon is present in *B. subtilis*, whereas the *cysJI* operon coding for the sulfide reductase is found elsewhere (Guillouard et al. 2002; Mansilla et al. 2000). The sulfate assimilation pathway is missing in several microorganisms including *Staphylococcus aureus*, Streptococci and Listeria.

2.3 Assimilation of Sulfonates

Several microorganisms can use sulfonates as their sole sulfur source under sulfate or sulfur-limiting conditions. Aliphatic sulfonates are first taken up using ABC transporters (SsuABC or TauABC). In *E. coli*, taurine is then converted into sulfite by TauD, an α-ketoglutarate-dependent dioxygenase (EC 1.14.11.17). Other aliphatic sulfonates are mostly broken down by SsuD, a monooxygenase (EC 1.14.14.5) that requires FMNH2, which is delivered by an FMN reductase, SsuE (van der Ploeg et al. 2001b). Monooxygenases are also involved in the degradation of aromatic sulfonates by *Pseudomonas* species (Kertesz and Wietek 2001).

3 Cysteine Biosynthetic Pathways

Cysteine is produced from the carbon backbone of serine in different ways depending on the organism that incorporates sulfur. The thiolation pathway directly incorporates sulfide or thiosulfate into *O*-acetyl-L-serine (OAS) whereas the transsulfuration pathways interconvert homocysteine and cysteine via intermediary formation of cystathionine (Fig. 1).

3.1 Synthesis of Cysteine from Sulfide or Thiosulfate

The biosynthesis of cysteine from serine in bacteria and plants is carried out by a two-step pathway starting with the *O*-acetylation of serine, followed by the β-replacement of the acetyl group by sulfide or thiosulfate.

3.1.1 The Serine Acetyltransferase, CysE

The first step in cysteine biosynthesis involves the serine acetyltransferase (SAT; EC 2.3.1.30), which is encoded by *cysE* (Becker et al. 1969). CysE catalyzes the reaction of serine and acetyl-coenzyme A to give OAS (Fig. 1). Cysteine inhibits CysE activity in different bacteria (Denk and Bock 1987; Haitani et al. 2006; Johnson et al. 2004), in *Saccharomyces cerevisiae* (Takagi et al. 2003), in protists and in plants. The natural insensitivity of SAT to feedback inhibition by cysteine has only been seen in endosymbiotic bacteria (Lai and Baumann 1992) and in some plants (Kawashima et al. 2005). The molecular basis of feedback inhibition by cysteine has been studied in CysE from different organisms. *E. coli* CysE activity inhibition by cysteine ($K_i = 1\ \mu M$) arises from competition between cysteine and serine at the serine-binding site, which inhibits *O*-acetylation of serine (Hindson 2003). Strains with mutations in *cysE* that lead to the synthesis of a feedback inhibition insensitive SAT, overproduce cysteine because of the loss of the cysteine control (see Sect. 7).

3.1.2
The OAS-Sulfhydrylases, CysK and CysM

The second step of cysteine synthesis involves an OAS-sulfhydrylase, also called OAS-thiol-lyase, (OASS, EC 2.5.1.47). OASS converts OAS and sulfide into cysteine and acetate (Fig. 1). OASS uses pyridoxal-5′-phosphate (PLP) as cofactor and belongs to the β-family of the PLP-dependent enzymes. OASS and cystathionine β-synthase, which is also involved in cysteine biosynthesis, belong to this family. Two different OAS-sulfhydrylases have been characterized in *E. coli* and *S. typhimurium* (Kredich 1996): OASS-A (CysK) and OASS-B (CysM), which are 43% identical. Both are homodimeric enzymes with one tightly bound PLP per subunit attached to a lysine residue. The *cysK cysM* double mutant of *S. typhimurium* requires cysteine to grow whereas strains lacking either *cysK* or *cysM* are cysteine prototrophs, showing that the two isoenzymes catalyze the same step in cysteine biosynthesis (Hulanicka et al. 1979). Both OAS-sulfhydrylases are inhibited by cysteine and sulfide but at concentrations that are too high to be physiologically significant (Kredich 1996). As observed in higher plants (Wirtz and Hell 2006), CysE and CysK from *S. typhimurium*, *E. coli* and *Haemophilus influenzae* form a complex, called cysteine synthase (Campanini et al. 2005; Kredich 1996). Although its role is not well understood, the complex formation may modulate OASS-A activity (Huang et al. 2005). By contrast, OASS-B cannot complex with CysE (Zhao et al. 2006). Moreover, CysM may play a more important role during anaerobic growth and for thiosulfate utilization (Kredich 1996). Accordingly, CysM is a S-sulfocysteine synthase that catalyzes the reaction of OAS and thiosulfate to give S-sulfocysteine (Fig. 1). The further conversion of S-sulfocysteine into cysteine is still uncharacterized (Nakamura et al. 1984). This reaction is a convenient way of incorporating sulfur from thiosulfate without the need for sulfate reduction, which consumes two ATP molecules (Kredich 1996). Two isoenzymes have been detected in a variety of bacteria and three OASS-type proteins are present in *B. subtilis*, (encoded by *cysK*, *ytkP* and *yrhA*) and *Staphylococcus aureus* (encoded by *cysM*, *cysK* and SA0112). CysK seems to be the main enzyme involved in the synthesis of cysteine from sulfate in *B. subtilis*, whereas CysM is required for the growth of *S. aureus* with thiosulfate (Lithgow et al. 2004; van der Ploeg et al. 2001a). Two OASS-like proteins are present in *Lactococcus lactis*: CysM may be an OASS whereas CysK is involved in the conversion of methionine to cysteine (Sperandio et al. 2005). An enzyme with broad specificity is present in the hyperthermophilic archaeon *Aeropyrum pernix*. This enzyme has *O*-acetylserine sulfhydrylase, S-sulfocysteine synthase and cystathionine β-synthase activity in vitro (Mino and Ishikawa 2003a). Surprisingly, it also uses *O*-phospho-L-serine to synthesize cysteine (*O*-phosphoserine sulfhydrylase, EC 2.5.1.65) (Mino and Ishikawa 2003b). Finally, a new pathway for the cysteine synthesis has been described in *M. tuberculosis* (Burns et al. 2005) involving the products of the *cysMO-mec* cluster. CysM is an OASS-B that uses the thiocarboxylate form of the CysO protein (CysO-SH) as a sulfide donor to form a CysO-cysteine adduct in the presence of OAS. CysO-cysteine is then cleaved by the Mec protease to release cysteine.

3.2 Conversion of Methionine to Cysteine

Several microorganisms including *B. subtilis*, *Klebsiella aerogenes*, *P. putida* or *P. aeruginosa*, *M. tuberculosis* and *S. cerevisiae* can use methionine as sole sulfur source indicating that this compound can be efficiently converted into cysteine (Seiflein and Lawrence 2001; Sekowska and Danchin 1999; Thomas and Surdin-Kerjan 1997; Vermeij and Kertesz 1999; Wheeler et al. 2005). By contrast, *E. coli* is unable to grow with methionine.

3.2.1 The Reverse Transsulfuration Pathway

The conversion of methionine to cysteine first involves the synthesis of homocysteine via S-adenosyl-methionine (SAM). Homocysteine is then converted to cysteine by the reverse transsulfuration pathway, via a cystathionine intermediate (Fig. 1). SAM is synthesized from methionine and ATP by SAM-synthase (EC 2.5.1.6), which is encoded by the *metK* gene. The by-product of SAM-dependent methylation reactions is S-adenosyl-homocysteine (SAH), which is degraded to homocysteine either by an SAH hydrolase (EC 3.3.1.1) or by the successive action of an SAH nucleosidase (EC 3.2.2.9) and an S-ribosyl-homocysteinase (LuxS) (EC 4.4.1.21). The LuxS enzyme is also involved in the production of AI-2 (autoinducter 2), a universal signaling factor for interspecies communication (Vendeville et al. 2005). Homocysteine can also be produced directly from sulfide by an *O*-acylhomoserine sulfhydrylase (EC 2.5.1.49) (Fig. 1). The reverse transsulfuration pathway requires the sequential action of cystathionine β-synthase (EC 4.2.1.22) and cystathionine γ-lyase (EC 4.4.1.1). Cystathionine β-synthase catalyzes the condensation of L-homocysteine and serine to give cystathionine whereas cystathionine γ-lyase catalyzes the conversion of cystathionine to ammonia, α-ketobutyrate and cysteine. Cystathionine β-synthase and cystathionine γ-lyases function with PLP. Experimental evidence has been found for cystathionine γ-lyases in *Streptomyces phaeochromogenes*, *M. tuberculosis*, *L. lactis*, *Lactobacillus fermentum*, *K. pneumoniae*, *P. aeruginosa* and *S. cerevisiae* (Bruinenberg et al. 1996; Seiflein and Lawrence 2001; Sperandio et al. 2005; Vermeij and Kertesz 1999; Wheeler et al. 2005; Yamagata et al. 1993). The cystathionine β-synthase of *S. cerevisiae* is a homotetramer made from a 55 kDa subunit and is encoded by *CYS4*. *CYS4* mutants are cysteine auxotrophs and thus cysteine can only be synthesized by reverse transsulfuration in *S. cerevisiae* as observed in mammals. By contrast, very little is known about cystathionine β-synthases in bacteria. Mutants with inactive genes encoding the putative cystathionine β-synthases in *L. lactis* and *Streptomyces venezuelae* cannot synthesize cysteine from methionine although the corresponding enzymes have not been characterized (Chang and Vining 2002; Sperandio et al. 2005). A non-typical reverse transsulfuration pathway has been proposed in *Methanococcus janaschii*, in which the cystathionine β-synthase uses *O*-phospho-serine as a homocysteine acceptor rather than serine (White 2003).

3.2.2
Conversion of Methionine to Cysteine Via Methanethiol Formation

Microorganisms can catabolize L-methionine by different pathways (Fig. 1). Methionine can be directly converted to methanethiol, α-ketobutyrate and ammonium by a methionine γ-lyase (E.C.4.4.1.11), an enzyme that is found in many bacteria (Amarita et al. 2004; Inoue et al. 1997). The use of methionine may also occur via a two-step degradation pathway initiated by an aminotransferase (Yvon et al. 2000). This enzyme requires the presence of an acceptor, such as α-ketoglutarate, to give α-keto-γ-methyl-thiobutyric acid (KMBA), which is subsequently degraded to methanethiol (Bonnarme et al. 2000). In *K. aerogenes* and *P. putida*, the methanethiol may be oxidized to sulfite via a methanesulfonate intermediate (Seiflein and Lawrence 2001; Vermeij and Kertesz 1999).

3.3
Cysteine Production from Glutathione or Peptides

Glutathione, which is a cysteine-containing tripeptide (γ-glu-cys-gly; GSH), can be used as a sulfur source by different organisms. In *E. coli*, GSH is taken up by the products of the *ybiK-yliABCD* gene cluster, which encodes an ABC transporter TC 3.A.1.207.1 (Suzuki et al. 2005). GSH is then hydrolyzed by γ-glutamyltranspeptidase (EC 2.3.2.2) (encoded by *ggt*) to give cysteinyl-glycine and glutamate (Suzuki et al. 1993). The cysteine-containing dipeptide is then cleaved by peptidases (aminopeptidases A, B, and N and dipeptidase D) to give cysteine and glycine (Suzuki et al. 2001). Two γ-glutamyltranspeptidase proteins are present in *B. subtilis*, but only one is important for using glutathione as a sulfur source (Minami et al. 2003). Finally, all the environmental proteins can serve as a cysteine source for cells. For example, lactic acid bacteria have a complex proteolytic system that allows the cells to degrade extracellular milk proteins, to then take up the protein breakdown products and finally to further hydrolyze the peptides into their corresponding amino acids (Savijoki et al. 2006).

4
Cysteine Degradation by Cysteine Desulfhydrases

Cysteine can be broken down by cysteine desulfhydrases to give sulfide, pyruvate and ammonia. In various microorganisms, different enzymes that participate in sulfur metabolism (cystathionine-β-lyases, cystathionine-γ-lyases, OAS-sulfhydrylases) have cysteine desulfhydrase activity in vitro (Auger et al. 2005; Awano et al. 2005; Wada et al. 2002). In addition, the tryptophanase (TnaA) from *E. coli*, the synthesis of which is induced in the presence of cysteine, is also a cysteine desulfhydrase (Awano et al. 2003). Five different cysteine desulfhydrases (MetC, MalY, CysK, CysM, TnaA) have been characterized in *E. coli*, whereas only four have been in *B. subtilis* (MetC, PatB, YrhB, CysK). Cysteine desulfhydrase activity induced by the addition of cysteine to the growth medium can still be detected in a quintet *E. coli* mutant suggesting that other enzymes with this activity exist (Awano et al. 2005). The hydrogen sulfide produced by microorganisms in the oral cavity (*Streptococci*, *Fusobacterium nucleatum*, *Treponema denticola*) from the degradation of cysteine is one of the predominant volatile sulfur compounds that is responsible for bad breath (Chu et al. 1997; Fukamachi et al. 2002; Yoshida et al. 2003).

5
Transport of Cystine or Cysteine

In environmental oxidizing conditions, cysteine dimerizes to form the disulfide-linked cystine, which is normally the compound transported and a source of cysteine for the cell. To limit the toxicity of cysteine at high concentration, bacteria have also developed systems for cysteine excretion.

5.1
Cystine Uptake

Cystine transport has mostly been investigated in enterobacteria and in *B. subtilis* (Burguiere et al. 2004; Kredich 1996). The L-cystine uptake systems that have so far been identified belong to two different transporter families: the polar amino acid uptake family of the ABC transporters (TC 3.A.1.3) and the dicarboxylate amino acid:cation (Na^+ and/or H^+) symporter family (TC 2.A.23). In *E. coli*, there are two kinetically identifiable L-cystine transport systems. One also transports diaminopimelic acid and several cystine analogues, whereas the second is more specific but has not yet been identified (Berger and Heppel 1972). The first system corresponds to an ABC transporter ($K_T = 0.3\ \mu M$). The synthesis of FliY, the periplasmic L-cystine-binding protein from *E. coli*, increases during sulfate starvation (Quadroni et al. 1996). The *fliY* gene forms an operon with *fliA*, encoding σF, which is involved in flagellar synthesis. The YecS and YecC proteins are probably the permease and ATP binding protein for this transporter (Hosie and Poole 2001). In *S. typhimurium*, L-cystine is taken up by three different systems, CTS-1 ($K_T = 2\ \mu M$), CTS-2 ($K_T = 0.1\ \mu M$) and CTS-3 (non saturable) (Baptist and Kredich 1977). In gram-positive bacteria such as *L. fermentum*, the high affinity L-cystine solute binding protein, BspA or CyuC, ($K_d = 0.2\ \mu M$) is required for L-cystine uptake (Hung et al. 2005; Turner et al. 1999). Two ABC transporters, TcyABC and TcyJKLMN and a (Na^+/H^+) symporter, TcyP, have recently been characterized in *B. subtilis*. A triple *ΔtcyP ΔtcyJKLMN ΔtcyA* mutant does not grow in the presence of cystine and cannot take this compound up (Burguiere et al. 2004). The L-cystine uptake by TcyP ($K_T = 0.6\ \mu M$) is strongly inhibited by seleno-cystine and the transport due to the TcyJKLMN system ($K_T = 2.5\ \mu M$) is also considerably reduced in the presence of cystathionine, djenkolic acid or S-methyl-L-cysteine. The expression of *tcyJKLMN* and *tcyP* increases during cysteine starvation while the level of *tcyA* expression remains low under all conditions tested. Cystine uptake by TcyP is inhibited by a 10-fold excess of cysteine, indicating that this symporter can also transport cysteine. All the ABC transporters described above take up cystine but not cysteine. Recently, the crystal structure of a solute binding protein, CjaA of *Campylobacter jejuni*, revealed that cysteine is a bound ligand (Muller et al. 2005). CjaA corresponds to the first solute receptor protein with a high affinity for L-cysteine ($K_d = 0.1\ \mu M$). In yeast, seven different permeases, which also transport other amino acids, participate in cysteine uptake (During-Olsen et al. 1999).

5.2 Cysteine Efflux

The SH group of cysteine makes it a highly reactive compound and is toxic at high concentrations (Kari et al. 1971). Three systems for cysteine excretion have recently been characterized in *E. coli*. YdeD, a highly hydrophobic membrane protein with 9 or 10 transmembrane segments and YfiK, a member of the RhtB family of amino acid exporters, are involved in the excretion of OAS and cysteine or its derivatives (Dassler et al. 2000; Franke et al. 2003). Neither a *ydeD* nor a *yfiK* mutant shows a phenotype. However, some strains that overproduce YdeD or YfiK excrete considerable amounts of

OAS and cysteine, which can condense with pyruvate to give 2-methyl-2,4-thiazolidinedicarboxylic acid. Also, a strain overproducing YdeD cannot grow in minimal medium without a reduced sulfur compound because the cysteine regulon is not induced (see Sect. 6.1) (Dassler et al. 2000). CydDC is an ABC transporter required for cytochrome assembly. A *cydD* mutant is hypersensitive to cysteine and accumulates high levels of cytoplasmic cysteine. In everted membrane vesicles, an ATP and CydD-dependent import of cysteine is observed indicating that CydDC is a cysteine exporter in vivo (Pittman et al. 2002). Proteins similar to these cysteine exporters are present in several bacteria.

6 Regulation of Cysteine Biosynthesis and Transport

There is a large variety of molecular mechanisms that participate in fine-tuning the regulation of cysteine metabolism in bacteria as is seen for methionine metabolism (Rodionov et al. 2004).

6.1 The CysB Regulon of *E. Coli*

In *E. coli* and *S. typhimurim*, the genes required for the transport of sulfate, thiosulfate, glutathione and cystine, for sulfate reduction and its incorporation into cysteine are positively controlled by CysB, a LysR-type transcriptional activator (LTTR) (Kredich 1996; Ostrowski and Kredich 1989; Parry and Clark 2002). In addition to CysB, the full expression of the *cys* regulon, requires an inducer (*N*-acetyl-L-serine; NAS) and a limiting amount of reduced sulfur (Fig. 2). The regulation mechanism by CysB has been extensively characterized by DNA-binding assays, footprinting and transcription studies for the *cysJ*, *cysK* and *cysP* genes of *S. typhimurium* (Kredich 1996). The interaction of CysB with regulated promoter regions appears complex. The CysB DNA-binding sites, which have low sequence similarities, correspond to imperfect dyad symmetry between 19-bp half-sites. The number of half-sites and their arrangement vary among promoters, but in general they contain one major activation site (a pair of convergently oriented half-sites) located just upstream from the – 35 region and several accessory sites. In absence of the inducer, CysB binds to DNA and covers a large region that includes accessory sites (Hryniewicz and Kredich 1991; Kredich 1996; Lochowska et al. 2004). NAS is thought to cause a conformational change in CysB that stimulates the binding of CysB to activation sites. This stable interaction is essential for transcriptional activation and may be required for the recruitment of RNA polymerase. CysB also acts as a repressor of its own transcription by interacting with a DNA region that overlaps the RNA polymerase-binding site

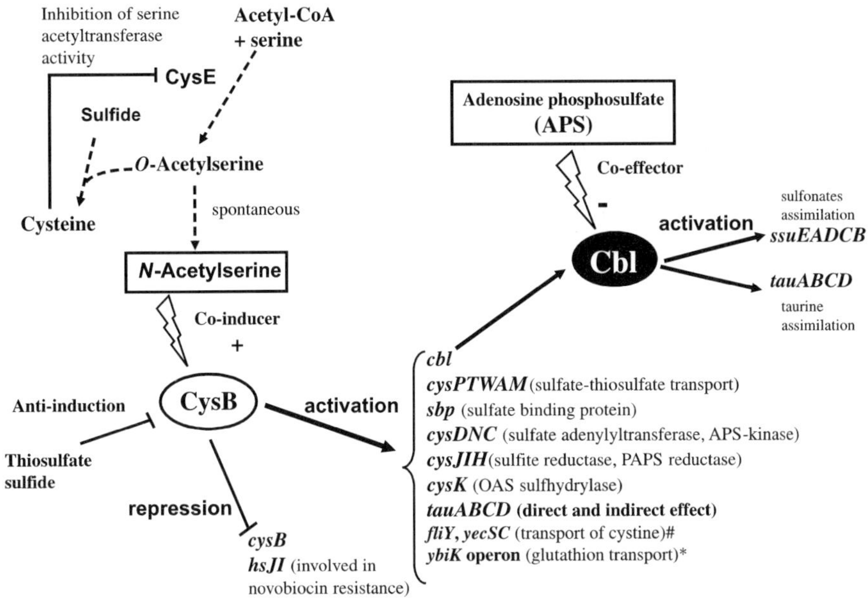

Fig. 2 The CysB regulon from *E. coli*. The targets of CysB and Cbl that have been characterized are indicated. The effectors of these two activators are boxed. The reactions involved in metabolic pathways are indicated by dotted arrows. CysE is the serine acetyltransferase. $*$, evidence for a direct binding of CysB to the promoter region is lacking. #, cystine transport is part of the *cys* regulon in *S. typhimurium*. The *adiA* gene encoding the arginine decarboxylase is also positively controlled by CysB but the mechanism remains unknown

(Kredich 1996). In this case, NAS reduces this binding. Limiting the amount of available sulfur is necessary for derepressing the *cys* regulon (Kredich 1996). There are two main levels of control (Fig. 2). The first is the inhibition by cysteine of the CysE-dependent formation of OAS, the precursor of NAS. By controlling the synthesis of the CysB inducer, cysteine decreases the expression level of the *cys* regulon. Mutant strains, in which CysE is insensitive to cysteine feedback inhibition, overproduce cysteine showing that the allosteric control is a key regulation point in sulfur metabolism. The second level of control is mediated by two other sulfur sources. Sulfide and thiosulfate act as anti-inducers by competing with NAS for the CysB activator (Hryniewicz and Kredich 1991; Kredich 1996). In *S. typhimurium*, anti-induction seems to be the main regulatory factor in vivo responsible for reducing *cys* gene expression (Oppezzo 1998). A second LTTR, Cbl, controls the expression of genes involved in sulfonate metabolism in *E. coli* (van der Ploeg et al. 2001b). Cbl directly activates the transcription of the *ssu* and *tau* operons by binding upstream from the – 35 region of their promoters (Bykowski et al. 2002; van der Ploeg et al. 2001b). In contrast to *cys* genes, the expression of these genes is fully repressed in the presence of sulfate and Cbl activity is negatively regulated by APS, which is a metabolite of the sulfate assimilatory pathway and is the signaling molecule for excess sulfate (Bykowski et al. 2002). CysB indirectly regulates the expression of the *ssu* and *tau* operons by controlling Cbl synthesis (Iwanicka-Nowicka and Hryniewicz 1995). CysB also binds to the *tau* and *ssu* promoter regions although a direct effect of CysB in vivo has only been observed for the *tau* operon (van der Ploeg et al. 2001b). The expression of other genes such as *cysK*, *fliY* and *sbp* may also be positively controlled by Cbl (van der Ploeg et al. 2001b).

6.2
The CymR Regulon in *B. Subtilis*

CymR (YrzC) belongs to the Rrf2 family of regulators. It has recently been identified as a master regulator of cysteine metabolism in *B subtilis* (Even et al. 2006). A transcriptome analysis identified 24 genes that are repressed in the presence of sulfate or cysteine (Auger et al. 2002; Burguiere et al. 2004) as being derepressed in a *cymR* mutant. The CymR repressor controls several pathways leading to cysteine formation (Fig. 3), including the OAS-thiol-lyase

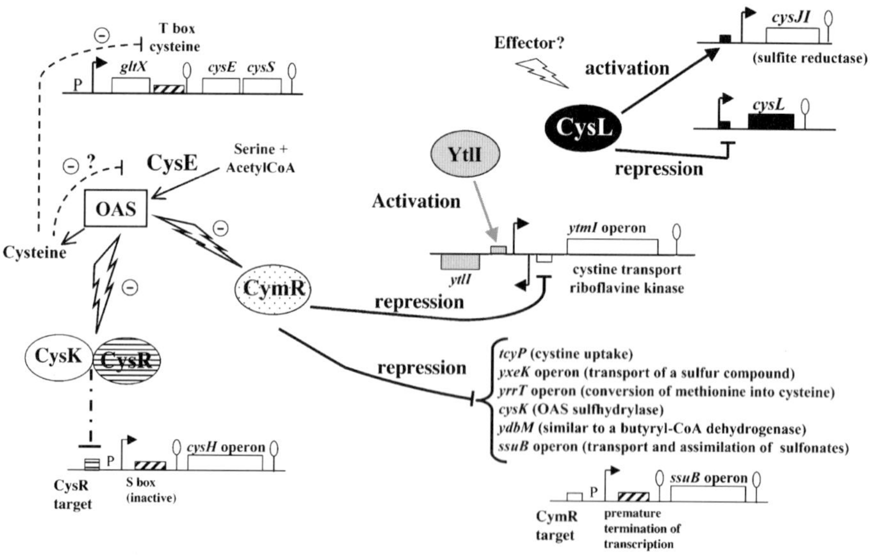

Fig. 3 Regulation of cysteine metabolism in *B. subtilis*. The targets of CymR, CysL, CysR and YtlI that have been characterized are indicated. The effectors of these regulators are boxed. The reactions involved in metabolic pathways are indicated by *thin arrows*. *CysE*, *cysH*, *cysS* and *gltX*, encode the serine acetyltransferase, the APS/PAPS reductase, the cysteinyl and glutamyl tRNA synthetases, respectively. Promoters and terminators are indicated. Motifs involved in premature termination of transcription are indicated by *striped boxes*. OAS, *O*-acetylserine

(CysK), L-cystine transporters (TcyP and TcyJKLMN), sulfonate assimilation (SsuABCD) and the methionine to cysteine conversion that involves YrhA and YrhB (unpublished results, Burguiere et al. 2004; van der Ploeg et al. 1998, 2001a). By contrast, the sulfate assimilation pathway does not seem to be regulated by CymR. CymR binds to the promoter region of seven genes or operons (*ytlI*, *ssuB*, *tcyP*, *yrrT*, *yxeK*, *cysK* or *ydbM*). A 27-bp CymR-binding motif TAAWNCN2ANTWNAN3ATMGGAATTW has been identified by DNase I footprinting or through deletions or point mutations (Even et al. 2006). This motif is either found downstream from the promoter or overlaps the – 35 box. Physiological data suggests that OAS, the direct precursor of cysteine (Fig. 3) plays a key role in the signaling pathway that controls *ytlI* and *ssuB* expression (Burguière et al. 2005; van der Ploeg et al. 2001b). The addition of OAS prevents CymR-dependent binding to its DNA target in vitro. Cysteine probably limits the size of the intracellular pool of OAS leading to CymR-dependent repression. As observed for *E. coli*, CysE activity may be feedback-inhibited by cysteine (Kredich 1996). In *B. subtilis*, *cysE* expression is also regulated by transcription antitermination at a cysteine-specific T-box (Gagnon et al. 1994). Thus, the level of OAS in microorganisms may be correlated with the level of uncharged cysteinyl-tRNA, which signals the cysteine status in the cell. Interestingly, CysK, the OASS of *B. subtilis* is also a global regulator of cysteine biosynthesis (Albanesi et al. 2005). Both CymR and CysK control a large set of genes. However, the relative role of these two proteins in this control remains to be determined. For the *cysH* operon, a more precise mechanism for CysK-dependent regulation has been proposed (Albanesi et al. 2005). The CysK protein may interact with an uncharacterized repressor, CysR (Mansilla et al. 2000), to promote its binding to the *cysH* promoter. OAS may prevent binding of CysR to its target by modulating the interaction between CysK and CysR in response to cysteine availability (Albanesi et al. 2005; Mansilla et al. 2000). The CymR regulon and the *cysH* operon are repressed in the presence of both sulfate and cysteine whereas the level of expression of the *cysJI* operon, which encodes the sulfite reductase, is only reduced by cysteine. An LTTR, CysL, positively regulates the *cysJI* operon by binding to its promoter region upstream from the – 35 box (Fig. 3) (Guillouard et al. 2002). The effector that modulates CysL binding remains uncharacterized.

In the CymR regulon, the expression of the *ssuB* and *ytmI* operon is derepressed several hundred-fold during cysteine starvation by a more complex regulation mechanism. In addition to the CymR-dependent control, the *ssuB* operon is also regulated by premature transcription termination. In both cases, OAS plays a central role in the signalling pathway (van der Ploeg et al. 2001b). The control of the *ytmI* operon, which encodes a cystine ABC transporter, involves a regulation cascade. In the presence of sulfate or cysteine, CymR represses the expression of the *ytlI* gene encoding an LTTR (Burguière et al. 2005). YtlI directly activates the transcription of the *ytmI* operon by binding to its promoter region. The sulfur-dependent regulation mostly occurs in YtlI synthesis (Fig. 3) (Burguière et al. 2005).

6.3 Global Control of Sulfur Metabolism in *L. Lactis* and *C. Glutamicum*

In *L. lactis* and *C. glutamicum*, a unique regulator controls most of the genes of the cysteine and methionine biosynthetic pathways (Rey et al. 2005; Sperandio et al. 2005). A global transcriptional approach has shown that FhuR, also called CmbR, broadly controls the sulfur amino acid metabolism in *L. lactis* (Sperandio et al. 2005). FhuR belongs to the LTTR family and positively controls the transcription of most of the genes involved in the biosynthesis and uptake of cysteine or methionine and their interconversion. Only the genes encoding serine acetyl-transferase and methionine synthase are not under FhuR control (Sperandio et al. 2005). FhuR directly regulates the expression of these genes by binding to a 13-bp box in their promoter region that follows the general rules for LTTR boxes. OAS, which is probably the cofactor of FhuR, induces the FhuR regulon expression in vivo and increases the affinity of the regulator for its targets in vitro (Golic et al. 2005; Sperandio et al. 2005). The FhuR-regulon expression is also repressed in the presence of cysteine (Sperandio et al. 2005). As in *E. coli*, cysteine may control FhuR activity by modulating the intracellular concentration of OAS (Sect. 3.1.1). Surprisingly, FhuR controls the synthesis of metabolic pathways that function in opposite directions, leading to apparent futile cycles. This strange behavior may be explained by the presence of other uncharacterized controls. In *C. glutamicum*, McbR, which belongs to the TetR-family of repressors, has been isolated by DNA affinity chromatography through its binding to the *metY* promoter region (Rey et al. 2003). Comparison of the proteome and the transcriptome of the wild-type strain and an *mcbR* mutant shows that McbR controls the transcription of genes involved in sulfate or sulfonate assimilation, in the synthesis of cysteine and methionine and in sulfate or methionine transport (Rey et al. 2003, 2005; Ruckert et al. 2005). In *silico* analysis suggested that the McbR binding site is an inverted repeat TAGAC-N6-GTCTA present upstream from 22 genes or operons. A purified McbR has been shown to bind to DNA fragments containing this motif in the promoter region of the *hom*, *cysI*, *cysK*, *metK* and *mcbR* genes (Rey et al. 2005). Three genes encoding regulators are found among the direct targets of McbR; There are: (i) the *mcbR* gene itself indicating a negative autoregulation that ensures a constant level of repressor in the cell; (ii) *cg0156*, which encodes a NagC family regulator, the function of which remains to be determined; (iii) *cg0012*, now called *ssuR* for sulfonate sulfur utilization regulator. SsuR, a NagC-family regulator, activates the transcription of four groups of genes involved in the uptake and degradation of aliphatic sulfonates and sulfonate esters in the absence of sulfate (Koch et al. 2005a,b). Purified SsuR can interact with the *ssuI*, *seuA*, *ssuD2* and *ssuD1* promoter regions. A 21 bp conserved SsuR-binding motif is present upstream from the – 35 box of the corresponding promoters. Consistent with the in vivo repression of the controlled genes by sulfate, SsuR binding decreases in the presence of sulfate, APS, sulfite or sulfide (Koch et al. 2005a). The global regulator McbR responds to a different signaling molecule. Indeed, S-adenosylhomocysteine, a major product of SAM-dependent trans-methylation prevents McbR binding to its targets (Rey et al. 2005). In *C. glutamicum*, a more specific control of cysteine metabolism in response to cysteine availability has yet to be found.

7
Production Process Development

Apart from its physiological importance, cysteine is also biotechnologically interesting as a nutritional supplement, a pharmaceutical (antidote) or a precursor for drugs. At present, most cysteine is produced by extraction from acid hydrolysates of keratine. L-cystine obtained from protein hydrolysis is converted into cysteine by electrochemical reduction (Leuchtenberger et al. 2005). An industrially used enzymatic process also exists in which the DL-2-amino-D2-thiazolin-4-carbonic acid (DL-ATC) is a substrate for the production of cysteine. This bioconversion involves three enzymes from Pseudomonas sp.: ATC racemase, L-ATC-hydrolase, and S-carbamoyl-L-cysteine hydrolase (Leuchtenberger et al. 2005). To meet a growing worldwide demand, fermentation processes with *E. coli* or *C. glutamicum* are currently being developed (Dassler et al. 2000). The biosynthesis of L-cysteine is regulated through feedback inhibition of serine acetyl-transferase by cysteine (Sects. 3.1.1 and 6.1). Thus, the supply of OAS is the limiting factor for cysteine production. Mutants, in which the methionine at position 256 is replaced by several other amino acids, are partially insensitive to feedback inhibition (Nakamori et al. 1998). Strains with serine acetyl-transferase insensitive to feedback inhibition by cysteine produce high quantities of cysteine and cystine (200 mg L^{-1}) (Nakamori et al. 1998). However, cysteine production is very unstable probably due to degradation of this compound by cysteine desulfhydrases. Five cysteine desulfhydrases have recently been characterized in *E. coli*: CysK, CysM, MetC, MalY and tryptophanase (TnaA) (Awano et al. 2003, 2005). The inactivation of *cysM*, *metC*, *malY* or *tnaA* doubles cysteine production. The export of amino acids out of the overproducing cells also limits production. The *ydeD* and *yfiK* genes that encode exporters of cysteine have been identified as genes that increase cysteine production when overexpressed in an industrial *E. coli* strain (Dassler et al. 2000; Franke et al. 2003). In the presence of thiosulfate, cysteine production in the medium by a strain overproducing YdeD reaches 75 mg L^{-1}. The overproduction of YfiK also leads to cysteine accumulation out of the cell (120 mg L^{-1}), but only in a *cysE* mutant encoding a serine acetyl-transferase insensitive to feedback inhibition by cysteine.

L-α-amino acids with unnatural substituents may also be produced by metabolic engineering of the cysteine biosynthetic pathway (Maier 2003; Zhao et al. 2004). These compounds can be incorporated into peptides to produce peptide-mimicking molecules that modulate interactions between proteins of pharmaceutical or agrochemical interest and their target polypeptides. OASS catalyzes β-substitutions on OAS with a broad range of substrates but purified CysK and CysM have different substrate preferences. CysK has a higher reactivity with N-heterocycles whereas CysM is more efficient with thiols (Maier 2003). In vivo, a strain carrying a plasmid with a modified *cysE* gene that encodes CysE-T167A-G245S, a serine acetyltransferase with decreased inhibition by L-cysteine, produces large amounts of an unnatural amino acid (up to $13\,g\,L^{-1}$) by fermentation and $3.8\,g\,L^{-1}$ of cysteine using thiosulfate as a sulfur source (Maier 2003). The unnatural amino acids are efficiently excreted although the mechanism involved remains unknown. A two-phase fermentation process has been developed to incorporate toxic precursors such as azide or cyanide. A first strain is used for producing and excreting OAS followed by a biotransformation step with permeabilized CysM-containing cells in the presence of a nucleophile substrate (Maier 2003).

Acknowledgements We thank Pierre Burguière and Brice Sperandio whose PhD theses have been extremely helpful for the writing of this paper. We are grateful to Sandrine Auger and Olga Soutourina for critical reading of the paper. This research was supported by grants from the "Ministère de l'Education Nationale de la Recherche et de la Technologie", the "Institut National de la Recherche argronomique" (UR895), the "Centre National de la Recherche Scientifique" (URA 2171), the "Institut Pasteur" and the "Université Paris 7".

References

Albanesi D, Mansilla MC, Schujman GE, de Mendoza D (2005) *Bacillus subtilis* cysteine synthetase is a global regulator of the expression of genes involved in sulfur assimilation. J Bacteriol 187:7631–7638

Amarita F, Yvon M, Nardi M, Chambellon E, Delettre J, Bonnarme P (2004) Identification and functional analysis of the gene encoding methionine-γ-lyase in *Brevibacterium linens*. Appl Environ Microbiol 70:7348–7354

Auger S, Danchin A, Martin-Verstraete I (2002) Global expression profile of *Bacillus subtilis* grown in the presence of sulfate or methionine. J Bacteriol 184:5179–5186

Auger S, Gomez MP, Danchin A, Martin-Verstraete I (2005) The PatB protein of *Bacillus subtilis* is a C – S-lyase. Biochimie 87:231–238

Awano N, Wada M, Kohdoh A, Oikawa T, Takagi H, Nakamori S (2003) Effect of cysteine desulfhydrase gene disruption on L-cysteine overproduction in *Escherichia coli*. Appl Microbiol Biotechnol 62:239–243

Awano N, Wada M, Mori H, Nakamori S, Takagi H (2005) Identification and functional analysis of *Escherichia coli* cysteine desulfhydrases. Appl Environ Microbiol 71:4149–4152

Baptist EW, Kredich NM (1977) Regulation of L-cystine transport in *Salmonella typhimurium*. J Bacteriol 131:111–118

Becker MA, Kredich NM, Tomkins GM (1969) The purification and characterization of O-acetylserine sulfhydrylase-A from Salmonella typhimurium. J Biol Chem 244:2418–2427

Berger EA, Heppel LA (1972) A binding protein involved in the transport of cystine and diaminopimelic acid in *Escherichia coli*. J Biol Chem 247:7684–7694

Berndt C, Lillig CH, Wollenberg M, Bill E, Mansilla MC, de Mendoza D, Seidler A, Schwenn JD (2004) Characterization and reconstitution of a 4Fe – 4S adenylyl sulfate/phosphoadenylyl sulfate reductase from *Bacillus subtilis*. J Biol Chem 279:7850–7855

Bick JA, Dennis JJ, Zylstra GJ, Nowack J, Leustek T (2000) Identification of a new class of 5′-adenylylsulfate (APS) reductases from sulfate-assimilating bacteria. J Bacteriol 182:135–142

Bonnarme P, Psoni L, Spinnler HE (2000) Diversity of L-methionine catabolism pathways in cheese-ripening bacteria. Appl Environ Microbiol 66:5514–5517

Bruinenberg PG, de Roo G, Limsowtin GK (1996) Purification and characterization of cystathionine γ-lyase from *Lactococcus lactis* subsp. *cremoris* SK11: possible role in flavor compound formation during cheese maturation. Appl Environ Microbiol 63:561–566

Burguiere P, Auger S, Hullo MF, Danchin A, Martin-Verstraete I (2004) Three different systems participate in L-cystine uptake in *Bacillus subtilis*. J Bacteriol 186:4875–4884

Burguière P, Fert J, Guillouard I, Auger S, Danchin A, Martin-Verstraete I (2005) Regulation of the *Bacillus subtilis ytmI* operon, involved in sulfur metabolism. J Bacteriol 187:6019–6030

Burns KE, Baumgart S, Dorrestein PC, Zhai HL, McLafferty FW, Begley TP (2005) Reconstitution of a new cysteine biosynthetic pathway in Mycobacterium tuberculosis. J Am Chem Soc 127:11602–11603

Bykowski T, van der Ploeg JR, Iwanicka-Nowicka R, Hryniewicz MM (2002) The switch from inorganic to organic sulphur assimilation in Escherichia coli: adenosine 5′-phosphosulphate (APS) as a signalling molecule for sulphate excess. Mol Microbiol 43:1347–1358

Campanini B, Speroni F, Salsi E, Cook PF, Roderick SL, Huang B, Bettati S, Mozzarelli A (2005) Interaction of serine acetyltransferase with O-acetylserine sulfhydrylase active site: evidence from fluorescence spectroscopy. Protein Sci 14:2115–2124

Chang Z, Vining LC (2002) Biosynthesis of sulfur-containing amino acids in *Streptomyces venezuelae* ISP5230: roles for cystathionine β-synthase and transsulfuration. Microbiology 148:2135–2147

Chu L, Ebersole JL, Kurzban GP, Holt SC (1997) Cystalysin, a 46-kilodalton cysteine desulfhydrase from *Treponema denticola*, with hemolytic and hemoxidative activities. Infect Immun 65:3231–3238

Dassler T, Maier T, Winterhalter C, Bock A (2000) Identification of a major facilitator protein from *Escherichia coli* involved in efflux of metabolites of the cysteine pathway. Mol Microbiol 36:1101–1112

Denk D, Bock A (1987) L-cysteine biosynthesis in Escherichia coli: nucleotide sequence and expression of the serine acetyltransferase (cysE) gene from the wild-type and a cysteine-excreting mutant. J Gen Microbiol 133:515–525

During-Olsen L, Regenberg B, Gjermansen C, Kielland-Brandt MC, Hansen J (1999) Cysteine uptake by *Saccharomyces cerevisiae* is accomplished by multiple permeases. Curr Genet 35:609–617

Even S, Burguière P, Auger S, Soutourina O, Danchin A, Martin-Verstraete I (2006) Global control of cysteine metabolism by CymR in *Bacillus subtilis*. J Bacteriol 188:2184–2197

Franke I, Resch A, Dassler T, Maier T, Bock A (2003) YfiK from *Escherichia coli* promotes export of *O*-acetylserine and cysteine. J Bacteriol 185:1161–1166

Fukamachi H, Nakano Y, Yoshimura M, Koga T (2002) Cloning and characterization of the L-cysteine desulfhydrase gene of *Fusobacterium nucleatum*. FEMS Microbiol Lett 215:75–80

Gagnon Y, Breton R, Putzer H, Pelchat M, Grunberg-Manago M, Lapointe J (1994) Clustering and co-transcription of the *Bacillus subtilis* genes encoding the aminoacyl-tRNA synthetases specific for glutamate and for cysteine and the first enzyme for cysteine biosynthesis. J Biol Chem 269:7473–7482

Golic N, Schliekelmann M, Fernandez M, Kleerebezem M, van Kranenburg R (2005) Molecular characterization of the CmbR activator-binding site in the metC-cysK promoter region in Lactococcus lactis. Microbiology 151:439–446

Guillouard I, Auger S, Hullo MF, Chetouani F, Danchin A, Martin-Verstraete I (2002) Identification of *Bacillus subtilis* CysL, a regulator of the *cysJI* operon, which encodes sulfite reductase. J Bacteriol 184:4681–4689

Haitani Y, Awano N, Yamazaki M, Wada M, Nakamori S, Takagi H (2006) Functional analysis of L-serine O-acetyltransferase from Corynebacterium glutamicum. FEMS Microbiol Lett 255:156–163

Hindson VJ (2003) Serine acetyltransferase of Escherichia coli: substrate specificity and feedback control by cysteine. Biochem J 375:745–752

Hosie AHF, Poole PS (2001) Bacterial ABC transporters of amino acids. Res Microbiol 152:259–270

Hryniewicz MM, Kredich NM (1991) The cysP promoter of Salmonella typhimurium: characterization of two binding sites for CysB protein, studies of in vivo transcription initiation, and demonstration of the anti-inducer effects of thiosulfate. J Bacteriol 173:5876–5886

Huang B, Vetting MW, Roderick SL (2005) The active site of O-acetylserine sulfhydrylase is the anchor point for bienzyme complex formation with serine acetyltransferase. J Bacteriol 187:3201–3205

Hulanicka MD, Hallquist SG, Kredich NM, Mojica AT (1979) Regulation of O-acetylserine sulfhydrylase B by L-cysteine in Salmonella typhimurium. J Bacteriol 140:141–146

Hung J, Turner MS, Walsh T, Giffard PM (2005) BspA (CyuC) in *Lactobacillus fermentum* BR11 is a highly expressed high-affinity L-cystine-binding protein. Curr Microbiol 50:33–37

Inoue H, Inagaki K, Eriguchi SI, Tamura T, Esaki N, Soda K, Tanaka H (1997) Molecular characterization of the *mde* operon involved in L-methionine catabolism of *Pseudomonas putida*. J Bacteriol 179:3956–3962

Iwanicka-Nowicka R, Hryniewicz MM (1995) A new gene, cbl, encoding a member of the LysR family of transcriptional regulators belongs to Escherichia coli cys regulon. Gene 166:11–17

Johansson P, Hederstedt L (1999) Organization of genes for tetrapyrrole biosynthesis in gram-positive bacteria. Microbiology 145:529–538

Johnson CM, Huang B, Roderick SL, Cook PF (2004) Kinetic mechanism of the serine acetyltransferase from Haemophilus influenzae. Arch Biochem Biophys 429:115–122

Kari C, Nagy Z, Kovacs P, Hernadi F (1971) Mechanism of the growth inhibitory effect of cysteine on *Escherichia coli*. J Gen Microbiol 68:349–356

Kawashima CG, Berkowitz O, Hell R, Noji M, Saito K (2005) Characterization and expression analysis of a serine acetyltransferase gene family involved in a key step of the sulfur assimilation pathway in Arabidopsis. Plant Physiol 137:220–230

Kertesz MA (2001) Bacterial transporters for sulfate and organosulfur compounds. Res Microbiol 152:279–290

Kertesz MA, Wietek C (2001) Desulfurization and desulfonation: applications of sulfur-controlled gene expression in bacteria. Appl Microbiol Biotechnol 57:460–466

Koch DJ, Ruckert C, Albersmeier A, Huser AT, Tauch A, Puhler A, Kalinowski J (2005a) The transcriptional regulator SsuR activates expression of the *Corynebacterium glutamicum* sulphonate utilization genes in the absence of sulphate. Mol Microbiol 58:480–494

Koch DJ, Ruckert C, Rey DA, Mix A, Puhler A, Kalinowski J (2005b) Role of the *ssu* and *seu* genes of *Corynebacterium glutamicum* ATCC 13032 in utilization of sulfonates and sulfonate esters as sulfur sources. Appl Environ Microbiol 71:6104–6114

Kredich NM (1996) Biosynthesis of Cysteine. In: Neidhardt FC (ed) *Escherichia coli* and *Salmonella*, cellular and molecular biology. ASM Press, Washington DC, p 514–527

Lai CY, Baumann P (1992) Sequence analysis of a DNA fragment from Buchnera aphidicola (an endosymbiont of aphids) containing genes homologous to dnaG, rpoD, cysE, and secB. Gene 119:113–118

Leuchtenberger W, Huthmacher K, Drauz K (2005) Biotechnical production of amino acids and derivatives: current status and prospects. Appl Microbiol Biotechnol 69:1–8

Lillig CH, Prior A, Schwenn JD, Aslund F, Ritz D, Vlamis-Gardikas A, Holmgren A (1999) New thioredoxins and glutaredoxins as electron donors of 3′-phosphoadenylylsulfate reductase. J Biol Chem 274:7695–7698

Lithgow JK, Hayhurst EJ, Cohen G, Aharonowitz Y, Foster SJ (2004) Role of a cysteine synthase in Staphylococcus aureus. J Bacteriol 186:1579–1590

Lochowska A, Iwanicka-Nowicka R, Zaim J, Witkowska-Zimny M, Bolewska K, Hryniewicz MM (2004) Identification of activating region (AR) of Escherichia coli LysR-type transcription factor CysB and CysB contact site on RNA polymerase alpha subunit at the cysP promoter. Mol Microbiol 53:791–806

Maier THP (2003) Semisynthetic production of unnatural L-α-aminoacids by metabolic engineering of the cysteine biosynthetic pathway. Nat Biotechnol 21:422–427

Mansilla MC, Albanesi D, de Mendoza D (2000) Transcriptional control of the sulfur-regulated *cysH* operon, containing genes involved in L-cysteine biosynthesis in *Bacillus subtilis*. J Bacteriol 182:5885–5892

Mansilla MC, de Mendoza D (1997) L-cysteine biosynthesis in *Bacillus subtilis*: identification, sequencing, and functional characterization of the gene coding for phosphoadenylylsulfate sulfotransferase. J Bacteriol 179:976–981

Mansilla MC, de Mendoza D (2000) The *Bacillus subtilis cysP* gene encodes a novel sulphate permease related to the inorganic phosphate transporter (Pit) family. Microbiology 146:815–821

Mendoza-Cozatl D, Loza-Tavera H, Hernandez-Navarro A, Moreno-Sanchez R (2005) Sulfur assimilation and glutathione metabolism under cadmium stress in yeast, protists and plants. FEMS Microbiol Rev 29:653–671

Minami H, Suzuki H, Kumagai H (2003) A mutant Bacillus subtilis gamma-glutamyltranspeptidase specialized in hydrolysis activity. FEMS Microbiol Lett 224:169–173

Mino K, Ishikawa K (2003a) Characterization of a novel thermostable O-acetylserine sulfhydrylase from Aeropyrum pernix K1. J Bacteriol 185:2277–2284

Mino K, Ishikawa K (2003b) A novel O-phospho-L-serine sulfhydrylation reaction catalyzed by O-acetylserine sulfhydrylase from Aeropyrum pernix K1. FEBS Lett 551:133–138

Muller A, Thomas GH, Horler R, Brannigan JA, Blagova E, Levdikov VM, Fogg MJ, Wilson KS, Wilkinson AJ (2005) An ATP-binding cassette-type cysteine transporter in *Campylobacter jejuni* inferred from the structure of an extracytoplasmic solute receptor protein. Mol Microbiol 57:143–155

Nakamori S, Kobayashi SI, Kobayashi C, Takagi H (1998) Overproduction of L-cysteine and L-cystine by *Escherichia coli* strains with a genetically altered serine acetyltransferase. Appl Environ Microbiol 64:1607–1611

Nakamura T, Iwahashi H, Eguchi Y (1984) Enzymatic proof for the identity of the S-sulfocysteine synthase and cysteine synthase B of Salmonella typhimurium. J Bacteriol 158:1122–1127

Oppezzo OJ (1998) In vivo effects of anti-inducers of the cysteine regulon in Salmonella typhimurium. FEMS Microbiol Lett 163:143–148

Ostrowski J, Kredich NM (1989) Molecular characterization of the cysJIH promoters of Salmonella typhimurium and Escherichia coli: regulation by cysB protein and *N*-acetyl-L-serine. J Bacteriol 171:130–140

Parry J, Clark DP (2002) Identification of a CysB-regulated gene involved in glutathione transport in Escherichia coli. FEMS Microbiol Lett 209:81–85

Pinto R, Tang QX, Britton WJ, Leyh TS, Triccas JA (2004) The Mycobacterium tuberculosis cysD and cysNC genes form a stress-induced operon that encodes a tri-functional sulfate-activating complex. Microbiology 150:1681–1686

Pittman MS, Corker H, Wu G, Binet MB, Moir AJ, Poole RK (2002) Cysteine is exported from the *Escherichia coli* cytoplasm by CydDC, an ATP-binding cassette-type transporter required for cytochrome assembly. J Biol Chem 277:49841–49849

Quadroni M, Staudenmann W, Kertesz M, James P (1996) Analysis of global responses by protein and peptide fingerprinting of proteins isolated by two-dimensional gel electrophoresis. Application to the sulfate-starvation response of *Escherichia coli*. Eur J Biochem 239:773–781

Rey DA, Nentwich SS, Koch DJ, Ruckert C, Puhler A, Tauch A, Kalinowski J (2005) The McbR repressor modulated by the effector substance S-adenosylhomocysteine controls directly the transcription of a regulon involved in sulphur metabolism of *Corynebacterium glutamicum* ATCC 13032. Mol Microbiol 56:871–887

Rey DA, Puhler A, Kalinowski J (2003) The putative transcriptional repressor McbR, member of the TetR-family, is involved in the regulation of the metabolic network directing the synthesis of sulfur containing amino acids in *Corynebacterium glutamicum*. J Biotechnol 103:51–65

Rodionov DA, Vitreschak AG, Mironov AA, Gelfand MS (2004) Comparative genomics of the methionine metabolism in Gram-positive bacteria: a variety of regulatory systems. Nucleic Acids Res 32:3340–3353

Ruckert C, Koch DJ, Rey DA, Albersmeier A, Mormann S, Puhler A, Kalinowski J (2005) Functional genomics and expression analysis of the *Corynebacterium glutamicum fpr2-cysIXHDNYZ* gene cluster involved in assimilatory sulphate reduction. BMC Genomics 6:121

Savijoki K, Ingmer H, Varmanen P (2006) Proteolytic systems of lactic acid bacteria. Appl Microbiol Biotechnol 71:394–406

Seiflein TA, Lawrence JG (2001) Methionine-to-cysteine recycling in *Klebsiella aerogenes*. J Bacteriol 183:336–346

Sekowska A, Danchin A (1999) Identification of *yrrU* as the methylthioadenosine nucleosidase gene in *Bacillus subtilis*. DNA Res 6:255–264

Sirko A, Zatyka M, Sadowy E, Hulanicka D (1995) Sulfate and thiosulfate transport in *Escherichia coli* K-12: evidence for a functional overlapping of sulfate- and thiosulfate-binding proteins. J Bacteriol 177:4134–4136

Sperandio B, Polard P, Ehrlich DS, Renault P, Guedon E (2005) Sulfur amino acid metabolism and its control in *Lactococcus lactis* IL1403. J Bacteriol 187:3762–3778

Suzuki H, Hashimoto W, Kumagai H (1993) Escherichia coli K-12 can utilize an exogenous gamma-glutamyl peptide as an amino acid source, for which gamma-glutamyltranspeptidase is essential. J Bacteriol 175:6038–6040

Suzuki H, Kamatani S, Kim ES, Kumagai H (2001) Aminopeptidases A, B, and N and dipeptidase D are the four cysteinylglycinases of Escherichia coli K-12. J Bacteriol 183:1489–1490

Suzuki H, Koyanagi T, Izuka S, Onishi A, Kumagai H (2005) The yliA, -B, -C, and -D genes of Escherichia coli K-12 encode a novel glutathione importer with an ATP-binding cassette. J Bacteriol 187:5861–5867

Takagi H, Yoshioka K, Awano N, Nakamori S, Ono B (2003) Role of Saccharomyces cerevisiae serine O-acetyltransferase in cysteine biosynthesis. FEMS Microbiol Lett 218:291–297

Thomas D, Surdin-Kerjan Y (1997) Metabolism of sulfur amino acids in *Saccharomyces cerevisiae*. Microbiol Mol Biol Rev 61:503–532

Turner MS, Woodberry T, Hafner LM, Giffard PM (1999) The *bspA* locus of *Lactobacillus fermentum* BR11 encodes an L-cystine uptake system. J Bacteriol 181:2192–2198

van der Ploeg JR, Barone M, Leisinger T (2001a) Functional analysis of the Bacillus subtilis cysK and cysJI genes. FEMS Microbiol Lett 201:29–35

van der Ploeg JR, Cummings NJ, Leisinger T, Connerton IF (1998) Bacillus subtilis genes for the utilization of sulfur from aliphatic sulfonates. Microbiology 144:2555–2561

van der Ploeg JR, Eichhorn E, Leisinger T (2001b) Sulfonate-sulfur metabolism and its regulation in *Escherichia coli*. Arch Microbiol 176:1–8

Vendeville A, Winzer K, Heurlier K, Tang CM, Hardie KR (2005) Making sense of metabolism: Autoinducer-2, LuxS and pathogenic bacteria. Nat Rev Microbiol 3:383–396

Vermeij P, Kertesz MA (1999) Pathways of assimilative sulfur metabolism in *Pseudomonas putida*. J Bacteriol 181:5833–5837

Wada M, Awano N, Haisa K, Takagi H, Nakamori S (2002) Purification, characterization and identification of cysteine desulfhydrase of *Corynebacterium glutamicum*, and its relationship to cysteine production. FEMS Microbiol Lett 217:103–107

Wheeler PR, Coldham NG, Keating L, Gordon SV, Wooff EE, Parish T, Hewinson RG (2005) Functional demonstration of reverse transsulfuration in the *Mycobacterium tuberculosis* complex reveals that methionine is the preferred sulfur source for pathogenic mycobacteria. J Biol Chem 280:8069–8078

White RH (2003) The biosynthesis of cysteine and homocysteine in *Methanococcus jannaschii*. Biochim Biophys Acta 1624:46–53

Williams SJ, Senaratne RH, Mougous JD, Riley LW, Bertozzi CR (2002) 5′-Adenosylphosphosulfate lies at a metabolic branchpoint in Mycobacteria. J Biol Chem 277:32606–32615

Wirtz M, Hell R (2006) Functional analysis of the cysteine synthase protein complex from plants: structural, biochemical and regulatory properties. J Plant Physiol 163:273–286

Wooff E, Michell SL, Gordon SV, Chambers MA, Bardarov S, Jacobs WR Jr, Hewinson RG, Wheeler PR (2002) Functional genomics reveals the sole sulphate transporter of the *Mycobacterium tuberculosis* complex and its relevance to the acquisition of sulphur in vivo. Mol Microbiol 43:653–663

Yamagata S, D'Andrea RJ, Fujisaki S, Isaji M, Nakamura K (1993) Cloning and bacterial expression of the *CYS3* gene encoding cystathionine γ-lyase of *Saccharomyces cerevisiae* and the physicochemical and enzymatic properties of the protein. J Bacteriol 175:4800–4808

Yoshida Y, Negishi M, Amano A, Oho T, Nakano Y (2003) Differences in the β C – S lyase activities of viridans group streptococci. Biochem Biophys Res Commun 300:55–60

Yvon M, Chambellon E, Bolotin A, Roudot-Algaron F (2000) Characterization and role of the branched-chain aminotransferase (BcaT) isolated from *Lactococcus lactis* subsp. cremoris NCDO 763. Appl Environ Microbiol 66:571–577

Zhao C, Moriga Y, Feng B, Kumada Y, Imanaka H, Imamura K, Nakanishi K (2006) On the interaction site of serine acetyltransferase in the cysteine synthase complex from Escherichia coli. Biochem Biophys Res Commun 341:911–916

Zhao CH, Ohno K, Sogoh K, Imamura K, Sakiyama T, Nakanishi K (2004) Production of nonproteinaceous amino acids using recombinant *Escherichia coli* cells expressing cysteine synthase and related enzymes with or without the secretion of *O*-acetyl-L-serine. J Biosci Bioeng 97:322–328

Microbiol Monogr (5)
V. F. Wendisch: Amino Acid Biosynthesis
DOI 10.1007/7171_2006_061/Published online: 9 November 2006

Microbial Arginine Biosynthesis: Pathway, Regulation and Industrial Production

Nicolas Glansdorff[1] (✉) · Ying Xu[2]

[1]JM Wiame Research Institute for Microbiology, 1 ave E. Gryzon, 1070 Brussels, Belgium
nglansdo@vub.ac.be

[2]Marine Sciences Research Center, SUNY at Stony Brook, Stony Brook, NY 11794-5000, USA

Abstract A vast number of prokaryotic and eukaryotic microorganisms can synthesize arginine de novo from glutamate. The first committed step of this pathway is acetylation of L-glutamate at the *N*-position. A surprising variety of proteins were found to catalyze this reaction in prokaryotes: (i) the classical, two-domain N-acetylglutamate synthase (NAGS) originally found in γ-Proteobacteria, (ii) shorter NAGS of the GNAT acetyltransferase family, either independent or fused with the ArgH protein (argininosuccinase), (iii) bifunctional ornithine acetyltransferases (OAT), i.e. able to acetylate glutamate with both acetyl-CoA and acetylornithine. In many organisms, including most Archaea, the enzyme acetylating glutamate remains elusive; possible connections with lysine biosynthesis may be envisaged. In fungi, NAGS appears only distantly related to its prokaryotic analog and requires association with acetylglutamate kinase (NAGK) to be functional. In most organisms, the acetyl group of acetylornithine is either split by an acetylornithinase (AO) or recycled on glutamate by OAT; in both cases, one of the products is ornithine which is carbamoylated into citrulline. In some Proteobacteria however, acetylornithine is carbamoylated into acetylcitrulline. These discoveries on arginine precursor acetylation have important metabolic and evolutionary implications.

Among Bacteria, regulation of arginine biosynthetic genes was analyzed intensively in *Escherichia coli*, *Salmonella typhimurium*, *Bacilli* and more recently in *Pseudomonas* sp. Archaea remain to be investigated. Among Eukarya, the yeast *Saccharomyces cerevisiae* was studied in great detail. The comparison of the mechanisms found to operate in these very different organisms is interesting from several points of view: (i) the occurrence of repressor-operator interactions on both sides of the prokaryote/eukaryote divide and the first evidence for a Jacob—Monod regulatory mechanism in eukaryotes, (ii) the coordination of carbamoyl phosphate synthesis with the two pathways that depend on this metabolite (arginine and pyrimidine biosyntheses), (iii) the coordination of arginine biosynthesis with arginine catabolism and the first indication ever that a repressor (the *E. coli* ArgR protein) may also function as a gene activator, (iv) the extensive conservation of the ArgR/AhrC transcription control system throughout the bacterial domain with the notable exception of *Pseudomonas* sp. and related Bacteria, (v) the functional and possible evolutionary relationship between proteins involved in arginine metabolic control systems and proteins controlling DNA replication and partition.

Arginine is an important "nutraceutical". Knowledge of the regulatory mechanisms controlling the function or synthesis of arginine biosynthetic enzymes in prokaryotes has been used to engineer arginine-overproducing strains amenable to industrial exploitation.

1
Introductory Remarks

Before examining the succession of enzymatic steps involved in arginine biosynthesis and their regulation, it should be emphasized that the pathway is interconnected in several ways (Figs. 1 and 2): (i) the intermediate ornithine is a constituent of certain hydroxamate siderophores (Winkelmann, 2002); in *E. coli*, when the flow of arginine intermediates is curtailed by feedback inhibition of acetylglutamate synthase, a transient iron depletion may ensue; this initiates a chain of reactions which increases the flow of aromatic amino acid precursors and leads to the production of the siderophore enterochelin (see Charlier and Glansdorff, 2004). Adding arginine to a growing culture can therefore be considered as a metabolic stress; (ii) L-Ornithine is a constituent of major membrane lipids in many Bacteria (Gao et al. 2004) while D- or L-ornithine is found in cell-walls of Gram-positive Bacteria; (iii) as a substrate for ornithine decarboxylase, ornithine is a source of putrescine and other polyamines, which are vital metabolites; when the flow of ornithine precursors is curtailed by the addition of arginine to the growth medium, putrescine must be synthesized from arginine via agmatine; (iv) ornithine is a limiting amino acid in the biosynthesis of several antibiotics (clavulanic acid, bacitracin, gramicidin S; Romero et al. 1986); (v) carbamoylphosphate (CP) is a substrate for both ornithine carbamoyltransferase (OTC) and aspartate carbamoyltransferase (ATC); it is therefore a precursor of both arginine and pyrimidines, a situation which calls for adequate regulatory mechanisms coordinating the corresponding pathways; (vi) in addition

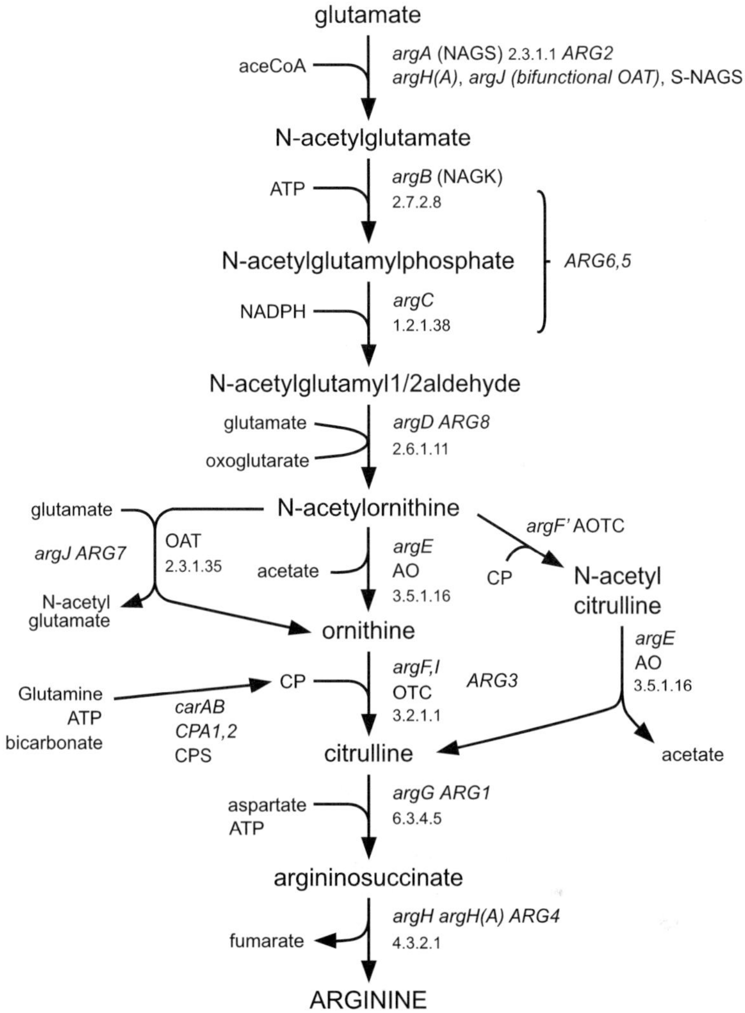

Fig. 1 Outline of pathways for de novo arginine biosynthesis from glutamate. Prokaryotic genetic symbols are in small italics, yeast genetic symbols in capital italics. EC numbers are indicated. Nonstandard abbreviations: NAGS, N-acetylglutamate synthase; S-NAGS, short N-acetylglutamate synthase; NAGK, N-acetylglutamate kinase; AO, N-acetylornithinase; OTC, ornithine carbamoyltransferase; AOTC, N-acetylornithine carbamoyltransferase; OAT, ornithine acetyltransferase; CP, carbamoylphosphate; CPS, carbamoylphosphate synthase

to these anabolic interconnections, arginine can be metabolized by various routes and the intermediates citrulline and ornithine, or their succinylated versions appear in certain arginine catabolic pathways; (vii) ornithine can be the intracellular source of proline; this happens in most Archaea that do not have the classical pathway for proline biosynthesis but generate proline

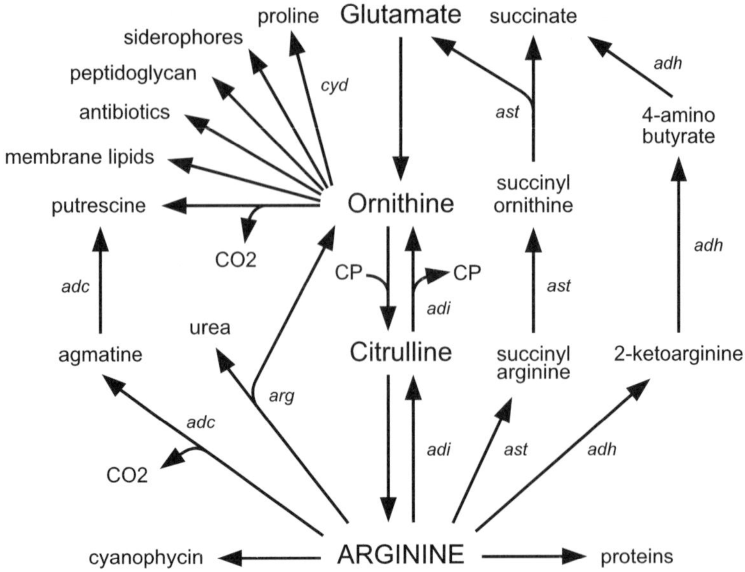

Fig. 2 Schematic representation of the main metabolic interconnections of arginine biosynthesis. Only key intermediates are mentioned. adi: arginine deiminase pathway; ast: arginine succinylase pathway; adh: arginine oxidase/dehydrogenase/transaminase pathway (Lu, 2006); arg: arginase pathway; adc: arginine decarboxylase pathway; cyd: ornithine cyclodeaminase; CP: carbamoylphosphate

by cyclization of ornithine (Graupner and White, 2001; Higuchi et al. 1999); (viii) in most cyanobacteria and in some heterotrophic bacteria, arginine itself is a component of cyanophycin (multi-L-arginyl-poly-L-aspartic acid) a polymer forming inclusion bodies that can be regarded as a nitrogen- and energy-storage compound (Maheswaran et al. 2006; Obst and Steinbuchel, 2006).

A detailed discussion of the metabolic consequences of these multiple interconnections is beyond the scope of this review but the coupling of arginine biosynthesis genetic regulation with the control of catabolism in various bacteria and yeast is briefly discussed in Sects. 3.2.1 and 3.2.2. The diverse arginine catabolic pathways of Bacteria and their control mechanisms were reviewed recently by Lu (2006).

It would have been impossible to cover the full range of information available on the enzymology and regulation of arginine biosynthesis in a single piece of work of reasonable size. Choices had to be made and the emphasis was put mainly on recent data and discoveries. Perhaps surprisingly, most of the novelties deal with the pathway itself. Recent reviews by Caldovic and Tuchman (2003) and Charlier and Glansdorff (2004) provide references to much of the early work, including the path-breaking discoveries of Gorini L, Maas W and Vogel H in the 1950s, and to more detailed accounts of certain aspects. Key papers and reviews concerning fungi are indicated in Sect. 3.2.2. The role of acetylglutamate in animal metabolism was reviewed by Caldovic and Tuchman (2003) and arginine biosynthesis in plants by Slocum (2005).

2
The de novo Pathway for Arginine Biosynthesis

De novo arginine biosynthesis occurs in many prokaryotes, fungi and plants but not in animals (Caldovic and Tuchman, 2003; Slocum, 2005). The present review focuses on prokaryotes and eukaryotic microorganisms.

2.1
Prokaryotes

In the classical de novo pathway for arginine biosynthesis, the intermediate ornithine is carbamoylated into citrulline. Recent findings, however, brought to light a novel pathway where the ornithine precursor—acetylornithine—is carbamoylated into acetylcitrulline. Moreover, the synthesis of acetylglutamate, the first committed intermediate of the pathway, is carried out by a previously unsuspected variety of enzymes. These findings have profound metabolic and evolutionary implications.

2.1.1
Biosynthesis of Arginine via Ornithine

In contrast to de novo proline biosynthesis, where glutamate is successively converted into glutamyl phosphate and the corresponding semi aldehyde that spontaneously cyclizes into pyrroline carboxylate, a direct proline precursor, the first intermediates of the arginine pathway are acetylated at the N-position (Fig. 1). This prevents the cyclization of the semi aldehyde intermediate. Most of the enzymology of the pathway was reviewed by Caldovic and Tuchman (2003) and Charlier and Glansdorff (2004). This review emphasizes more recent developments.

The first committed step of the pathway is catalyzed in many Proteobacteria (Table 1) by a two-domain enzyme: N-acetylglutamate synthase (NAGS, EC 2.3.1.1, ArgA). The N-terminal domain contains a carbamate kinase fold and is homologous to the next enzyme, acetylglutamate kinase (NAGK, EC 2.7.2.8, ArgB, Ramon-Maiques et al. 2002); this N-terminal domain may contain a glutamate binding site. The C-terminal domain contains an acetyl-CoA binding fold that is related to the vast family of Gcn5-related acetyltransferases (GNAT).

The same reaction can also be catalyzed by certain ornithine acetyltransferases (OAT, EC 2.3.1.35, ArgJ) which can use both acetyl-CoA and acetylor-

Table 1 Synthesis of acetylglutamate in prokaryotes

	GROUP [a]	OAT [b]	NAGS [c]	S-NAGS [d]	ARGH(A) [e]
BACTERIA					
ECOLI	γ Proteob.	–	+	–	–
VIBCH	"	–	+	–	–
VIBFI	"	–	+	–	+
VIBPA	"	–	+	+	+
VIBVU	"	–	+	–	+
PHOPR	"	–	+	–	+
COLPS	"	+	+	–	+
MORAB	"	nd	nd	nd	+
IDILO	"	–	–	–	+
PSEHA	"	–	–	–	+
PSEAE	"	m	+	–	–
NEIGO	β Proteob.	b	+	–	–
RHIME	α Proteob.	+	–	+	–
GEOSL	δ Proteob.	+	–	+	–
DESPS	"	+	–	+	–
DESVH	"	+	–	+	–
HELHE	ε Proteob.	+	–	+	–
WOLSU	"	+	–	+	–
CHLTE	Bact/Chlor.	+	–	+	–
AQUAE	Aquif.	+	–	+	–
PIRBA	Planctom.	+	–	+	–
LEPIN	Spiroch.	+	–	+	–
THETE	Ther/Dein.	m	–	+	–
DEIRA	"	+	–	+	–
LEIXY	Firmicutes	+	–	+	–
STRCO	"	m	–	+	–
STRAV	"	+	–	+	–
NOCFA	"	+	–	+	–
MYCTU	"	m	–	+	–
MYCLE	"	+	–	+	–
SYMTH	"	+	–	+	–
BACSU	"	b	–	–	–
GEOST	"	b	nd	nd	nd
GEOKA	"	+	–	+	–
BACCE	"	+	–	+	–
BACHA	"	+	–	+	–
THEMA	Thermot.	+	–	–	–
THENE	"	b	nd	nd	nd

Table 1 (continued)

	GROUP [a]	OAT [b]	NAGS [c]	S-NAGS [d]	ARGH(A) [e]
ARCHAEA	Euryarch.				
HALMA	"	–	–	+	–
METAC	"	+	–	+	–
METMA	"	+	–	+	–
METJA	"	m	–	–	–
ARCFU	"	+	–	–	–

[a] groups of organisms are listed in the order Proteobacteria (γ, β, α, δ, ε), Bacteroidetes/Chlorobium, Aquificales, Planctomycetes, Spirochaetes, Thermus/Deinococcus, Firmicutes (Actinobacteria and Bacilli), Thermotogales, Archaea (Euryarchaeota)
[b] ornithine acetyltransferase (m: monofunctional; b: bifunctional)
[c] acetylglutamate synthase
[d] short NAGS
[e] fusion between the gene for argininosuccinase and S–NAGS
nd: not determined. Species names abbreviated as in http://www.expasy.ch/cgi-bin/speclist

nithine as the acetyl donor for the synthesis of acetylglutamate (Fig. 1) and are therefore called "bifunctional" (Marc et al. 2000; Weigel et al. 2002). OAT is responsible for the conversion of acetylornithine into ornithine in many Bacteria; only a few were characterized as bifunctional (in *Bacillus subtilis, Geobacillus,* formerly *Bacillus stearothermophilus*, and *Neisseria gonorrheae*) despite widespread but invalid "bifunctional" genomic annotations based on mere sequence similarity (see Xu et al. 2006). Many OATs can not use acetyl-CoA as a substrate and are therefore called "monofunctional" (Table 1). Not surprisingly, a number of organisms using a bifunctional OAT have no NAGS, though some functional redundancies have been observed (f. ex. *N. gonorrheae*), that has both a bifunctional OAT and a NAGS.

There is no significant sequence similarity between NAGS and bifunctional OAT. This may not be surprising since contrary to NAGS, OAT operates by ping pong bi-bi mechanisms involving the formation of an acetyl-enzyme intermediate and the CoA moiety of acetyl-CoA does not appear to enter the OAT catalytic site (Weigel et al. 2002). Active OAT is produced by autoproteolysis of the ArgJ gene product at a conserved T residue that is also involved in catalysis (Marc et al. 2001; see also further yeast *ARG7*).

A specialized, monofunctional OAT takes part in the synthesis of the antibiotic clavulanic acid in *Streptomyces clavuligerus* (Kershaw et al. 2002; de la Fuente et al. 2004; Elkins et al. 2005).

As OAT recycles the acetyl group on glutamate, it provides a more economical pathway for ornithine synthesis than the alternative route via acetylornithinase (AO, EC 3.5.1.16, ArgE) which simply splits acetylornithine into ornithine and acetate. OAT and AO epitomize the so-called "cyclic" and "linear" pathways for ornithine synthesis, respectively. AO appears restricted to a few groups of Proteobacteria (including enterics where it was first characterized) and may operate in Archaea (see further). However, as AO can deacetylate different acetylated amino acids and is homologous to succinyldiaminopimelate desuccinylase (DapE, EC 3.5.1.18, a lysine biosynthetic enzyme) and certain carboxypeptidases, genomic annotation of a deacetylase as ArgE (as in *P. aeruginosa*, which has an OAT) must be regarded with caution in the absence of functional evidence (Charlier and Glansdorff, 2004). Moreover, genes for amidohydrolases only distantly related to ArgE could be cloned by complementation of *E. coli argE* mutants (Sakanyan et al. 1993b). Some of these genes complement only weakly, such as the hippurate hydrolase *hipO* determinant from *C. jejuni*; it is interesting that *C. jejuni hipO* weakly complements *argA* mutants as well (Hani et al. 1999).

Surprisingly, the genomes of numerous Bacteria from widely different lineages were found to possess genes coding for putative GNAT-related enzymes (from 150 to 170 amino acids, the length of the C-terminal, acetyl-CoA domain of classical NAGS) able to complement *argA E. coli* mutants. When the cognate organism has no classical NAGS and no OAT or only a monofunctional OAT (see Table 1 and Xu et al. 2006), the presumption that it depends on this shorter NAGS version for glutamate acetylation is very strong. Among these short NAGS (S-NAGS), we find *C. jejuni* ArgO (Hani et al. 1999) and a distantly related, large group of putative acetyltransferases that are clearly homologous to the C-terminal domain of NAGS (Xu et al. 2000, 2006); the latter either are independent proteins (the only example characterized biochemically is from *Mycobacterium tuberculosis*, Errey and Blanchard, 2005) or are fused with ArgH (argininosuccinase) that catalyzes the last step of the pathway. The *argH(A)* fusion pattern appears restricted to marine Bacteria of the *Alteromonas-Vibrio* group (Xu et al. 2000, 2006). It is possible that these short NAGS (S-NAGS) require association with a protein partner providing an efficient glutamate binding site (Errey and Blanchard, 2005; Xu et al. 2006). This would be reminiscent of fungal NAGS that show very little similarity with classical prokaryotic NAGS but require association with NAGK to be functional (Abadjieva et al. 2001; Pauwels et al. 2003).

The enzymes catalyzing the second, third and fourth steps of ornithine biosynthesis (ArgB, C and D, respectively) are ubiquitous among Bacteria; in *E. coli*, the transaminase ArgD was shown to be identical with DapC, an enzyme of the lysine pathway (Ledwidge and Blanchard, 1999).

The situation in Archaea is much less documented than in Bacteria and eukaryotic microorganisms. NAGS (ArgA) appears absent from this domain. ArgJ (uncharacterized functionally except in *Methanocaldococcus jannaschii* where it appears to be monofunctional, Marc et al. 2000) is found in the genomes of some Euryarchaeota, where putative short NAGS were found as well. Early experiments reported enzymatic activities associated with the cyclic arginine pathway in methanogens at a time no genetic correlation could be attempted (Meile and Leisinger, 1984).

An interesting hypothesis (Nishida et al. 1999) suggests that the genes annotated *argB, C, D* and *E* in most Archaea and in *Pyrococci* in particular (Makarova et al. 1999; Brinkman et al. 2002; Cohen et al. 2003) code for bifunctional enzymes involved in both arginine biosynthesis and a new pathway for synthesis of lysine from aminoadipic acid (AAA) discovered in *Thermus thermophilus.* This hypothesis remains unsubstantiated however; moreover, at least some of the genes for the diaminopimelate lysine biosynthetic pathway are present in *Pyrococci* (Velasco et al. 2002) and the lysine and arginine pathways remain separate in *Thermus*, even though *Thermus* ArgD and E actually are active on both arginine and lysine intermediates (see Brinkman et al. 2002; Miyasaki et al. 2001, 2002).

Of potential interest for glutamate acetylation in Archaea is the protein LysX which was shown by genetic disruption to catalyze the synthesis of the first intermediate in the conversion of AAA to lysine in *Thermus* (Nishida et al. 1999; Sakai et al. 2003); it is an enzyme of the ATP-dependent carboxylate-amino ligase RimK family. In *Thermus*, LysX may function exclusively in lysine biosynthesis (though this remains to be demonstrated) since acetylglutamate appears to be synthesized by the short version of ArgA (and not by ArgJ, as erroneously assumed in Nishida et al. 1999, since *Thermus* ArgJ is monofunctional, Baetens et al. 1998). As most Archaea lack an obvious enzyme to synthesize acetylglutamate the question arises whether a LysX homologue could be involved in glutamate acetylation in these organisms; *lysX* homologues have been detected in several archaeal genomes in close linkage with annotated *arg* genes, but their exact function remains to be defined.

Whatever the actual relationship between the arginine and lysine pathways, the pattern of glutamate acetylation in Archaea may thus turn out to be as varied as it is in Bacteria: short NAGS, bifunctional OATs (yet to be identified in this domain however), LysX homologues and other putative acetyltransferases are all candidates for this metabolic function.

In most organisms investigated until now, ornithine is carbamoylated into citrulline by ornithine carbamoyltransferase (OTC, EC 2.1.3.3). The gene is commonly called *argF*; actually *E. coli K12* was found to contain two paralogous genes, *argF* and *argI* whose products interact to form hybrid trimeric OTCs (see Charlier and Glansdorff 2004). Though the symbol *argF* is most frequently used to designate OTC in various organisms, the genuine resident gene of enterics is *argI*; *argF* features as one of the earliest cases of horizontal transfer to have been described among Bacteria: the gene has a GC content higher than the *E. coli* average and is flanked by two IS1 elements. It is unfortunate that the symbol *argI* has been repeatedly used in genomic annotations to designate the gene for arginase.

OTC has been particularly well studied in extremophiles. OTC from the hyperthermophilic archaeon *P. furiosus* is an eloquent example of the role played by hydrophobic interactions in the stabilization of a high order quaternary structure at temperatures close to 100 °C (Villeret et al. 1998; Massant et al. 2005); on the other hand, OTC from the strictly psychrophilic bacterium *Moritella abyssi* illustrates the trade off between substrate affinity and catalytic velocity that accompanies adaptation to low temperature, suggesting that enzyme adaptation to cold may be constrained by natural limits to optimization of catalytic efficiency (Xu et al. 2003a).

The second substrate of the reaction catalyzed by OTC is carbamoylphosphate (CP). Depending on the microorganism, CP synthesis is catalyzed by a single carbamoylphosphate synthase (CPS) or by two distinct enzymes integrated in the arginine and pyrimidine regulons, respectively (see Sect. 3 and below); the amino donor is glutamine. The enzyme consists of two subunits, a small one (*E. coli* CarA, alias PyrAA in certain organisms, a glutaminase) and a larger one CarB (alias PyrAB), that can use NH_3 as a low affinity nitrogen donor for the synthesis of CP. The CarB protein synthesizes carboxyphosphate from bicarbonate and ATP, and carbamoylphosphate from carbamate and a second molecule of ATP; ammonia from the amino group of glutamine is channeled through the enzyme towards the carboxyphosphate active site, to allow the formation of carbamate (Thoden et al. 1997). The *carB* gene was shown to consist of two modules probably resulting from the duplication of an ancestral gene. The N-terminal part is the carboxyphosphate-forming part, whereas the C-terminal half is responsible for CP formation (Kothe et al. 2005). Structural aspects and details of the catalytic mechanism were reviewed by Charlier and Glansdorff, (2004) and Holden et al. (1999). Available databases show that the two genes are usually clustered in one *carAB* operon but in some organisms they are unlinked and in others still—notably Archaea but also *Aquifex aeolicus* (Ahuja et al. 2001)—the two halves of the CarB subunit are encoded by separate genes.

The single-CPS situation is typical of Gram-negative Bacteria such as *E. coli* and *P. aeruginosa* (Charlier et al. 2004; Kwon et al. 1994). In *P. aeruginosa* the *car* operon contains an unidentified open reading frame and is terminated by the gene for a transcriptional elongation factor, a situation of probable physiological significance (Lu et al. 1997). In some Gram-positive bacteria and in fungi, CP is synthesized by arginine- and pyrimidine-specific CPS, respectively (see Sect. 3.1.1); organisms devoid of CPS may synthesize CP from arginine by the catabolic arginine deiminase pathway (Cunin et al. 1986; Nicoloff et al. 2001).

In the hyperthermophilic Archaea *Pyrococci*, CP synthesis is catalyzed by a carbamate kinase (Legrain et al. 1995; Purcarea et al. 1996), an enzyme that may be related to the ancestor of CPS (Baur et al. 1989; Durbecq et al. 1997). Some debate developed regarding the number of ATP molecules consumed in the reaction using different preparations of enzymes from *Pyrococcus furiosus* and *P. abyssi* (Purcarea et al. 1996, 2001; Durbecq et al. 1997) but structural analysis leaves no doubt that the enzyme from *P. furiosus* is a carbamate kinase and crystal-pure preparations were reported to consume only one ATP molecule during the reaction, as would be expected from a true carbamate kinase (Uriarte et al. 1999, 2001). Kinetic arguments suggested that *P. furiosus* could protect the thermolabile CP from hydrolytic thermodegradation by channeling it between the kinase and OTC (Legrain et al. 1995); the two enzymes were actually shown to form in vivo and in vitro a loose complex capable of channeling CP toward the synthesis of citrulline (Massant et al. 2002); with aspartate carbamoyltransferase (ATC) a functionally similar but structurally quite different complex channels CP toward carbamoylaspartate, the first intermediate in the pyrimidine pathway (Massant et al. 2004, 2005). Kinetic evidence for CP channeling was also obtained for *P. abyssi* (Purcarea et al. 1999) and for *T. aquaticus*, where, however, a true CPS and not a carbamate kinase is responsible for CP synthesis (Van de Casteele, 1997). There are *carA* and *carB* homologues in the genome of *P. furiosus* but not of *P. horikoshii* nor of *P. abyssi* (Cammarano et al. 2002; Cohen et al. 2003), but their metabolic function is presently not known. It would appear, however, that the environmental conditions under which CP synthesis is expected to proceed in *Pyrococcus* are particularly adequate for a carbamate kinase (Uriarte et al. 1999).

The intermediate citrulline is converted by argininosuccinate synthase (EC 6.3.4.5 ArgG) into argininosuccinate, which is then split into arginine and fumarate by argininosuccinase (syn. argininosuccinate lyase, EC 4.3.2.1, ArgH). We have already mentioned above the unexpected gene fusion discovered in certain marine bacteria between *argH* and the short NAGS gene (Xu et al. 2006). The impact of this structure on the kinetics of glutamate acetylation remains to be investigated.

2.1.2
Biosynthesis of Arginine via Acetylornithine

The discovery of an acetylornithine carbamoyltransferase (AOTC) is one of the most recent and least expected additions to the repertoire of arginine biosynthetic functions. The cognate gene (*argF′*, Morizono et al. 2006) is homologous to *argF* but lacks the SMG conserved motif. It appears to replace *argF* in *Xylella fastidiosa* and *Xanthomonas campestris* (two Proteobacteria pathogenic for plants), in certain Flavobacteria and Bacteroidetes (ibid.). Acetylornithine carbamoylation was studied on pure *X. campestris* AOTC (Morizono et al. 2006). In *Bacteroides fragilis*, *argF′* was shown to be essential though the activity could not be detected (Shi et al. 2002), but recent findings show that *Bacteroides argF′* in fact codes for a novel N-succinyl-L-ornithine transferase (Shi et al. 2006). *Bacteroides* arginine biosynthesis therefore would appear to be different from the cognate pathway of most organisms.

The AOTC gene family appears to be ancient (Naumoff et al. 2004) and it will be interesting to see how close to the Last Universal Common Ancestor (LUCA) it can be traced. The phylogeny of carbamoyltransferases is certainly complex since previous studies by Labedan et al. (1999, 2004) suggested that the LUCA contained at least two exemplars of both OTC and ATC that segregated in different prokaryotic and eukaryotic cell lines according to a pattern of differential extinction.

AOTC catalyzes the synthesis of acetylcitrulline which is then deacetylated into citrulline. The enzyme responsible for the deacetylation appears to be an ArgE homolog. The fact that the same enzyme can also deacetylate acetylornithine (Morizono et al. 2006) and is therefore not a novel acetylcitrulline specific deacetylase (Shi et al. 2005) is presumably of prime metabolic importance since ornithine itself may be required for a variety of cellular functions (see introductory remarks).

2.1.3
Evolutionary Considerations

The classical two-domain NAGS appears to be present in some Proteobacteria only and absent from Archaea. On the other hand, OAT is found in many if not most Bacteria, in some Euryarchaeota and in lower Eukaryotes; for most of them it is, however, not known whether it is mono or bifunctional. Furthermore, a large number of Bacteria and some Archaea display the type of S-NAGS gene that was discovered recently; most significant are those instances where S-NAGS is present in an organism devoid of both a classical NAGS and a bifunctional OAT. Despite the many gaps in our information, it would appear that the most likely candidate for a primeval acetylglutamate synthase in the LUCA is S-NAGS. Bifunctional OAT may have become involved at some later stage; the fact that OATs occur in some Euryarchaeota but appear absent from other Archaea may suggest that it has been acquired by horizontal gene transfer though it can not be excluded that it was present in the LUCA and was lost in all but a few lineages.

The organization of classical NAGS in two domains suggests that it originated by fusion between a S-NAGS gene and a copy of the gene presently recognized as *argB* (Xu et al. 2006). It may very well be, however, that more than one type of S-NAGS was recruited from the vast array of GNAT acetyltransferases as already suggested by *C. jejuni* ArgO and the putative acetyltransferase annotated ArgA in the genome of *Xanthomonas campestris*, that appear only remotely related to other S-NAGS. This may also be the case in organisms where candidates for glutamate acetylation remain at present elusive. Besides, we must keep the possibility that LysX homologs may fulfill this function, particularly in Archaea.

A peculiarity of S-NAGS is the absence of any obvious glutamate-binding domain. This is corroborated by the extremely high Km (> 600 mM) for glutamate of *Mycobacterium* S-NAGS (Errey and Blanchard, 2005). As already mentioned above, the protein may have to associate with another one (perhaps NAGK, ArgB, as the fungal situation suggests) to achieve metabolic efficiency. The role of the ArgH protein in the organisms harboring *argH(A)* fusions remains to be determined.

An observation of as yet unknown significance is the presence of not functionally identified genes co-regulated with arginine operons in *Lactobacilli* and in *Thermus* (Arsene-Ploetze et al. 2005; Sanchez et al. 2000) and of such a gene in the *P. aeruginosa car* operon (Lu et al. 1997).

2.2
Eukaryotes

The most intensively studied eukaryotic microorganisms are fungi (mainly Ascomycetes). For early work on the basidiomycete *Coprinus* see Makoff and Radford (1978).

Compared with Bacteria, the gene enzyme relationship for ornithine synthesis in Ascomycetes displays specific features. In *Saccharomyces cerevisiae* and *Neurospora crassa*, the synthase and the kinase associate into a complex whose integrity is essential for activity and effective inhibition of the synthase (see Sect. 3.1.1). Yeast Arg5p (ArgC) and Arg6p (ArgB) are produced from a single polypeptide that is maturated in the mitochondrion (Boonchird et al. 1991; Gessert et al. 1994; Para-Gessert et al. 1998). These organisms use an OAT which, as in prokaryotes (Marc et al. 2001), is produced from a precursor protein by a self-cleavage mechanism operating at a conserved T residue that is also involved in the formation of the acetyl-enzyme intermediate (Abadjeva et al. 2000). Yeast OAT can use both acetylornithine and acetyl-CoA as the acetyl donor albeit only acetylornithine is used efficiently (Crabeel et al. 1997). The green algae *Chlorella* also appears to synthesize a bifunctional OAT since the two activities could not be separated in the course of purification (Morris and Thompson, 1975).

In *S. cerevisiae*, the first five steps of arginine biosynthesis take place in the mitochondrion (see Abadjeva et al. 2001) and ornithine is exported to the cytoplasm by transporter Ort1p (Crabeel et al. 1996; Palmieri et al. 1997). The identification of this transporter by Crabeel's group solved a long-standing mystery: mutants in the cognate gene, previously denominated *ARG11*, are arginine bradytrophs in which no enzymatic function had been found to be impaired. Mutations in a putative mitochondrial basic amino acid carrier encoded by the arg-13 gene of *N. crassa* give a similar phenotype (Liu and Dunlap, 1996).

3
Regulation of Arginine Biosynthesis

Arginine biosynthesis became a paradigm for the study of metabolic regulation very early (reviewed by Charlier and Glansdorff, 2004). As in other metabolic pathways, regulation of enzyme activity has to be distinguished from regulation of enzyme synthesis.

3.1 Regulation of Enzyme Activity

The flow of arginine precursors is controlled at the steps that limit the rate of formation of acetylated intermediates and at the level of carbamoylphosphate (CP) synthesis. In addition, in some yeasts and in *B. subtilis*, that catabolize arginine by the arginase pathway, a remarkable enzyme–enzyme activity control takes place: in the presence of arginine and ornithine, an OTC-arginase complex is formed, where OTC activity is inhibited but arginase remains catalytically active (see Sect. 3.1.2).

3.1.1 Activity Control by Metabolites

In *E. coli* and the other microorganisms that use an acetylornithinase but no acetyl recycling enzyme, NAGS catalyzes the first committed step of the pathway; it is feed-back-inhibited by arginine. When the acetyl group is recycled by an OAT, whether a NAGS is present or not, the reaction catalyzed by NAGK can become flow-controlling and is the actual target of feedback inhibition by arginine in a number of organisms (Udaka, 1965; Caldovic and Tuchmann, 2003; Charlier and Glansdorff, 2004; Ramon-Maiquez et al. 2006 and other references below). For example in *T. maritima*, where there is an OAT which is probably bifunctional, as suggested by the properties of its homologue from *T. neapolitana*, there is no NAGS nor any obvious S-NAGS (Xu et al. 2006) and it is NAGK that is inhibited by arginine (Fernandez-Murga et al. 2004; Ramon-Maiques et al. 2006). In *Corynebacterium glutamicum*, where there is a monofunctional OAT and the glutamate acetylase is not yet identified, NAGK is feedback inhibited (Sakanyan et al. 1996).

The transacetylation step itself is, however, also a "logical" target to control the flow of acetylated intermediates; in *Geobacillus stearothermophilus* NAGK is not inhibited by arginine (Sakanyan et al. 1993c), but both activities of the bifunctional OAT are strongly inhibited by ornithine (Sakanyan et al. 1993a). In *Thermus aquaticus* OAT is inhibited by arginine (Van de Casteele et al. 1990). In *P. aeruginosa*, where both NAGS and OAT are present, NAGS assumes an anaplerotic function only (i.e. primes the cycle of acetylated precursors with a first molecule of acetylglutamate) but both NAGS and NAGK are nevertheless inhibited by arginine (Haas et al. 1972). In *M. tuberculosis*, where a S-NAGS operates in conjunction with a monofunctional OAT, S-NAGS is inhibited by arginine (Errey and Blanchard, 2005).

A recent paper by Ramon-Maiquez et al. (2006) brings to light a structural basis for feedback inhibition by arginine. *E. coli* NAGK is an arginine-insensitive homodimer whereas *T. maritima* and *P. aeruginosa* arginine-sensitive NAGKs are ring-like homohexamers where arginine binds to each dimeric subunit at a site flanking the interdimeric junction. The fact that *E. coli* arginine-sensitive NAGS also is a hexamer and that NAGK is homologous to the N-terminal domain of NAGS led Ramon-Maiques et al. (2006) to localize the molecular signature for arginine inhibition in that N-terminal domain. It should be noted, however, that the activity of *Mycobacterium* S-NAGS, where this domain is absent, is nevertheless inhibited by arginine (Errey and Blanchard, 2005). In S-NAGS, both the glutamate and arginine binding sites thus remain undefined.

The sensitivity towards arginine of NAGKs from photosynthetic organisms (including cyanobacteria) is modulated by the nitrogen signal transduction protein P_{II}. The two proteins form a complex in which NAGK activity is enhanced and the concentration of arginine required for inhibition is increased; this regulation allows the accumulation of arginine when nitrogen is abundant (Maheswaran et al. 2004; Herrero, 2004); arginine is incorporated in a polymer of arginine and aspartic acid, cyanophycin (ibid and Sect. 4).

The situation in fungi was studied in great detail in *S. cerevisiae* (Abadjeva et al. 2001; Pauwels et al. 2003). Yeast NAGS (Arg2p) bears little amino acid sequence similarity to its prokaryotic counterpart but the same regulatory logic as in *P. aeruginosa* is maintained: both NAGS and NAGK are inhibited by arginine. However, to be both functional and sensitive to arginine inhibition, NAGS must be associated with NAGK. Knocking out arginine inhibition of NAGK makes NAGS arginine insensitive, whereas deleting the NAGS gene decreases the arginine sensitivity of the kinase. In *Neurospora crassa*, the situation would appear to be similar (Weiss and Lee,1980; Yu et al. 1996).

CP is a precursor of both arginine and pyrimidines, and it is therefore no surprise that the biosynthesis of this compound is regulated at the enzyme level by metabolites of both pathways. In *E. coli* and *Salmonella*, where the mechanisms were studied in great detail in the wild type and different types of mutants (see Charlier and Glansdorff, 2004) a single CPS supplies the two pathways. CPS activity is inhibited by UMP; this inhibition is antagonized by IMP and, most importantly, by ornithine. Since addition of arginine curtails the flow of ornithine precursors, this control mechanism amounts to a double feedback inhibition by arginine and pyrimidines. In the presence of saturating concentrations of ornithine, UMP does not inhibit the activity by more than about half and in the absence of ornithine, the inhibition becomes total. Allosteric control of CPS activity is, however, not perfectly balanced between the two pathways and regulatory interferences between them have been observed (Charlier and Glansdorff, 2004).

It will be interesting to determine how CPS is regulated in organisms where CP is used to carbamoylate acetylornithine rather than ornithine.

Because of the complexity of the reaction catalyzed by CPS and of its allosteric control mechanisms, a variety of phenotypes (for example sensitivity towards arginine or uracil, or sensitivity to both) may result from mutations affecting the *carB* gene (Delannay et al. 1999).

Another pattern is observed in some Gram-positive bacteria and fungi (Cunin et al. 1986; Davis, 1986; Messenguy and Dubois, 2000; Nicoloff, 2001, below and Sect. 3.2.2) where CP is provided by two different CPS, one repressed by arginine, the other inhibited and repressed by a pyrimidine nucleotide; a uracil-sensitive phenotype arises from mutations that impair the enzyme controlled by arginine. In bacteria with two CPS, the cognate genes are respectively integrated in a cluster of arginine biosynthetic genes (*carA* and *B*) and in a cluster of pyrimidine biosynthetic genes (where *carA* is sometimes annotated *pyrAA* according to a previous nomenclature). In Bacteria, the two-CPS situation is characteristic of certain *Bacilli*, such as *B. subtilis* and *Geobacillus stearothermophilus* (Quinn et al. 1991; Yang et al. 1997); the genomes of *B. licheniformis* and *B. halodurans* also appear to contain two CPS whereas only one, pyrimidine-specific CPS is annotated in the genome of *B. anthracis* and *B. cereus*. A similar disparity was noted among *Lactobacilli* (Bringel et al. 1997; Nicoloff et al. 2000, 2001; Kilstrup et al. 2005). In fungi the pyrimidine-specific CPS gene is part of the composite *URA2* gene encoding a multifunctional protein containing also aspartate carbamoyltransferase and a dihydroorotase-like protein (Davis, 1986; Souciet et al. 1986).

3.1.2 Enzyme–Enzyme Activity Control

The yeast *S. cerevisiae* has the potential to operate a urea cycle: arginine can be degraded into urea and ornithine by an arginase. It was recognized very early that in the presence of arginine and ornithine, OTC and arginase, both trimeric proteins, can form a one-to-one complex in which arginase remains active but OTC activity is inhibited (the so-called epi-arginase mode of regulation, Messenguy and Wiame, 1969). This regulation prevents the formation of a futile ornithine cycle when the biosynthesis of arginine is repressed and the catabolism of arginine activated by this amino acid. The residues responsible for the specificity of this protein–protein interaction have been identified (El Alaoui et al. 2003). The occurrence of this type of regulation is linked to the pattern of intracellular compartmentation of the enzymes concerned: it operates in yeasts where both OTC and arginase are cytosolic but not when OTC is segregated in the mitochondrion (ibid.).

The same type of regulation was found to operate in *B. subtilis* but has not been investigated in comparable detail (Issaly and Issaly 1974).

3.2
Regulation of Enzyme Synthesis

At the time the concept of the operon surged onto the biological scene (Jacob and Monod, 1961), the *E. coli* arginine pathway already appeared as a model system to study control of gene expression at a higher level of complexity. Indeed, the *arg* genes were not clustered into a single functional unit. Moreover, the occurrence of an intermediate common to the biosyntheses of arginine and the pyrimidines—carbamoyl phosphate—raised very early the question of gene regulation at metabolic branching points. Since these early discoveries, regulation of arginine biosynthesis has been studied in considerable detail in a variety of prokaryotes and in a few eukaryotic microorganisms, mainly fungi.

3.2.1
Prokaryotes

As regards regulation of enzyme synthesis in the arginine pathway, the archaeal domain is still unchartered territory, whereas three types of Bacteria have been studied in depth: the enterics (*E. coli*, *Salmonella* and *Moritella*), the *Bacilli* (*B. subtilis* and *G. stearothermophilus*) and *Pseudomonas aeruginosa*. To these molecular studies, the explosion of genomic data witnessed in the last few years has added much information on the phylogenetic distribution of regulatory mechanisms and their determinants.

3.2.1.1
The ArgR/AhrC System

Studies on arginine biosynthesis in *E. coli* gave rise to the concept of regulon (Maas and Clark, 1964; Maas et al. 1964; Maas, 1994): a set of structural genes controlled by a common transcriptional regulator. The system is now known in considerable detail (see Charlier and Glansdorff, 2004 and Charlier, 2004). The *E. coli* arginine repressor (ArgR) controls gene expression (including its own) by steric exclusion of RNA polymerase binding. Repressor binds at operators consisting of so-called ARG boxes: 18 bp-long imperfect palindromes overlapping the promoter elements. In most genes these boxes are present in pairs separated by 3 bp (2 for the *argR* gene itself).

In addition to arginine biosynthetic genes, *E. coli* ArgR also represses genes involved in an arginine specific transport system (encoded by the *artPIQMJ* cluster; TC 3.A.1.3.3) and exerts a modest repression on the expression of the *hisJQMP* histidine uptake system (TC 3.A.1.3.1; Caldara, Charlier and Cunin, personal communication).

At the time these molecular features were being unraveled it came as a surprise to many that the arginine system was not controlled by attenuation, as many other amino acid biosyntheses. Perhaps stretches of adjacent arginine codons would impede ribosome translocation because of the intrinsic instability of codon-anticodon interactions involving inosine or for other reasons linked to the usage of arginine codons (see Charlier and Glansdorff, 2004). It should also be noted that, despite the existence of RNA aptamers able to recognize individual molecules of arginine or arginine-rich peptides, no arginine-specific "riboswitch" control mechanism has been reported in prokaryotes, in contrast to the involvement of the chemically related amino acid lysine in such regulatory mechanisms (ibid.; see further for a putative arginase riboswitch in the fungus *Aspergillus*).

E. coli ArgR is a DNA binding protein belonging to the wHTH (winged helix-turn-helix) family, by far the most abundant and perhaps most ancient one in prokaryotes (Perez-Rueda et al. 2004). The ArgR family, however, appears absent from Archaea. It remains to be determined whether this absence is in anyway related to the above-mentioned possibility that in Archaea arginine and lysine biosyntheses might be part of the same system. Among Bacteria, ArgR and its mode of regulation is widespread (see Charlier, 2004; Makarova et al. 2001; Perez-Rueda et al. 2004, and below). In *Pseudomonas* species ArgR is replaced by a regulatory factor of the AraC/XylS family (Park et al. 1997 and below).

Extensive DNA-protein interaction studies and structural analyses brought to light fundamental similarities between *E. coli* ArgR and its homolog in *B. subtilis* (AhrC) and *G. stearothermophilus* (Van Duyne et al. 1996; Sunnerhagen et al. 1997; Ni et al. 1999; Dennis et al. 2002; Charlier, 2004; Charlier and Glansdorff, 2004 and additional references therein) despite relatively low sequence identity (between 25 and 30%). ArgR consists of a basic N-terminal DNA-domain connected through a flexible linker to an acidic C-terminal domain responsible for arginine binding and oligomerization of the subunit in a hexamer composed of two trimers; six arginine molecules bind at the interface between trimers. The conformational change induced by arginine binding increases the affinity of the hexameric repressor for DNA and orients the recognition helices of the wHTH motifs in such a way that they dock into successive major grooves on the same face of the DNA helix (Ni et al. 1999). The minimal target is a single ARG box, where base-specific contacts occur between two successive major groove segments and the intervening minor groove (Wang et al. 1988). Binding of a hexameric ArgR molecule to a tandem pair of ARG boxes exhibits a chelating effect provided the boxes are correctly spaced (Charlier et al. 1992; Tian et al. 1992; Burke et al. 1994).

ArgR blocks transcription by steric exclusion of RNA polymerase but is unable to block an elongating polymerase. This feature is particularly important to understand the role of ArgR in the regulation of the CPS genes (reviewed by Charlier and Glansdorff, 2004). The *carAB* operon is cumulatively repressed by arginine and the pyrimidines. The *carAB* promoter region consists of two consecutive promoters: P1, upstream and P2, downstream, overlapping a pair of ARG boxes. P1 is controlled by pyrimidines through a complex and architecturally dependent mechanism involving the general regulatory factor IHF, the PepA protein and UMP kinase, the pyrimidine sensor (PyrH); P1 is also subject to moderate repression by the PurR repressor and to stringent control of nucleic acid metabolism (Charlier and Glansdorff, 2004; Devroede et al. 2004, 2006); it is however clear, both in *E. coli* and *Salmonella*, where the system has also been studied in detail (Lu et al. 1992), that ArgR is not involved in P1 control and, most importantly, that transcription initiated at P1 is not blocked by P2-bound repressor, both in vivo and in vitro (Charlier, personal communication). P1 does not present the attenuation features found in the control region of the *pyrBI* operon, coding for aspartate carbamoyltransferase.

In some *Bacilli* and other Gram-positive Bacteria, multiple exemplars of potential arg repressor genes have been found; in *L. plantarum,* and *L. lactis* where the situation has been studied in detail, two ArgR proteins appear to interact in regulating arginine genes, perhaps by forming heterohexamers (Nicoloff et al. 2004; Larsen et al. 2005). The possible ecological significance of this gene multiplicity has been discussed by Nicoloff et al. (2004).

3.2.1.2
ArgR in Pseudomonas and Acinetobacter

In *P. aeruginosa* the transcription of only three of the arginine biosynthetic operons is repressible by arginine: *argF*, *carAB* and *argG* (Kwon et al. 1994; Lu et al. 2004). Ornithine biosynthesis is thus not repressible by arginine though it is feed-back inhibited by it (see above, Sect. 3.1). The repression is mediated by a regulatory molecule ArgR that is unrelated to ArgR/AhrC, both functionally and structurally (Park et al. 1997); the protein is dimeric and belongs to the AraC/XylS family of transcriptional regulators; it possesses two HTH DNA-binding domains at the C-terminus of each subunit. This ArgR is also involved as an activator in the regulation of the arginine-succinyltransferase (AST catabolic pathway (see below). In vitro, arginine does not enhance the affinity of *P. aeruginosa* ArgR for its target sites, but the protein is auto-induced from the arginine-responsive promoter of an operon coding for elements of an arginine/ornithine uptake system and for ArgR, the last gene of the operon (Nishijyo et al. 1998). In contrast to *E. coli*, the *P. aeruginosa car* operon is expressed from a single promoter; arginine modulates initiation of transcription and pyrimidines control *car* expression by a superimposed attenuation mechanism (Kwon et al. 1994).

Despite the absence of repression by arginine of ornithine biosynthetic enzymes in *P. aeruginosa*, it should be mentioned that Hebert and Houghton (1997) isolated mutants in a transcriptional regulatory gene (*oruR*) that would appear to control ornithine catabolism at the level of OAT synthesis by preventing the production of glutamate from ornithine (the reverse of the biosynthetic reaction).

The nutritionally versatile *Acinetobacter* strain ADP1 is related to *P. aeruginosa*, but nevertheless presents an interesting departure from the pattern outlined above: the genome contains a cluster regrouping genes coding for cyanophycin synthesis (see Sect. 4), the enzymes of the AST catabolic pathway and a regulatory protein (ArgR) which belongs to the AsnC/Lrp family; this family was originally thought to comprise only global regulators such as the archetypal Leucine-responsive Regulatory Protein (Lrp) but may in fact comprise specific regulators as well (Brinkman et al. 2003; Elbahloul and Steinbuchel, 2006). Since archaeal arginine-responsive regulators still remain to be identified, investigations in this direction might prove rewarding.

3.2.1.3 Evolutionary Considerations

Detailed functional studies have led Morin et al. 2003 (see also Charlier, 2004) to classify arginine repressors of the wHTH type in three major types: class I is typical of γ-Proteobacteria (such as *E. coli, S. typhimurium* and *M. profunda*); it is markedly arginine-dependent and highly sequence-specific, hardly binding to operators of organisms with repressors of class II and III; ideally, class I arginine repressors recognize pairs of ARG boxes separated by 3 bp; they exist as hexamers even at low protein concentration and in the absence of arginine; class II ArgR molecules (as in *Bacilli*, probably also *Lactobacilli*—see above—and *Streptomyces*, Rodriguez-Garcia et al. 2000) display a more moderate arginine-dependence and a broader target specificity since they can recognize class I and class III operators; they also interact with tandem pairs of ARG boxes; they exist in a concentration-dependent equilibrium between the trimeric and hexameric forms, the hexamer being stabilized by arginine (Dion et al. 1997; Holtham et al. 1999). Class III comprises the ArgR molecules from the extreme thermophilic bacteria *Thermotoga maritima* and *neapolitana*; arginine influences their binding to DNA only marginally and strong contacts are established with only one ARG box in a pair (Dimova et al. 2000; Song et al. 2002). *Thermotoga* ArgR behaves as a super-repressor in *E. coli* and it is possible that it behaves as a more global regulator in its native host. It is conceivable that ArgR proteins evolved from a general regulator with poor target specificity (class III) to more metabolically specialized molecules (class II and III).

Studies on the ArgR molecules of *M. profunda* (a strict psychrophile, Xu et al. 2003b) and *Thermotoga*, an extreme thermophile (Song et al. 2002; Morin et al. 2003) brought to light properties (stability, DNA binding) that are in agreement with the thermal profiles of their hosts.

From the point of view of evolution, it is particularly interesting that the ArgR protein, that appears to belong to an ancient class of DNA recognizing proteins, was also found (under the independently coined denomination XerA) to play a role in a totally unrelated cellular process: the site specific recombination that resolves ColEl dimers into monomers (Stirling et al. 1988). Similarly, the *pepA* gene (alias *xerB*) which takes part in the control of the *carAB* operon (see above) is involved in the same resolution process. Moreover, the ArgP protein, a transcriptional activator that controls both the *abpS* gene (encoding an arginine and ornithine binding protein involved in the low-affinity arginine/ornithine membrane transport system) and the synthesis of the ArgO (ex YggA) arginine exporter, is identical to IciA, an inhibitor of initiation of chromosomal replication (Celis, 1999; Nandineni and Gowrishankar, 2004; Caldara, Charlier and Cunin, personal communication). These observations strongly suggest that a number of metabolic regulatory proteins may have been recruited from ancestral DNA binding proteins involved in the control of essential cellular functions such as replication and partition of DNA molecules (see Charlier and Glansdorff, 2004).

As already stated above, the elements of the arginine regulatory system appear conserved in a large number of Bacteria. On the other hand, different types of genes are being identified for the glutamate acetylation function (see above). It will be interesting to determine if these genes are regulated and, assuming they are, whether they fit into the general pattern or present specific features.

3.2.1.4
Regulatory Coupling between Arginine Biosynthesis and various Arginine Metabolic Pathways in Bacteria

Since the emphasis of this review is on arginine biosynthesis, the reader is referred to Lu (2006) for a comprehensive treatment of arginine catabolism and connections with global control mechanisms. Figure 2 outlines major metabolic interconnections as presently documented. Several pathways for arginine degradation were identified among Bacteria (Abdelal, 1979; Cunin et al. 1986; Lu, 2006). The identification in *E. coli* of the arginine-succinyltransferase pathway (AST), that converts arginine to glutamate (Schneider et al. 1998) solved the long-standing mystery of the cryptic, arginine-inducible acetylornithine aminotransferase (ArgM, homologous to ArgD) brought to light by Bacon and Vogel (1963) and now found to operate in the AST pathway as a succinylornithine aminotransferase under the denomination AstC or CstC (see Charlier and Glansdorff, 2004). Since ArgM induction is controlled by ArgR (Bacon and Vogel, 1963), this observation provided the very first indication ever that a particular repressor could also be involved in gene activation. The AST pathway is indeed controlled by the *argR* product, which plays here the role of a transcriptional co-activator. In *E. coli*, the *ast* operon is under complex expression control, ArgR being necessary to achieve maximal expression but not essential (Kiupakis and Reitzer,

2002). In *Salmonella*, however, ArgR is required for *ast* expression (Lu and Abdelal, 1999).

The AST pathway was actually discovered in pseudomonads (Vander Wauven and Stalon, 1985) and appears widespread among related Bacteria as well as enterics and *Vibrionaceae*, but has not been identified so far in Gram-positive Bacteria (Lu, 2006); in *P. aeruginosa*, AST induction by arginine was shown to require an ArgR factor that is also involved in repression of some arginine biosynthetic genes but is different in structure and function from ArgR/AhrC proteins (Lu et al. 1999 and above). A *P. aeruginosa* transcriptome analysis showed that ArgR and arginine also induce an arginine decarboxylase and various transport functions, whereas they repress two operons involved in glutamate biosynthesis (Lu et al. 2004); on the other hand, ArgR and arginine activate the gene for a catabolic glutamate dehydrogenase (Lu and Abdelal, 2001).

The arginase pathway of *Bacilli* presents another example of an ambivalent repressor: in *B. subtilis*, the arginine biosynthetic AhrC repressor protein (see above) and the *rocR* gene product, a transcriptional activator of the NtrC/NifA family (Gardan et al. 1997) are required for induction of the arginase pathway. The arginase pathway is the only arginine catabolic route in *B. subtilis*. In *B. licheniformis*, a facultative anaerobic organism, arginase is induced under good aeration, but an additional arginine degradative pathway is induced under anaerobic conditions: the arginine deiminase pathway (ADI) that functions as a source of ATP at the level of the carbamate kinase catalyzed step (Fig. 1). Interestingly, induction of this pathway also involves an ArgR/AhrC homolog; the same type of regulatory protein thus appears to have been recruited for different pathways. Protein ArcR, an activator of the Crp/Fnr family is required as well (Maghnouj et al. 1998, 2000). ArcR is oxidation-sensitive (Wolkonig et al. 2004). A similar ArcR protein is involved in induction of the arginine deiminase pathway in *Lactobacillus sakei* (Zuniga et al. 2002). In *Enterococcus faecalis*, three putative transcriptional regulators (two of the Arg/AhrC type and one ArcR homologue) are involved in the regulation of the ADI pathway (Barcelona-Andres et al. 2002).

3.2.2
Fungi

At the time the basic features of the bacterial arginine regulon were still being unraveled in *E. coli*, it was already becoming clear that the mechanism controlling arginine biosynthesis in the yeast *S. cerevisiae* would provide the first instance of a eukaryotic regulatory system comprising the key elements postulated by Jacob and Monod (1961): a trans-acting component genetically identified as a repressor by recessive regulatory mutations and *cis*-acting control elements of the operator type. The former kind of mutant was obtained very early (Bechet and Wiame 1970) and operator mutants somewhat later

(Messenguy, 1976), but both were retrieved by the procedures that had proven successful with *E. coli*: selection for mutants resistant to the toxic arginine analog canavanine (trans-acting mutations of the ArgR type) and specific selection for increased rate of OTC synthesis (operator mutations of the OTC structural gene, *ARG3*), using a CPS deficient mutant forced to provide CP for pyrimidine synthesis from the unfavored reverse OTC catalyzed reaction (citrullinolysis).

Today, the system is understood in great detail; as in some prokaryotes, an interesting aspect is the coordination between the biosynthesis and the catabolism of arginine; in addition, the control of arginine biosynthesis is linked to cellular mechanisms of more general scope, making the overall picture rather complex (see Messenguy and Dubois, 2000 and below).

The expression of yeast *arg* genes *ARG1* (*argG*), *ARG3* (*argF*), *ARG5,6* (*argCB*) and *ARG8* (*argD*) is specifically controlled at the level of DNA transcription (Crabeel et al. 1990; de Rijcke et al. 1992) by a multi-component repressor system, the ArgR/Mcm1 complex; the formation of functional repressor requires three nonessential proteins (Arg80p, Arg81p and Arg82p) and the essential protein Mcm1 (reviewed by Messenguy and Dubois, 2000, 2003). Mcm1 belongs to the family of so-called MADS box proteins, eukaryotic combinatorial transcription factors that associate with other transcription factors into regulatory complexes deriving their specificity from this association (ibid). Arg80p is also a MADS box protein, very similar to Mcm1 (Jamai et al. 2002). Arg81p is a transcription factor of the Zn_2C_6-cluster family (Messenguy et al. 1986); Arg82p is not a DNA-binding protein, but rather a pleiotropic kinase involved in diverse cellular processes; it turned out to be identical with Ipk2p, an inositol polyphosphate kinase (Odom et al. 2000). Arg82p/Ipk2p is thought to facilitate the association between Arg80p and Mcm1 since the requirement for Arg82p can be bypassed by overproducing either of these proteins (El Bakkoury et al. 2000). Arg82 kinase activity is not required for transcriptional regulation of arginine genes (Dubois et al. 2000).

It should be noted that yeast ArgB (NAGK, Arg6p) may be implicated in the regulation of gene expression since the protein was found to bind defined mitochondrial and nuclear DNA sequences and a *ARG5-ARG6* deletion mutant displayed altered levels of cognate gene products (Hall et al. 2004). The physiological link between these effects and arginine (or possibly nitrogen) metabolism, however, remains to be clarified.

The component of the ArgR/Mcm1 complex that is comparable to bacterial ArgR is Arg81p; the protein interacts with Arginine Control (ARC) elements located in the promoters of *arg* genes; Arg81p (formerly ARGRII) is most probably the arginine sensor in the complex since cooperative binding of Arg81p/Mcm1p or Arg81p/Arg80p to ARC elements is stimulated by arginine and since part of the DNA-binding, N-terminal domain of Arg81p presents some sequence similarity with the N-terminal domain of bacterial ArgR (Amar et al. 2000). The last three bp of the ARC element constitute the consensus CGR motif present in the target sites of members of the Zn_2C_6 cluster family (de Rijcke et al. 1992). Similar ARC sequences are found in the control regions of arginine catabolic genes (CAR1 and CAR2, coding for arginase and ornithine aminotransferase, respectively); there, however, the ArgR/Mcm1 complex functions as an activator, so that mutations altering the ArgR/Mcm1 complex prevent the induction of the arginine catabolic genes (Bechet et al. 1970; Dubois and Messenguy, 1985; Dubois, Bercy and Messenguy, 1987). Regulation of CAR1 and CAR2 is complex however: several *trans*- and *cis*-acting elements are involved (Kovari et al. 1990, 1993; Messenguy et al. 1991, 2000; Viljoen et al. 1992; Messenguy and Dubois, 2000). For the identification of corresponding elements in the ascomycete *Aspergillus nidulans*, the reader is referred to Dzikowska et al. (2003) and Empel et al. (2001).

It is not yet clear by which mechanism the ArgR/Mcm1p complex modulates DNA transcription; sequence-swapping experiments performed with the control elements of the ARG3 gene suggested that repression of transcription would require the ARC element to be located close to the TATA box or downstream from it, whereas activation would require an upstream located ARC sequence (de Rijcke et al. 1992); in keeping with this view, the ARC elements of arginine catabolic genes, that are induced by arginine, are found in an upstream position (ibid., Kovari et al. 1990; Messenguy et al. 1991). However, analysis of the *ARG1* and *CPA1* genes (coding for the small subunit of arginine-specific CPS) indicated that repression could also be mediated by an upstream ARC sequence (Crabeel et al. 1995). In fact, understanding the specific repression mechanism in molecular detail will require unraveling the interactions that take place at each particular promoter between two sets of elements: those involved in repression itself and those involved in the so-called "general control of amino acid biosynthesis" (also known as "cross-pathway control") that is mediated by the transcriptional activator Gcn4p (Cpc-1 in *Neurospora*, Luo and Sachs, 1996; CpaA in *Aspergillus*, Krappmann et al. 2004) and is superimposed to repression (Delforge et al. 1975; Hinnebusch, 2005). Most amino acid biosynthetic genes become induced in yeast cells starved for any amino acid. Amino acid deprivation triggers translational induction of *GCN4* (see Natarajan et al. 2001); this leads to enhanced expression of amino acid biosynthetic genes. Gcn4p binds at specific upstream activating sequences present in the promoters of these genes. Studies of gene expression from the *ARG1* promoter showed that Gcn4p binding to DNA, followed by recruitment of co-activators SWI/SNF, SAGA and Srb Mediator, stimulates assembly of the pre-initiation complex (Govind et al. 2005). Surprisingly, Gcn4 binding was found to recruit the Mcm1p/Arg80p heterodimer independent from Arg81p and arginine; only when arginine was in excess did Gcn4p recruit the full ArgR/Mcm1 complex to the *ARG1* promoter (Yoon et al. 2004). Thus, at low intracellular arginine concentrations, recruitment of a fully operative repressor complex is impaired whereas under excess of arginine full repression becomes possible. This coupling mechanism between the elements respectively involved in specific repression and in general control allows fine tuning of the Gcn4p activation function in response to arginine availability (ibid.).

The observations of Yoon et al. (2004) thus indicate that *ARG1*-bound Gcn4p stimulates binding of the Mcm1p/Arg80p dimer to ARC sequences independent from Arg81p and arginine, a result that was unexpected since in vitro binding of Mcm1p/Arg80p was found to depend on both Arg81p and arginine (Amar et al. 2000). It will be interesting to see how the recruitment process operates at the *ARG3* promoter where de Rijcke et al. (1992) carried out their positional analysis. It is not excluded that in different promoter contexts an ARC element might become involved in different Gcn4p/repressor/DNA architectures and thereby behave in opposite ways.

The synthesis of *S. cerevisiae* arginine-specific CPS is negatively regulated by arginine both at the translational and transcriptional levels. The mRNA small subunit of the *CPA1* gene (that codes for the small subunit of the enzyme) contains a short leader sequence that must be translated for translational repression to occur; this effect is composition-dependent and the leader sequence confers arginine-dependent repression when fused to a reporter gene (Werner et al. 1987; Delbecq et al. 1994). A similar mechanism may operate in *Neurospora crassa* (Luo et al. 1996). Mutations in gene *CPAR*, unlinked to *CPA1*, impair repression; Messenguy et al. (2002) found *CPAR* to be identical to *UPF1* that codes for a protein involved in the premature termination step of the mechanism responsible for RNA surveillance by nonsense-mediated mRNA decay (Czaplinski et al. 1995; Ruiz-Echevarria and Pelz, 2000). The half-life of *CPA1* mRNA had indeed been found to be reduced twofold in excess arginine (Crabeel et al. 1990). The current model assumes that in the presence of arginine, the leader regulatory peptide would block ribosome scanning upstream from *CPA1* ORF; ribosome pausing would allow assembly of the elements of the surveillance pathway and initiate *CAP1* mRNA degradation (see Messenguy et al. 2002, and Gaba et al. 2005 for further insight into how translation of the leader is involved in controlling *CPA1* mRNA stability through the surveillance mechanism). *CPA1* is also partly regulated at the transcriptional level by the ArgR/Mcm1p system (Crabeel et al. 1990).

Arginine metabolism appears much less regulated in some other fungi than in *S. cerevisiae*. In *N. crassa*, *CPA1* is the only arginine biosynthetic gene known to be repressed by arginine (Davis, 1986; Davis and Weiss, 1988; Luo and Sachs, 1996) and no significant repression occurs in the yeast *Schizosaccharomyces pombe* though the arginase catabolic pathway is induced by arginine (Van Huffel et al. 1994). In both organisms, compartmentation plays a major role in controlling metabolic fluxes. In *A. nidulans*, where early indications had been obtained for a regulation common to the biosynthesis and catabolism of arginine (Cybis et al. 1972), more recent work focused on regulation of the arginase pathway. A regulatory system with interesting similarities to yeast ArgR/Mcm1p was studied in some detail (Empel et al. 2001; Dzikowska et al. 2003). Moreover, a putative arginine binding aptamer was identified in the 5′ untranslated region of the arginase structural gene (Borsuk et al. 1999). This putative arginase "riboswitch" is cleaved by a resident RNase P, but arginine does not influence the rate of cleavage (Altman et al. 2005); the possible regulatory significance of this mechanism thus remains to be evaluated.

4 Arginine as a Nutraceutical and Arginine Microbial Production

In human nutrition arginine is considered as a conditionally essential amino acid since it is produced endogenously in adults but may become necessary as a dietary supplement under conditions of increased demand such as growth or tissue repair; in addition, arginine exerts an immunosupportive effect (see de Jonge et al. 2002). It is also the source of nitrogen oxide, a vasodilatory messenger (Ignarro et al. 1987) that is also involved in a variety of cellular communications (Traylor and Sharma, 1992). Arginine is therefore regarded as an important "nutraceutical".

Microbial arginine production can be boosted to high levels by applying strategies based on the knowledge of enzyme properties and regulatory mechanisms or by repeated mutagenesis and screening. Utagawa (2004) described an example of the latter procedure applied to *Corynebacterium glutamicum* subsp. *flavum*, a natural producer of glutamic acid, a precursor of arginine; the multiple mutant produced as much as 25.3 g/L arginine. Grigorievna et al. (US patent 6 984 514, 2006) used a L-isoleucine auxotroph of *E. coli* to produce glutamic acid, proline and arginine in the presence of isoleucine. The arginine producing strain (up to 4.5 g/L) was also resistant to 6-azauracil, a condition resulting from a deficiency in the *upp* gene (coding for uracil pyrophosphorylase) and known to partly derepress arginine and pyrimidine biosyntheses (Pierard et al. 1972). An improved procedure was achieved using a derivative of that strain able to use acetate more efficiently as a C source (acetate is a product of the reaction catalyzed by acetylornithinase); up to 19.3 g/L arginine was produced in the presence of glucose as a C source (Gusyatiner et al. US patent 6 841 365, 2005).

More elaborate strategies were devised. Sakanyan et al. (US patent 6 897 048, 2005) obtained their best results (13 to 14 g/L) from engineered *E. coli* strains containing an *argR* mutation as well as plasmids containing an additional *argA* gene and a heterologous *argJ* gene coding for a bifunctional OAT (not inhibited by arginine). Rajagopal et al. (1998) constructed *E. coli* strains containing a plasmid harboring a gene coding for a feedback resistant NAGS (in order to circumvent the limitation imposed on ornithine biosynthesis by feedback inhibition), several arginine operators (to titrate the repressor) and the genes coding for CPS and OTC, respectively. The expression of these genes was made inducible by IPTG. When cells were suspended in minimal medium supplemented with glutamine in the presence of IPTG, arginine production amounted to about 100 nM/hr/mg dry weight; this would correspond approximately to 7 g/L during the three-hour incubation period. It should be stressed, however, that all these experiments are difficult to compare quantitatively; media compositions and conditions of incubation are quite different.

Aside from arginine production, the work of Rajagopal et al. (1998; see also Tuchman et al. 1997) presents the additional interest of providing a system that could be used for enhanced incorporation of ammonia into arginine in the intestine of hyperammonemic patients.

Further improvements in microbial arginine production could also result from the suppression of catabolic pathways (such as the AST pathway in *E. coli*, Lu, 2006) and/or monitoring membrane transport systems. For example, Nandineni and Gowrishankar (2004) recently identified *E. coli* ORF *yggA* as coding for a protein involved in arginine efflux and constructed a strain combining an *argR* mutation with an increased level of YggA (renamed ArgO) in order to achieve arginine overproduction (patent filed 12/22/05, USPTO class 435, #20050282258).

Cyanophycin is a polymer of arginine and aspartate (multi-L-arginyl-L-aspartic acid) where the backbone consists of aspartate residues linked by their β-carboxyl group to the α-amino group of arginine by an amide bond. This substance was initially detected as membraneless granules produced in cyanophyceae but genomic studies showed that homologs to the cyanobacterial gene coding for cyanophycin synthase, the enzyme assembling the polymer, are present in a number of Gram-positive and Gram-negative Bacteria (see Elbahloul and Steinbuchel, 2006; Obst and Steinbuchel, 2006). The present interest in this substance appears to reside mostly in the possibility to obtain poly(aspartic acid) as a substitute for nonbiodegradable polyacrylates (ibid and references therein); however, since it constitutes a form of arginine accumulation in autotrophic organisms, thus with minimal growth requirements, we thought useful to mention it in this section. Moreover, in the metabolically versatile *Acinetobacter* strain ADP1 (the first noncyanobacterial bacterium shown to produce cyanophycin), arginine synthesis was identified as a rate-limiting step in cyanophycin production. It is therefore interesting that Elbahloul and Steinbuchel (2006) were able to engineer derivatives of *Acinetobacter* ADP1 with improved arginine biosynthetic capacity and impaired arginine catabolism that increased their cyanophycin content under defined growth conditions.

Acknowledgements We are thankful to Dr. Danny Charlier, Dr. Marjolaine Crabeel and Dr. Raymond Cunin for critical reading of the manuscript, and to Bart Scherens for help with the artwork. NG gratefully acknowledges Professor G. Taylor's hospitality at Stony Brook MSRC during the preparation of the manuscript.

References

Abadjieva A, Hilven P, Pauwels K, Crabeel M (2000) The yeast *ARG7* gene product is autoproteolyzed to two subunit peptides, yielding active ornithine acetyltransferase. J Biol Chem 275:11361–11367

Abadjieva A, Pauwels K, Hilven P, Crabeel M (2001) A new yeast metabolon involving at least the two first enzymes of arginine biosynthesis. J Biol Chem 276:42869–42880

Abdelal A (1979) Arginine catabolism by microorganisms. Annu Rev Micrbiol 33:139–168

Ahuja A, Purcarea C, Guy HI, Evans DR (2001) A novel carbamoyl-phosphate synthetase from *Aquifex aeolicus*. J Biol Chem 276:45694–45703

Altman S, Wesolowski, Guerrier-Takada, Li Y (2005) RNase P cleaves transient structures in some riboswitches. Proc Natl Acad Sci USA 102:11284–11289

Amar N, Messenguy F, El Bakkoury M, Dubois E (2000) ArgRII, a component of the ArgR-Mcm1 complex involved in the control of arginine metabolism in *Saccharomyces cerevisiae*, is the sensor of arginine. Mol Cell Biol 20:2087–2097

Arsene-Ploetze F, Nicoloff H, Bringel F (2005) *Lactobacillus plantarum ccl* gene is non-essential, arginine-repressed and codes for a conserved protein in Firmicites. Arch Microbiol 183:307–316

Bacon DF, Vogel HJ (1963) A regulatory gene simultaneously involved in repression and induction. Cold Spring Harbor Symp Quant Biol 28:437–438.

Baetens M, Legrain C, Boyen A, Glansdorff N (1998) Genes and enzymes of the acetyl cycle of arginine biosynthesis in the extreme thermophilic bacterium Thermus thermophilus. Microbiology UK 144:479–492

Barcelona-Andres B, Marina A, Rubio V (2002) Gene structure, organization, expression and potential regulatory mechanisms of arginine catabolism in *Enterococcus faecalis*. J Bacteriol 184:6289–6300

Baur H, Luethi E, Stalon V, Mercenier A, Haas D (1989) Sequence analysis and expression of the arginine-deiminase and carbamate-kinase genes of Pseudomonas aeruginosa. Eur J Biochem 179:53–60

Bechet J, Grenson M, Wiame JM (1970) Mutations affecting the repression of arginine biosynthetic enzymes in *Saccharomyces cerevisiae*. Eur J Biochem 12:31–39

Boonchird C, Messenguy F, Dubois E (1991) Determination of amino acid sequences involved in the processing of the AR5/ARG6 precursor in *Saccharomyces cerevisiae*. Eur J Biochem 199:325–335

Borsuk P, Dzikowska A, Empel J, Grzelak A, Grzeskowiak R, Weglenski P (1999) Structure of the arginase coding gene and its transcript in *Aspergillus nidulans*. Acta Biochim Pol 46:391–403

Bringel F, Frey L, Boivin S, Hubert JC (1997) Arginine biosynthesis and regulation in *Lactobacillus plantarum*: the *carA* gene and the *argCJBDF* cluster are divergently transcribed. J Bacteriol 179:2697–2706

Brinkman AB, Bell SD, Lebbink RJ, de Vos WM, van der Ost (2002) The Sulfolobus solfataricus Lrp-like protein LysM regulates lysine biosynthesis in response to lysine availability. J Biol Chem 277:29537–29549

Brinkman AB, Ettema TJG, de Vos, van der Oost (2003) The Lrp family of transcriptional regulators. Mol Microbiol 48:287–294

Brochier C, Philippe H (2002) Phylogeny: a non-hyperthermophilic ancestor for bacteria. Nature 417:244

Burke M, Merican AF, Sherratt DJ (1994) Mutant *Escherichia coli* arginine repressor proteins that fail to bind L-arginine yet retain the ability to bind their normal DNA-binding sites. Mol Microbiol 13:609–618

Caldovic L, Morizono H, Yu X, Thompson M, Shi D, Gallegos R, Allewell NM, Malamy MH, Tuchman M (2002) Identification, cloning and expression of the mouse N-acetylglutamate synthase gene. Biochem J 364:825–831

Caldovic L, Tuchman M (2003) N-acetylglutamate and its changing role through evolution. Biochem J 372:279–290

Cammarano P, Gribaldo S, Johann A (2002) Updating carbamoylphosphate synhase (CPS) phylogenies: occurrence and phylogenetic identity of archaeal CPS genes. J Mol Evol 55:153–160

Celis R (1999) Repression and activation of arginine transport genes in *Escherichia coli* K-12 by the ArgP protein. J Mol Biol 294:1087–1095.

Charlier D (2004) Arginine regulation in *Thermotoga neapolitana* and *Thermotoga maritima*. Biochem Soc Trans 32:310–313

Charlier D, Roovers M, Van Vliet F, Boyen A, Cunin R, Nakamura Y (completer) (1992) Arginine regulon of *Escherichia coli K12*. A study of repressor-operator interactions and of in vitro binding affinities versus in vivo repression. J Mol Biol 226:387–397

Charlier D, Glansdorff N (2004) Biosynthesis of arginine and polyamines. In R. Curtiss III (Editor in Chief), EcoSal-Escherichia coli and Salmonella: Cellular and Molecular Biology. Online. http://www.ecosal.org. ASM Press, Washington, DC; Module 3.6.1.10

Cohen GN, Barbe V, Flament D, Galperin M, Heilig R, Lecompte O, Poch O, Prieur D, Querellou J, Ripp R, Thierry JC, Van der Oost J, Weissenbach J, Zivanovic Y, Forterre P (2003) An integrative analysis of the genome of the hyperthermophilic archaeon *Pyrococcus abyssi*. Mol Microbiol 47:1495–1512

Crabeel M, Lavalle R, Glansdorff N (1990) Arginine-specific repression in *Saccharomyces cerevisiae*: kinetic data on *ARG1* and *ARG3* mRNA transcription and stability support a transcriptional control mechanism. Mol Cell Biol 10:1226–1233

Crabeel M, de Rijcke M, Seneca S, Heimberg H, Pfeiffer I, Matisova A (1995) Further definition of the sequence and position requirements of the arginine control element that mediates repression and induction by arginine in *Saccharomyces cerevisiae*. Yeast 11:1367–1380

Crabeel M, Soetens O, de Rijcke M, Pratiwi R, Pankiewicz R (1996) The *ARG11* gene of Saccharomyces cerevisiae encodes a mitochondrial integral membrane protein required for arginine biosynthesis. J Biol Chem 57:231–239

Crabeel M, Abadjieva A, Hilven P, Desimpelaere J, Soetens O (1997) Characterization of the *Saccharomyces cerevisiae ARG7* gene encoding ornithine acetyltransferase, an enzyme also endowed with acetylglutamate synthase activity. Eur J Biochem 250: 232–241

Cunin R, Glansdorff N, Pierard A, Stalon V (1986) Biosynthesis and metabolism of arginine in bacteria. Microbiol Rev 50:314–352

Cybis J, Piotrowska M, Weglenski P (1972) Genetic control of the arginine pathways in Aspergillus nidulans. Common regulation of anabolism and catabolism. Mol Gen Genet 118:273–277

Czaplinski K, Weng Y, Hagan KW, Peltz SW (1995) Purification and characterization of the Upf1 protein: a factor involved in translation and mRNA degradation. RNA 1:610–613

Davis RH (1986) Compartmental and regulatory mechanisms in the arginine pathways of *Neurospora crassa* and *Saccharomyces cerevisiae*. Microbiol Rev 50:280–313

Davis RH, Weiss RL (1988) Novel mechanisms controlling arginine metabolism in *Neurospora*. Trends Biochem Sci 13:101–104

De Jonge WJ, Kwikkers KL, te Velde AA, van Deventer SJH, Nolte MA, Mebius RE, Ruijter JM, Lamers MC, Lamers WH (2002) Arginine deficiency affects early B cell maturation and lymphoid organ development in transgenic mice. J Clin Invest 110:1539–1548

De la Fuente A, Martin JF, Rodriguez-Garcia A, Liras P (2004) Two proteins with ornithine acetyltransferase activity show different functions in *Streptomyces clavuligerus*: Oat2 modulates clavulanic acid biosynthesis in response to arginine. J Bacteriol 186:6501–6507

Delannay S, Charlier D, Tricot C, Villeret V, Pierard A, Stalon V (1999) Serine 948 and threonine 1042 are crucial residues for allosteric regulation of *Escherichia coli* carbamoylphosphate synthase and illustrate coupling effects of activation and inhibition pathways. J Mol Biol 286:1217–1228

Delbecq P, Werner M, feller A, Filipkowski RK, Messenguy F, Pierard A (1994) A segment of mRNA encoding the leader peptide of the CPA1 gene confers repression by arginine on a heterologous yeast gene transcript. Mol Cell Biol 14:2378–2390

Delforge J, Messenguy F, Wiame JM (1975) The regulation of arginine biosynthesis in saccharomyces cerevisiae. The specificity of argR-mutations and the general control of amino-acid biosynthesis. Eur J Biochem 57:231–239

Dennis CA, Glykos NM, Rarsons MP, Phillips SEV (2002) The structure of AhrC, the arginine repressor/activator protein from *Bacillus subtilis*. Acta Crystallog Sect D 58:421–430

de Rijcke M, Seneca S, Punyammalee B, Glansdorff N, Crabeel M (1992) Characterization of the DNA target site for the yeast ARGR regulatory complex, a sequence able to mediate repression or induction by arginine. Mol Cell Biol 12:68–81

Devroede N, Thia-Toong TL, Gigot D, Charlier D (2004) Purine and pyrimidine-specific repression of the *Escherichia coli carAB* operon are functionally and structurally coupled. J Mol Biol 336:25–42

Devroede N, Huysveld N, Charlier D (2006) Mutational analysis of intervening sequences connecting the binding sites for integration host factor, PepA, PurR, and RNA polymerase in the control region of *Escherichia coli carAB* operon, encoding carbamoylphosphate synthase. J Bacteriol 188:3236–3245

Dimova D, Weigel P, Takahashi M Marc F, Van Duyne GD, Sakanyan V (2000) Thermostability, oligomerization and DNA-binding properties of the regulatory protein ArgR from the hyperthermophilic bacterium *Thermotoga neapolitana*. Mol Gen Genet 263:119–130

Dion M, Charlier D, Wang H, Gigot D, Savchenko A, Hallet J-N, Glansdorff N, Sakanyan V (1997) The highly thermostable arginine repressor of *Bacillus stearothermophilus*: gene cloning and repressor-operator interactions. Mol Microbiol 25:385–398

Dubois E, Bercy J, Messenguy F (1987) Characterization of two genes, ARGRI and ARGRIII required for specific regulation of arginine metabolism in yeast. Mol Gen Genet 207:142–148

Dubois E, Dewaste V, Erneux C, Messenguy F (2000) Inositol polyphosphate kinase activity of Arg82/ArgRIII is not required for the regulation of the arginine metabolism in yeast. FEBS Lett 486:300–304

Durbecq V, Legrain C, Roovers M, Pierard A, Glansdorff N (1997) The carbamate kinase-like carbamoylphosphate synthase of the hyperthermophilic archaeon *Pyrococcus furiosus*, a missing link in the evolution of carbamoylphosphate biosynthesis. Proc Natl Acad Sci USA 94:12803–12808

Dzikowska A, Kacprzak M, Tomecki R, Koper M, Scazzoccchio C, Weglenski P (2003) Specific induction and carbon/nitrogen repression of arginine catabolism gene of *Aspergillus nidulans*-functional in vivo analysis of the *otaA* promoter. Fungal Genet Biol 38:175–186

El Alami M, Messenguy F, Scherens B, Dubois E (2003) Arg82p is a bifunctional protein whose inositol polyphosphate kinase activity is essential for nitrogen and *PHO* gene expression but not for Mcm1p chaperoning in yeast. Mol Microbiol 49:457–468

El Alaoui M, Dubois E, Oudjama Y, Tricot C, Wouters J, Stalon, Messenguy F (2003) Yeast epiarginase regulation, an enzyme–enzyme activity control. Identification of residues of ornithine carbamoyltransferase and arginase responsible for enzyme catalytic and regulatory properties. J Biol Chem 278:21550–21558

Elbahloul Y, Steinbuchel (2006) Engineering the genotype of *Acinetobacter* sp. strain ADP1 to enhance biosynthesis of cyanophicin. Appl Environ Microbiol 72:1410–1419

El Bakkoury M, Dubois E, Messenguy F (2000) Recruitment of the yeast MADS-box proteins, ArgRI and Mcm1 by the pleiotropic factor ArgIII is required for their stability. Mol Microbiol 35:15–31

Elkins JM, Kershaw NJ, Schofield CJ (2005) X-ray crystal structure of ornithine acetyltransferase clavulanic acid biosynthesis gene cluster. Biochem J 385:565–573

Empel J, Sietkiewicz I, Andrukiewicz A, Lasocki K, Borsuk, Weglenski P (2001) *arcA*, the regulatory gene for the arginine catabolic pathway in *Aspergillus nidulans*. Mol Genet Genomics 266:591–597

Errey JC, Blanchard JS (2005) Functional characterization of a novel ArgA from *Mycobacterium tuberculosis*. J Bacteriol 187:3039–3044

Fernandez-Murga ML, Gil-Ortiz F, Llacer J, Rubio V (2004) Arginine biosynthesis in *Thermotoga maritima*: characterization of the arginine-sensitive N-acetyl-L-glutamate kinase. J Bacteriol 186:6142–6149

Gaba A, Jacobson A, Sachs MS (2005) Ribosome occupancy of the yeast *CPA1* upstream open reading frame termination codon modulates nonsense-mediated mRNA decay. Mol Cell 20:449–460

Gao JL, Weissenmayer B, Taylor AM, Thomas-Oates J, Lopez-Lara IM, Geiger O (2004) Identification of a gene required for the formation of lyso-ornithine lipid, an intermediate in the biosynthesis of ornithine-containing lipids. Mol Microbiol 53:1757–1770

Gardan R, Rapoport G, Debarbouille M (1995) Expression of the *rocDEF* operon involved in arginine catabolism in *Bacillus subtilis*. J Mol Biol 23:843–856

Gessert S, Kim JH, Nargang F, Weiss RL (1994) A polyprotein precursor of two mitochondrial enzymes in *Neurospora crassa*. Gene structure and precursor processing. J Biol Chem 269:8129–8203

Govind CK, Toon S, Qiu H, Govind S, Hinnebusch AG (2005) Simultaneous recruitment of coactivators by Gcn4p stimulates multiple steps of transcription in vivo. Mol Cell Biol 25:5626–5638

Graupner M, White RH (2001) *Methanococcus jannaschii* generates L-proline by cyclization of L-ornithine. J Bacteriol 183:5203–5205

Haas D, Kurer V, Leisinger T (1972) N-acetylglutamate synthetase of *Pseudomonas aeruginosa*. An assay in vitro and feed-back inhibition by arginine. Eur J Biochem 31:290–295

Hall DA, Zhu H, Zhu X, Royce T, Gerstein M, Snyder M (2004) Regulation of gene expression by a metabolic enzyme. Science 306:482–484

Hani EK, Ng D, Chan V-L (1999) Arginine biosynthesis in *Campylobacter jejuni* THG9011: determination of the *argCOBD cluster*. Can J Microbiol 45:959–969

Hashim S, Kwon D-H, Abdelal A, Lu C-D (2004) The arginine regulatory protein mediates repression by arginine of the operons encoding glutamate synthase and anabolic glutamate dehydrogenase in *Pseudomonas aeruginosa*. J Bacteriol 186:3848–3854

Hebert MD, Houghton JE (1997) Regulation of ornithine utilization in *Pseudomonas aeruginosa* is mediated by a transcriptional regulator OruR. J Bacteriol 179:7834–7842

Herrero A (2004) New targets of the P_{II} signal transduction protein identified in cyanobacteria. Mol Microbiol 52:1225–1228

Higuchi S, Kawashima T, Suzuki M (1999) Comparison of pathways for amino acid biosynthesis in archaebacteria using their genomic DNA sequences. Proc Jpn Acad Ser B 75:241–245

Hinde RW, Jacobson JA, Weiss RL, Davis RH (1986) N-acetyl-L-glutamate synthase of *Neurospora crassa*. Characteristics, localization, regulation, and genetic control. J Biol Chem 261:5848–5852

Hinnebusch AG (2005) Translational regulation of *GCN4* and the general amino acid control of yeast. Annu Rev Microbiol 59:407–450

Holden HM, Thoden JB, Raushel FM (1999) Carbamoyl phosphate synthetase: an amazing odyssey from substrate to product. Cell Mol Life Sci 56:507–522

Holtham CA, Jumel K, Miller CM, Harding SE, Baumberg S, Stockley PG (1999) Probing activation of the prokaryotic arginine transcriptional regulator using chimeric proteins. J Mol Biol 289:707–727

Ignarro LJ, Buga GM, Wood KS, Byrns RE, Chaudhuri G (1987) Endothelium-derived relaxing factor produced and released from artery and vein is nitric oxide. Proc Natl Acad Sci USA 84:9265–9269

Issaly IM, Issaly AS (1974) Control of ornithine carbamoyltransferase activity by arginase in *Bacillus subtilis*. Eur J Biochem 49:485–495

Jacob F, Monod J (1961) Genetic regulatory mechanisms in the synthesis of proteins. J Mol Biol 3:318–356

Jamai A, Dubois E, Vershon AK, Messenguy F (2002) Swapping functional specificity of a MADS box protein: residues required for Arg80 regulation of arginine metabolism. Mol Cell Biol 22:5741–5752

Kershaw NJ, McNaughton HJ, Hewitson KS, Hernandez H, Griffin J, Hughes C, Greaves P, Barton B, Robinson CV, Schofield CJ (2002) ORF6 from the clavulanic acid gene cluster of *Streptomyces clavuligerus* has ornithine acetyltransferase activity. Eur J Biochem 269:2052–2059

Kim JH, Weiss RL (1995) Genetic analysis of interactions between arg-14 and arg-6 products in Neurospora crassa. Mol Cells 5:461–466

Kiupakis A, Reitzer L (2002) ArgR-independent induction and ArgR-dependent superinduction of the *astCABDE* operon in *Escherichia coli*. J Bacteriol 184:2940–2950

Kothe M, Purcarea C, Guy HI, Evans DR, Powers-Lee SG (2005) Direct demonstration of carbamoylphospkate formation on the C-terminal domain of carbamoyl phosphate synthetase. Protein Sci 14:37–44

Kovari LZ, Fourie M, Park HD, Kovari IA, Van Vuuren HJ, Cooper TJ (1993) Analysis of the inducer-responsive CAR1 upstream activation sequence (UAS1) and the factors required for its operation. Yeast 9:835–845

Kovari L, Sumrada R, Kovari I, Cooper TG (1990) Multiple positive and negative *cis*-acting elements mediate induced arginase *(CAR1)* gene expression in *Saccharomyces cerevisiae*. Mol Cell Biol 10:5087–5097

Krappmann S, Bignell EM, Reichard U, Rogers T, Haynes K, Braus GH (2004) The *Aspergillus fumigatus* transcriptional activator CpcA contributes significantly to the virulence of this fungal pathogen. Mol Microbiol 52:785–799

Kwon D-H, Lu C-D, Nalthall DA, Brown TM, Houghton JE, Abdelal A (1994) Structure and regulation of the carAB operon in *Pseudomonas aeruginosa* and *Pseudomonas stutzeri*. J Bacteriol 176:2532–2542

Labedan B, Boyen A, Baetens M, Charlier D, Pingguo C, Cunin R, Durbecq V, Glansdorff N, Herve G, Legrain C, Liang Z, Purcarea C, Roovers M, Sanchez R, Toong TL, Van de Casteele M, Van Vliet F, Xu Y, Zhang YF (1999) The evolutionary history of carbamoyltransferases: a comolex set of paralogous genes was already present in the last common ancestor. J Mol Evol 49:461–473

Labedan B, Xu Y, Naumoff DG, Glansdorff N (2004) Using quaternary structures to assess the evolutionary history of proteins: the case of the aspartate carbamoyltransferase. Mol Biol Evol 21:364–373

Larsen R, Kok J, Kuipers OP (2005) Interactions between ArgR and AhrC controls regulation of arginine metabolism in *Lactobacillus lactis*. J Biol Chem 280:19319–19330

Ledwidge R, Blanchard JS (1999) The dual biosynthetic capability of N-acetylornithine aminotransferase in arginine and lysine biosynthesis. Biochem 38:3019–3024

Legrain C, Demarez M, Glansdorff N, Pierard A (1995) Ammonia-dependent synthesis and metabolic channeling of carbamoylphosphate in the hyperthermophilic archaeon *Pyrococcus furiosus*. Mibrobiogy UK 141:1093–1099

Liu Q, Dunlap JC (1996) Isolation and analysis of the arg-13 gene of *Neurospora crassa*. Genetics 143:1163–1174

Lu C-D (2006) Pathways and regulation of bacterial arginine metabolism and perspectives for obtaining arginine overproducing strains. Appl Environ Microbiol 70:261–272

Lu C-D, Houghton JE, Abdelal A (1992) Characterization of the arginine repressor from *Salmonella typhimurium* and its interactions with the carAB operator. J Mol Biol 225:11–24

Lu CD, Kwon DH, Abdelal A (1997) Identification of *greA* encoding a transcriptional elongation factor as a member of the *carA-orf-carB-greA* operon of *P. aeruginosa* PAO1. J Bacteriol 179:3043–3046.

Lu C-D, Abdelal A (1999) Role of ArgR in activation of the *ast* operon, encoding enzymes of the arginine succinyltransferase pathway in *Salmonella typhimurium*. J Bacteriol 181:1934–1938

Lu C-D, Abdelal A (2001) The gdhB gene of Pseudomonas aeruginosa encodes an arginine-inducible NAD(+)-dependent glutamate dehydrogenase which is subject to allosteric regulation. J Bacteriol 183:490–499

Lu C-D, Yang Z, Li W (2004) Transcriptome analysis of the ArgR regulon in *Pseudomonas aeruginosa*. J Bacteriol 186:3855–3861

Luo Z, Sachs MS (1996) Role of an upstream open reading frame in mediating arginine-specific translational control in *Neurospora crassa*. J Bacteriol 178:2172–2177

Maas WK (1994) The arginine repressor of *Escherichia coli*. Microbiol Rev 58:631–640

Maas WK, Maas R, Wiame JM, Glansdorff N (1964) Studies on the mechanism of repression of arginine biosynthesis in *Escherichia coli* I. Dominance of repressibility in zygotes. J Mol Biol 8:359–364

Maas WK, Clark AJ (1964) Studies on the mechanisms of repression of arginine biosynthesis in *Escherichia coli* II. Dominance of repressibility in diploids. J Mol Biol 8:365–370

Maghnouj A, Abu-Bakr AAW, Baumberg, Stalon V, Vander Wauven (2000) Regulation of anaerobic arginine catabolism in *Bacillus licheniformis* by a protein of the Crp/Fnr family. FEMS Microbiol Lett 191:227–234

Maghnouj A, de Sousa Cabral TF, Stalon V, Vander Wauven C (1998) The arcABDC gene cluster, encoding the arginine deiminase pathway of *Bacillus licheniformis*, and its inactivation by the arginine repressor ArgR. J Bacteriol 180:6468–6475

Maheswaran M, Urbanke C, Forchammer K (2004) Complex formation and catalytic activation by the P_{II} signaling protein of N-acetyl-L-glutamate kinase *from Synechococcus elongatus* strain PCC7942. J Biol Chem 279:55202–55210

Maheswaran M, Ziegler K, Lockau W, Hagemann, Forchammer K (2006) P_{II}- regulated arginine synthesis controls accumulation of cyanophycin in *Synechocystis* sp. Strain PCC 6803. J Bacteriol 188:2730–2734

Makarova KS, Mironov AA, Gelfand MS (2001) Conservation of the binding site for the arginine repressor in all bacterial lineages. Genome Biol 2:0013.1

Makoff AJ, Radford A (1978) Genetics and biochemistry of carbamoylphosphate biosynthesis and its utilization in the pyrimidine biosynthetic pathway. Microbiol Rev 42:307–328

Marc F, Weigel P, Legrain C, Almeras Y, Santrot M, Glansdorff N, Sakanyan V (2000) Characterization and kinetic mechanism of mono- and bifunctional ornithine acetyltransferases from thermophilic organisms. Eur J Biochem 267:5217–5226

Marc F, Weigel P, Legrain C, Glansdorff N, Sakanyan V (2001) An invariant threonine is involved in self-catalyzed cleavage of the precursor protein for ornithine acetyltransferase. J Biol Chem 276:25404–25410

Martin PR, Mulks MH (1992) Sequence analysis and complementation studies of the *argJ* gene encoding ornithine acetyltransferase from *Neisseria gonorrhoeae*. J Bacteriol 174:2694–2701

Massant J, Verstreken P, Durbecq V, Kholti A, Legrain C, Beeckmans S, Cornelis P, Glansdorff N (2002) Metabolic channeling of carbamoylphosphate, a thermolabile intermediate: evidence for physical interactions between carbamate kinase-like carbamoylphosphate synthetase and ornithine carbamoyltransferase from the hyperthermophile *Pyrococcus furiosus*. J Biol Chem 277:18517–18522

Massant J, Wouters J, Glansdorff N (2003) Refined structure of *Pyrococcus furiosus* ornithine carbamoyltransferase at 1.87 A. Acta Cryst D Biol Crystallogr 59:2140–2149

Massant J, Glansdorff N (2004) Metabolic channeling of carbamoylphosphate in the hyperthermophilic archaeon *Pyrococcus furiosus*: dynamic enzyme–enzyme interactions involved in the formation of the channeling complex. Biochem Soc Trans 32:306–309

Massant J, Glansdorff N (2005) New experimental approaches for investigating interactions between Pyrococcus furiosus carbamate kinase and carbamoyltransferases, enzymes involved in the channeling of thermolabile carbamoylphosphate. Archaea 1:365–373

Meile L, Leisinger T (1984) Enzymes of arginine biosynthesis in methanogenic bacteria. Experientia 40:899–900

Messenguy F (1976) Regulation of arginine biosynthesis in *Saccharomyces cerevisiae*: isolation of a cis-dominant, constitutive mutant for ornithine carbamoyltransferase synthesis. J Bacteriol 128:49–55

Messenguy F, Dubois E (2000) Regulation of arginine metabolism in *Saccharomyces cerevisiae*: a network of specific and pleiotropic proteins in response to multiple environmental signals. Food Technol Biotechnol 38:277–285

Messenguy F, Dubois E (2003) Role of MADS box proteins and their cofactors in combinatorial control of gene expression and cell development. Gene 316:1–21

Messenguy F, Wiame JM (1969) The control of ornithinetranscarbamylase activity by arginase in *Saccharomyces cerevisiae*. FEBS Lett 3:47–49

Messenguy F, Dubois E, Descamps J (1986) Nucleotide sequence of the ARGRII regulatory gene and amino acid sequence homologies between ARGRII, PPRI and GAL4 regulatory proteins. Eur J Biochem 15:77–81

Messenguy F, Dubois E, Boonchird C (1991) Determination of the DNA-binding sequences of ARGR proteins to arginine anabolic and catabolic promoters. Mol Cell Biol 11:2852–2863

Messenguy F, Vierendeels F, Scherens B, Dubois E (2000) In *Saccharomyces cerevisiae*, expression of catabolic genes CAR1 and CAR2 in response to exogenous nitrogen availability is mediated by the Ume6 (CargRI)-Sin3 (CargRII)-Rpd3 (CargRIII) complex. J Bacteriol 182:3158–3164

Messenguy F, Vierendeels F, Pierard A, Delbecq P (2002) Role of RNA surveillance proteins Upf1/CpaR, Upf2 and Upf3 in the translational regulation of yeast CPA1 gene. Curr Genet 41:224–231

Miyasaki J, Kobashi N, Nishiyama M, Yamane H (2001) Functional and evolutionary relationship between arginine biosynthesis and prokaryotic lysine biosynthesis through α-aminoadipate. J Bacteriol 183:5067–5073

Miyasaki J, Kobashi N, Nishiyama M, Yamane H (2002) Charcterization of a lysK gene as an argE homolog in *Thermus thermophilus* HB27. FEBS Lett 512:269–274

Morin A, Huysveld N, Braun F, Dimova D, Sakanyan V, Charlier D (2003) Hyperthermophilic *Thermotoga* arginine repressor binding to full-length cognate and heterologous arginine operators and to half-site targets. J Mol Biol 332:537–553

Morizono H, Caldovic L, Shi D, Tuchman M (2004) Mammalian N-acetylglutamate synthase. Mol Genet Metab 81(1):4–11

Morizono H, Cabrera-Luque J, Shi D, Gallegos R, Yamaguci S, Yu X, Allewell NM, Malamy MH, Tuchman M (2006) Acetylornithine transcarbamylase: a novel enzyme in arginine biosynthesis. J Bacteriol 188:2974–2982

Morris CJ, Thompson JF (1975) Acetyl coenzyme A-glutamate acetyltransferase and N^2-acetylornithine-glutamate acetyltransferase of *Chlorella*. Plant Physiol 55:960–967

Nandineni MR, Gowrishankar J (2004) Evidence for an arginine exporter encoded by yggA (argO) that is regulated by the LysR-type of transcriptional regulator ArgP in *Escherichia coli*. J Bacteriol 186:3539–3546

Natarajan K, Meyer MR, Jackson BM, Slade D, Roberts C, Hinnebusch AG, Marton MJ (2001) Transcriptional profiling shows that Gcn4p is a master regulator of gene expression during amino acid starvation in yeast. Mol Cell Biol 21:4347–4368

Naumoff DG, Xu Y, Glansdorff N, Labedan B (2004) Retrieving sequences of enzymes experimentally characterized but erroneously annotated: the case of the putrescine carbamoyltransferase. BMC Genomics 5:52

Ni J, Sakanyan V, Charlier D, Glansdorff N, Van Duyne (1999) Structure of the arginine repressor from *Bacillus stearothermophilus*. Nature Struct Biol 6:427–432

Nicoloff H, Hubert JC, Bringel F (2000) In *Lactobacillus plantarum*, carbamoylphosphate is synthesized by two carbamoylphosphate synthetases (CPS): carbon dioxide differentiates the arginine-repressed from the pyrimidine-regulated CPS. J Bacteriol 182:3416–3422

Nicoloff H, Hubert JC, Bringel F (2001) Carbamoylphosphate synthetases (CPS) in lactic acid bacteria and other Gram-positive bacteria. Lait 81:1–2

Nicoloff H, Arsene-Ploetze F, Malandain C, Kleerebezem M, Bringel F (2004) Two arginine repressors regulate arginine biosynthesis in *Lactobacillus plantarum*. J Bacteriol 186:6059–6069

Nishida H, Nishiyama M, Kobashi N, Kosuge T, Hoshino T, Yamane H (1999) A prokaryotic gene cluster involved in synthesis of lysine though the amino adipate pathway: a key to the evolution of amino macid biosynthesis. Genome Res 9:1175–1183

Nishijyo T, Park SM, Lu CD, Itoh Y, Abdelal A (1998) Molecular characterization and regulation of an operon encoding a system for transport of arginine and ornithine and the ArgR regulatory protein in *Pseudomonas aeruginosa*. J Bacteriol 180:5559–5566

Obst M, Steinbuchel A (2006) Cyanophycin—an ideal Bacterial Nitrogen Storage Material with Unique Chemical properties. In: Microbiology Monographs, vol 1. Springer, Berlin Heidelberg New York, p 167–193 (DOI 10.1007/7171_007)

Odom AR, Stahlberg A, Wente SR, York JD (2000) A role for nuclear inositol 1,4,5-trisphosphate kinase in transcriptional control. Science 287:1937–1939

Palmieri L, De Marco V, Iacobazzi V, Palmieri F, Runswick MJ, Walker JE (1997) Identification of the yeast ARG-11 gene as a mitochondrial ornithine carrier involved in arginine biosynthesis. FEBS Lett 410:447–451

Para-Gessert L, Koo K, Fajardo J, Weiss RL (1998) Processing and function of a polypeptide precursor of two mitochondrial proteins in *Neurospora crassa.* J Biol Chem 273:7972–7980

Park S-M, Lu C-D, Abdelal A (1997) Purification and characterization of an arginine regulatory protein ArgR, from *Pseudomonas aeruginosa* and its interactions with the control regions for the *car*, *argF* and *aru* operons. J Bacteriol 179:5309–5317

Pauwels K, Abadjieva A, Hilven P, Stankiewicz, Crabeel M (2003) The N-acetylglutamate synthase/N-acetylglutamate kinase metabolon of Saccharomyces cerevisiae allows coordinated feedback regulation of the first two steps in arginine biosynthesis. Eur J Biochem 270:1014–1024

Perez-Rueda E, Collado-Vides, Segovia L (2004) Phylogenetic distribution of DNA-binding transcription factors in bacteria and archaea. Comput Biol Chem 28:341–350

Pierard A, Glansdorff N, Yashphe (1972) Mutations affecting uridine monophosphate pyrophosphorylase or the argR gene in *Escherichia coli.* Effects on carbamoylphosphate and pyrimidine biosynthesis and on uracil uptake. Mol Gen Genet 118:235–245

Purcarea C, Simon V, Prieur D, Herve G (1996) Purification and characterization of carbamoylphosphate synthetase from the deep-sea hyperthermophilic archaebacterium *Pyrococcus abyssi.* Eur J Biochem 236:189–199

Purcarea C, Evans DR, Herve G (1999) Channeling of carbamoylphosphate to the pyrimidine and arginine biosynthetic pathways in the deep sea hyperthermphilic archaeon *Pyrococcus abyssi.* J Biol Chem 274:6122–6129

Purcarea C, Herve G, Cunin R, Evans DR (2001) Cloning, expression and structural analysis of carbamate-like carbamoyl phosphate synthetase from *Pyrococcus abyssi.* Extremophiles 5:229–239

Quinn CL, Stephenson BT, Switzer RL (1991) Functional organization and nucleotide sequence of the *Bacillus subtilis* pyrimidine biosynthetic operon. J Biol Chem 266:9113–9127

Rajagopal BS, De Ponte III J, Tuchman M, Malamy MH (1998) Use of inducible feedback-resistant N-acetylglutamate synthetase (*argA*) genes for enhanced arginine biosynthesis by genetically engineered *Escherichia coli* K-12 strains. Appl Environ Microbiol 64:1805–1811

Ramon-Maiques S, Marina A, Gil-Ortiz F, Fita I, Rubio V (2002) Structure of acetylglutamate kinase, a key enzyme for arginine biosynthesis and a prototype for the amino acid kinase enzyme family, during catalysis. Structure 10:329–342

Ramon-Maiques S, Fernandez-Murga ML, Gil-Ortiz F, Vagin A, Fita I, Rubio V (2006) Structural bases of feed-back control of arginine biosynthesis, revealed by the structures of two hexameric N-acetylglutamate kinases, from *Thermotoga maritima* and *Pseudomonas aeruginosa.* J Mol Biol 356:695–713

Rodriguez-Garcia A, de la Fuente A, Perez-Redondo R, Martin JF, Liras P (2000) Characterization and expression of the arginine biosynthesis gene cluster of *Streptomyces clavuligerus.* J Mol Microbiol Biotechnol 2:543–550

Romero J, Liras P, Martin JF (1986) Utilization of ornithine and arginine as specific precursors of clavulanic acid. Appl Environ Microbiol 52:892–897

Ruiz-Echevarria MJ, Peltz SM (2000) The RNA binding protein Pub1 modulates the stability of transcripts containing upstream open reading frames. Cell 23:741–751

Sakai H, Vassylyeva MN, Matsuura T, Sekine S-I, Gotoh K, Nishiyama M, Terada T, Shirouzu M, Kuramitsu S, Vassylyev DG, Yokoyama (2003) Crystal structure of a lysine biosynthetic enzyme, LysX, from *Thermus thermophilus* HB8. J Mol Biol 332:729–740

Sakanyan V, Charlier D, Legrain C, Koxhikyan A, Mett I, Pierard A, Glansdorff N (1993a) Primary structure, partial purification and regulation of key enzymes of the acetyl cycle of arginine biosynthesis in *Bacillus stearothermophilus*: dual function of ornithine acetyltransferase. J Gen Microbiol 139:393–402

Sakanyan V, Desmarez L, Legrain C, Charlier D, Mett I, Kochikyan A, Savchenko A, Boyen A, Falmagne P, Pierard A, Glansdorff N (1993b) Gene cloning, sequence analysis, purification, and characterization of a thermostable aminoacylase from *Bacillus stearothermophilus*. Appl Environ Microbiol 59:3878–3888

Sakanyan V, Legrain C, Charlier D, Kochikian AV (1993c) N-acetylglutamate-5-phosphotransferase of the thermophilic bacterium *Bacillus stearothermophilus*: nucleotide sequence of the gene and enzyme characterization (in Russian). Genetika 29:556–570

Sakanyan V, Petrosyan P, Lecocq M, Boyen A, Legrain C, demarez M, Hallet JN, Glansdorff N (1996) Genes and enzymes of the acetyl cycle of arginine biosynthesis in Cyanobacterium glutamicum: enzyme evolution in the early steps of the arginine pathway. Microbiology UK 144:479–492

Sanchez R, Roovers M, Glansdorff N (2000) Organization and expression of a Thermus thermophilus arginine cluster: presence of unidentified open reading frames and absence of a Shine-Dalgarno sequence. J Bacteriol 182:5911–5915

Schneider BL, Kiupakis AK, Reitzer LJ (1998) Arginine catabolism and the arginine succinyltransferase pathway in *Escherichia coli*. J Bacteriol 180:4278–4286

Shi D, Gallegos R, DePonte J 3rd, Morizono H, Yu X, Allewell NM, Malamy M, Tuchman M (2002) Crystal structure of a transcarbamylase-like protein from the anaerobic bacterium *Bacteroides fragilis* at 2.0 A resolution. J Mol Biol 320:899–908

Shi D, Morizono H, Xiaolin Y, Roth L, Calcovic L, Allewell NM, Malamy MH, Tuchman M (2005a) Crystal structure of a N-acetylornithine transcarbamylase from *Xanthomonas campestris*: a novel enzyme in a new arginine biosynthetic pathway found in several eubacteria. J Biol Chem 280:14366–14369

Shi D, Xiaolin Y, Roth L, Morizono H, Hathout Y, Allewell NM, Tuchman M (2005b) Expression, purification, crystallization and preliminary X-ray crystallographic studies of a novel acetylcitrulline deacetylase from *Xanthomonas campestris*. Acta Cryst F61:676–679

Shi D, Morizono H, Cabrera-Luque J, Xiaolin Y, Roth L, Malamy MH, Allewell NM, Tuchman M (2006) Structure and catalytic mechanism of a novel N-succinyl-L-ornithine transcarbamylase in arginine biosynthesis of *Bacteroides fragilis*. J Biol Chem 281:20623–20631

Slocum RD (2005) Genes, enzymes and regulation of arginine biosynthesis in plants. Plant Physiol Biochem 43:729–745

Song H, Wang H, Gigot D, Dimova D, Sakanyan V, Glansdorff N, Charlier D (2002) Transcription regulation in thermophilic bacteria: high resolution contact probing of *Bacillus stearothermophilus* and *Thermotoga neapolitana* arginine repressor-operator interactions. J Mol Biol 315:255–274

Souciet JL, Nagy M, Le Gouar M, Lacroute F, Potier S (1989) Organization of the yeast URA2 gene: identification of a defective dihydroorotase-like domain in the multifunctional carbamoylphosphate synthetase-aspartate transcarbamylase complex. Gene 79:59–70

Steger DJ, Haswell ES, Miller AL, Wente SR, O'Shea EK (2003) Regulation of chromatin remodeling by inositol polyphosphates. Science 299:114–116

Stirling CJ, Szatmari G, Stewart G, Smith MC, Sherratt DJ (1988) The arginine repressor is essential for plasmid-stabilizing site-specific recombination at the ColE1 cer locus. EMBO J 7:4389–4395

Sunnerhagen M, Nilges M, Otting G, Carey J (1997) Solution structure of the DNA binding domain and model for the complex multifunctional hexameric arginine repressor with DNA. Nature Struct Biol 4:819–826

Thoden JB, Holden HM, Wesenberg G, Raushel FM, Rayment I (1997) Structure of carbamoylphosphate synthetase: a journey of 96 A from substrate to product. Biochem 36:6305–6316

Tian G, Lim D, Carey J, Maas WK (1992) How does the arginine repressor of *Escherichia coli* K12 bind to its operator sites. J Mol Biol 226:387–397

Traylor TG, Sharma VS (1992) Why NO? Biochem 31:2847–2849

Tuchman M, rajagopal BS, McCann MT, Malamy M (1997) Enhanced production of arginine and urea by genetically engineered *Escherichia coli* K-12 strains. Appl Environ Microbiol 63:33–38

Udaka S (1965) Pathway-specific pattern of control of arginine biosynthesis in bacteria. J Bacteriol 91:617–621

Uriarte M, Marina A, Ramon-Maiques S, Fita I, Rubio V (1999) The carbamoylphosphate synthetase of *Pyrococcus furiosus* is enzymologically and structurally a carbamate kinase. J Biol Chem 274:16295–16303

Uriarte M, Marina A, Ramon-Maiques S, Rubio V, Durbecq V, Legrain C, Glansdorff N (2001) Carbamoylphosphate synthesis: carbamate kinase from *Pyrococcus furiosus*. Methods Enzymol 331:236–247

Utagawa T (2004) Production of arginine by fermentation. J Nutr 134:2854-2857

Van de Casteele M, Demarez M, Legrain C, Glansdorff N, Pierard A (1990) Pathways of arginine biosynthesis in extreme thermophilic archaeaobacteria and eubacteria. J Gen Microbiol 136:1177–1183

Van de Casteele M, Legrain C, Demarez L, Chen PG, Pierard A, Glansdorff N (1997) Molecular physiology of carbamoylation under extreme conditions: what can we learn from extreme thermophilic microorganisms. Comp Biochem Physiol 188A:463–473

Vander Wauven C, Stalon V (1985) Occurrence of succinyl derivatives in the catabolism of arginine in Pseudomonas cepacia. J Bacteriol 164:882–886

Van Duyne GD, Ghosh G, Maas WK, Sigler (1996) Structure of the oligomerization and L-arginine domain of the arginine repressor of *Escherichia coli*. J Mol Biol 256:377–391

Van Huffel C, Dubois E, Messenguy F (1994) Cloning and sequencing of *Schizosaccharomyces pombe car1* gene, encoding arginase. Expression of the arginine anabolic and catabolic genes in response to arginine and related metabolites. Yeast 10:923–933

Velasco AM, Leguina JI, Lazcano A (2002) Molecular evolution of the lysine biosynthetic pathways. J Mol Evol 55:445–459

Viljoen M, Kovari LZ, Kovari IA, Park HD, van Vuuren HJJ, Cooper TG (1992) Tripartite structure of the *Saccharomyces cerevisiae* arginase *(CAR1)* gene inducer-responsive upstream activation sequence. J Bacteriol 174:6831–6839

Villeret V, Clantin B, Tricot C, Legrain C, Roovers M, Stalon V, Glansdorff, Van Beeumen J (1998) The crystal structure of *Pyrococcus furiosus* ornithine carabamoyltransferase reveals a key role for oligomerization in enzyme stability at extremely high temperatures. Proc Natl Acad Sci USA 95:2801–2806

Wang H, Glansdorff N, Charlier D (1998) The arginine repressor of *Escherichia coli* K-12 makes direct contacts to minor and major groove determinants of the operators. J Mol Biol 277:805–824.

Weigel P, Marc F, Simon S, Sakanyan V (2002) Ornithine N-acetyltransferase and arginine biosynthesis in thermophilic bacteria. Rec Res Devel Microbiol 6:95–106

Weiss RL, Lee C (1980) Isolation and characterization of *Neurospora crassa* mutants impaired in feedback control of ornithine synthesis. J Bacteriol 141:1305–1311

Werner M, Feller A, Messenguy F, Pierard A (1987) The leader peptide of yeast gene CPA1 is essential for the translational repression of its expression. Cell 19:805–813

Winkelmann G (2002) Microbial siderophore-mediated transport. Biochem Soc Trans 30:691–696

Woese C (1998) The universal ancestor. Proc Natl Acad Sci USA 95:6854–6859

Wohlkonig A, Stalon, Vander Wauven C (2004) Purification of ArcR, an oxidation-sensitive regulatory protein from *Bacillus licheniformis*. Prot Expr Purif 37:32–38

Xu Y, Liang Z, Legrain C, Ruger H-J, Glansdorff N (2000) Evolution of arginine biosynthesis in the bacterial domain: novel gene-enzyme relationships from psychrophilic *Moritella* strains (*Vibrionaceae*) and the evolutionary significance of N-α-acetylornithinase. J Bacteriol 182:1609–1615

Xu Y, Feller G, Gerday C, Glansdorff N (2003a) Metabolic enzymes from psychrophilic bacteria: challenge of adaptation to low temperature in ornithine carbamoyltransferase from *Moritella abyssi*. J Bacteriol 185:2161–2168

Xu Y, Sun Y, Huysveld N, Gigot D, Glansdorff N, Charlier D (2003b) regulation of arginine biosynthesis in the psychropiezophilic bacterium *Moritella profunda*: in vivo repressibility and in vitro repressor-operator contact probing. J Mol Biol 326:353–369

Xu Y, Glansdorff N, Labedan B (2006) Bioinformatic analysis of an unusual gene-enzyme relationship in the arginine biosynthetic pathway among marine gamma proteobacteria: implications concerning the formation of N-acetylated intermediates in prokaryotes. BMC Genomics 7:4

Yang H, Park SM, Nolan WG, Lu CD, Abdelal A (1997) Cloning and characterization of the arginine-specific carbamoyl-phosphate synthetase from *Bacillus stearothermophilus*. Eur J Biochem 249:443–449

Yoon S, Govind CK, Qiu H, Kim S-J, Dong J, Hinnebusch AG (2004) Recruitment of the ArgR/Mcm1p repressor is stimulated by the activator Gcn4p: a self-checking activation mechanism. Proc Natl Acad Sci USA 101:11713–11718

Yu YG, Turner GE, Weiss RL (1996) Acetylglutamate synthase from *Neurospora crassa*: structure and regulation of expression. Mol Microbiol 22:545–554

Zuniga M, Miralles Md Mdel C, Perez-Martinez G (2002) The product of *arcR*, the sixth gene of the *arc* operon of *Lactobacillus sakei*, is essential for expression of the arginine deiminase pathway. Appl Environ Microbiol 68:6051–6058

Microbiol Monogr (5)
V. F. Wendisch: Amino Acid Biosynthesis
DOI 10.1007/7171_2006_068/Published online: 18 January 2007

L-Serine and Glycine

Lothar Eggeling

Institute for Biotechnology 1, Research Centre Juelich, 52425 Juelich, Germany
l.eggeling@fz-juelich.de

Abstract The biosynthesis of glycine and L-serine is closely connected, and both amino acids are produced in industry. However, whereas glycine is made chemically, L-serine production relies largely on microbial processes. These include conversions of added glycine by C-1 utilizing microorganisms. But such precursor conversions usually suffer from low yields, as did previous attempts to produce L-serine from glucose. As more recent molecular and physiological studies have shown, microorganisms like *Corynebacterium glutamicum* have a high L-serine degradation capacity corresponding to an apparent key position of this amino acid in metabolism. Considering this key position, deletion of a serine dehydratase gene and prevention of folate synthesis to reduce serine hydroxymethyltransferase activity together with increased biosynthesis resulted in L-serine producers of *C. glutamicum* with excellent production characteristics and maximal specific productivities of 1.45 mmol g^{-1} h^{-1} accumulating more than 50 g l^{-1} L-serine.

1 Features and Use

L-Serine is an uncharged polar amino acid. It is not essential for humans, but synthesized either from 3-phosphoglycerate or glycine, the latter reaction preferably occurring in the fetal liver (Cetin et al. 1991).

It is estimated that approximately 300 t of L-serine are produced each year (Kumagai 2000). The amino acid is used for instance in the pharmaceutical industry in infusion solutions, in the feed industry as an additive to drinks, or in the cosmetics industry as a moisturizing agent in skin lotions. Also, a still substantial amount of L-serine is required for its enzymatic conversion together with indole to L-tryptophan (Eggeling et al. 2006), although competitive processes for the total cellular synthesis of L-tryptophan by either *Escherichia coli*, *Corynebacterium glutamicum*, or *Bacillus subtilis* exist (Ikeda 2006).

Glycine is a nonpolar amino acid. It is the simplest of the 20 standard proteinogenic amino acids as its side chain is a hydrogen atom. Because there is a second hydrogen atom at the α carbon, glycine is not optically active, which is a unique feature among the proteinogenic amino acids.

As estimated by the Japan Amino Acid Association, in 1996 about 22 000 t of glycine were produced each year (Kumagai 2000). Due to its sweet taste and buffering properties, glycine is used as a food and feed additive, for instance, to improve the flavor and taste of vinegar, fruit juices, and salted vegetables. It is also used as a solvent for removing CO_2 in the fertilizer industry, and as an additive to galvanizing solutions. Although formation of up to $4\,g\,l^{-1}$ glycine as a byproduct during L-histidine production with *C. glutamicum* ssp. *lactofermentum* was observed and thus microbial production was in principle demonstrated (Araki and Nakayama 1971), glycine is exclusively made chemically since it is not optically active. It is made via the classical Strecker synthesis of amino acids using ammonium chloride, formaldehyde, and hydrogen cyanide as educts.

The microbial synthesis of L-serine and glycine is closely linked. The synthesis of L-serine starts from 3-phosphoglycerate by a short sequence of three reactions and L-serine is converted by serine hydroxymethyltransferase (SHMT, EC 2.1.2.1) to glycine. The SHMT reaction also involves pyridoxal-5′-phosphate and 5,6,7,8-tetrahydrofolate (THF). The latter serves as an acceptor for the C-1 unit generated as a consequence of L-serine cleavage to yield 5,10-methylene-THF. Thus, SHMT provides glycine plus an activated C-1 compound. The 5,10-methylene-THF in turn can be converted into 5,10-methenyl-THF, and other activated C-1 units like 10-formyl-THF, 5-formyl-THF, and 5-methyl-THF to serve different demands in metabolism. For example, 5-methyl-THF is used in L-methionine synthesis, 5,10-methenyl-THF for D-pantothenate synthesis, or 10-formyl-THF for N-formylmethionyl-tRNA and purine synthesis. It is obvious that only due to the requirement for purine synthesis and provision of $tRNA^{fMet}$ for bacterial translation initiation, a high flux via the SHMT reaction occurs, which is estimated to make up 15% of total glycolytic flux in *E. coli* (Pizer and Potochny 1964). Therefore, L-serine has to be regarded as a metabolite in central metabolism rather than as the end product of a biosynthesis pathway. In the following, the attempts to produce L-serine by classical approaches as well as that by recent metabolic engineering will be summarized.

2
L-Serine from Glycine

Engineering of microorganisms accumulating L-serine from simple carbon sources was difficult. This is very different from L-lysine, L-arginine, or L-histidine, for instance, where already at the early times of strain engineering the successful microbial production of these and further amino acids was achieved. This process is reviewed in *The Microbial Production of Amino Acid* edited by Yamada et al. (1972) and in *Biotechnology of Amino Acid Production* edited by Aida et al. (1986) and *Recent Progress in Microbial Production of Amino Acids* edited by Enei et al. (1989). The successful engineering for the amino acids mentioned above was largely based on the use of amino acid analogues to derive feedback-resistant key enzymes and the use of auxotrophic mutants. However, similar attempts to derive L-serine producers were not convincing. Such attempts yielded a *Corynebacterium glutamicum* strain accumulating 3.8 $g\,l^{-1}$ L-serine during growth on glucose (Yoshida and Nakayama 1974), and an *Arthrobacter paraffineus* strain accumulating 1.5 $g\,l^{-1}$ L-serine during growth on paraffin (Kase and Nakayama 1974).

Therefore, development of L-serine production strains was largely dependent on the use of cheap precursors like glycine, methanol and formaldehyde making use of the serine hydroxymethyltransferase (SHMT) reaction, which catalyzes the interconversion of 5,10-methylenetetrahydrofolate + glycine + H_2O = tetrahydrofolate + L-serine (see above), with the interconversion and dissociation of reactants within the same order of magnitude (Schirch et al. 1977).

Glycine addition to numerous bacteria resulted in observable L-serine formation, however, at the same time glycine also inhibited growth, thus reducing space-time yields (Kubota et al. 1971). Therefore, a glycine-resistant *Corynebacterium glycinophilum* strain was isolated from putrefied banana (Kubota et al. 1972). This strain accumulated up to 9.1 $g\,l^{-1}$ L-serine from added glycine (together with glucose) at a molar yield of 53%. Using undirected mutagenesis, a mutant unable to use L-serine as a nitrogen source was isolated, which exhibited an improved L-serine accumulation of up to 13.9 $g\,l^{-1}$, at the cost of a reduced glycine yield (33 mol %) (Kubota 1985). This mutant was shown to be devoid of serine dehydratase (s. below) pointing to the importance of L-serine degrading activities (Kubota et al. 1989). The SHMT of *C. glycinophilum* was characterized and exhibited the expected comparable affinities for its substrates glycine, formaldehyde and L-serine with a K_m (in mM) of 1.85, 0.29, and 1.64, respectively (Kubota and Yokozeki 1979). In an approach to further improve the performance of *C. glycinophilum* by immobilization in calcium alginate-entrapped cells, productivities of up to 1.5 $g\,l^{-1}\,day^{-1}$ were obtained (Tanaka et al. 1989).

3 L-Serine from Glycine plus C-1 Units

It is obvious that the limited conversion of glycine as a precursor stimulated studies on the supply of the additionally required C-1 compound to improve production of L-serine from glycine. Consequently, the natural use of the SHMT reaction in some methylotrophic bacteria for assimilation of the C-1 compound was exploited. In these bacteria, the SHMT reaction fuels the serine cycle (Laukel et al. 2004) to yield C-3 compounds for anabolism and for further C-1 fixation (Fig. 1). The acceptor glycine is generated via the joint glyoxylate regeneration cycle. The idea of an advantageous use of the serine cycle was substantiated by the fact that L-serine formation occurred with *Pseudomonas* 3ab upon glycine addition during growth on methanol, but not on succinate or pyruvate (Keune et al. 1976). With *Pseudomonas* 3ab high L-serine accumulations of $7\,g\,l^{-1}$ at a 50% molar yield were obtained under fed-batch culture conditions (Behrendt et al. 1984). Another *Pseudomonas* strain isolated, *Pseudomonas* MS31, accumulated up to $2.1\,g\,l^{-1}$ L-serine from glycine during growth on methanol (Morinaga et al. 1981a).

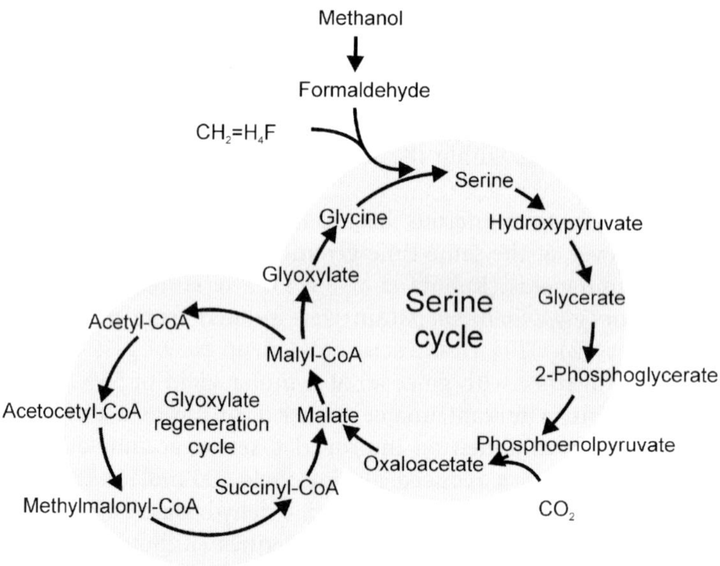

Fig. 1 Methanol and formaldehyde fixation via the serine cycle in methylotrophs. The activated C-1 unit is condensed with glycine via SHMT activity to L-serine, which is converted to phospho*enol*pyruvate. From phospho*enol*pyruvate oxaloacetate is generated, which, after rearrangement, is cleaved into glyoxylate and acetyl CoA. Two acetyl CoA molecules are used in the glyoxylate regeneration cycle for glyoxylate regeneration to fuel the serine cycle. Only part of the reactions of the glyoxylate cycle are given, which is in part still hypothetically

When cultivated at an elevated temperature of 42 °C its L-serine degradation capacity seemed to be reduced, since a higher product concentration was obtained (Morinaga et al. 1981b). A mutant of *Pseudomonas* MS31 with 1.7-fold-increased SHMT activity was able to accumulate up to 23.9 $g\,l^{-1}$ L-serine (Watanabe et al. 1987). Another methanol-utilizing bacterium able to accumulate L-serine is *Hyphomicrobium methylovorum* (Izumi et al. 1982). Resting cells of mutant GM2 of *H. methylovorum* with increased methanol dehydrogenase and SHMT activities accumulated 34 $g\,l^{-1}$ L-serine (Yamada et al. 1986). A further optimization of the feeding regime resulted in accumulation of 45 $g\,l^{-1}$ L-serine with a glycine yield of 50% (Izumi et al. 1993). Interestingly, an almost 100% molar conversion was obtained in the presence of 1 mM $CdCl_2$ possibly by directly preventing the aldolytic cleavage of L-serine (Yoshida et al. 1993). Other methanol-utilizing strains elaborated were *Arthrobacter globiformis* (Tani et al. 1978), and *Sarcina albida* accumulating up to 21 $g\,l^{-1}$ L-serine (Ema et al. 1979; Omori et al. 1983). However, in almost all cases the conversion yields of glycine were unsatisfying. Therefore, only recently a new screening of methanol-utilizing bacteria was initiated (Hagishita et al. 1996) and resulted in a *Methylobacterium* strain exhibiting an exceptional high glycine conversion of 93%, with final L-serine concentrations of 65 $g\,l^{-1}$, when cultures were fed repeatedly with methanol.

Due to limiting precursor conversion and toxicity of C-1 substrates in fermentative L-serine production, another development assayed enzymatic routes towards L-serine production. The catalyst was *Klebsiella aerogenes* overexpressing *glyA* of *E. coli* encoding the SHMT (Hamilton and Hsiao 1985). Expression of *glyA* on pGX139 under control of its own promoter accounted for more than 10% of the cellular protein as SHMT in the recombinant *Klebsiella* strain. With this strain titers of 30 000 units l^{-1} at cell densities of 30 $g\,l^{-1}$ were achieved. L-serine formation was performed in bioreactors with permeabilized cells or cell lysates and 8.8 mM tetrahydrofolate and 1 mM pyridoxal-5′-phosphate. The formaldehyde feeding was critical since the SHMT is very sensitive to free formaldehyde. Therefore, a control system was necessary. This used the rapid and non-enzymatic conversion of formaldehyde plus tetrahydrofolate to 5,10-methylenetetrahydrofolate resulting in very low levels of free formaldehyde. In case of a reduced conversion of 5,10-methylenetetrahydrofolate and glycine to L-serine, the increasing free formaldehyde reacted with amino groups (e.g., glycine or L-serine) to produce a Schiff base and a proton, thus lowering the pH, which was used as a control signal to reduce the formaldehyde feed. With this process L-serine titers of 450 $g\,l^{-1}$ with an 88% molar conversion of glycine at a volumetric productivity of 8.9 $g\,l^{-1}\,h^{-1}$ were achieved (Hsiao and Wei 1985).

An entirely other enzymatic L-serine synthesis used D-glycerate as substrate. D-glycerate conversion required glyoxylate reductase (EC 1.1.1.26), which catalyzes the NAD-linked oxidation of D-glycerate into β-hydroxypyruvate. The following reductive amination of β-hydroxypyruvate to

L-serine under oxidation of the NADH formed in the first reaction step was catalysed by L-alanine dehydrogenase (EC 1.4.1.1) (Furuyoshi 1989). Using 2 kU glyoxylate reductase and 100 kU L-alanine dehydrogenase the formation of 20 $g\,l^{-1}$ L-serine from 21 g D-glycerate within 25 h at a molar yield of 96% was possible.

4
L-Serine from Glucose

As is evident, the traditional development of microorganisms by undirected methods or enzymological processes was not convincing. However, modern molecular engineering techniques in conjunction with physiological knowledge, genome-wide approaches, and determination of intracellular fluxes provide the tools to analyze L-serine metabolism in more detail. The consequent application of such techniques to the “work horse” of amino acid production, *Corynebacterium glutamicum* (Eggeling and Bott 2005), is an example on the use of such modern techniques for strain development, which will be described in the following section.

4.1
Biosynthesis and 3-Phosphoglycerate Dehydrogenase

The biosynthesis of L-serine occurs in three steps initiating from the glycolytic intermediate 3-phosphoglycerate (Fig. 2). Based on the genome sequence and enzyme assays, the corresponding genes *serA, serB* and *serC* were identified in *C. glutamicum* (Peters-Wendisch et al. 2002). The *serA* encoded 3-phosphoglycerate dehydrogenase (PGDH, EC 1.1.1.95) generates 3-phosphohydroxypyruvate in an NAD dependent reaction (Fig. 2), and the *serC* encoded phosphoserine aminotransferase (PSAT, EC 2.6.1.52) catalyzes the amino transfer from L-glutamate to generate 3-phosphoserine. The last step is a phosphotransferase reaction catalysed by the *serB* encoded phosphoserine phosphatase (PSP, EC 3.1.3.3) probably by involvement of an enzyme-bound phosphoaspartate as an intermediate (Collet et al. 1999).

Furthermore, it was shown that *serA*-encoded PGDH is allosterically controlled by L-serine. After incubation in the presence of 10 mM L-serine the activity was substantially reduced to 38% (Fig. 2). A detailed sequence analysis of the *serA* encoded polypeptide of 530 aminoacyl residues (aa), together with a comparison of crystal structures of PGDH’s revealed a clear domain organization, with the C-terminus likely to represent the domain involved in allosteric control. This is reminiscent of the situation of the aspartate kinase (EC 2.7.2.4), homoserine dehydrogenase (EC 1.1.1.3) and threonine dehydratase (EC 4.3.1.19) of *C. glutamicum* that also contain domains involved in allosteric control located in their C-termini. Based on the domain

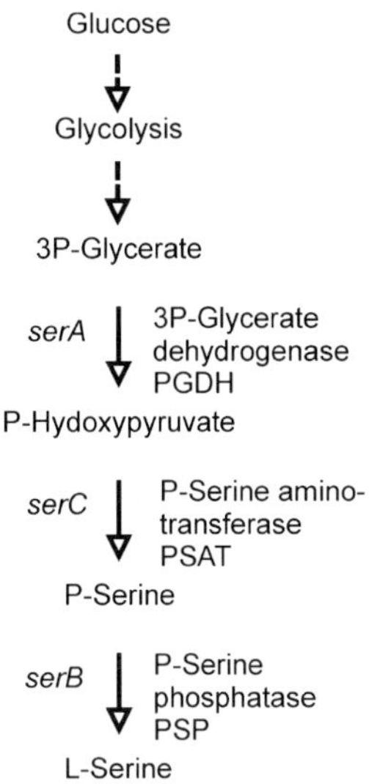

Construct	Strain	Specific activity	Residual activity (%)
	WT pZ1	0.13	38
	WT p*serA*	2.15	37
	WT p*serA*Δ79	0.19	15
	WT p*serA*Δ188	0.44	99
	WT p*serA*Δ197	1.32	92
	WT p*serA*Δ205	0.39	< 1
	WT p*serA*Δ211	0.15	< 1

Fig. 2 L-serine synthesis and regulation of 3-phosphoglycerate dehydrogenase muteins. In the table, the specific activity is given in μmol min^{-1} mg(protein)$^{-1}$, and the strains are all recombinant wild-type derivatives of *C. glutamicum* (WT) carrying plasmid-encoded *serA* alleles (Peters-Wendisch et al. 2002). For details, see text. On the left, a sketch on the relevant reactions and genes for L-serine synthesis is given

organization, a comprehensive set of truncated *serA* versions was studied (Peters-Wendisch et al. 2002). Most prominent was the deletion of as much as 197 aa from the C-terminus, as present in *serA*Δ197, resulting in an activity almost unsensitive to L-serine inhibition, but with its catalytic activity largely retained (Fig. 2). Slightly shorter or longer deletions just differing by 8 aa from *serA*Δ197 negatively influenced total activity, control, or stability. The notion that the C-terminus of SerA of the PGDH enzyme is involved in the allosteric transition from the tight inactive to the relaxed active state of the enzyme is furthermore substantiated by a mutation identified in *C. glutamicum* resistant to serine homologs such as β-(-2-thienyl)-D,L-alanine or azaserine, where Glu in position 325 was substituted by Lys and the corresponding enzyme was resistant to L-serine inhibition (Suga et al. 1999). Also in *E. coli*, short truncations in the C-terminus of SerA led to abolishment of feedback inhibition, which was beneficial for L-cysteine production due to an increased L-serine flux (Wacker 1991). Although relieved of inhibition by the product L-serine, overexpression of the mutant allele *serA*Δ197 in *C. glutamicum* either alone or in combination with overexpression of *serC* and *serB* did not result in significant L-serine accumulation (Peters-Wendisch et al. 2002, 2005) pointing to an intracellular conversion of L-serine. This meets the expectations already derived from the poor L-serine accumulations obtained with the classically made strains.

4.2
L-Serine Degradation and L-Serine Dehydratase

Bacteria like *E. coli* and *C. glutamicum* have a high capacity to utilize L-serine present in the external environment (Prüß et al. 1994; Netzer et al. 2004). Thus, in the complex medium CGIII containing peptone and yeast extract L-serine is an amino acid rapidly utilized by *C. glutamicum* as is also the case for L-glutamine and L-methionine (Fig. 3), whereas L-lysine, L-alanine or L-phenylalanine are hardly utilized, if at all (Zimmermann 2001). On the mineral salt medium CGXII (Eggeling and Bott 2005) up to 100 mM L-serine are consumed in presence of 80 mM glucose with uptake rates of up to 20 nmol min^{-1} $mg(dryweight)^{-1}$ (Netzer et al. 2004). This cometabolism of glucose with L-serine by *C. glutamicum* results in higher growth yields as compared to growth on glucose alone and, since utilization also occurred in the stationary phase, it might also help to survive under conditions when substrates are limiting. Tracing the label of L-[U-^{13}C]serine revealed that L-serine as an entity is converted to pyruvate and that an *sdaA* encoded serine dehydratase (EC 4.3.1.17) is present in *C. glutamicum* (Netzer et al. 2004). The serine dehydratase contains an [4Fe – 4S] cluster involved in the pyridoxal-5′-phosphate independent deamination of L-serine to pyruvate. When *sdaA* is deleted the L-serine uptake rate of *C. glutamicum* was reduced to about 9 nmol min^{-1} $mg(dryweight)^{-1}$, showing that serine dehydratase contributes significantly to L-serine consumption. The absence of serine dehydratase might explain why a *C. glutamicum* mutant obtained by undirected mutagenesis has reduced serine degradation capacity (Suga et al. 1999).

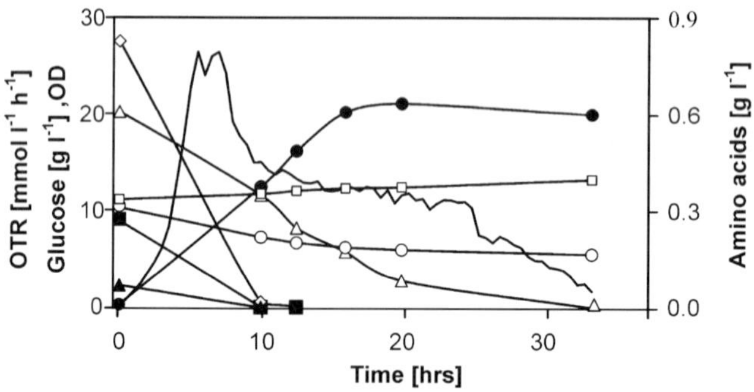

Fig. 3 Fermentation of *Corynebacterium glutamicum* ATCC13032 in medium CGIII containing 4% glucose. Shown is the oxygen transfer rate (OTR, *solid line without symbols*), growth (•), and the residual concentrations of glucose (△) and of the amino acids glutamine (◇), lysine (□), phenylalanine (○), serine (■) and methionine (▲) in the complex medium. (Courtesy of HF Zimmermann, Zimmermann 2001)

4.3
Serine Hydroxymethyltransferase and C-1 Metabolism

The NMR study with labelled L-[U-^{13}C]serine confirmed that L-serine is also metabolized to glycine by *C. glutamicum*. This reaction is catalysed by the serine hydroxymethyltransferase (SHMT) and serves to provide glycine plus 5,10-methylene-tetrahydrofolate (see above). Since activated C-1 units are, amongst others, required for the synthesis of purines, thymine, L-methionine and N-formylmethionyl-tRNA, it is not surprising that the total carbon flux towards L-serine during growth on glucose minimal medium is as much as 7.5% of the glucose uptake rate, as estimated for *C. glutamicum* (Marx et al. 1996). This cellular demand cannot be entirely bypassed by external metabolite addition, since SHMT is essential in *C. glutamicum* (Simic et al. 2001).

However, to nevertheless enable SHMT activity control and to create conditions of reduced L-serine degradation a special recombinant strain was made. For this purpose, a non-replicative plasmid was constructed containing the 5′end of *glyA* under control of the *tac* promoter P-*tac* including *lac* operator (Simic et al. 2001). In addition, the plasmid carried a kanamycin resistance gene and *lacI*^{Q} for production of *E. coli* Lac repressor, which is required for IPTG-dependent control of P-*tac*. Transformation of *C. glutamicum* to kanamycin resistance in the presence of IPTG yielded a recombinant strain, which carried an intact copy of *glyA* under control of P-*tac* and a second inactive *glyA* fragment under control of the natural promoter in the chromosome. When this strain was cultivated without IPTG, growth was retarded and *glyA* expression strongly reduced (Peters-Wendisch et al. 2005). The specific SHMT activity was 10 nmol min^{-1} mg(protein)$^{-1}$ instead of 40 nmol min^{-1} mg(protein)$^{-1}$. The wild-type derivative with reduced *glyA* expression transiently accumulated up to 1 mM of L-serine. Additional overexpression of *serA*, *serC* and *serB* resulted in accumulation of 15 mM L-serine, which could be further improved by deletion of *sdaA* (accumulation of 100 mM L-serine). However, strains with *glyA* under control of P-*tac* were difficult to handle and notoriously unstable in larger cultivations, which became apparent in loss of L-serine accumulation. Analysis of clones derived from such cultures revealed, amongst others, mutations in *lacI*^{Q}, which apparently resulted in the inability of the encoded Lac repressor to interact with the *lac* operator in front of P-*tac*, thus making *glyA* expression independent from the presence of IPTG (Stolz et al. 2006).

4.4
Tetrahydrofolate Synthesis and Serine Hydroxymethyltransferase Control

To overcome the inconvenient *glyA* expression control by the *tac* promoter, an entirely different approach was chosen based on the requirement of SHMT activity on the availability of 5,6,7,8-tetrahydrofolate (THF). Although the aldol

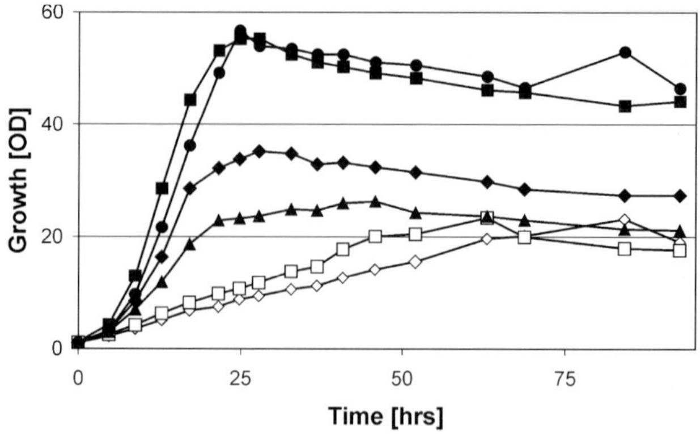

Fig. 4 Growth of *C. glutamicum* 13032Δ*sdaA*Δ*pabABC* p*serACB* in minimal medium containing different folate concentrations (0 mM, ◊; 0.01 mM, □; 0.1 mM, ▲; 0.25 mM, ♦; 1 mM, •). The control strain 13032Δ*sdaA* p*serACB* did not receive folate (■). (Stolz et al. 2006)

cleavage reaction of L-threonine, which is also catalysed by SHMT, is THF-independent (Simic et al. 2002), the cleavage of L-serine requires THF. Inspection of the genome of *C. glutamicum* detected 2 genes adjacent to *glyA* encoding aminodeoxychorismate synthase (EC 6.3.5.8) and aminodeoxychorismate lyase (EC 4.1.3.38) catalyzing steps of THF formation. The aminodeoxychorismate synthase is a bifunctional enzyme with a glutamine amidotransferase and a chorismate aminating activity. In *C. glutamicum* both polypeptides are fused and encoded by *pabAB* (Stolz et al. 2006), while separate polypeptides exist in *E. coli* or *Mycobacterium tuberculosis*, for instance (Cole et al. 1998). The lyase is encoded by *pabC*. Deletions of either *pabAB* or *pabC* revealed the requirement of THF (or p-aminobenzoate) of the corresponding mutants, albeit the phenotype of the *pabC* mutant was less pronounced probably due to decomposition of the labile intermediate 4-amino-4-deoxychorismate. In liquid culture growth of a *pabABC* deletion mutant on mineral salt media responded in a dose-dependent manner to external folate supply, with 1 mM folate satisfying the demand for growth with 4% glucose enabling growth undistinguishable from that of the wild-type (Fig. 4).

4.5 Fermentative Production of L-Serine

To generate a strain suitable for glucose conversion to L-serine, the serine dehydratase gene as well as *pabABC* were deleted in *C. glutamicum* and the resulting strain transformed with a plasmid containing the genes for feedback resistant 3-phosphoglycerate dehydrogenase, phosphoserine aminotransferase, and phosphoserine phosphatase to generate strain 13032Δ*sdaA* Δ*pabABC* p*serACB* (Stolz et al. 2006). Under industrially relevant conditions, the performance of this strain was evaluated in a 20-l reactor based on corn steep liquor (CSL) medium. The medium contained 35 g l^{-1} solid CSL plus initially 15 g l^{-1} glucose and 15 g l^{-1} fructose, and salts. The minimum dissolved oxygen concentration was set to 50% saturation to ensure no oxygen limitation. As can be seen in Fig. 5, inoculation of the reactor with cells derived from the preculture CGXII enabled rapid growth up to a maximum specific growth rate of 0.25 h^{-1}. L-serine formation occurred already from the beginning, suggesting a suitable folate supply in the culture due to CSL, which can be assumed to contain at least traces of this vitamin. The maximum oxygen uptake rate, OURmax, was about 110 mol l^{-1} h^{-1}, which was observed at the end of the exponential growth phase of the culture. The maximal specific productivity was 1.45 mmol g^{-1} h^{-1}, and the volumetric productivity was about 1.4 g l^{-1} h^{-1}. In this experiment, a final concentration of 345 mM L-serine was reached, but significantly higher concentrations could be obtained.

As the continued work on L-serine synthesis in *C. glutamicum* has shown, the key to derive substantial concentrations of this amino acid is the adequate circumvention of its catabolism. Limitation of the vitamin folate is an ideal tool for this purpose. It is known from other examples that growth limitation, and vitamin limitation in particular, is advantageous for metabolite overproduction. For instance, control of D-pantothenate availability is favorable to improve L-valine formation, whose carbon skeleton is derived from two pyruvate molecules (Radmacher et al. 2002). The basis is that D-pantothenate is a constituent of coenzyme A, and reduced coenzyme A availability results in reduced activity of the pyruvate dehydrogenase thus limiting pyruvate decarboxylation. It should be noted that vitamin limitations in strain constructions are entirely different from the well-established "pathway tailoring" by removing competing reactions or removing bottlenecks (Eggeling and Sahm 1999).

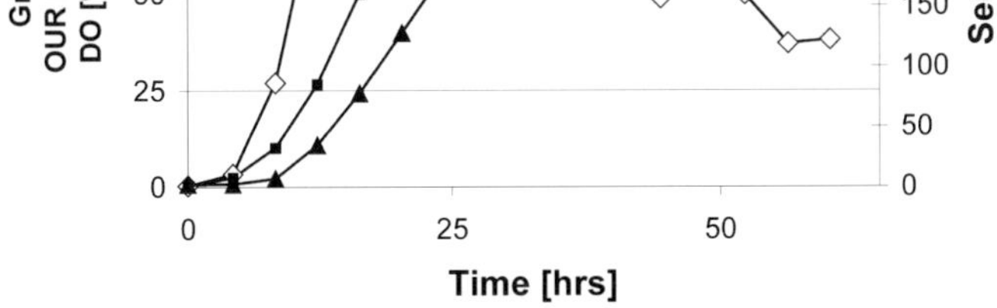

Fig. 5 Performance of *C. glutamicum* 13032Δ*sdaA*Δ*pabABC* p*serACB* in a 20-l reactor showing growth (■), the L-serine concentration in the medium (▲), the dissolved oxygen DO (□), and the oxygen uptake rate, OUR (◇) (Stolz et al. 2006)

The reason is that vitamin limitation enables a gradual control of growth thus probably leading to increased intracellular supply of reducing power and precursors for the biosynthesis pathways. Another example is a favorable *leuC* mutation of *C. glutamicum*, resulting in a leucine leaky phenotype and increased L-lysine synthesis (Hayashi et al. 2006). In this case, increased expression of lysine biosynthesis genes by unknown mechanisms has been demonstrated to occur. It cannot be excluded that under the severe growth limitations by folate auxotrophy such mechanisms also come into play to contribute to the excellent production properties of such strains.

References

Aida K, Chibata I, Nakayama K, Takinami K, Yamada H (eds) (1986) Biotechnology of amino acid production. Kodansha Ltd, Tokyo. Elsevier Science Publishers BV, Amsterdam

Araki K, Nakayama K (1971) Studies on histidine fermentation. 1. Histidine production by histidine analog-resistant mutants from several bacteria. Agric Biol Chem 5:2081–2084

Behrendt U, Bang W-G, Wagner F (1984) The production of L-serine with a methylotrophic microorganism using the L-serine pathway and coupling with an L-tryptophan-producing process. Biotech Bioeng 26:308–314

Cetin I, Fennessey PV, Quick AN Jr, Marconi AM, Meschia G, Battaglia FC Sparks JW (1991) Glycine turnover and oxidation and hepatic serine synthesis from glycine in fetal lambs. Am J Physiol Endocrinol Metab 260:E371–E378

Cole ST, Brosch R, Parkhill J, Garnier T, Churcher C, Harris D, Gordon SV, Eiglmeier K, Gas S, Barry CE 3rd, Tekaia F, Badcock K, Basham D, Brown D, Chillingworth T, Connor R, Davies R, Devlin K, Feltwell T, Gentles S, Hamlin N, Holroyd S, Hornsby T, Jagels K, Krogh A, McLean J, Moule S, Murphy L, Oliver K, Osborne J, Quail MA, Rajandream MA, Rogers J, Rutter S, Seeger K, Skelton J, Squares R, Squares S, Sulston JE, Taylor K, Whitehead S, Barrell BG (1998) Deciphering the biology of *Mycobacterium tuberculosis* from the complete genome sequence. Nature 393:537–544

Collet JF, Stroobant V, Van Schaftingen E (1999) Mechanistic studies of phosphoserine phosphatase, an enzyme related to P-type ATPases. J Biol Chem 274:33985–33990

Eggeling L, Bott M (2005) Handbook of *Corynebacterium glutamicum*. CRC Press, Taylor Francis Group, Boca Raton, FL

Eggeling L, Pfefferle W, Sahm H (2006) Amino acids. In: Radledge C, Kristiansen B (eds) Basic biotechnology, 3rd edn. Cambridge University Press, Cambridge, pp 335–358

Eggeling L, Sahm H (1999) Amino acid production: principles of metabolic engineering. In: Lee SY et al., Papoutsakis (eds) Metababolic Engineering. Marcel Dekker Inc., New York, Basel, pp 153–176

Ema M, Kakimoto T, Chibata I (1979) Production of L-serine by *Sarcina albida*. Appl Environ Microbiol 37:1053–1058

Enei H, Yokozeki K, Akashi K (1989) Recent progress in microbial production of amino acids. Gordon and Breach Science Publishers, Amsterdam

Furuyoshi S, Kawabata N, Nagata S, Tanaka H, Soda K (1989) Enzymatic synthesis of L-serine from D-glycerate. Agr Biol Chem 53:3075–3076

Hagishita T, Yoshida T, Isumi Y, Mitsunaga T (1996) Efficient L-serine production from methanol and glycine by resting cells of *Methylobacterium* sp. strain MN43. Biosci Biotech Biochem 60:1604–1607

Hamilton BK, Hsiao H (1985) Manufacture of L-amino acids with bioreactors. Trends Biotechnol 3:64–69

Hayashi M, Mizoguchi H, Ohnishi J, Mitsuhashi S, Yonetani Y, Hashimoto S, Ikeda M (2006) A *leuC* mutation leading to increased L-lysine production and *rel*-independent global expression changes in *Corynebacterium glutamicum*. Appl Microbiol Biotechnol 72:783–789

Hsiao H, Wei T (1985) Enzymatic production of L-serine with a feedback control system for formaldehyde addition. Biotechnol Bioeng 28:1510–1518

Ikeda M (2006) Towards bacterial strains overproducing L-tryptophan and other aromatics by metabolic engineering. Appl Microbiol Biotechnol 69:615–626

Izumi Y, Takizawa M, Tani Y, Yamada H (1982) L-Serine production by resting cells of a methanol-utilizing bacterium. J Ferment Technol 60:269–276

Izumi Y, Yoshida T, Miyazaki SS, Mitsunaga T, Ohshiro T, Shimao M, Miyata A, Tanabe T (1993) L-Serine production by a methylotroph and its related enzymes. Appl Microbiol Biotechnol 39:427–432

Kase H, Nakayama K (1974) L-Serine production by mutants of *Corynebacterium glutamicum* and *Athrobacter paraffineus*. Nippon-Nogeikagaku-Kaishi 48:209–213

Keune H, Sahm H, Wagner F (1976) Production of L-serine by the methanol utilizing bacterium *Pseudomonas* 3ab. Eur J Appl Microbiol 2:175–184

Kubota K (1985) Improved production of L-serine by mutants of *Corynebacterium glycinophylum* with less serine dehydratase activity. Agr Biol Chem 49:7–12

Kubota K, Kageyama K, Maeyashiki I, Yamada K, Okumura S (1972) Fermentative production of L-serine. Production of L-serine from glycine by *Corynebacterium glycinophilum* Nov. SP. J Gen Appl Microbiol 18:365–375

Kubota K, Kageyama T, Shiro S, Okumura S (1971) Fermentative production of L-serine. J Gen Appl Microbiol 17:167–168

Kubota K, Yokozeki K (1979) Production of L-serine from glycine by *Corynebacterium glycinophylum* and properties os serine hydroxymethyltransferase, a key enzyme in L-serine production. J Ferment Bioeng 67:387–390

Kubota K, Yokozeki K, Ozaki H (1989) Effects of L-serine dehydratase activity on L-serine production by *Corynebacterium glycinophilum* and an examination of the properties of the enzyme. J Ferment Bioeng 67:391–394

Kumagai H (2000) Microbial production of amino acids in Japan. In: Scheper T (ed) Adv in Biochem Eng, vol 69. Springer, Berlin, Heidelberg, New York, pp 71–85

Laukel M, Rossignol M, Borderies G, Volker U, Vorholt JA (2004) Comparison of the proteome of *Methylobacterium extorquens* AM1 grown under methylotrophic and non-methylotrophic conditions. Proteomics 4:1247–1264

Marx A, deGraaf AA, Wiechert W, Eggeling L, Sahm H (1996) Determination of the fluxes in the central metabolism of *Corynebacterium glutamicum* by nuclear magnetic resonance spectroscopy combined with metabolite balancing. Biotechnol Bioeng 49:111–129

Morinaga Y, Yamanaka S, Takinami K (1981a) L-Serine production by methanol-utilizing bacterium *Pseudomonas* MS31. Agr Biol Chem 45:1419–1424

Morinaga Y, Yamanaka S, Takinami K (1981b) L-Serine production by temperature-sensitive mutants of methanol-utilizing bacterium *Pseudomonas* MS31. Agr Biol Chem 45:1425–1430

Netzer R, Peters-Wendisch P, Eggeling L, Sahm H (2004) Cometabolism of a nongrowth substrate: L-serine utilization by *Corynebacterium glutamicum*. Appl Environ Microbiol 70:7148–7155

Omori K, Kakimoto T, Chibata I (1983) L-serine production by a mutant of *Sarcina albida* defective in L-serine degradation. Appl Environ Microbiol 45:1722–1726

Peters-Wendisch P, Netzer R, Eggeling L, Sahm H (2002) 3-Phosphoglycerate dehydrogenase from *Corynebacterium glutamicum*: the C-terminal domain is not essential for activity but is required for inhibition by L-serine. Appl Microbiol Biotechnol 60:437–441

Peters-Wendisch P, Stolz M, Etterich H, Kennerknecht N, Sahm H, Eggeling L (2005) Metabolic engineering of *Corynebacterium glutamicum* for L-serine production. Appl Environ Microbiol 71:7139–7144

Pizer LI, Potochny ML (1964) Nutritional and regulatory aspects of serine metabolism in *Escherichia coli*. J Bacteriol 88:611–619

Prüß BM, Nelms JM Park C, Wolfe AJ (1994) Mutations in NADH:ubiquinone oxidoreductase of *Escherichia coli* affect growth on mixed amino acids. J Bacteriol 176:2143–2150

Radmacher E, Vaitsikova A, Burger U, Krumbach K, Sahm H, Eggeling L (2002) Linking central metabolism with increased pathway flux: L-valine accumulation by *Corynebacterium glutamicum*. Appl Environ Microbiol 68:2246–2250

Schirch LV, Tatum CM Jr, Benkovic SJ (1977) Serine transhydroxymethylase: evidence for a sequential random mechanism. Biochemistry 16:410–419

Simic P, Willuhn J, Sahm H, Eggeling L (2002) Identification of *glyA* (encoding serine hydroxymethyltransferase) and its use together with the exporter ThrE to increase L-threonine accumulation by *Corynebacterium glutamicum*. Appl Environ Microbiol 68:3321–3327

Stolz M, Peters-Wendisch P, Etterich H, Gerharz T, Faurie R, Sahm H, Fersterra H, Eggeling L (2006) Reduced folate supply as a key to enhanced L-serine production with *Corynebacterium glutamicum*. Appl Environ Microbiol, DOI: 10.1128/AEM.02208-06

Suga M, Sugimoto M, Osumi T, Nakamatsu T, Hibino W, Ito M (1999) Method for producing L-serine by fermentation. EP 0 943 687 A2

Tanaka T, Yamamoto K, Towpprayoon S, Nakajima H, Sonomoto K, Yokozeki K, Kubota K, Tanaka A (1989) Continuous production of L-serine by immobilized growing *Corynebacterium glycinophilum* cells. Appl Microbiol Biotech 30:564–568

Tani Y, Kanagawa T, Hanpongkittikun A, Ogata K, Yamada H (1978) Production of L-serine by a methanol-utilizing bacterium *Arthrobacter globiformis* SK-200. Agr Biol Chem 42:2275–2279

Wacker F (1991) Materials and methods for biosynthesis of serine and serine-related products. PCT WO 93/12235

Watanabe M, Morinaga Y, Enei H (1987) L-Serine production by a methionine-auxotrophic mutant of methylotrophic *Pseudomonas*. J Ferment Technol 65:617–620

Yamada H, Miyazaki SS, Izumi Y (1986) L-Serine production by a glycine-resistant mutant of methylotrophic *Hyphomicrobium methylovorum*. Agr Biol Chem 50:17–21

Yamada K, Kinoshita S, Tsunoda T, Aida K (eds) (1972) The microbial production of amino acid. Kodansha Ltd, Tokyo. John Wiley & Sons, New York, London, Sidney, Toronto

Yoshida H, Nakayama K (1974) Production of L-serine by L-serine analog-resistent mutants from various bacteria and the effect of L-threonine and L-homoserine on the production of L-serine. Nippon-Nogeikagaku-Kaishi 48:201–208

Yoshida T, Mitsunaga T, Izumi Y (1993) L-serine production using a resting cell system of *Hyphomycobium* strains. J Ferment Technol 75:405–408

Zimmermann HF (2001) Online-Messung der Sauerstoff-Transferrate in geschüttelten Bioreaktoren zur Beschreibung der Stoffwechselaktivitäten von *Corynebacterium glutamicum*. Diplomarbeit, Universität Stuttgart, Stuttgart

Microbiol Monogr (5)
V. F. Wendisch: Amino Acid Biosynthesis
DOI 10.1007/7171_2006_062/Published online: 9 December 2006

Alanine, Aspartate, and Asparagine Metabolism in Microorganisms

Tadao Oikawa

Department of Biotechnology, Faculty of Engineering, Kansai University, 3-3-35 Yamate-Cho, Suita, 564-8680 Osaka-Fu, Japan
oikawa@ipcku.kansai-u.ac.jp

Abstract L-Alanine, L-aspartate, and L-asparagine are non-essential amino acids that can be produced in microorganisms with various enzymes. L-Alanine is produced by alanine transaminase (EC 2.6.1.2), alanine:oxo-acid aminotransferase (EC 2.6.1.12), β-alanine-pyruvate aminotransferase (EC 2.6.1.18), alanine-glyoxylate transaminase (EC 2.6.1.44), aspartate 1-decarboxylase (EC 4.1.1.11), aspartate 4-decarboxylase (EC 4.1.1.12), alanine racemase (EC 5.1.1.1), and alanine-tRNA ligase (EC 6.1.1.7). L-Aspartate is produced by aspartate transaminase (EC 2.6.1.1), asparaginase (EC 3.5.1.1), aspartate ammonia-lyase (EC 4.3.1.1), aspartate racemase (EC 5.1.1.13), adenylosuccinate synthase (EC 6.3.4.4), argininosuccinate synthetase (EC 6.3.4.5), and asparagine synthase (EC 6.3.5.4). L-Alanine, L-aspartate, and L-asparagine are converted to each other: L-aspartate is converted to L-alanine and L-asparagine directly by L-aspartate 4-decarboxylase and

L-aspartate:ammonia ligase (EC 6.3.1.1), respectively, whereas L-asparagine is converted to L-aspartate by L-asparagine amidohydrolase (EC 3.5.1.1). Unusual amino acids such as D-alanine and D-aspartate are produced by alanine racemase (EC 5.1.1.1) and aspartate racemase (EC 5.1.1.13), respectively. The α-amino group of L-alanine and L-aspartate is transferred to L-glutamate by L-alanine:2-oxo-acid aminotransferase and L-aspartate: 2-oxoglutarate aminotransferase and is subsequently released as ammonia by glutamate dehydrogenase (EC 1.4.1.2). When these enzymes catalyze the reversed reactions, L-alanine and L-aspartate are synthesized via L-glutamate from ammonia and pyruvate and from ammonia and oxaloacetate, respectively. L-Alanine, L-aspartate, L-asparagine, D-alanine, and D-aspartate are useful as ingredients or starting materials for the industrial production of foods, cosmetics, medicines, and other products. For example, L-alanine is widely used as a natural moisture balancer in various cosmetics to keep the skin moist. L-Aspartate is one of the important raw materials for production of the artificial sweetener, aspartame. The *N*-acyl derivatives of D-alanine and D-aspartate show antibotulinal activity in the presence of sodium nitrite, and are expected to be antibacterial agents (Paquet and Rayman 1987).

1 Alanine

Alanine ($C_3H_7NO_2$; MW 89.09; 2-aminopropionic acid) was first isolated from the hydrolyzate of fibroin by Schutzenberger and Bourgeois in 1850. L-Alanine is a non-polar and hydrophobic amino acid, and is found in almost all bacteria in both free and bound form. D-Alanine is one of the essential components in bacterial cell walls.

1.1 Enzymes Involved in the Biosynthetic Pathway of Alanine and Their Regulation

In this chapter, the enzymes involved in the biosynthetic pathway of alanine in microorganisms (see Fig. 1) will be described in terms of reaction, enzymatic characteristics, function, and their regulation in order to understand the alanine flux totally.

1.1.1 Alanine Transaminase

Alanine transaminase (L-alanine:2-oxoglutarate aminotransferase, EC 2.6.1.2) catalyzes the transfer of the α-amino group of L-alanine to 2-oxoglutarate and forms L-glutamate and pyruvate in the presence of pyridoxal 5′-phosphate (PLP) as a cofactor. L-Alanine is mainly synthesized from pyruvate and L-glutamate by the reaction of L-alanine:2-oxoglutarate aminotransferase. Most of the enzymes have been purified from plants and mammals; there have been only a few reports of their occurrence in microorganisms. The

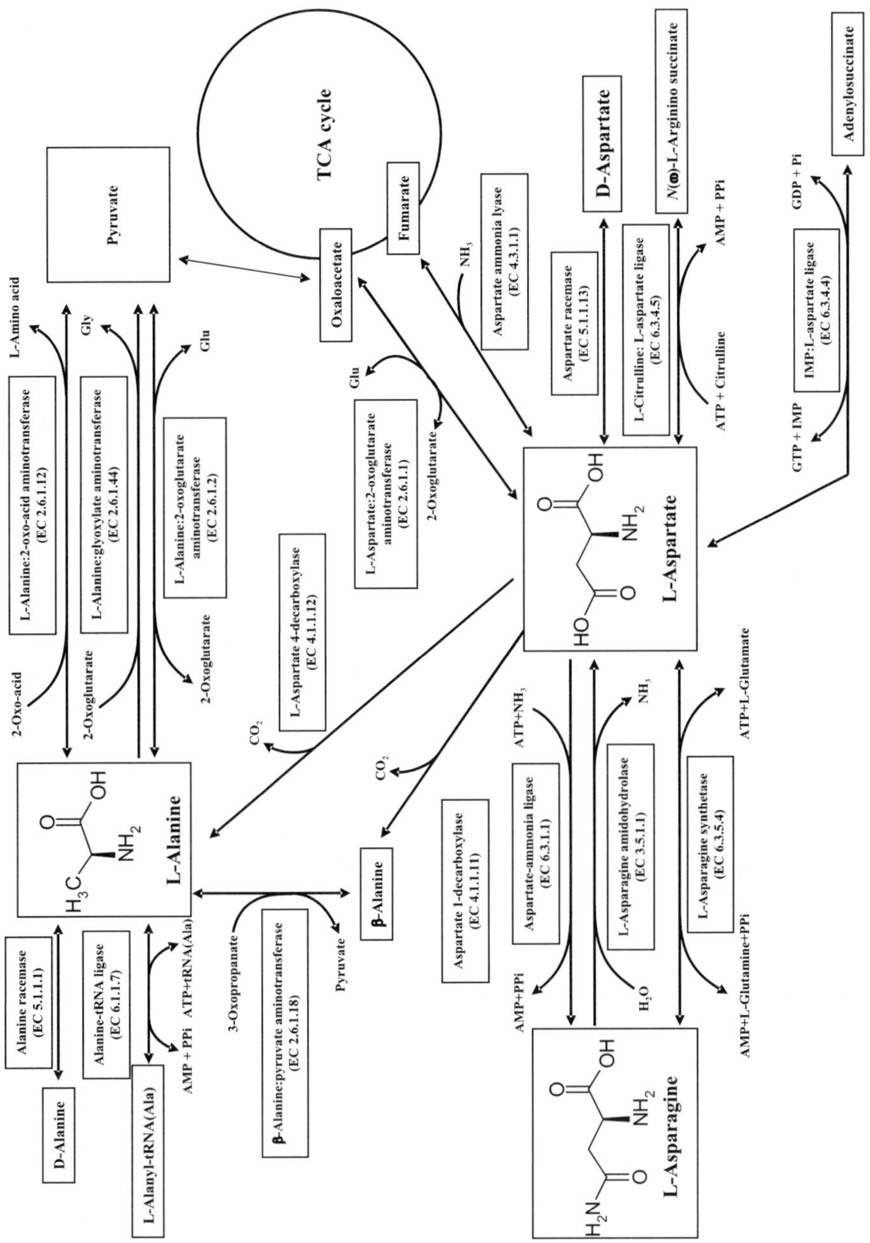

Fig. 1 Alanine, aspartate, and asparagine metabolism in microorganisms

Candida maltosa enzyme, a homodimer with a molecular weight of 99 000 (Umemura et al. 1994), contains 2 mol of PLP per mol of enzyme and is inhibited by aminooxyacetate, hydroxylamine, and *p*-hydroxymercuribenzoate. The enzyme uses L-alanine and L-glutamate as amino donors and 2-oxoglutarate and 2-oxobutanoate as amino acceptors. It is active in the temperature range 25–55 °C, with maximum activity at 55 °C. The enzyme is stable between pH 5.5 and 9.5, with an optimum pH of 6. The *Pyrococcus furiosus* enzyme, a homodimer with a molecular weight of 93 400 (Ward et al. 2000), shows broader substrate specificity than other enzymes from eukaryal sources and exhibits significant activity with L-alanine, L-glutamate, and L-aspartate with either 2-oxoglutarate or pyruvate as the amino acceptor. Optimal activity occurs within a pH range of 6.5 to 7.8 and at temperatures over 95 °C. The genes encoding alanine transaminase and glutamate dehydrogenase of *P. furiosus* are co-regulated at the transcriptional level, and the expression of both genes is induced when the cells are grown on pyruvate.

1.1.2 Alanine:Oxo-acid Aminotransferase

Alanine:oxo-acid aminotransferase (L-alanine:2-oxo-acid aminotransferase, EC 2.6.1.12) catalyzes reactions similar to those of the enzymes classified as EC 2.6.1.2, but it uses 2-oxopentanoate, 2-oxobutanoate, 2-oxohexanoate, 2-oxo-3-phenylpropanoate, 3-methyl-2-oxobutanoate, and 3-methyl-2-oxopentanoate in addition to 2-oxoglutarate as the amino acceptor for L-alanine. The enzyme has not been purified from any organism and has been found only in *Pseudomonas* sp. (Koide et al. 1977) and *Brucella abortus* (Altenbern and Housewright 1953). The enzymological properties of the enzyme have not been reported in detail.

1.1.3 β-Alanine-Pyruvate Aminotransferase

β-Alanine-pyruvate aminotransferase (L-alanine:3-oxopropanoate aminotransferase, EC 2.6.1.18) catalyzes the transfer of L-alanine to malonic semialdehyde and forms β-alanine and pyruvate in the presence of PLP as a cofactor. Accordingly, the enzyme catalyzes the conversion of L-alanine to β-alanine. The enzyme has been found only in bacteria, and the enzymological studies have been carried out mainly with the enzymes from *Pseudomonas fluorescens* (Hayashi et al. 1961) and *Bacillus cereus* (Nakano et al. 1977). The *Pseudomonas* enzyme uses 2-oxobutanoate, 2-oxohexanoate, 2-oxoisohexanoate, acetoacetate, and glyoxylic acid in addition to pyruvate as the amino acceptor. In the *Bacillus* enzyme, the amino acceptors are 2-oxobutanoate, oxoisopentanoate, oxopentanoate, and oxaloacetic acid. Both enzymes are strongly inhibited by *p*-chloromercuribenzoate. The optimum temperature for both the *Pseudomonas* and *Bacillus* enzymes is 35 °C, although the *Bacillus* enzyme is stable up to 40 °C. The optimum pH values of the *Pseudomonas* and *Bacillus* enzymes are pH 9.2 and 10, respectively. The specific activity of the *Pseudomonas* enzyme is 56.6 μmol/min/mg. No structural data for the enzymes are available.

1.1.4 Alanine-Glyoxylate Transaminase

Alanine-glyoxylate transaminase (L-alanine:glyoxylate aminotransferase, EC 2.6.1.44) catalyzes the transfer of the α-amino group of L-alanine to glyoxylate and forms glycine and pyruvate in the presence of PLP as a cofactor. The enzyme occurs widely in eukaryotes, and there have been only a few reports of its occurrence in microorganisms. Growth experiments with different background mutations were carried out in *S. cerevisiae*, and the enzyme was found to be one of three enzymes for glycine synthesis (Schlösser et al. 2004). Recently, a novel archaeal L-alanine:glyoxylate aminotransferase was found in *Thermococcus litoralis* and was characterized enzymologically (Sakuraba et al. 2004). The molecular weight of the enzyme is 170 000, and it exists as a homotetramer with a subunit molecular weight of 42 000. The optimum pH is about pH 7.5. The enzyme activity increases with increasing temperatures from 37 to 90 °C and is stable at temperatures up to 80 °C. The enzyme is strongly inhibited by 10 mM L-penicillamine and 10 mM hydroxylamine, and it is highly specific for L-alanine when glyoxylate is used as the amino acceptor. While the amino acid sequence of the enzyme does not show a similarity to any L-alanine:glyoxylate aminotransferases previously reported, the homologs of the enzyme seem to be present in most hyperthermophilic archaea whose genome analyses have been completed. The physiological function of the enzyme is still unclear.

1.1.5 Alanine Racemase

Alanine racemase catalyzes the racemization of L- and D-alanine and is classified into the fold-type III group of the PLP-dependent enzymes (Alexander et al. 1994). The enzyme occurs widely in eukaryote and prokaryote and is important for bacteria since D-alanine is one of the essential components of the peptidoglycan for bacterial cell walls (Walsh 1989). Generally, alanine racemase is divided into two groups. Two alanine racemases have been identified in *Escherichia coli* and *Salmonella typhimurium*. The alanine racemase (biosynthetic alanine racemase) encoded by the *alr* gene is constitutive, whereas the *dadX* gene (catabolic alanine racemase) is inducible and is essential only for L-alanine catabolism to provide a substrate for D-alanine dehydrogenase encoded by the *dadA* gene. Previous studies on the alanine racemase have focused mainly on *Geobacillus stearothermophilus* (Inagaki et al. 1986) and *E. coli* (Wang and Walsch 1981), and very few studies have focused on the coryneform bacteria (Strych et al. 2001). Recently, the *alr* gene from *Corynebacterium glutamicum* was cloned into *E. coli*, and the gene product, alanine racemase, was characterized enzymologically (Oikawa et al. 2006). The enzyme is a dimer with a molecular weight of 78 000. It requires PLP as a coenzyme and contains 2 mol of PLP per mol of the enzyme. The holoenzyme shows maximum absorption at 420 nm, while the reduced form of the enzyme shows it at 310 nm. The enzyme is specific for alanine, and the optimum pH is observed at about 9. The enzyme is relatively thermostable, and its half-life at 60 °C is estimated to be 26 min. The K_m and V_{max} values have been determined as follows: L-alanine to D-alanine, K_m (L-alanine) 5.01 mM and V_{max} 306 U/mg; D-alanine to L-alanine, K_m (D-alanine) 5.24 mM and V_{max} 345 U/mg. The K_{eq} value has been calculated to be 1.07, which is in good agreement with the theoretical value for the racemization reaction.

1.1.6 Alanine-tRNA Ligase

Alanine-tRNA ligase (L-alanine:$tRNA_{Ala}$, EC 6.1.1.7) catalyzes the synthesis of L-alanyl-$tRNA_{Ala}$ from L-alanine and $tRNA_{Ala}$ in the presence of ATP. Accordingly, the enzyme is distributed widely in living organisms and is essential for the biosynthesis of L-alanine-containing polypeptides and proteins. In microorganisms, the enzyme from *E. coli* has been extensively studied enzymologically (Putney et al. 1981). The enzyme requires ATP and Mg^{2+} as cofactors, and Zn^{2+} is tightly bound to the enzyme. The elongation factor Tu strongly inhibits the aminoacylation of $tRNA_{Ala}$ since it interferes with the formation of the canonical ternary complex among the elongation factor Tu, the nucleotide cofactor, and aminoacyl-tRNA. The K_m values for L-alanine, ATP, and $tRNA_{Ala}$ are 0.34, 0.083, and 0.0024 mM, respectively. The optimum pH is 7.5, and maximum activity is observed at 37 °C.

2 Aspartate and Asparagine

L-Aspartate ($C_4H_7NO_4$ MW 133.10; 2-aminobutanedioic acid) was first identified as the acid hydrolysis product of L-asparagine ($C_4H_8N_2O_3$ MW 132.12; 2-aminosuccinamic acid) by Plisson in 1827 (Plisson 1827). Aspartate is classified as the amino acid with acidic and non-polar side chain, while asparagine is that with neutral and polar side chain. L-Aspartate is the precursor of L-lysine, L-methionine, L-threonine, and purine and pyrimidine bases in microorganisms.

2.1
Enzymes Involved in Biosynthetic Pathway of Aspartate and Asparagine and Their Regulation

In this chapter, the enzymes involved in the biosynthetic pathway of aspartate and asparagine in microorganisms will be described in terms of reaction, enzymatic characteristics, function, and their regulation in order to understand the aspartate and asparagine flux totally.

2.1.1
Aspartate Transaminase

Aspartate transaminase (L-aspartate:2-oxoglutarate aminotransferase, EC 2.6.1.1) catalyzes the transfer of the α-amino group of L-aspartate to 2-oxoglutarate and forms L-glutamate and oxaloacetate in the presence of PLP as cofactor. L-Aspartate is mainly synthesized from oxalacetate and L-glutamate by the reaction of L-aspartate:2-oxoglutarate aminotransferase. In contrast to L-alanine:2-oxoglutarate aminotransferase (EC 2.6.1.2), the enzyme exists widely in eukaryotes and prokaryotes, and the enzymes from *E. coli* (Yagi et al. 1979) and *Thermus thermophilus* (Makai et al. 1999) have been extensively studied enzymologically. The enzymes are divided into two subgroups. The enzyme from *E. coli* belongs to subgroup Ia, whereas the *T. thermophilus* enzyme is in subgroup Ib. X-ray crystallographic analysis has revealed that the overall and active site structures are well conserved, but two differences apparently exist between the subgroups. In subgroup Ia, substrate binding induces a large movement of the small domain close to the active site, whereas in subgroup Ib, only the N-terminal region of the small domain interacts with the substrate. The amino acid residue that interacts with the side-chain carboxylate of the substrate is Arg292 in subgroup Ia, but in subgroup Ib, Lys 109 instead of Arg292 forms a salt bridge with the side-chain carboxylate.

2.1.2
Asparaginase

Asparaginase (L-asparagine amidohydrolase, EC 3.5.1.1) catalyzes the hydrolysis of the amido bond of L-asparagine and irreversibly produces L-aspartate and ammonia. The enzyme is widely distributed in microorganisms, animals, and plants. The bacterial enzymes from *Acinetobacter calcoaceticus* (Joner et al. 1973), *Bacillus coagulans* (Law and Wriston 1971), *E. coli* (Peterson 1977), and *Vibrio succinogenes* (Distasio et al. 1976) also show enzymatic activity on D-asparagine. The enzyme from *E. coli* has been used for the industrial production of L-asparagine. *Saccharomyces cerevisiae* produces the enzyme both intracellularly and extracellularly (Dunlop et al. 1978). The synthesis of the enzyme is stimulated by nitrogen starvation, requires an available energy source, and is prevented by cycloheximide. The intracellular enzyme appears to be constitutive. The extracellular activity is relatively insensitive to *p*-hydroxymercuribenzoate inhibition, whereas the intracellular activity is highly inhibited by this compound.

2.1.3 Aspartate 1-Decarboxylase

Aspartate 1-decarboxylase (L-aspartate 1-carboxy-lyase, EC 4.1.1.11) catalyzes the α-decarboxylation of L-aspartate to form β-alanine and carbon dioxide. The enzyme is found mainly in microorganisms, but enzymological studies of the enzyme have not been well done. The enzyme from *Escherichia coli* is relatively well-characterized (Williamson 1979). The enzyme is translated as an inactive pro-protein, i.e., a pi-protein, which undergoes intramolecular self-cleavage at Gly24/Ser25, producing α- and β-subunits. The molecular self-processing mechanism proceeds slowly. The active enzyme is a tetramer composed of three α- and β-subunits and an incompletely processed pi-protein. The molecular weight of the enzyme has been estimated to be about 58 000 by gel filtration.

2.1.4 Aspartate 4-Decarboxylase

Aspartate 4-decarboxylase (L-aspartate 4-carboxy-lyase, EC 4.1.1.12) catalyzes the β-decarboxylation of L-aspartate to form L-alanine and carbon dioxide and requires PLP as a coenzyme. The enzyme was found in *Alcaligenes faecalis* (Novogrodsky and Meister 1964), *Achromobacter* sp. (Wilson and Komberg 1963), *Clostridium perfringens*, *Desulfovibrio desulfuricans*, *Nocadia globerula*, *Pseudomonas dacunhae*, and even in *Bombyx mori* and Lobster (Tate and Meister 1971). The enzyme from *Achromobacter* sp. has been crystallized and well characterized. The PLP bound to the enzyme is probably in a reduced form. The enzyme is inhibited by β-chloro-L-alanine, DL-β-methylaspartate, DL-erythro-β-hydroxyaspartate, DL-threo-β-hydroxyaspartate, L-alanine, and maleate, and most strongly by phosphate buffers above pH 5.0. The enzyme is activated by 2-oxoglutarate, acetaldehyde, glyoxal, glyoxylate, and pyruvate. The K_m value for L-aspartate is 0.08 mM. L-Alanine is produced industrially with *Pseudomonas dacunhae* by the addition of L-aspartate to the medium (Takamatsu et al. 1981).

2.1.5 Aspartate Ammonia-lyase

Aspartate ammonia-lyase (aspartase, aspartate ammonia-lyase, EC 4.3.1.1) catalyzes the reversibe deamination of L-aspartate to fumarate and ammonia and serves as a turning point between amino acid metabolism and organic acid metabolism. The enzyme has been found mainly in microorganisms and plants but scarcely in animals. The enzyme was purified from a psychrotolerant, *Cytophaga* sp. KUC-1 (Kazuoka et al. 2003), recently re-identified as *Flavobacterium frigidimaris* KUC-1, mesophiles, *E. coli* (Suzuki et al. 1973), *Pseudomonas fluorescens* (Takagi et al. 1984), *Bacillus subtilis* (Sun and Setlow 1991), and a thermophile, *Bacillus* sp. YM55-1 (Kawata et al. 1999). The *E. coli* enzyme, which has been studied extensively, has a molecular weight of about 200 000, with four identical subunits. The enzyme requires Mg^{2+}, and its optimum pH is 8.2 in the presence of Mg^{2+}. The substrate saturation curve shows positive cooperation over pH 7.4 and negative cooperation under pH 7.4. The K_m value at pH 7.4 for L-aspartate is 1.0 mM. The substrate specificity is very strict, and only L-aspartate serves as a substrate. However, in the reverse reaction, hydroxylamine has been cited as a substrate to produce hydroxyaspartate (Suzuki et al. 1973). The *Cytophaga* enzyme is more thermostable than the *E. coli* enzyme even though it is from a psychrotolerant. The *E. coli* enzyme loses about 83% of the initial activity after incubation at 50 °C for 45 min, but the *Cytophaga* enzyme retains about 80% of its activity after incubation at 50 °C for 60 min.

2.1.6 Aspartate Racemase

Aspartate racemase (aspartate racemase, EC 5.1.1.13) catalyzes racemization of L- and D-aspartate. The enzyme is not very widely distributed in nature, but its extensive study has been possible because it can be obtained from lactic acid bacteria such as *Streptococcus faecalis* (Lamont et al. 1972) and *Streptococcus thermophilus* (Yamaguchi et al. 1992), from a hyperthermophile such as *Pyrococcus horikoshii* OT3 (Liu et al. 2002), and from the blood shell *Scapharca broughtonii* (Shibata et al. 2003). The *S. faecalis* enzyme requires PLP as coenzyme, but the *S. thermophilus* and *P. horikoshii* OT3 enzymes contain no bound PLP. Only the crystal structure of the *P. horikoshii* OT3 enzyme has been reported, and two conserved cysteine residues (Cys82 and Cys192), which have been shown to act as catalytic acid and base, are located on both sides of a cleft between two domains. Accordingly, the enzyme reaction proceeds through a two-base catalytic mechanism and is a typical PLP-independent amino acid racemase.

2.1.7 Adenylosuccinate Synthetase

Adenylosuccinate synthetase (IMP: L-aspartate ligase (GDP-forming), EC 6.3.4.4) reversibly catalyzes the conversion of adenylosuccinate, GDP, and Pi to L-aspartate, IMP, and GTP. The equilibrium of the enzyme reaction tends toward adenylosuccinate synthesis. The enzyme catalyzes the first committed step in the de novo biosynthesis of AMP. Adenylosuccinate synthetase has been obtained from various sources ranging from Archeae and bacteria to mammals. The enzyme from *E. coli* is the best characterized enzyme and is known as a model for a *ter*-reactant enzyme (Poland and Fromm 1996). In their pioneering work, Rudolph and Fromm (1969) showed that the enzyme binds its substrates sequentially and completely randomly. The *E. coli* enzyme requires divalent metal ions such as Ba^{2+}, Ca^{2+}, Mg^{2+}, and Mn^{2+}. The best activator is Mg^{2+}, which exists in the active site. Adenylosuccinate synthetase is a target of herbicides and drugs. The herbicidal action of hydantocidin is shown through the inhibition of the enzyme by the 5′-phosphorylated compounds (Rudolph and Fromm 1969).

2.1.8
Argininosuccinate Synthetase

Argininosuccinate synthetase (L-citrulline: L-aspartate ligase (AMP-forming), EC 6.3.4.5) catalyzes reversibly the conversion of argininosuccinate, AMP, and PPi to L-aspartate, L-citrulline, and ATP. The enzyme has been found in microorganisms and animals and catalyzes the seventh step of the arginine biosynthesis process (the second step of the urea cycle). Recently, the structures of the enzyme from *Thermus thermophilus* HB8 (Goto et al. 2003) complexed with intact ATP, aspartate, and citrulline (substrates) as well as with AMP and arginosuccinate (products) have been determined, as has the mechanism of the enzyme reaction. The protonated amino group of the bound aspartate interacts with the α-phosphate of ATP and the ureido group of citrulline.

2.1.9
Asparagine Synthase

Asparagine synthase (L-aspartate:L-glutamine amido-ligase (AMP-forming), EC 6.3.5.4) catalyzes reversibly the conversion of L-asparagine, L-glutamate, AMP, and PPi to L-aspartate, L-glutamine, and ATP. The enzyme is distributed in microorganisms, animals, and plants and has been studied mainly in plants such as *Helianthus annuus* (sunflower), *Hordeum vulgare* L. (barley), and *Oryza sativa* (rice). In microorganisms, the enzyme from *E. coli* has been extensively studied. The enzyme also produces β-aspartylhydroxamate from L-aspartate, hydroxylamine, and ATP. The enzyme is highly activated by Mg^{2+}. Recently, the three-dimensional structure of the *E. coli* enzyme was solved, and the important amino acid residues for substrate binding were determined (Larsen et al. 1999). The N-terminal region contains two layers of antiparallel β-sheets, with each layer containing six strands. Wedged between these layers is the active site responsible for the hydrolysis of glutamine. The key residues for binding the glutamine substrates in the active site are Arg 49, Asn 74, Glu 76, and Asp 98. The C-terminal domain is responsible for the binding of Mg^{2+}, ATP, and aspartate, and Ser 346, Val 272, Leu 232, and Gly 347 have been used to anchor the AMP moiety.

2.1.10
Asparagine Synthetase

Asparagine synthetase (L-aspartate:ammonia ligase (AMP-forming), EC 6.3.1.1) catalyzes the reversible conversion of L-aspartate, NH_4^+, and ATP to L-asparagine, AMP, and PPi. The enzyme is distributed widely in nature, but its enzymological properties have not been studied in detail. Pioneering studies have been made on the enzymes from lactic acid bacteria. The enzyme from *Lactobacillus arabinosus* can be stored at 4 °C for 3 weeks but not at −20 °C (Meister 1974). The optimum pH is 8.2, and the optimum temperature is about 40 °C. The enzyme is specific for L-aspartate and does not act on L-glutamate. β-L-Aspartyl hydroxamate is synthesized when hydroxamate is added to the reaction mixture instead of NH_4^+. The enzyme requires Mg^{2+} and is activated by Mn^{2+}. No activation of Mg^{2+} was observed for the *E. coli* (Sugiyama et al. 1992) and *Streptococcus bovis* enzymes.

3
Biotechnological Applications

Aspartate 4-decarboxylase from *Pseudomonas dacunhae* (Takamatsu et al. 1981) is useful for the industrial production of L-alanine from L-aspartate (Fig. 2). The enzyme is constitutive, but its production is stimulated twofold by the addition of L-glutamate and is repressed by one-tenth by L-serine, L-glutamine, L-proline, L-leucine, glycine, and L-threonine. The intermediates from L-glutamate metabolism show no effect, and this induction is observed only for L-glutamate. Several enzymes linked to L-glutamate metabolism, such as γ-glutamyl transpeptidase, L-glutamate dehydrogenase, glutamate-oxalacetate transaminase, and glutamate decarboxylase, have no relation to this induction. The industrial production of L-alanine was

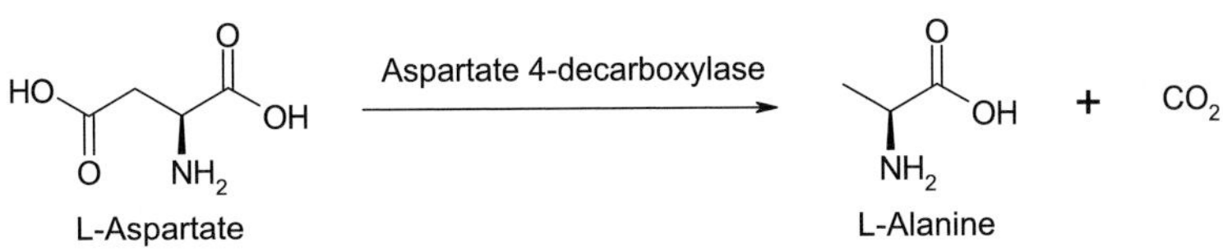

Fig. 2 Enzymatic reaction catalyzed by aspartate 4-decarboxylase from *Pseudomonas dacunhae*

achieved previously from a 40-g feed of L-aspartate to 100 mL of culture broth (Chibata et al. 1965). With the addition of L-glutamate, L-alanine can be produced from a > 100-g feed of L-aspartic acid to 100 mL of culture broth. Accordingly, *P. dacunhae* grown in L-glutamate medium with high aspartate 4-decarboxylase activity is advantageous for the industrial production of L-alanine from L-aspartate. In Japan, L-alanine has been produced industrially since 1965 from L-aspartate by the batch method with aspartate 4-decarboxylase from *P. dacunhae*. Furthermore, L-alanine has been produced continuously using immobilized *P. dacunhae* since 1980, after the development of immobilization with κ-carrageenan and the stabilization of enzyme activity, reaction under pressure, and conditions for cell preparation. Subsequently, L-alanine has been produced from fumarate and ammonia by using immobilized *P. dacunhae* and *E. coli* with aspartate ammonia-lyase (aspartase) since 1982.

Alanine racemase from *Corynebacterium glutamicum* ATCC 13032 is expected to be a biocatalyst for D-alanine production from the L-counterpart since it is highly specific for alanine (Fig. 3). *Corynebacterium fascians* ATCC 21950 secretes D-alanine stereospecifically through the cell membrane and produces D-alanine from pyruvate in the extracellular medium (Yamada et al. 1976). Both *C. glutamicum* ATCC 13032 and *C. fascians* ATCC 21950 belong to *Corynebacterianeae*, and the transcription regulation system of two closely related organisms is expected to be similar. Accordingly, the expression of the *alr* gene from *C. glutamicum* ATCC 13032 into *C. fascians* ATCC 21950 is expected to be easier than that from other organisms. By the transformation of the *alr* gene from *C. glutamicum* ATCC 13032 into *C. fascians* ATCC 21950, the clone obtained probably produces D-alanine outside of the cells with high optical purity, and the produced D-alanine might be easily separated from the medium by conventional ion exchange column chromatography. Another possibility for using the alanine racemase from *C. glutamicum* ATCC 13032 for D-alanine production is by applying it to a bioreactor consisting of two columns of sequentially connected immobilized alanine racemase and L-alanine dehydrogenase (EC 1.4.4.4, from *Bacillus subtilis*) (Oikawa et al. 1999). L-Alanine is racemized by the alanine racemase column, and the remaining L-alanine is dehydrogenated by the L-alanine dehydrogenase column. Consequently, only D-alanine remains, and

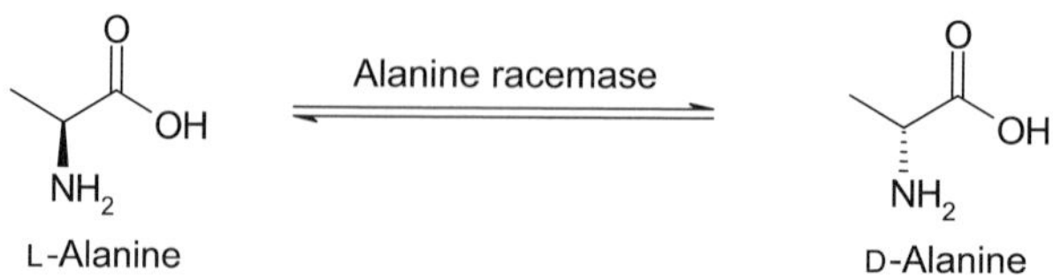

Fig. 3 Proposed D-alanine production catalyzed by alanine racemase from *Corynebacterium glutamicum*

can be easily separated from pyruvate formed by cation-exchange column chromatography since pyruvate has no positive charge.

Several recent microbial studies have also sought to generate D- and L-alanine through fermentation. For example, strain DAN 75 of the genus *Arthrobacter*, selected for an inability to grow on D-alanine, accumulated L-alanine to 75.6 g/L (with 1.2 g/L D-alanine) in 120 h with a mass yield of over 50% from glucose (Hashimoto and Katsumata 1998). Anaerobically, *Zymomonas mobilis* transformed with the *alaD* gene from *Bacillus sphaericus* produced 7.5 g/L L-alanine when the essential cofactor thiamine PPi was limited (Uhlenbusch et al. 1991). Limiting this cofactor reduced carbon flux through pyruvate decarboxylase, a pathway that competes with alanine generation in this organism. A study using *E. coli* transformed with the *Arthrobacter* sp. HAP1 L-alanine dehydrogenase gene resulted in the accumulation of 2.9 g/L DL-alanine under aerobic conditions and 8.1 g/L under oxygen-limited conditions in shake flasks (Katsumata and Hashimoto 1996).

The enzymes from hyperthermophilic archaea are highly thermostable and are generally useful for the industrial production of various compounds and for gene engineering. DNA polymerase from *Thermococcus aquaticus* is an example of the enzymes from hyperthermophilic archaea enabling an automatic polymerase chain reaction (PCR) (Tindall and Kunkel 1988). Alanine transaminase from hyperthermophilic archaeon *Pyrococcus furiosus*, which is also thermostable, shows broad substrate specificity and acts on L-alanine, L-aspartate, L-valine, L-isoleucine, and L-leucine when 2-oxoglutarate is used as an amino acceptor (Ward et al. 2000). Accordingly, L-alanine, L-aspartate, L-valine, L-isoleucine, and L-leucine can be produced from L-glutamate and the corresponding keto acids by the reversed reaction of the enzyme without denaturation for a long time (Fig. 4).

The pioneering study of L-aspartate production was done in Japan by Chibata et al. (1960). Based on the findings of this study, L-aspartate is produced industrially with aspartase of *E. coli* from fumarate and ammonia (Fig. 5). Since 1973, L-aspartate has been produced continuously with polyacryl amide gel-immobilized cells. Since 1982, with the development of the fumarase-lacking *E. coli* mutant with high aspartase activity, L-aspartate has been

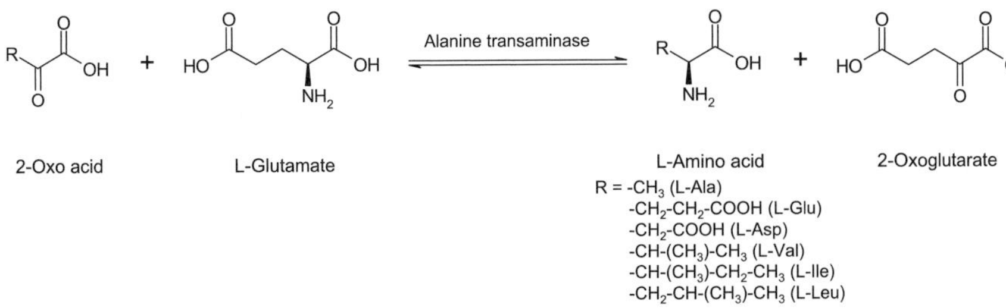

Fig. 4 Enzymatic reaction catalyzed by alanine transaminase from *Pyrococcus furiosus*

Fig. 5 Enzymatic reaction catalyzed by aspartase from *E. coli*

produced industrially with κ-carrageenan-immobilized cells of the *E. coli* mutant. Although old, this is still the major method of L-aspartate production in industry. The thermostable aspartases from *Cytophaga* sp. KUC-1 and *Bacillus* sp. YM55-1 were reported by Kazuoka et al. (2003) and Kawata et al. (1999), respectively. These enzymes would be useful for L-aspartate production and could be applied to L-aspartate production in the same way as in the *E. coli* aspartate production system.

References

Alexander FW, Sandmeier E, Mehata PK, Christen P (1994) Evolutionary relationships among pyridoxal-5′-phosphate-dependent enzymes. Regio-specific alpha, beta and gamma families. Eur J Biochem 219:953–960

Altenbern RA, Housewright RD (1953) Transaminases in smooth *Brucella abortus*, strain 19. J Biol Chem 204:159–167

Chibata IT, Kakimoto J, Kato J (1965) Enzymatic production of L-alanine by *Pseudomonas dacunhae*. Appl Microbiol 13:638–645

Distasio JA, Niederman RA, Kafkewitz D, Goodman D (1976) Purification and characterization of L-asparaginase with antilymphoma activity from *Vibrio succinogenes*. J Biol Chem 251:6929–6933

Dunlop PC, Meyer GM, Ban D, Roon RJ (1978) Characterization of two forms of asparaginase in *Saccharomyces cerevisiae*. J Biol Chem 253:1297–1304

Goto M, Omi R, Miyahara I, Sugahara M, Hirotsu K (2003) Structures of arginosuccinate synthetase in enzyme-ATP substrates and enzyme-AMP product forms. J Biol Chem 278:22964–22971

Hashimoto S, Katsumata R (1998) L-Alanine fermentation by an alanine racemase-deficient mutant of the DL-alanine hyperproducing bacterium *Arthrobacter oxydans* HAP-1. J Ferment Bioeng 86:385–390

Hayashi O, Nishizuka Y, Tatibana M, Takeshita M, Kuno S (1961) Enzymatic studies on the metabolism of β-alanine. J Biol Chem 236:781–790

Inagaki K, Tanizawa K, Badet B, Walsh CT, Tanaka H, Soda K (1986) Thermostable alanine racemase from *Bacillus stearothermophilus*: molecular cloning of the gene, enzyme purification, and characterization. Biochemistry 25:3268–3274

Joner PE, Kristiansen T, Einasson M (1973) Purification and properties of L-asparaginase A from *Acinetobacter calcoaceticus*. Biochim Biophys Acta 327:146–456

Katsumata R, Hashimoto S (1996) Process for producing alanine. US Patent 5 559 016

Kazuoka T, Masuda Y, Oikawa T, Soda K (2003) Thermostable aspartase from a marine psychrophile, *Cytophaga* sp. KUC-1: molecular characterization and primary structure. J Biochem 133:51–58

Kawata Y, Tamura K, Yano S, Mizobata T, Nagai J, Esaki N, Soda K, Tokushige M (1999) Purification and characterization of thermostable aspartase from *Bacillus* sp. YM55-1. Arch Biochem Biophys 366:40–46

Larsen TM, Boehlein SK, Schuster SM, Richards NG, Thoden JB, Holden HM, Rayment I (1999) Three-dimensional structure of *Escherichia coli* asparagines synthetase B: a short journey from substrate to product. Biochemistry 38:16146–16157

Yumoto N (1999) Purification and characterization of thermostable aspartase from *Bacillus* sp. YM55-1. Arch Biochem Biophys 366:40–46

Koide Y, Honma M, Shimomura T (1977) L-alanine-α-keto acid aminotransferase of *Pseudomonas* sp. Agric Biol Chem 41:781–784

Lamont HC, Staudenbauer WL, Strominger JL (1972) Partial purification and characterization of an aspartate racemase from *Streptococcus faecalis*. J Biol Chem 267:18361–18364

Law AS, Wriston JC (1971) Purification and properties of *Bacillus coagulans* L-asparaginase. Arch Biochem Biophys 147:744–752

Liu L, Iwata K, Kita A, Kawarabayasi Y, Yohda M, Miki K (2002) Crystal structure of aspartate racemase from *Pyrococcus horikoshii* OT3 and its implications for molecular mechanism of PLP-independent racemization. J Mol Biol 319:479–489

Meister A (1974) Asparagine synthesis. In: Boyer PD (ed) The Enzymes, 3rd edn, vol 10. Academic, New York, pp 561–580

Nakai T, Okada K, Akutsu S, Miyahara I, Kawaguchi S, Kato R, Kuramitsu S, Hirotsu K (1999) Structure of *Thermus thermophilus* HB8 aspartase aminotransferase and its complex with maleate. Biochemistry 38:2413–2424

Nakano Y, Tokunaga H, Kitaoka S (1977) Two ω-amino acid transaminases from *Bacillus cereus*. J Biochem 81:1375–1381

Novogrodsky A, Meister A (1964) Control of aspartate β-decarboxylase activity by transamination. J Biol Chem 239:879–888

Oikawa T, Tauch A, Schaffer S, Fujioka T (2006) Expression of *alr* gene from *Corynebacterium glutamicum* ATCC 13032 in *Escherichia coli* and molecular characterization of the recombinant alanine racemase. J Biotech 125:503–512

Oikawa T, Watanabe M, Makiura H, Kusakabe H, Yamade K, Soda K (1999) Production of D-glutamate from L-glutamate with glutamate racemase and L-glutamate oxidase. Biosci Biotech Biochem 63:2168–2173

Paquet A, Rayman K (1987) Aome *N*-acyl-D-amino acid derivatives having antibotulinal properties. Can J Microbiol 33:577–582

Peterson RG, Richards FF, Handschumacher RE (1977) Structure of peptide from active site region of *Escherichia coli* L-asparaginase. J Biol Chem 252:2072–2076

Plisson A (1827) Non-essential amino acid for human development. First identified as the acid hydrolysis product of asparagines. J Pharm 13:477–492

Poland BW, Fromm HJ (1996) Crystal structure of adenylosuccinate synthetase. J Mol Biol 264:1013–1027

Rudolph FB, Fromm HJ (1969) Initial rate studies of adenylosuccinate synthetase with product and competitive inhibitors. J Biol Chem 244:3832–3839

Putney SD, Sauer RT, Schimmel PR (1981) Purification and properties of alanine tRNA synthetase from *Escherichia coli* A tetramer of identical subunits. J Biol Chem 256:198–204

Sakuraba H, Kawakami R, Takahashi H, Ohshima T (2004) Novel archaeal alanine:glyoxylate aminotransferase from *Thermococcus litoralis*. J Bacteriol 186:5513–5518

Schlösser T, Gätgens C, Weber U, Stahmann KP (2004) Alanine: glyoxylate aminotransferase of *Saccharomyces cerevisiae*-encoding gene *AGXI* and metabolic significance. Yeast 21:63–73

Shibata K, Watanabe T, Yoshikawa H, Abe K, Takahashi S, Kera Y, Yamada RH (2003) Purification and characterization of aspartate racemase from the bivalve mollusk *Scapharca broughtonii*. Comp Biochem Physiol B 134:307–314

Strych U, Penland RL, JImsnez M, Krause KL, Benedik MJ (2001) Characterization of the alanine racemases from two mycobacteria. FEMS Microbiol Lett 196:93–98

Sugiyama A, Kato H, Nishioka T, Oda J (1992) Overexpression and purification of asparagines synthetase from *Escherichia coli*. Biosci Biotechnol Biochem 56:376–379

Sun DX, Setlow P (1991) Cloning, nucleotide sequence, and expression of the *Bacillus subtilis* ans operon, which codes for L-asparaginase and L-aspartase. J Bacteriol 173:3831–3845

Suzuki S, Yamaguchi J, Tokushige M (1973) Purification and molecular properties of aspartase from *Escherichia coli*. Biochim Biophys Acta 321:369–381

Takagi JS, Fukunaga R, Tokushige M, Katsuki H (1984) Purification, crystallization, and molecular properties of aspartase from *Pseudomonas fluorescens*. J Biochem 96:545–552

Takamatsu S, Yamamoto K, Tosa T, Chibata I (1981) Stabillization of L-aspartate β-decarboxylase activity of *Pseudomonas dacunhae* immobilized with carrageenan. J Ferment Technol 59:489–493

Tate S, Meister A (1971) L-aspartate β-decarboxylase. Structure, catalytic activities, and allosteric regulation. Adv Enzymol Relat Areas Mol Biol 35:503–543

Tindall KR, Kunkel TA (1988) Fidelity of DNA synthesis by the *Thermus aquaticus* DNA polymerase. Biochemistry 27:6008–6013

Uhlenbusch I, Sahm H, Sprenger GA (1991) Expression of an L-alanine dehydrogenase gene in *Zymomonas mobilis* and excretion of L-alanine. Appl Environ Microbiol 57:1360–1366

Umemura I, Yanagiya K, Komatsubara S, Sato T, Tosa T (1994) Purification and some properties of alanine aminotransferase from *Candida maltosa*. Biosci Biotechnol Biochem 58:283–287

Walsh CT (1989) Enzymes in the D-alanine branch of bacterial cell wall peptidoglycan assembly. J Biol Chem 264:2393–2396

Wang EA, Walsch C (1981) Characteristics of β-difluoroalanine and β,β,β-trifluoroalanine as suicide substrates for *Escherichia coli* B alanine racemase. Biochemistry 20:7539–7546

Ward DE, Kengen SW, van Der Oost J, de Vos WM (2000) Purification and characterization of the alanine aminotransferase from the hyper thermophilic Archaeon *Pyrococcus furiosus* and its role in alanine production. J Bacteriol 182:2559–2566

Williamson JM and Brown GM (1979) Purification and properties of L-aspartate-α-decarboxylase, an enzyme that catalyzes the formation of β-alanine in *Escherichia coli*. J Biol Chem 254:8074–8082

Wilson EM, Komberg HL (1963) Properties of crystalline L-aspartate 4-carboxy-lyase from *Achromobacter* sp. Biochem J 88:578–587

Yagi T, Kagamiyama H, Motosugi K, Nozaki M, Soda K (1979) Crystallization and properties of aspartate aminotransferase from *Escherichia coli* B. FEBS Lett 100:81–84

Yamada S, Wada M, Izuno N, Chibata I (1976) Mechanism of D-alanine production by *Corynebacterium fascians*. Appl Environ Microbiol 32:1–6

Yamauchi T, Choi SY, Okada H, Yohda M, Kumagai H, Esaki N, Soda K (1992) Properties of aspartate racemase, a pyridoxal 5′-phosphate-independent amino acid racemase. J Biol Chem 267:18361–1836

Microbiol Monogr (5)
V. F. Wendisch: Amino Acid Biosynthesis
DOI 10.1007/7171_2006_069/Published online: 22 February 2007

Amino Acid Transport Systems in Biotechnologically Relevant Bacteria

Kay Marin · Reinhard Krämer (✉)

Institut für Biochemie, Universität zu Köln, Zülpicher Str. 47, 50674 Cologne, Germany
r.kraemer@uni-koeln.de

Electronic Supplementary Material

The electronic reference of this chapter is http://dx.doi.org/10.1007/7171_7171_2006_069. The online version of this chapter includes electronic supplementary material. This material is available to authorised users and can be accessed by means of the ESM button beneath the abstract. To cite or link to this chapter you can use the above reference.

Abstract Besides metabolic pathways and regulatory networks, transport reactions are also pivotal for understanding amino acid metabolism and production in bacteria. Apart from substrate uptake, this refers to product (amino acid) excretion as well as product re-uptake. Both the mechanistic (kinetic and energetic) as well as structural properties of these transport systems are relevant for understanding their significance and for providing a basis for rational metabolic design. Transport systems have been classified into numerous different carrier families, according to their structural properties and their putative evolutionary relation. The diversity of amino acid uptake and excretion systems in two biotechnologically relevant organisms, namely *Escherichia coli* K12 and *Corynebacterium glutamicum* ATCC 13032 is described in this review on the basis of their relation to these different transporter families. Particular functional and molecular properties of specific amino acid excretion systems in these two organisms, in particular those responsible for efflux of lysine (and arginine), threonine, branched chain amino acids, cysteine (and cysteine derivatives) and glutamate are described. A complete list of all secondary and primary transport systems in *C. glutamicum* putatively related to amino acid transport is provided.

1 Introduction

Cells exchange matter, energy and information with their surroundings and membrane-bound solute transport systems are essential for these processes. Bacteria are equipped with a broad variety of transport systems, with some of them, e.g., phosphotransferase systems and binding protein dependent ABC transporters, being exclusively present in prokaryotic organisms. Amino acid transport systems, which are found in bacteria like in all other living cells, include members of many different transporter families and are thus in principle not distinct from other substrate categories of transport systems, e.g., organic acids or carbohydrates. From a historic point of view, however, they are remarkable in a sense that the significance of specific solute excretion mechanisms was recognized earlier in the case of amino acids than for most other solutes except cytotoxic compounds. This refers both to the physiological and biochemical description of export systems as well as to their molecular definition, for reviews see: (Krämer 1994; Burkovski and Krämer 2002; Eggeling and Sahm 2003; Eggeling 2005). Beside amino acid uptake, however, amino acid excretion was discovered already early in the history of modern biotechnology (Kinoshita et al. 1957). At that time, solute transport in general and solute export in particular was not acknowledged to be a central step within the network of biochemical pathways. In the meantime it is obvious that, besides the sum of metabolic and regulatory events within the bacterial cell, transport reactions are also of core importance.

In this contribution, due to space restrictions, we will focus on a few central aspects of this broad topic. Based on their importance in biotechnology as well as on the availability of detailed knowledge of genomic and biochemical data, amino acid transport systems of *E. coli* K12 and *C. glutamicum* ATCC 13032 will be reviewed mainly. An inventory of amino acid transport systems of these two organisms will be presented, organized in those being proven by both genetic and biochemical data, and those being suggested only on the basis of gene annotation and/or similarity screen. Where available, they will be combined with physiological and biochemical data on amino acid uptake and excretion.

2
Basic Properties of Bacterial Amino Acid Transport Systems

Before listing particular protein families of amino acid transport systems as well as properties of individual transport systems, some basic characteristics of transport systems in general and amino acid transport systems in particular will be summarized. This includes a brief outline of available structural and functional data of this type of carrier systems and mechanisms.

2.1
Structure of Transport Systems

Most amino acid transporters belong to the class of secondary carriers (see below) and are thus polytopic integral membrane proteins, comprising one or several subunits with different numbers of α-helical transmembrane segments. Although still few in number, several high resolution 3D structures are now available for a small number of integral membrane transport proteins (www.tcdb.org, www.membranetransport.org, www.mpibp-frankfurt.mpg.de/michel/public/memprotstruct) also including a few amino acid transporters, e.g., the glutamate carrier of *Pyrococcus horikoshii* (Yernool et al. 2004) or the leucine carrier from *Aquifex aeolicus* (Yamashita et al. 2005). Several representatives of amino acid transport proteins belong to primary active ABC-type carriers (see below), where a number of structures (Chang and Roth 2001; Locher et al. 2002; Dawson and Locher 2006) are available, however, not from entire amino acid transporting ABC carrier systems.

2.2
Transport Mechanisms: Kinetics and Energetics

The formal treatment of transport kinetics follows the well-established formulation of classical enzyme kinetics and will not be discussed here. In this context, V_{max} and K_t (equivalent to an apparent K_m value for transport reactions) and K_i values are used to describe the biochemical properties of transport systems with respect to activity, specificity and inhibitory action. For practical considerations, transport energetics is of high significance for the description of the diversity of amino acid transport systems in one single organism in the presence of a multitude of systems for one particular substrate (amino acid). In addition, knowledge on transport energetics is helpful for a mechanistic description of efflux systems (see below and Fig. 1) as well as for a combination of uptake and efflux mechanisms for the same substrate (Sect. 2.3).

In bacteria, the majority of amino acid transport systems functions as secondary systems, i.e., their driving force is an electrochemical ion potential across the plasma membrane, which, in general, means the electrochemical H^+ or Na^+ potential (Fig. 1). Alternatively, bacterial amino acid uptake systems may follow a primary transport mechanism, i.e., depending on ATP as the energy source and involving an external binding protein (ABC transport systems). So far, with the exception of the CydDC system in *E. coli* (Sect. 3.20), amino acid export systems exclusively follow secondary mechanisms. However, a first report on the involvement of a channel type of transport mechanism in the case of glutamate export has appeared recently (Sect. 4.2.5).

2.3
Uptake versus Excretion

Besides the variety of sophisticated mechanisms for protein excretion, a significant number of export systems for solutes has been described in bacteria, mainly falling into the group of MDR (multidrug resistance) proteins. These systems are in general responsible for the excretion of cytotoxic compounds or waste products. *C. glutamicum* is the paradigm for which the simultaneous presence of both uptake and export systems specific for the same solute (amino acid) has been described first, later also followed by similar combinations mainly in *E. coli*. Typical examples are the amino acids lysine, leucine/isoleucine, threonine and glutamate, which will be described below in more detail. It is obvious that the presence of both uptake and efflux systems for the same amino acid poses an extra problem for the cell, both in terms of energetics and regulation, because of the possible occurrence of energy wasting futile cycling (Krämer 1994, 1996; Krämer and Hoischen 1994).

3
Repertoire of Amino Acid Transporters in *C. glutamicum* and *E. coli*

Within this chapter we will present data illustrating the equipment of the biotechnological important bacteria *C. glutamicum* and *E. coli* with carriers for amino acids. We included known and characterized transport proteins as well as predicted candidates for particular transport functions of different transporter families.

3.1
Carrier Identification, Nomenclature and Occurrence

Although the unequivocal identification of a transport protein requires the biochemical characterization of its transport function, carrier proteins are frequently predicted based on sequence data. The overall amino acid sequence similarity, the occurrence of domains specific for a certain transporter class, and the topology derived from the number and position of predicted transmembrane helical segments (TMS) provide useful hints on the putative carrier function of a membrane protein. The increasing number of known bacterial genome sequences provides a large pool of putative transporters to be used for development of prediction methods and identification of specific domains or motif sequences. However, in view of the fact that the prediction of substrate specificity from sequence data is not possible for the majority of transport systems, a detailed characterization by biochemical methods is still indispensable for the correct assignment of a carrier. Furthermore, only by biochemical studies so far unknown carriers of new transporter classes can be discovered.

A large variety of transport systems was classified by several systematic approaches. The most common classification system, the TC nomenclature, developed by the Saier group, is similar to the enzyme nomenclature (Busch and Saier 2004). The system is accessible online at http://www.tcdb.org (Saier et al. 2006) and was applied for the genome-wide prediction of transporter repertoires of prokaryotic and eukaryotic organisms (Ren and Paulsen 2005) accessible at http://www.membranetransport.org. In this chapter functionally characterized as well as predicted amino acid carriers from *C. glutamicum* and *E. coli* are presented. Detailed information for *E. coli* K12 is available online in public databases like EchoBase (http://www.biolws1.york.ac.uk/echobase), ecogene (http://www.ecogene.org) or NCBI (http://www.ncbi.nlm.nih.gov). For *C. glutamicum*, however, open data sources of this quality are missing. A general comparison of the transporter equipment of *C. glutamicum* and

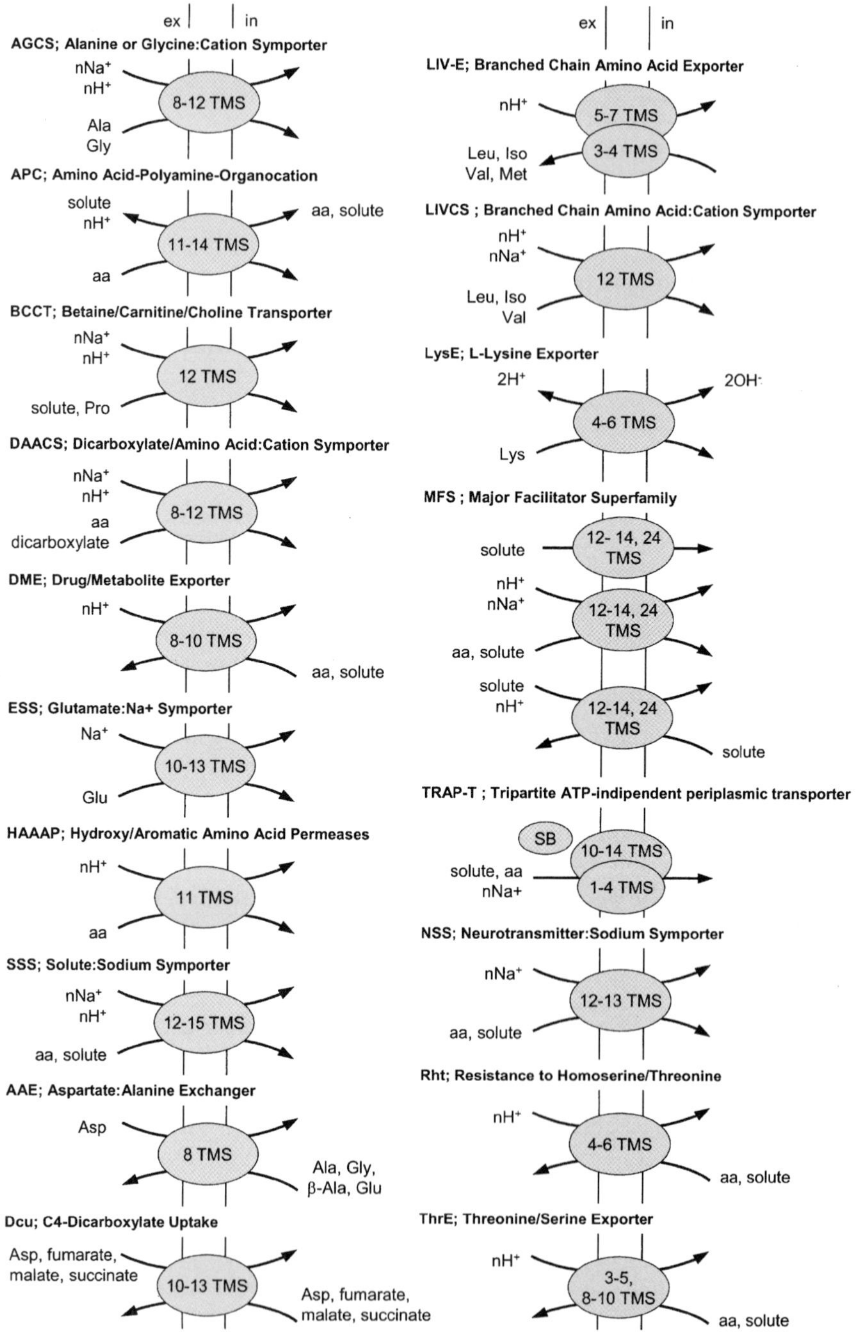

Fig. 1 Membrane topology as well as transport mechanism of secondary carriers (putatively) involved in amino acid transport. (ex: external, in: internal, aa: amino acid, TMS: transmembrane segments)

C. efficiens is available (Winnen et al. 2005). We provide for *C. glutamicum* detailed information on members of all transporter classes putatively involved in amino acid transport plus information on protein structure and function as well as experimental evidence for carrier function (see electronic supplementary material online Table S1 and S2).

The genome of *E. coli* comprises more than 4,200 protein encoding genes. For more than 1,000 of the respective gene products, the localization within the cytoplasmic membrane is predicted thus making the function as a carrier (-subunit) possible. Using bioinformatic tools, researchers identified 66 different transporter classes (Ren and Paulsen 2005). They comprise 52 different secondary transporter families, more than 68 ABC-type transporters and six PTS-type carriers. In total, more than 530 genes are predicted to encode for components of transport systems. In *E. coli* K12, members of 16 different transporter classes are shown or are predicted to be involved in amino acid transport (Table 1). At least one carrier system was found or predicted for the uptake of each amino acid, whereby the import of a number of amino acids is accomplished by several carriers. Beside a few systems involved in the exchange of amino acids, several systems have been found capable for the export of cysteine (and derivatives), threonine, homoserine, leucine, and arginine.

The genome of *C. glutamicum* carries more than 2,900 protein encoding genes. For more than 750 proteins, a membrane localization and function as a transporter (-subunit) is possible. Up to now 57 different transporter families have been predicted to exist. Among them are more than 130 secondary transporter (units), at least 57 primary transport systems of the ABC-type and four PTS-type carriers. The components of these carrier systems are encoded by 344 genes. Members of 17 transporter classes are shown or predicted to be involved in amino acid transport (Table 2). As in *E. coli*, the import of a number of amino acids is accomplished be several carriers of different transporter types. For some amino acids, the presence of carriers has only been derived from biochemical data, but until now the proteins are unknown. In particular, uptake carriers for glycine, cysteine, serine, threonine and other polar amino acids are not (yet) identified. In contrast to *E. coli*, *C. glutamicum* is characterized by a limited catabolism of several amino acids e.g., lysine (Nakayama 1985), isoleucine (Kennerknecht et al. 2002), threonine (Simic et al. 2001) or methionine (Trötschel et al. 2005), which may be related to a more pronounced capacity for amino acid export (see below).

In the following the variety of transporter families putatively comprising amino acid carriers are described and functionally characterized or predicted members of these families in *C. glutamicum* and *E. coli* are mentioned, using the TC nomenclature. For each carrier family, the general topology of TMS and a scheme of the transport reaction mechanism are listed in Figs. 1 and 2. For genes in *E. coli*, gene names as well as the locus tag (*b* number) for the MG1655 genome (Blattner et al. 1997) are listed. For genes in *C. glutamicum*, the *cg* nomenclature is used (Kalinowski et al. 2003). As supplementary mate-

Table 1 Transporter families (putatively) involved in amino acid transport in *E. coli* [a]

Transporter class		members in E.c.	Gly	Ala	Val	Leu	Ile	Cys	Met	Phe	Tyr	Trp	Pro	Ser	Thr	Asn	Gln	Asp	Glu	His	Lys	Arg	derivatives
MFS	TC 2.A.1	71						blue O [1]					red O [2]										blue O [1]
APC	TC 2.A.3	22	red O [3]	red O [3]						red O [4]	red O [4]		red × [5]	red O [6]				red × [5]	green O [7]		red O green O [8]	green O [9]	red O green O [10]
DME	TC 2.A.7	16										blue × [11]			blue O [12]								
Dcu	TC 2.A.13	2																green O [?]					
BCCT	TC 2.A.15	3																					green O [14]
SSS	TC 2.A.21	4											red O [15]										
NSS	TC 2.A.22	0																					
DAACS	TC 2.A.23	3												red O [16]					red O [17]				
AGCS	TC 2.A.25	1		red × [5]																			
LIVCS	TC 2.A.26	1			red O [18]	red O [18]	red O [18]																
ESS	TC 2.A.27	1																	red O [19]				
HAAAP	TC 2.A.42	8									red O [20]	red O [21]		red O [22]	red O [23]								
TRAP-T	TC 2.A.56	1																	red × [5]				
LysE	TC 2.A.75	1?																			blue ×? [24]	blue O [24]	
RhtB	TC 2.A.76	5				blue O [25]		blue O [26]							blue O [27]								blue O [27]
LIV-E	TC 2.A.78	1?			blue × [5]	blue × [5]	blue × [5]																
ThrE	TC 2.A.79	1																					
AAE	TC 2.A.81	2		green × [5]														green × [5]					
ABC	TC 3.A.1	>41			red O [28]	red O [28]	red O [28]	blue O [29]	red O [30]				red O [31]			red × [32]	red O [33]	red O [34]	red O [35]	red O [36]	red O [37]	red O [38]	red O [39]

(a) Beside the number of particular members of the transporter families (for abbreviations see text) the carrier identification by experimental characterization (O) or bioinformatic tools (×) is indicated. The presence of different carriers of the same transporter family for particular amino acids predicted either experimentally or by bioinformatic methods is also indicated (⊗). The function as importer (red), exporter (blue) or antiporter (green) is specified by the color and the particular references are: [1] (Yamada et al. 2006), (Dassler et al. 2000); [2] (Culham et al. 1998); [3] (Lee et al. 1975), (Steffes et al. 1992); [4] (Chye et al. 1986); [5] (Ren and Paulsen 2005); [6] (Pi et al. 1991); [7] (Metzer and Halpern 1990); [8] (Steffes et al. 1992), (Meng and Bennett 1992); [9] (Gong et al. 2003); [10] (Kashiwagi et al. 1992), (Meng and Bennett 1992), (Metzer and Halpern 1990), (Gong et al. 2003); [11] Ryback et al. 2006; [12] (Livshits et al. 2003); [13] (Andresen et al. 1988), (Jung et al. 2002); [14] (Jackowski and Alix 1990); [15] (Jung 1998); [16] (Ogawa et al. 1997), (Kim et al. 2002); [17] (Deguchi et al. 1989); [18] (Guardiola et al. 1974); [19] (Essenberg 1984); [20] (Whipp et al. 1980); [21] (Heatwole and Somerville 1991); (Sarsero et al. 1991); [22] (Kayahara et al. 1992); [23] (Sumantran et al. 1990); [24] (Nandineni and Gowrishankar 2004); [25] (Kutukova et al. 2005); [26] (Franke et al. 2003); [27] (Zakataeva et al. 1999); (Eggeling and Sahm 2003); [28] (Oxender et al. 1980), (Ovchinnikov et al. 1977), (Adams et al. 1990), (Sack et al. 1989); [29] (Pittman et al. 2002); [30] (Merlin et al. 2002); [31] (Checroun and Gutierrez 2004), (Dattananda and Gowrishankar 1989); [32] (Willis and Woolfolk 1975), (Ren and Paulsen 2005); [33] (Nohno et al. 1986); [34] (Willis and Furlong 1975); [35] (Liu and Ames 1997); [36] (Oh et al. 1993); [37] (Wissenbach et al. 1995); [38] (van der Ploeg et al. 1996), (Berger and Heppel 1972), (Leive and Davis 1965).

Table 2 Transporter families (putatively) involved in amino acid transport in *C. glutamicum* [a]

Transporter class		members in C.g.	Gly	Ala	Val	Leu	Ile	Cys	Met	Phe	Tyr	Trp	Pro	Ser	Thr	Asn	Gln	Asp	Glu	His	Lys	Arg	derivatives
			Substrate																				
MFS	TC 2.A.1	49						× [1]					⊗ [2]										
APC	TC 2.A.3	4		O [3]	O [3]	O [3]				O [4]	O [4]	O [4]									O [3]		
DME	TC 2.A.7	4																					
Dcu	TC 2.A.13	0																					
BCCT	TC 2.A.15	4											O [5]										O [6]
SSS	TC 2.A.21	2											O [7]										
NSS	TC 2.A.22	1																					
DAACS	TC 2.A.23	3												× [8]									
AGCS	TC 2.A.25	1		× [1]																			
LIVCS	TC 2.A.26	1			O [9]	O [9]	O [10]																
ESS	TC 2.A.27	1																	O [11]				
HAAAP	TC 2.A.42	1								× [1]	× [1]	× [1]											
TRAP-T	TC 2.A.56	1?																	× [1]				
LysE	TC 2.A.75	2													× [1]						O [12]	O [12]	
RhtB	TC 2.A.76	3				×		×															
LIV-E	TC 2.A.78	2			⊗	⊗	⊗		O														
ThrE	TC 2.A.79	1												O	O								
AAE	TC 2.A.81	2		×														×					
ABC	TC 3.A.1	>57			×	×	×	×	×								×	×	O	×	×	×	×
YggB	???	1																	O				

(a) Beside the number of particular members of the transporter families (for abbreviations see text) the carrier identification by experimental characterization (O) or bioinformatic tools (×) is indicated. The presence of different carriers of the same transporter family for particular amino acids predicted either experimentally or by bioinformatic methods is also indicated (⊗).The function as importer (red), exporter (blue) or antiporter (green) is specified by the color and the particular references are: [1] (Ren and Paulsen 2005); [2] (Peter et al. 1998); (Ren and Paulsen 2005); [3]; (Bröer and Krämer 1991a); [4]; (Wehrmann et al. 1995); [5] (Peter et al. 1998), (Steger et al. 2004); [6] (Peter et al. 1998); [7] (Peter et al. 1997); [8] This work; [9] This work, (Ebbighausen et al. 1989); [10] (Tauch et al. 1998); [11] (Trötschel et al. 2003); [12] (Bellmann et al. 2001); [13] This work; [14] (Kennerknecht et al. 2002), This work; [15] (Trötschel et al. 2005); [16] (Simic et al. 2001); [17] (Ren and Paulsen 2005), This work; [18] (Kronemeyer et al. 1995); [19] (Nakamura et al. 2006).

rial a listing of all genes of *C. glutamicum* encoding known or putative, amino acid transporters is given according to the transport mechanism (secondary transporters in supplementary Table S1, primary transporters in supplementary Table S2) according to the particular transporter class. The tables are accessible at: http://www.springer.com.

3.2
The Major Facilitator Superfamily (MFS) (TC 2.A.1)

The MFS is a large and diverse superfamily including more than 6,900 sequenced members found in all kingdoms of live. They catalyze uniport, solute:cation (H^+ or Na^+) symport and/or solute:H^+ or solute:solute antiport and possess either 12, 14 or 24 TMS, whereby the proteins with 24 TMS are most likely fusion proteins of two homologous but distinct MFS permeases (Fig. 1). MFS permeases transport sugars, polyols, drugs, neurotransmitters, Krebs cycle metabolites, phosphorylated glycolytic intermediates, amino acids, peptides, osmolytes, siderophores (efflux), iron-siderophores (uptake), nucleosides, organic anions, inorganic anions, etc. (Busch and Saier 2004). For the MFS carriers GlpT, LacY (*E. coli*) and OxlT (*Oxalobacter formigenes*) high resolution structures are available (Abramson et al. 2003; Hirai et al. 2003; Huang et al. 2003). As the only amino acid substrate of an MFS uptake carrier proline was found (Culham et al. 1993; Peter et al. 1998). In addition, the Bcr protein was shown to be involved in efflux of cysteine and other amino acids or derivatives (Yamada et al. 2006).

In *C. glutamicum*, 49 MFS systems are predicted to exist. Among them, the ProP protein encoded by *cg3395* was shown to be capable of proline uptake (Peter et al. 1998). The gene *cg3403* encodes a carrier of high similarity but smaller size and proline was proposed as substrate (Ren and Paulsen 2005). In *E. coli*, 71 MFS systems are found, among them are the proline uptake carrier ProP, encoded by the gene *b*4111 (Culham et al. 1993), as well as the proteins YhjE and YdfJ encoded by the genes *b*3523 and *b*1543, respectively, for which proline was proposed as putative substrate. The efflux carrier Bcr (*b*2182) has already been mentioned above. Because of the variety of substrates known for MFS transporters in *E. coli* and *C. glutamicum*, the potential participation in uptake as well as export of further amino acids has to be considered.

3.3
The Amino Acid-Polyamine-Organocation (APC) Superfamily (TC 2.A.3)

The APC superfamily of transport proteins currently includes more than 1400 members, which occur in bacteria, archaea, and eukaryotes. They function as solute:cation symporters and solute:solute antiporters and vary in length and topology (Saier 2000), (Fig. 1). Some members possess only 10 TMS and rather function as amino acid receptors (Jack et al. 2000; Cabrera-Martinez et al. 2003). Some animal homologs associate with a small membrane chaperone, which is essential for insertion and/or activity of the permease and linked by a disulfide bridge. In general, APC carriers were found or proposed to be involved in the transport of a variety of amino acids, namely glycine, alanine, leucine, isoleucine, valine, tyrosine, tryptophan, histidine, S-methylmethionine, phenylalanine, proline, serine, lysine and asparagine.

In *C. glutamicum*, four genes code for members of the APC family, including the LysI system (*cg1105*), which catalyzes exchange of lysine against lysine, alanine, valine, or leucine with very low activity (Bröer and Krämer 1991a; Seep-Feldhaus et al. 1991). The gene *cg1257* encodes AroP, the only known uptake system for aromatic amino acids in *C. glutamicum* (Wehrmann et al. 1995) and the genes *cg0555* and *cg1305* code for carriers of unknown substrates. In *E. coli*, 22 members of the APC family were found. The AroP (*b0112*) system is capable of phenylalanine and tyrosine uptake (Chye et al. 1986), the PheP (*b0576*) protein represents an uptake system for phenylalanine (Pi et al. 1991), the CycA (*b4208*) system is the main alanine carrier in *E. coli* but also competent for serine and glycine uptake (Lee et al. 1975; Schneider et al. 2004) and the LysP (*b2156*) protein acts as a lysine importer (Steffes et al. 1992). Moreover, three systems involved in pH regulation, the GabP (*b2663*) transporter responsible for exchange of glutamate and γ-aminobutyrate (Metzer and Halpern 1990), the CadB (*b4132*) system exchanging cadaverine and lysine (Meng and Bennett 1992), and the arginine agmatine antiporter AdiC (YjdE, *b4115*) (Gong et al. 2003) belong to this class. As transporters for amino acid derivatives, PotE (*b0692*) exchanging putrescine and ornithine (Kashiwagi et al. 1992) and MmuP (*b0260*) taking up S-methylmethionine (Thanbichler et al. 1999) are present. Further APC transporter of *E. coli* are YeeF (*b2014*), YifK (*b3795*), YjeH (*b4141*), YjeM (*b4156*), YhfM (*b3370*), YgjI (*b3078*), XasA (*b1492*), ProY (*b0402*), YbaT (*b0486*), YcaM (*b0899*), AnsP (*b1453*), YdgI (*b1605*) and YcjJ (*b1296*), including the putative proline carrier ProY and the possible asparagine uptake system AnsP (Ren and Paulsen 2005). These data indicate that members of this family are capable of transporting a large variety of substrates and further members may be supposed to represent still missing amino acids carriers.

3.4
The Drug/Metabolite Transporter (DMT) Superfamily (TC 2.A.7)

The DMT Superfamily currently consists of more than 1,800 members ordered in 18 recognized families, each with a characteristic function, size and topology. Within these subfamilies amino acid carriers are identified only for the drug/metabolite exporter subgroup (2.A.7.3, DME). Proteins of the DME family range from 287 to 310 amino acyl residues and carry 8–10 putative α-helical TMS (Busch and Saier 2004). In *C. glutamicum*, four members of the DME subfamily are predicted to be encoded by the genes *cg0168*, *cg0701*, *cg2339* and *cg2356*. None of them was functionally characterized but they are supposed to be involved in amino acid export. In *E. coli*, 16 members of the DME subfamily were identified, among them, the YddG (*b1473*) carrier is supposed to be involved in tryptophan export (see below). The YdeD carrier (*b1533*, EamD) was shown to represent an exporter of cysteine and cysteine derivatives (Dassler et al. 2000). The RhtA (YbiF, *b0813*) protein has been related to efflux of threonine and homoserine based on an increased resistance against these amino acids after overexpression of the *rhtA* gene (Livshits et al. 2003).

3.5 The C4-Dicarboxylate Uptake (Dcu) Family (TC 2.A.13)

More than 100 members of the Dcu family are found. They consist of approx. 440 amino acyl residues in length and possess 10–12 TMS. In *E. coli*, the DcuA (*b4138*) and DcuB (*b4123*) proteins are known as exchangers of aspartate, malate, fumarate and succinate under anaerobic growth conditions (Engel et al. 1994). Two additional genes encoding the dicarboxylate transporters DcuC (*b0621*) and DcuD (*b3227*) are present showing weak sequence similarity to DcuA and DcuB and they were, therefore, assigned to a distinct family (TC 2.A.61). While in *C. diphtheriae* DcuA and DcuB homologs are found, no member of the Dcu family is known in *C. glutamicum*.

3.6 The Betaine/Carnitine/Choline Transporter (BCCT) Family (TC 2.A.15)

More than 140 proteins of the BCCT family are found in bacteria and archaea. In general, they transport solutes with a quaternary ammonium group $[R-N^{+}(CH_3)_3]$ (Busch and Saier 2004). The only amino acid substrate of BCCT carriers is proline. Some of these transporters exhibit inherent osmosensory and osmoregulatory properties (Rübenhagen et al. 2000). In *C. glutamicum*, BetP encoded by *cg1016*, EctP encoded by *cg2539* and LcoP encoded *cg2563* were characterized (Peter et al. 1998; Steger et al. 2004). All three are capable of uptake of the amino acid derivatives glycine betaine and ectoine, but only EctP and LcoP accept proline. In *E. coli*, three BCCT transporters, BetT, CaiT and YeaV that are encoded by *b0314*, *b0040*, and *b1801*, respectively, are present. None of these systems seems to be competent for the transport of proline or another amino acid (Andresen et al. 1988; Jung et al. 2002).

3.7 The Solute:Sodium Symporter (SSS) Family (TC 2.A.21)

More than 420 members of the SSS family have been identified in bacteria, archaea and animals possessing 12-15 TMS and catalyzing solute: Na^{+} symport

(Fig. 1). A variety of solutes such as sugars, amino acids, organic cations (e.g., choline), nucleosides, inositols, vitamins, urea or anions are transported. For some members regulatory domains are found homologous to histidine kinases (Jung 2001; Busch and Saier 2004). In *E. coli*, four members of the SSS family are present. Besides the PutP (*b1015*) system catalyzing proline uptake (Jung 1998), the PanF protein, encoded by *b3258*, involved in pantothenate uptake, the acetate uptake system ActP (*b4067*), as well as the putative myo-inositol transporter YidK (*b3679*) are members of this family (Jackowski and Alix 1990; Gimenez et al. 2003; Ren and Paulsen 2005). In *C. glutamicum*, two SSS-type carriers are found including a PutP protein (*cg1314*) as the main proline uptake system at low external osmolality (Peter et al. 1997). The gene *cg0953* encodes an additional SSS transporter with high sequence similarity to the acetate transporter ActP of *E. coli*, but the substrate is still unknown.

3.8
The Neurotransmitter:Sodium Symporter (NSS) Family (TC 2.A.22)

Members of the NSS family catalyze uptake of a variety of neurotransmitters, amino acids, osmolytes and related nitrogenous substances by a solute:Na^+ symport mechanism (Fig. 1). More than 190 members are mostly found in animals, but bacterial and archaeal homologues have also been identified. TnaT of *Symbiobacterium thermophilum* has been shown to be a Na^+-dependent tryptophan uptake permease (Androutsellis-Theotokis et al. 2003). For the LeuT protein from *Aquifex aeolicus* facilitating leucine sodium co-transport, the crystal structure has been determined (Yamashita et al. 2005). Glycine and tyrosine were shown or proposed as additional substrates of eukaryotic NSS carriers (Morrow et al. 1998; Ren and Paulsen 2005). Whereas for *C. glutamicum* one member of the NSS family with unknown substrate specificity encoded by *cg1169* was found, in *E. coli* no transporter of this family is present.

3.9
The Dicarboxylate/Amino Acid : Cation (Na^+ or H^+) Symporter (DAACS) Family (TC 2.A.23)

The more than 340 members of the DAACS family catalyze Na^+ and/or H^+ symport together with a Krebs cycle dicarboxylate or with an amino acid as indicated in Fig. 1. The 3D structure of a member of the DAACS family, the glutamate carrier from *Pyrococcus horikoshi*, has been determined (Yernool et al. 2004). DAACS carriers were found or proposed to be involved in the transport of acidic, small zwitterionic, as well as basic amino acids. In *C. glutamicum*, three members were predicted to be encoded by the genes *cg2810*, *cg2870* and *cg3356*. The gene *cg2870* encodes a DctA homolog, making a function as dicarboxylate transporter likely. Furthermore, all three proteins display a weak similarity to the serine uptake system SstT of *E. coli*. In *E. coli*, three members of the DAACS family were found, the dicarboxylate transporter DctA (Lo and Bewick 1978), the glutamate carrier GltP (Deguchi et al. 1989) and the serine and threonine transporter SstT (YgjU) (Ogawa et al. 1997; Kim et al. 2002) encoded by *b3528*, *b4077* and *b3089*, respectively.

3.10
The Alanine or Glycine:Cation Symporter (AGCS) Family (TC 2.A.25)

Members of the AGCS family in general transport alanine and/or glycine in symport with Na^+ and or H^+ and comprise 8–12 TMS (Fig. 1). They are found in bacteria and archaea and similarity with the APC family (TC 2.A.3) has been established (Busch and Saier 2004). Three members of the AGCS family have been functionally characterized as alanine uptake systems in different organisms (MacLeod and MacLeod 1992; Kanamori et al. 1999; Moore and Leigh 2005). The single member of this family in *C. glutamicum* is encoded by *cg0254* and alanine was proposed as substrate. In *E. coli*, also a single member of the AGCS family is present, the YaaJ protein encoded by *b0007*. Also in this case, alanine was proposed as substrate; however, CycA (DagA, *b4208*), a member of the APC carrier family, has been identified as the main alanine uptake system (see above).

3.11
The Branched Chain Amino Acid:Cation Symporter (LIVCS) Family (TC 2.A.26)

Characterized members of this family import all three branched chain amino acids. More than 110 members were found in bacteria; they function by a Na^+ or H^+ symport mechanism. In *E. coli* as well as in *C. glutamicum*, a single LIVCS system was found named BrnQ encoded by *b0401* and *cg3537*, respectively (Guardiola et al. 1974; Tauch et al. 1998).

3.12
The Glutamate : Na^+ Symporter (ESS) Family (TC 2.A.27)

This family comprises more than 50 members only known in bacteria and proteins from *E. coli*, *Salmonella*, and *C. glutamicum* have been functionally characterized (Essenberg 1984; Alvarez-Jacobs et al. 1986; Trötschel et al. 2005). In *C. glutamicum*, the GltS system is encoded by *cg3080* and was characterized at the molecular and biochemical level (Burkovski and Krämer 1995; Trötschel et al. 2005). In *E. coli*, the single member of this family was also named GltS and is encoded by *b3653* (Marcus and Halpern 1969). In both organisms, other glutamate uptake systems are present besides the GltS permease and belong to the ABC or DAACS family, respectively (see Sects. 3.20 and 3.9).

3.13
The Hydroxy/Aromatic Amino Acid Permease (HAAAP) Family (TC 2.A.42)

More than 150 homologues are present in bacteria and most of them were found or proposed to be involved in the transport of tyrosine, tryptophan, serine and threonine. In *E. coli*, the HAAAP family comprises eight proteins including the high affinity tryptophan permease, Mtr (*b3161*), the low affinity tryptophan carrier TnaB (*b3709*), the tyrosine-specific permease TyrP (*b1907*), the serine permease SdaC (*b2796*) as well as the threonine permease TdcC (*b3116*) (Whipp et al. 1980; Sumantran et al. 1990; Heatwole and Somerville 1991; Sarsero et al. 1991; Kayahara et al. 1992). Further members are encoded by *b3539* (YhjV), *b3110* (YhaO) and *b2845* (YgeG), but their transport substrates are unknown. In *C. glutamicum*, a single member of the HAAAP family is encoded by *cg0568*. Based on sequence similarity, aromatic amino acids are supposed as substrates of this carrier.

3.14
The Tripartite ATP-Independent Periplasmic (TRAP-T) Transporter Family (TC 2.A.56)

TRAP-T family carrier consists of three components, two integral membrane proteins of 12 and four TMS, respectively, as well as a periplasmic substrate binding protein, therefore representing the only known binding protein dependent secondary carriers. The substrate binding protein can be anchored in the membrane in gram negative and gram positive bacteria. As observed for ABC-type transporters, subunit/domain fusions and splicing occurred during evolution (Busch and Saier 2004). More than 300 members of this family are known in bacteria and archaea. The only report on amino acid transport by a TRAP-T carrier concerns glutamate uptake in *Rhodobacter sphaeroides* (Jacobs et al. 1996). In *E. coli*, a single TRAP-T-type carrier is known comprising the membrane bound subunits YiaM (*b3577*) and YiaN (*b3578*) as well as the substrate binding component YiaO (*b3579*). In *C. glutamicum*, the genes *cg2568* and *cg2569* encode the large and small membrane bound subunits, respectively and *cg2570* the substrate binding protein of a TRAP-T system of unknown function. Additionally, the orphan gene *cg2546* encodes a large membrane component.

3.15
The Lysine Exporter (LysE) Family (TC 2.A.75)

More than 70 LysE family members are found widely distributed in bacteria. Together with the cadmium resistance family (CadC) and the resistance to homoserine/threonine family (RhtB) the LysE family forms the LysE superfamily. The energy source for transport is proton motive force (H^+ antiport or OH^- symport, (Bröer and Krämer 1991b). Two members of the LysE family were functionally characterized, LysE of *C. glutamicum* encoded by *cg1424* and ArgO of *E. coli* encoded by the gene *b2923* (Vrljic et al. 1996; Nandineni and Gowrishankar 2004). Whereas for LysE of *C. glutamicum*, the specific export of lysine and arginine was demonstrated experimentally (Bellmann et al. 2001), for ArgO (YggA) of *E. coli* the specific efflux of arginine was proposed (Nandineni and Gowrishankar 2004), (Sect. 4.2.1). In *C. glutamicum*, an additional LysE-type carrier is encoded by *cg0183*, which is annotated as putative threonine carrier, but probably responsible for the export of so far unknown substrates.

3.16 The Resistance to Homoserine/Threonine (RhtB) Family (TC 2.A.76)

More than 420 proteins, derived from bacteria and archaea, comprise the RhtB family as subfamily of the LysE superfamily. The characterization of a few of them revealed a topology and function as amino acid exporters (Fig. 1). *E. coli* possesses five paralogues of about the same size and apparent topology. The genes *b3823*, *b3824*, *b1798*, *b0328*, and *b2578* encode the RhtC, RhtB, YeaS, YahN, and YfiK proteins, respectively. The overexpression of *yeaS*, *yfiK*, *rhtC* and *rhtB* provides resistance to threonine whereas an additional resistance to homoserine and homoserine lactone was observed by overexpression of *yeaS* and *rhtB* (Zakataeva et al. 1999; Eggeling and Sahm 2003). The overexpression of *yahN* had no positive effect on growth of *E. coli* cells exposed to higher concentrations of threonine, homoserine, and homoserine lactone. For the YfiK protein, the export of cysteine, O-acetylserine and azaserine was demonstrated as well (Franke et al. 2003), whereas YeaS was recently identified as leucine exporter and consequently named LeuE (Kutukova et al. 2005). In *C. glutamicum*, three members of the RhtB family are encoded by the genes *cg0183*, *cg2574* and *cg2941*, respectively, but for none of them a transport substrate could be identified and ThrE of the threonine/serine exporter family was found as the major threonine export system (Sect. 3.18). Based on sequence similarity, the *cg0183* and *cg2574* gene products could be assigned to homoserine export while *cg2941* might encode a leucine exporter.

3.17 The Branched Chain Amino Acid Exporter (LIV-E) Family (TC 2.A.78)

The LIV-E family consists of more than 104 members in a diverse group of bacteria and archaea. Pairs of integral membrane proteins comprise an efflux pump for branched chain amino acids. In *C. glutamicum*, two homologous systems are found. The BrnFE system (*cg0315*, *cg0314*) has been functionally characterized to catalyze export of methionine, isoleucine, leucine and valine (Kennerknecht et al. 2002; Trötschel et al. 2005). The second system was named AzlCD (*cg3412* and *cg3413*), but the substrate specificity is unknown. In *E. coli*, a similarity to the large subunits of LIV-E systems was found for YgaZ encoded by *b2682*; however, no counterpart of a small subunit is known until now. It remains unclear, therefore, whether a complete LIV-E system is present in *E. coli*, since efflux of leucine and other amino acids can probably be accomplished by the LeuE system (Sect. 3.16).

3.18 The Threonine/Serine Exporter (ThrE) Family (TC 2.A.79)

Around 60 members of the ThrE family are diverse in sequence and are ubiquitous in bacteria, archaea and eukaryotes. It is interesting to note that most of these transporters carry an extended N-terminal domain showing a weak sequence similarity to hydrolases, which has been supposed to be responsible for additional functions of this type of carriers (Eggeling and Sahm 2003). In *C. glutamicum*, ThrE (*cg2905*), the only member of this family, has been functionally characterized as proton motive force-dependent exporter of threonine and serine (Simic et al. 2001). In *E. coli*, the gene *b4363* was predicted to encode a small ThrE carrier, the YjjP protein, for which the substrate is still unknown.

3.19 The Aspartate:Alanine Exchanger (AAE) Family (TC 2.A.81)

A single functionally characterized protein, the aspartate:alanine exchanger AspT of the Gram-positive lactic acid bacterium *Tetragenococcus halophila* D10 served to define the AAE family (Abe et al. 2002; Busch and Saier 2004). Until now at least 40 further members of the AAE family have been predicted in many bacteria. In *C. glutamicum*, the genes *cg0683* and *cg2425* encode proteins that display similarity to the AspT protein sequence, whereas in *E. coli* the proteins encoded by *b3685* and *b0847* belong to the AAE family. In both organisms, however, the proteins have not been functionally characterized.

3.20 The ATP-binding Cassette (ABC) Superfamily (TC 3.A.1)

The ABC superfamily contains both uptake and efflux transport systems, and genes for subunits of individual members of these two groups frequently form gene clusters. According to their substrate specificity, numerous families within the ABC superfamily were classified. The members of the ABC superfamily consist of two integral membrane domains/proteins and two cytoplasmic domains/proteins that constitute a homo- or heterodimer each. In addition, the uptake systems possess extracytoplasmic solute-binding receptors (one or more per system), which in Gram-negative bacteria are found in the periplasm, whereas in Gram-positive bacteria they are in general tethered to the external surface of the cytoplasmic membrane by lipid anchors (Fig. 2). Some members possess receptor domains fused to either the N- or C-terminus of the translocating membrane protein (van der Heide and Poolman 2002). ABC-type uptake systems have not been identified in eukaryotes, but ABC-type efflux systems are widely found both in pro- and eukaryotes. Efflux systems frequently have four domains (two cytoplasmic and two membrane domains) fused into either one or two polypeptide chains. The three-dimensional structures of several entire bacterial drug export proteins (MsbA, BtuCD, and Sav1866) were solved (Chang and Roth 2001; Locher et al. 2002; Dawson and Locher 2006). Moreover, several substrate binding proteins were crystallized, see Table 3. According to the transporter classification system by Saier (Busch and Saier 2002), amino acid transporter subfamilies are the polar amino acid uptake transporter (PAAT, TC 3.A.1.3) family, the hydrophobic amino acid uptake transporter (HAAT, TC 3.A.1.4) family, the quaternary amine uptake transporter (QAT, TC 3.A.1.12) family, the methionine uptake transporter (MUT, TC 3.A.1.24) family and the cydDC cysteine exporter (CydDC-E, TC 3.A.1.129.1) family. In general, ABC-type carriers have been shown or proposed to be involved in the transport of glycine, valine, leucine, isoleucine, cysteine, methionine, proline, asparagine, glutamine, aspartate, glutamate, histidine, lysine, and arginine.

In *E. coli*, more than 68 ABC-type transport systems are present. Among them are 13 supposed to be involved in amino acid transport (Table 3). Members of the PAAT family are the lysine, arginine and ornithine uptake system ArgTHisPQM (Wissenbach et al. 1995) and the histidine uptake system HisJMPQ, which comprise the same permease and ATP binding proteins, but a different substrate binding protein (Liu and Ames 1997). ArtIMPQ catalyzes uptake of arginine (Wissenbach et al. 1993, 1995), GlnHPQ of glutamine (Nohno et al. 1986), GltIJKL is responsible for the uptake of glutamate and aspartate (Willis and Furlong 1975; Deguchi et al. 1989) and CysXYZ catalyzes

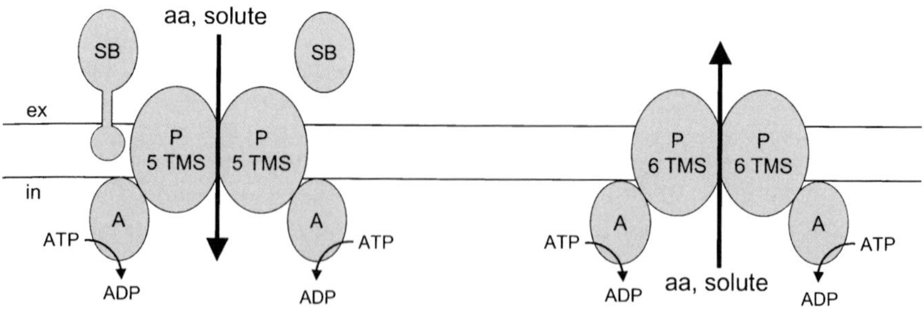

Fig. 2 Membrane topology as well as transport mechanism of primary active carriers (putatively) involved in amino acid transport (ex: external, in: internal, aa: amino acid, TMS: transmembrane segments)

Table 3 Known ABC-type amino acid transporters in *E. coli*[a]

ABC-system	TC class	Permease subunits	Substrate binding protein	ATP binding proteins	Substrate	Refs.
ArgTHisPQM	3.A.1.3.1	HisM (*b2307*) HisQ (*b2308*)	ArgT (*b2310*) S	HisP (*b2306*)	lysine arginine ornithine	1
HisJMPQ	3.A.1.3.1	HisM (*b2307*) HisQ (*b2308*)	HisJ (*b2309*) S	HisP (*b2306*)	histidine	2
GlnHPQ	3.A.1.3.2	GlnP (*b0810*)	GlnH (*b0811*) S	GlnQ (*b0809*)	glutamine	3
ArtIMPQ	3.A.1.3.3	ArtM (*b0861*) ArtQ (*b0862*)	ArtI (*b0863*)	ArtP (*b0864*)	arginine	4
GltIJKL	3.A.1.3.4	GltJ (*b0654*) GltK (*b0653*)	GltI (*b0655*)	GltL (*b0652*)	glutamate aspartate	5
CysXYZ	3.A.1.3.10	YecS (*b1917*)	FliY (*b1920*)	YecC (*b1918*)	cystine diamino- pimelic acid	6
LivFGHJM	3.A.1.4.1	LivM (*b3456*) LivH (*b3457*)	LivJ (*b3460*) S	LivF (*b3454*) LivG (*b3455*)	leucine iso-leucine valine	7
ProU	3.A.1.12.1	ProW (*b2678*)	ProX (*b2679*)	ProV (*b2677*)	glycine betaine proline	8
YehWXYZ	3.A.1.12.?	YehW (*b2128*) YehY (*b2130*)	YehZ (*b2131*)	YehX (*b2129*)	proline?	9
TauABC	3.A.1.17.1	TauC (*b0367*)	TauA (*b0365*)	TauB (*b0366*)	taurine	10
MetD	3.A.1.24.1	MetI (*b0198*)	MetQ (*b0197*)	MetN (*b0199*)	D-methio- nine	11
CydDC	3.A.1.129.1	CydD (*b0887*) CydC (*b0886*)			cysteine glutathione	12
YhdWXYZ	3.A.1.?	YhdY (*b3270*) YhdX (*b3269*)	YhdW (*b3268*)	YhdZ (*b3271*)	amino acid?	13

[a] ABC-type carriers and the particular subunits as well as identified substrates in *E. coli*. The 3D structure of several binding proteins was solved (S). The particular references are: [1] (Wissenbach et al. 1995), (Oh et al. 1993); [2]; (Liu and Ames 1997), (Yao et al. 1994); [3] (Nohno et al. 1986), (Sun et al. 1998); [4] (Wissenbach et al. 1993), (Wissenbach et al. 1995); [5] (Willis and Furlong 1975), (Deguchi et al. 1989); [6] (Berger and Heppel 1972), (Leive and Davis 1965); [7] (Oxender et al. 1980), (Ovchinnikov et al. 1977), (Adams et al. 1990), (Sack et al. 1989); [8] (May et al. 1986), (Dattananda and Gowrishankar 1989); [9] (Checroun and Gutierrez 2004); [10] (van der Ploeg et al. 1996); [11] (Merlin et al. 2002); [12] (Pittman et al. 2002); [13] (Ren and Paulsen 2005).

uptake of cystine and diaminopimelic acid (Leive and Davis 1965; Berger and Heppel 1972). The LivFGHJM system was described as an importer for hydrophobic amino acids (Ovchinnikov et al. 1977; Oxender et al. 1980; Adams et al. 1990). The QAT family consists of three members in *E. coli*, the ProU system, which is a proline carrier and a component of the osmotic stress response, (May et al. 1986, Dattananda and Gowrishankar 1989), the TauABC transporter as taurine uptake system (van der Ploeg et al. 1996) and the putative proline transporter YehWXYZ (Checroun and Gutierrez 2004). Uptake of D-methionine in *E. coli* is catalyzed by the MetD (MetINQ) system (Merlin et al. 2002), and YhdWXYZ was proposed as a further transporter for amino acids (Ren and Paulsen 2005). The only known primary active export system for amino acids in *E. coli* is the CydDC transporter, an ABC-type cysteine and glutathione exporter required for cytochrome assembly (Pittman et al. 2002). The dimeric protein is comprised of two subunits with a fused permease and ATP binding domain each. In this family of transporters, the structure of four substrate binding proteins of amino acid carriers (ArgT, HisJ, GlnH and LivJ) was solved (Sack et al. 1989; Oh et al. 1993; Yao et al. 1994; Sun et al. 1998).

In *C. glutamicum*, at least 57 ABC-type uptake systems are present and several orphan genes encoding single components of ABC-carriers are found. Until now, only one ABC-type uptake system for amino acids has been characterized, the GluABCD transporter (Kronemeyer et al. 1995), catalyzing uptake of glutamate. The substrate binding protein GluB shows similarity to the GltI protein of the Glt system in *E. coli*. An uptake system for polar amino acids is supposed to be encoded by the gene cluster *cg1502-04* because of the high sequence similarity of the binding protein (*cg1504*) to the GlnH, ArgT and HisJ proteins of *E. coli*. A high degree of sequence similarity also exists between the components MetI, MetN, and MetQ of the methionine uptake system MetD in *E. coli* and the corresponding *C. glutamicum* proteins encoded by the genes *cg0735*, *cg0736*, and *cg0737*. Furthermore, sequence similarity also suggests a taurine transporter binding protein, encoded by *cg1441*, however, a permease gene is not found in the adjacent gene cluster *cg1438* and *cg1440*. Because of sequence similarity, the protein encoded by the orphan gene *cg3045* was proposed to function as permease of an uptake system for glutamine or arginine (Ren and Paulsen 2005). A high sequence similarity to the CydDC system of *E. coli* is found for the proteins encoded by the genes *cg1298* and *cg1299* of *C. glutamicum*, indicating that these proteins may constitute an efflux carrier for cysteine and glutathione. As transport system for branched chain amino acids the ABC-type system encoded by the genes *cg1061-66* (Ren and Paulsen 2005) was proposed. It was recently shown, however, that these components constitute the urea uptake system of *C. glutamicum* (Beckers et al. 2004). Interestingly, no primary uptake system for proline is present in *C. glutamicum*, since proline uptake is facilitated by BCCT transporters (Sect. 3.6) only (Morbach and Krämer 2003).

3.21
So far Unidentified Carrier Systems

As a result of current efforts both on *C. glutamicum* and *E. coli* by using bioinformatics as well as molecular, biochemical and physiological tools the identification and functional characterization of still missing or even unknown carrier systems is expected. These proteins may belong to known transporter families or to new classes of transporters. For a number of substrates the presence of carrier systems can be predicted based on biochemical and physiological results combined with the fact that for many of the related putative substrates a limited rate of diffusion was proven. Recently, the 3D structure of the *Campylobacter jejuni* CjaA periplasmic substrate binding protein with a bound cysteine was solved, indicating the existence of ABC-type carriers for this particular amino acid (Müller et al. 2005). Examples for further expected carrier systems of this kind are mentioned in the chapters on specific amino acid uptake and excretion systems.

4
Particular Amino Acid Transport Systems in *C. glutamicum* and *E. coli*

For the same reasons as described above, we will restrict ourselves to the two biotechnologically highly relevant bacteria *E. coli* K12 and *C. glutamicum* in this chapter. This seems a bit arbitrary in the case of uptake systems; however, it is fully appropriate for amino acid export systems, since they have only been explored in these two organisms to a significant extent.

4.1
Amino Acid Uptake Systems

A list of amino acid uptake systems in *C. glutamicum* and *E. coli* K12 is shown in Table 1, and the corresponding mechanisms are listed in Fig. 1. A multiplicity of transport systems are observed for several amino acids both in the case of *E. coli* and *C. glutamicum*, at least when also predicted transport systems are included, and an even higher degree of multiplicity may turn out to be true if further putative carriers will be assigned in the future. Frequently, the observed multiplicity has a particular meaning in terms of regulation. This is true, for example, for proline in both organisms or for aromatic, as well as branched chain amino acids in *E. coli*. "Housekeeping systems" are constitutively expressed and frequently show a broad specificity and low affinity. On the other hand, additional transport systems with confined specificity and higher affinity are regulated on the level of transcription and are only synthesized, when a particular amino acid substrate is present extracellularly, or in case of special needs, e.g., under hyperosmotic conditions, where proline and amino acid derivatives are required for osmoprotection.

As mentioned in the first part of this review, the majority of amino acid uptake carriers function according to a secondary mechanism, i.e., they are driven by electrochemical ion potentials; however, in particular for *E. coli*, also a significant number of primary ABC-type systems that depend on ATP as driving force were found. The transport affinities are in general in the μm range for specific substrates and up to mM for the uptake of alternative substrates, the definition of which is of course subject to interpretation.

Because of the opposing direction of substrate flux, amino acid uptake systems may be of significance for amino acid production by bacteria, too. At least in the case of tryptophan and threonine production it has been shown that modification of the respective amino acid uptake activity by deletion or overexpression of genes coding for uptake systems may affect amino acid production (Ikeda and Katsumata 1995; Okamoto et al. 1997). Overexpression of the *aroP* gene, coding for an aromatic amino acid uptake system resulted in a drastic decrease of tryptophan production, whereas mutants impaired in tryptophan uptake were shown to be more effective in tryptophan production than the corresponding parent strains (Ikeda and Katsumata 1995).

Interestingly, a number of amino acid transport systems are still missing, at least with respect to their molecular definition, although many of them have been shown to be present by functional tests such as growth dependence or biochemical uptake measurements. This is true in *C. glutamicum* for alanine, for glutamine and asparagine, for the basic amino acids histidine and arginine, as well as for the polar amino acids serine, threonine, and cysteine. Uptake of glycine seems to be absent in *C. glutamicum* (Krämer, unpublished observation). For *E. coli*, the list of carrier systems seems to be much more complete and only uptake system(s) for cysteine have not yet been assigned to particular gene products.

4.2
Amino Acid Excretion Systems

In contrast to amino acid uptake, which is well studied in many bacterial species and for which numerous examples of uptake systems are described in terms of physiology, biochemistry and molecular biology, this is not true for amino acid excretion. There are a number of reviews available on this particular topic (Krämer 1994; Burkovski and Krämer 2002; Eggeling and Sahm 2003; Eggeling 2005), but these reviews only refer to the biotechnologically relevant organisms *E. coli* and *C. glutamicum*. It is an interesting and yet unsolved question, both in terms of basic microbial physiology as well as in terms of biotechnological application, whether amino acid excretion systems are wide spread among bacterial organisms or whether they are more restricted to particular species, like *E. coli* and *C. glutamicum*. Basically, the occurrence of members of carrier families, which are known to function as amino acid excretion systems in many organisms (e.g., members of the LysE, ThrE, and RhtB family are found in genomes of 72, 50, and 104 different species, respectively), clearly argues for a broad distribution. Moreover, the fact that for the production of other amino acids, e.g., glutamine, serine and alanine, besides the major "work horses" *C. glutamicum* and *E. coli*, also other bacteria are used, e.g., *Serratia marcescens* and *Bacillus subtilis*, indeed argues for a broader distribution of this kind of systems.

Another valid argument for this view is based on the physiological explanation for most of the observed amino acid excretion processes (Krämer 1994; Burkovski and Krämer 2002). Besides glutamate (see below), the presence of bacterial amino acid export systems can, in general, be explained by a putative function as emergency valves for situations of metabolic imbalance. An example is the uptake of peptides as sole or at least major source of carbon and energy and possibly also nitrogen. Since in *C. glutamicum*, pathways of amino acid catabolism are relatively limited (Nakayama 1985), a fact which is also the basis for application of the peptide feeding method for inducing amino acid efflux (Bröer and Krämer 1991a,b; Simic et al. 2001), particular amino acids may accumulate in the cytoplasm and are excreted in order to guarantee a high overall metabolic flux under these conditions. Although *E. coli* is better equipped with amino acid degradation pathways, they may still be kinetically limiting under particular metabolic conditions thus creating an evolutionary advantage of the presence of excretion systems.

All transporters that are known to catalyze amino acid export in *E. coli* and *C. glutamicum* are included in Tables 1–3. The first amino acid excretion systems have been described in physiological and biochemical terms about 15 years ago (Hoischen and Krämer 1989; Bröer and Krämer 1991a), at least in *C. glutamicum*, and the majority of them has been defined in molecular terms to a large extent by the work of L. Eggeling and his group (Eggeling and Sahm 2003; Eggeling 2005). Since a number of instructive reviews (see above) are available for the transport systems known so far, besides a general overview, details on these systems will not be given here.

4.2.1 Lysine and Arginine Export (LysE of *C. glutamicum* and ArgO of *E. coli*)

Lysine is the amino acid with the second largest biotechnological production capacity (Kelle et al. 2005). Furthermore, lysine export in *C. glutamicum* was the first amino acid excretion system to be described in mechanistic terms (Bröer and Krämer 1991b) and to be identified on the molecular level (Vrljic et al. 1996). In view of the impermeability of this amino acid by passive diffusion through the plasma membrane, on the one hand, and because of the energetically unfavorable charge movement (a cation would be moved to the positive side of the membrane), on the other, the requirement of a regulated and energy driven extrusion system for this amino acid seems to be obvious. The lysine export carrier is an integral membrane protein with five (Vrljic et al. 1996) to six (Haier and Krämer, unpublished) transmembrane segments, which most probably functions in the membrane as a dimer. Besides lysine, it also accepts arginine as a substrate, and it is energetically driven by antiport with hydroxyl ions, or cotransport with protons, respectively (Bröer and Krämer 1991b). Expression of the *lysE* gene is under the control of the LysR-type transcription factor LysG which, in the presence of the co-inducers lysine or arginine (or citrulline or histidine), increases expression of *lysE* (Bellmann et al. 2001). The ArgO transporter of *E. coli*, which has a relatively high similarity to LysE of *C. glutamicum*, accepts arginine as a substrate and the expression of *argO* is under control of the LysR-type transcriptional activator ArgP (Nandineni and Gowrishankar 2004).

It has been shown convincingly that loss of the lysine export carrier creates a serious problem in *C. glutamicum* by accumulation of excessive internal lysine concentrations when growing on peptide substrates (Vrljic et al. 1996). The direct consequence of this observation, namely that overexpression of the *lysE* gene should lead to enhanced excretion of lysine, has in fact been shown (Vrljic et al. 1996; Kelle et al. 2005), however, a successful application to lysine producing *C. glutamicum* strains is not known. It may, thus, be assumed that in these lysine production strains lysine export, although being carrier mediated, is most probably not rate limiting in the overall process. On the other hand, the presence of LysE may increase lysine production in other organisms (Gunji and Yasueda 2006).

4.2.2
Threonine and Homoserine Export (ThrE of *C. glutamicum* and the RhtB Family Transporter of *E. coli*)

Threonine export was biochemically characterized in *C. glutamicum* (Palmieri et al. 1996) and the prototypical ThrE exporter was identified in the same organism (Simic et al. 2001). At elevated internal threonine concentrations, ThrE was shown to be responsible for the majority of threonine efflux. The remaining export activity is due to other, yet unidentified excretion systems as well as to passive diffusion (Simic et al. 2001). For threonine export, it has been shown that overexpression of the *thrE* gene in fact may lead to a significantly increased external threonine accumulation (Simic et al. 2002).

For biotechnological threonine production, however, *E. coli* rather than *C. glutamicum* strains are used (Leuchtenberger et al. 2005), but the situation concerning threonine export systems is not as clear. Carriers of the RhtB family of transporters, comprising five members in *E. coli*, namely *yahN*, *yeaS*, *yfiK*, *rhtB*, and *rhtC*, have been made responsible for export of this amino acid (Aleshin et al. 1999; Zakataeva et al. 1999). Later on, a major contribution of RhtB and RhtC to threonine efflux has been excluded (Kruse et al.

2002), although an influence of these proteins on resistance to threonine, homoserine, and homoserine lactone has been shown (Eggeling and Sahm 2003). Moreover, RhtA, a transporter belonging to the DME family (DMT superfamily, see Sect. 3.4) has recently been made responsible for increasing the resistance to inhibitory concentrations of threonine and homoserine, which was taken as an indication for being involved in threonine efflux (Livshits et al. 2003).

4.2.3
Export of Branched Chain Amino Acids and Methionine (BrnFE of *C. glutamicum*)

Isoleucine was one of the first amino acids for which active export has been demonstrated in *C. glutamicum*, in addition to a basic flux due to passive diffusion (Zittrich and Krämer 1994; Hermann and Krämer 1996). When the exporter was identified, it turned out to be a novel two-component carrier encoded by the *brnFE* genes (Kennerknecht et al. 2002). Previously, a similar transport system has been found in *B. subtilis* to be related to 4-azaleucine resistance (Belitsky et al. 1997). The originally identified substrates for this proton motive force-driven secondary system were the three branched-chain amino acids. Recently, however, methionine was identified to be most probably the major export substrate of BrnFE in *C. glutamicum* (Trötschel et al. 2005). In this publication, it was also shown that *C. glutamicum* carries at least one further methionine export system in addition to BrnFE. In *E. coli*, the existence of an entire LIV-E system is unclear, but the *yeaS* gene product has recently been related to the efflux of leucine (Kutukova et al. 2005).

4.2.4
Export of Cysteine and Cysteine Derivatives in *E. coli*

In *E. coli*, three proteins from very different transporter families have previously been identified to catalyze export of cysteine or cysteine derivatives, YdeD from the DME family (Dassler et al. 2000), YfiK from the RhtB family (Franke et al. 2003), as well as the ABC-type system CydDC (Pittman et al. 2002). Additionally, a recent systematic study on the effect of transporter gene overexpression revealed altogether eight different transporters as putative cysteine exporters (Yamada et al. 2006). The sensitivity of an *E. coli* strain carrying a disrupted cysteine desulfhydrase gene, which leads to a block of cysteine degradation, was reversed by overexpression of these genes. Further transport assays indicated a major contribution of the Bcr protein, a member of the MFS family. One of the reasons for the difficulty in unequivocally assigning the correct export system in this case may be due to the fact that it is not yet clear whether cysteine itself or derivatives of it are transported across the plasma membrane, such as cystine or the condensation product with pyruvate, 2-methyl-2,4-thiazolidinecarboxylic acid (Dassler et al. 2000; Yamada et al. 2006).

4.2.5 Glutamate Export in *C. glutamicum*

Glutamate was the first amino acid to be produced by bacteria (Kinoshita et al. 1957), and it is by far the amino acid with the highest production capacity (Kimura 2005). It was also one of the first amino acids for which carrier-mediated export was demonstrated (Hoischen and Krämer 1989, 1990; Gutmann et al. 1992). Nevertheless, significant progress in defining the molecular basis for glutamate excretion has been achieved only very recently. In contrast to the excretion of other amino acids, such as lysine or isoleucine, glutamate excretion always seemed to be connected with a particular physiological situation called "overflow metabolism" (Tempest and Neijssel 1992), which describes surplus of energy, carbon and nitrogen in the presence of a particular limiting factor, leading in general to cessation of growth. A number of treatments have been developed in the course of the years, which ultimately lead to massive export of glutamate. It has to be taken into account that, even under normal physiological conditions, the cytoplasm of *C. glutamicum* shows a high steady-state concentration of glutamate, in the range of up to several 100 mM (Gutmann et al. 1992). Nevertheless, under these conditions, glutamate transport functions only in one direction, namely uptake, with extremely high efficiency of accumulation (Krämer and Lambert 1990), due to the presence of the GluABC uptake system (Kronemeyer et al. 1995).

The original concept explaining the physiological basis of glutamate excretion was the combination of two major aspects. On one hand, a strongly decreased or even missing activity of the Krebs cycle enzyme oxoglutarate dehydrogenase (OGDH) was assumed and the significance of this enzyme for glutamate production was shown by inactivation (Shiio et al. 1961; Kawahara et al. 1997). Later on, it turned out that this enzyme activity is rather difficult to measure in vitro because of its inherent instability (Kimura 2005). Nevertheless, in production strains it was found that in fact the OGDH was severely impaired (Shingu and Terui 1972). The second aspect obviously related to glutamate excretion was an altered state of the cell envelope. Upon application of different, highly diverse kinds of treatment, which include, among others, (1) biotin limitation, (2) addition of particular detergents, (3) addition of antibiotic substances with different modes of action, such as penicillin, ethambutol and cerulenin, and (4) use of fatty acid or glycerol auxotrophs, changed metabolic conditions in *C. glutamicum* cause continuous and efficient efflux of glutamate without a loss of basic viability (Kimura 2005).

Obviously, all these treatments are, in one way or the other, connected to the integrity of the cell envelope, either affecting directly the state of the plasma membrane and/or that of the cell wall. This has led to the suggestion that primarily the permeability properties of the cell wall may be responsible for glutamate efflux (Eggeling et al. 2001; reviewed in Kimura 2005), which, in the case of Corynebacteria, in fact, represents a second permeability barrier in addition to that of the plasma membrane. This concept, however, was challenged by the observation that at the same time, uptake of glutamate, which has to cross the same permeability barrier, is not increased after treatments leading to an increased efflux (Burkovski and Krämer 2002).

Interestingly, in a recent publication, the first hint to the molecular basis of a possible connection between OGDH activity and glutamate excretion was found (Niebisch et al. 2006). A novel mechanism for the regulation of this enzyme was identified, including a 15 kDa protein OdhI, which inhibits the OGDH when it is present in its unphosphorylated form, as well as the action of a soluble Ser/Thr protein kinase PknG on the OdhI protein. Thus, OdhI and PknG are interesting putative candidates for a regulatory linkage between treatments leading to glutamate excretion and metabolic responses.

According to results described in a recently published patent, it seems that the central question on the mechanism of glutamate excretion in *C. glutamicum* has been solved (Nakamura et al. 2006; Nakamura et al., submitted for publication). Based on the observation that glutamate producing strains of *C. glutamicum* in general carry a defective OGDH, it has been demonstrated that strains in which solely this enzyme was mutated did in fact excrete glutamate, however, at relatively low level. These strains now turned out to be genetically unstable, and secondary mutations were found triggering much higher glutamate excretion. When these mutations were analyzed, repeatedly a mutated *yggB* gene was identified, which could be interpreted in terms of an altered activity regulation or an altered specificity. The YggB protein in *C. glutamicum* has previously been described as a putative mechanosensitive channel (Nottebrock et al. 2003), similar to the *E. coli* MscS system (Levina et al. 1999). It was shown that point mutations in the *yggB* gene for modification of particular domains as well as overexpression led, even in the absence of an altered OGDH, to significantly enhanced glutamate excretion, which could further be stimulated by the well known membrane triggers, e.g., addition of tween or biotin limitation. Deletion of the *yggB* gene, on the other hand, reduced glutamate efflux both in the presence and absence of the triggers mentioned above. A connection between osmotic stress and glutamate excretion has in fact been noted earlier (Lambert et al. 1995). It has to be elucidated in the future (1) whether the main function of the *C. glutamicum* YggB protein is, in fact, glutamate excretion or rather unspecific solute export in case of hypoosmotic stress, and (2) how the known mechanisms triggering glutamate excretion are related to YggB function.

4.2.6 Further Amino Acid Excretion Systems

On the one hand, it is known that those amino acids, which are currently produced in *C. glutamicum*, can in general be produced in *E. coli*, too, and the other way round. Threonine, for example, is currently produced in *E. coli* strains, and methionine can also be produced in the same organism, (Bestel-Core et al. 2005), but, in contrast to *C. glutamicum*, the responsible export carriers are not clearly defined (threonine) or not known (methionine). *E. coli* is also able to excrete glutamate under particular conditions (Broda 1968), however, a glutamate export system has not been identified so far. Tryptophan can be produced by both organisms (Berry 1996; Ikeda 2005), but there is no clear evidence for a carrier-mediated export so far, although the *yddG* gene product has been related to tryptophan excretion in *E. coli* (Ryback et al. 2006). Whereas it seems plausible that phenylalanine may cross the membrane barrier effectively by passive diffusion, due to its hydrophobic nature, this is not very probable for the more polar tyrosine and tryptophan thus asking for the presence of efflux systems in these cases.

There are further observations of amino acid efflux, which are not related to known amino acid export systems. A general observation under a number of metabolic conditions is export of alanine, which is frequently found in significant concentrations in the medium, and the same is true for glycine. At least the former amino acid is supposed to have a significant rate of passive, diffusion-controlled permeability (Ruhrmann et al. 1994). Another example is proline, which is found to be excreted in *C. glutamicum* under situations when mechanosensitive efflux channels are not supposed to be active (Krämer, unpublished observations). Finally, in view of the rather unexpected mechanism suggested to be valid for glutamate efflux in *C. glutamicum* and in view of a significant residual efflux activity even in the absence of YggB, further mechanisms and/or carriers responsible for this glutamate efflux may be present.

5 Conclusions and Perspectives

Bacterial amino acid transport is a potentially important factor in biotechnological amino acid production. At least three different transport reactions, namely carbon and nitrogen substrate uptake, product excretion and product re-uptake, contribute to the overall net flux starting with the substrate removed from the surrounding medium by the production organism and ending with the product accumulating in the external medium. Even in the best studied bacterial organisms of biotechnological relevance, namely *E. coli* and *C. glutamicum*, a significant part of putative amino acid transport systems has not yet been identified and characterized. Detailed biochemical characterization is of particular significance in the case of transport systems because of the notoriously unreliable assignment of transport substrate specificity based on sequence similarity. These tasks, e.g., identification of substrate specificity, transport mechanism, as well as integration into regulatory networks, will still require extensive effort in terms of biochemical analysis in the future. The final goal for biotechnological application will be the availability of various kinds of transport systems with various kinds of mechanisms (e.g., energy dependence) and specificity (e.g., discrimination of major and minor substrates), in order to improve the number of valuable tools for rational strain design.

Acknowledgements We are grateful for information and help from Lothar Eggeling and Jörn Kalinowski during preparation of this review. Thanks to Jun Nakamura for information prior to publication. The contribution of Muhammad Ashfaq in establishing our in house data bank on *C. glutamicum* genes and proteins is gratefully acknowledged.

References

Abe K, Ohnishi F, Yagi K, Nakajima T, Higuchi T, Sano M, Machida M, Sarker RI, Maloney PC (2002) Plasmid-encoded asp operon confers a proton motive metabolic cycle catalyzed by an aspartate-alanine exchange reaction. J Bacteriol 184:2906–2913

Abramson J, Smirnova I, Kasho V, Verner G, Kaback HR, Iwata S (2003) Structure and mechanism of the lactose permease of *Escherichia coli*. Science 301:610–615

Adams MD, Wagner LM, Graddis TJ, Landick R, Antonucci TK, Gibson AL, Oxender DL (1990) Nucleotide sequence and genetic characterization reveal six essential genes for the LIV-I and LS transport systems of *Escherichia coli*. J Biol Chem 265:11436–11443

Aleshin VV, Zakataeva NP, Livshits VA (1999) A new family of amino-acid-efflux proteins. Trends Biochem Sci 24:133–135

Alvarez-Jacobs J, de la Garza M, Ortega MV (1986) Biochemical and genetic characterization of L-glutamate transport and utilization in *Salmonella typhimurium* LT-2 mutants. Biochem Genet 24:195–205

Andresen PA, Kaasen I, Styrvold OB, Boulnois G, Strom AR (1988) Molecular cloning, physical mapping and expression of the bet genes governing the osmoregulatory choline-glycine betaine pathway of *Escherichia coli*. J Gen Microbiol 134:1737–1746

Androutsellis-Theotokis A, Goldberg NR, Ueda K, Beppu T, Beckman ML, Das S, Javitch JA, Rudnick G (2003) Characterization of a functional bacterial homologue of sodium-dependent neurotransmitter transporters. J Biol Chem 278:12703–12709

Beckers G, Bendt AK, Krämer R, Burkovski A (2004) Molecular identification of the urea uptake system and transcriptional analysis of urea transporter- and urease-encoding genes in *Corynebacterium glutamicum*. J Bacteriol 186:7645–7652

Belitsky BR, Gustafsson MC, Sonenshein AL, Von Wachenfeldt C (1997) An *lrp*-like gene of *Bacillus subtilis* involved in branched-chain amino acid transport. J Bacteriol 179:5448–5457

Bellmann A, Vrljic M, Patek M, Sahm H, Krämer R, Eggeling L (2001) Expression control and specificity of the basic amino acid exporter LysE of *Corynebacterium glutamicum*. Microbiology 147:1765–1774

Berger EA, Heppel LA (1972) A binding protein involved in the transport of cystine and diaminopimelic acid in *Escherichia coli.* J Biol Chem 247:7684–7694
Berry A (1996) Improving production of aromatic compounds in *Escherichia coli* by metabolic engineering. Trends Biotechnol 14:250–256
Bestel-Core G, Chateau M, Figge RM, Raynaud C, Soucaille PNP (2005) WO 2005-108561
Blattner FR, Plunkett G III, Bloch CA, Perna NT, Burland V, Riley M, Collado-Vides J, Glasner JD, Rode CK, Mayhew GF, Gregor J, Davis NW, Kirkpatrick HA, Goeden MA, Rose DJ, Mau B, Shao Y (1997) The complete genome sequence of *Escherichia coli* K-12. Science 277:1453–1474
Broda P (1968) Ribonucleic acid synthesis and glutamate excretion in *Escherichia coli.* J Bacteriol 96:1528–1534
Bröer S, Krämer R (1991a) Lysine excretion by *Corynebacterium glutamicum.* 1. Identification of a specific secretion carrier system. Eur J Biochem 202:131–135
Bröer S, Krämer R (1991b) Lysine excretion by *Corynebacterium glutamicum.* 2. Energetics and mechanism of the transport system. Eur J Biochem 202:137–143
Burkovski A, Krämer R (1995) Functional expression of the glutamate uptake system from *Corynebacterium glutamicum* in *Escherichia coli.* FEMS Microbiol Lett 127:263–266
Burkovski A, Krämer R (2002) Bacterial amino acid transport proteins: occurrence, functions, and significance for biotechnological applications. Appl Microbiol Biotechnol 58:265–274
Busch W, Saier MH Jr (2002) The transporter classification (TC) system, 2002. Crit Rev Biochem Mol Biol 37:287–337
Busch W, Saier MH Jr (2004) The IUBMB-endorsed transporter classification system. Mol Biotechnol 27:253–262
Cabrera-Martinez RM, Tovar-Rojo F, Vepachedu VR, Setlow P (2003) Effects of overexpression of nutrient receptors on germination of spores of *Bacillus subtilis.* J Bacteriol 185:2457–2464
Chang G, Roth CB (2001) Structure of MsbA from *E. coli*: a homolog of the multidrug resistance ATP binding cassette (ABC) transporters. Science 293:1793–1800
Checroun C, Gutierrez C (2004) Sigma(s)-dependent regulation of *yehZYXW*, which encodes a putative osmoprotectant ABC transporter of *Escherichia coli.* FEMS Microbiol Lett 236:221–226
Chye ML, Guest JR, Pittard J (1986) Cloning of the *aroP* gene and identification of its product in *Escherichia coli* K-12. J Bacteriol 167:749–753
Culham DE, Dalgado C, Gyles CL, Mamelak D, MacLellan S, Wood JM (1998) Osmoregulatory transporter ProP influences colonization of the urinary tract by *Escherichia coli.* Microbiology 144:91–102
Culham DE, Lasby B, Marangoni AG, Milner JL, Steer BA, van Nues RW, Wood JM (1993) Isolation and sequencing of *Escherichia coli* gene *proP* reveals unusual structural features of the osmoregulatory proline/betaine transporter, ProP. J Mol Biol 229:268–276
Dassler T, Maier T, Winterhalter C, Böck A (2000) Identification of a major facilitator protein from *Escherichia coli* involved in efflux of metabolites of the cysteine pathway. Mol Microbiol 36:1101–1112
Dattananda CS, Gowrishankar J (1989) Osmoregulation in *Escherichia coli*: complementation analysis and gene-protein relationships in the *proU* locus. J Bacteriol 171:1915–1922
Dawson RJ, Locher KP (2006) Structure of a bacterial multidrug ABC transporter. Nature 443:180–185
Deguchi Y, Yamato I, Anraku Y (1989) Molecular cloning of *gltS* and *gltP*, which encode glutamate carriers of *Escherichia coli* B. J Bacteriol 171:1314–1319

Ebbighausen H, Weil B, Krämer R (1989) Transport of branched-chain amino acids in *Corynebacterium glutamicum*. Arch Microbiol 151:238–244

Eggeling L (2005) Export of amino acids and other solutes. In: Eggeling L, Bott M (eds) Handbook of *Corynebacterium glutamicum*. Taylor & Francis, Boca Raton, pp 187–214

Eggeling L, Krumbach K, Sahm H (2001) L-glutamate efflux with *Corynebacterium glutamicum*: why is penicillin treatment or Tween addition doing the same? J Mol Microbiol Biotechnol 3:67–68

Eggeling L, Sahm H (2003) New ubiquitous translocators: amino acid export by *Corynebacterium glutamicum* and *Escherichia coli*. Arch Microbiol 180:155–160

Engel P, Krämer R, Unden G (1994) Transport of C4-dicarboxylates by anaerobically grown *Escherichia coli*. Energetics and mechanism of exchange, uptake and efflux. Eur J Biochem 222:605–614

Essenberg RC (1984) Use of homocysteic acid for selecting mutants at the *gltS* locus of *Escherichia coli* K12. J Gen Microbiol 130:1311–1314

Franke I, Resch A, Dassler T, Maier T, Böck A (2003) YfiK from *Escherichia coli* promotes export of O-acetylserine and cysteine. J Bacteriol 185:1161–1166

Gimenez R, Nunez MF, Badia J, Aguilar J, Baldoma L (2003) The gene *yjcG*, cotranscribed with the gene *acs*, encodes an acetate permease in *Escherichia coli*. J Bacteriol 185:6448–6455

Gong S, Richard H, Foster JW (2003) YjdE (AdiC) is the arginine:agmatine antiporter essential for arginine-dependent acid resistance in *Escherichia coli*. J Bacteriol 185:4402–4409

Guardiola J, De Felice M, Klopotowski T, Iaccarino M (1974) Mutations affecting the different transport systems for isoleucine, leucine, and valine in *Escherichia coli* K-12. J Bacteriol 117:393–405

Gunji Y, Yasueda H (2006) Enhancement of l-lysine production in methylotroph *Methylophilus methylotrophus* by introducing a mutant LysE exporter. J Biotechnol 127:1–13

Gutmann M, Hoischen C, Krämer R (1992) Carrier-mediated glutamate secretion by *Corynebacterium glutamicum* under biotin limitation. Biochim Biophys Acta 1112:115–123

Heatwole VM, Somerville RL (1991) Cloning, nucleotide sequence, and characterization of *mtr*, the structural gene for a tryptophan-specific permease of *Escherichia coli* K-12. J Bacteriol 173:108–115

Hermann T, Krämer R (1996) Mechanism and Regulation of Isoleucine Excretion in *Corynebacterium glutamicum*. Appl Environ Microbiol 62:3238–3244

Hirai T, Heymann JA, Maloney PC, Subramaniam S (2003) Structural model for 12-helix transporters belonging to the major facilitator superfamily. J Bacteriol 185:1712–1718

Hirokawa T, Boon-Chieng S, Mitaku S (1998) SOSUI: classification and secondary structure prediction system for membrane proteins. Bioinformatics 14:378–379

Hoischen C, Krämer R (1989) Evidence for an efflux carrier system involved in the secretion of glutamate by *Corynebacterium glutamicum*. Arch Microbiol 151:342–347

Hoischen C, Krämer R (1990) Membrane alteration is necessary but not sufficient for effective glutamate secretion in *Corynebacterium glutamicum*. J Bacteriol 172:3409–3416

Huang Y, Lemieux MJ, Song J, Auer M, Wang DN (2003) Structure and mechanism of the glycerol-3-phosphate transporter from *Escherichia coli*. Science 301:616–620

Ikeda M (2005) L-Tryptophan production. In: Eggeling L, Bott M (eds) Handbook of *Corynebacterium glutamicum*. Taylor & Francis, Boca Raton, pp 489–510

Ikeda M, Katsumata R (1995) Tryptophan production by transport mutants of *Corynebacterium glutamicum*. Biosci Biotechnol Biochem 59:1600–1602

Itou H, Okada U, Suzuki H, Yao M, Wachi M, Watanabe N, Tanaka I (2005) The CGL2612 protein from Corynebacterium glutamicum is a drug resistance-related transcriptional repressor: structural and functional analysis of a newly identified transcription factor from genomic DNA analysis. J Biol Chem 280:38711–38719

Jack DL, Paulsen IT, Saier MH (2000) The amino acid/polyamine/organocation (APC) superfamily of transporters specific for amino acids, polyamines and organocations. Microbiology 146:1797–814

Jackowski S, Alix JH (1990) Cloning, sequence, and expression of the pantothenate permease (*panF*) gene of *Escherichia coli.* J Bacteriol 172:3842–3848

Jacobs MH, van der Heide T, Driessen AJ, Konings WN (1996) Glutamate transport in *Rhodobacter sphaeroides* is mediated by a novel binding protein-dependent secondary transport system. Proc Natl Acad Sci USA 93:12786–12790

Jager W, Kalinowski J, Pühler A (1997) A *Corynebacterium glutamicum* gene conferring multidrug resistance in the heterologous host *Escherichia coli.* J Bacteriol 179:2449–2451

Jung H (1998) Topology and function of the Na+/proline transporter of *Escherichia coli,* a member of the Na+/solute cotransporter family. Biochim Biophys Acta 1365:60–64

Jung H (2001) Towards the molecular mechanism of Na(+)/solute symport in prokaryotes. Biochim Biophys Acta 1505:131–143

Jung H, Buchholz M, Clausen J, Nietschke M, Revermann A, Schmid R, Jung K (2002) CaiT of *Escherichia coli,* a new transporter catalyzing L-carnitine/gamma -butyrobetaine exchange. J Biol Chem 277:39251–39258

Kalinowski J, Bathe B, Bartels D, Bischoff N, Bott M, Burkovski A, Dusch N, Eggeling L, Eikmanns BJ, Gaigalat L, Goesmann A, Hartmann M, Huthmacher K, Krämer R, Linke B, McHardy AC, Meyer F, Mockel B, Pfefferle W, Pühler A, Rey DA, Rückert C, Rupp O, Sahm H, Wendisch VF, Wiegrabe I, Tauch A (2003) The complete *Corynebacterium glutamicum* ATCC 13032 genome sequence and its impact on the production of L-aspartate-derived amino acids and vitamins. J Biotechnol 104:5–25

Käll L, Krogh A, Sonnhammer EL (2004) A combined transmembrane topology and signal peptide prediction method. J Mol Biol 338:1027–1036

Kanamori M, Kamata H, Yagisawa H, Hirata H (1999) Overexpression of the alanine carrier protein gene from thermophilic bacterium PS3 in *Escherichia coli.* J Biochem (Tokyo) 125:454–9

Kashiwagi K, Miyamoto S, Suzuki F, Kobayashi H, Igarashi K (1992) Excretion of putrescine by the putrescine-ornithine antiporter encoded by the *potE* gene of *Escherichia coli.* Proc Natl Acad Sci USA 89:4529–4533

Kawahara Y, Takahashi-Fuke K, Shimizu E, Nakamatsu T, Nakamori S (1997) Relationship between the glutamate production and the activity of 2-oxoglutarate dehydrogenase in *Brevibacterium lactofermentum.* Biosci Biotechnol Biochem 61:1109–1112

Kayahara T, Thelen P, Ogawa W, Inaba K, Tsuda M, Goldberg EB, Tsuchiya T (1992) Properties of recombinant cells capable of growing on serine without NhaB Na+/H+ antiporter in *Escherichia coli.* J Bacteriol 174:7482–7485

Kelle R, Herrmann T, Bathe B (2005) L-Lysine Production. In: Eggeling L, Bott M (eds) Handbook of *Corynebacterium glutamicum.* Taylor & Francis, Heidelberg, pp 465–488

Kennerknecht N, Sahm H, Yen MR, Patek M, Saier MH Jr, Eggeling L (2002) Export of L-isoleucine from *Corynebacterium glutamicum*: a two-gene-encoded member of a new translocator family. J Bacteriol 184:3947–3956

Kim YM, Ogawa W, Tamai E, Kuroda T, Mizushima T, Tsuchiya T (2002) Purification, reconstitution, and characterization of Na(+)/serine symporter, SstT, of *Escherichia coli.* J Biochem (Tokyo) 132:71–76

Kimura E (2005) L-Glutamate production. In: Eggeling L, Bott M (eds) Handbook of *Corynebacterium glutamicum*. Taylor & Francis, Boca Raton, pp 439–464

Kinoshita K, Udata S, Shimono M (1957) Studies on the amino acid fermentation. I. Production of L-glutamic acid by various microorganisms. J Gen Appl Microbiol 7:193–205

Koch DJ, Rückert C, Rey DA, Mix A, Pühler A, Kalinowski J (2005) Role of the *ssu* and *seu* genes of *Corynebacterium glutamicum* ATCC 13032 in utilization of sulfonates and sulfonate esters as sulfur sources. Appl Environ Microbiol 71:6104–6114

Krämer R (1994) Secretion of amino acids by bacteria: Physiology and mechanism. FEMS Microbiol Rev 13:75–94

Krämer R (1996) Genetic and physiological approaches for the production of amino acids. J Biotechnol 45:1–21

Krämer R, Hoischen C (1994) Futile cycling caused by the simultaneous presence of separate transport systems for uptake and secretion of amino acids in *Corynebacterium glutamicum*. In: Westerhoff H (eds) Biothermokinetics. Intercept Publ., Amsterdam, pp 19–26

Krämer R, Lambert C (1990) Uptake of glutamate in *Corynebacterium glutamicum*. 2. Evidence for a primary active transport system. Eur J Biochem 194:937–944

Krogh A, Larsson B, von Heijne G, Sonnhammer EL (2001) Predicting transmembrane protein topology with a hidden Markov model: application to complete genomes. J Mol Biol 305:567–580

Kronemeyer W, Peekhaus N, Krämer R, Sahm H, Eggeling L (1995) Structure of the *gluABCD* cluster encoding the glutamate uptake system of *Corynebacterium glutamicum*. J Bacteriol 177:1152–1158

Kruse D, Krämer R, Eggeling L, Rieping M, Pfefferle W, Tchieu JH, Chung YJ, Saier MH Jr, Burkovski A (2002) Influence of threonine exporters on threonine production in *Escherichia coli*. Appl Microbiol Biotechnol 59:205–210

Kutukova EA, Livshits VA, Altman IP, Ptitsyn LR, Zyiatdinov MH, Tokmakova IL, Zakataeva NP (2005) The *yeaS* (*leuE*) gene of *Escherichia coli* encodes an exporter of leucine, and the Lrp protein regulates its expression. FEBS Lett 579:4629–4634

Lambert C, Erdmann A, Eikmanns M, Krämer R (1995) Triggering Glutamate Excretion in *Corynebacterium glutamicum* by Modulating the Membrane State with Local Anesthetics and Osmotic Gradients. Appl Environ Microbiol 61:4334–4342

Lee M, Robbins JC, Oxender DL (1975) Transport properties of merodiploids covering the *dagA* locus in *Escherichia coli* K-12. J Bacteriol 122:1001–1005

Leive L, Davis BD (1965) The transport of diaminopimelate and cystine in *Escherichia coli*. J Biol Chem 240:4362–4369

Leuchtenberger W, Huthmacher K, Drauz K (2005) Biotechnological production of amino acids and derivatives: current status and prospects. Appl Microbiol Biotechnol 69:1–8

Levina N, Totemeyer S, Stokes NR, Louis P, Jones MA, Booth IR (1999) Protection of *Escherichia coli* cells against extreme turgor by activation of MscS and MscL mechanosensitive channels: identification of genes required for MscS activity. Embo J 18:1730–1737

Liu CE, Ames GF (1997) Characterization of transport through the periplasmic histidine permease using proteoliposomes reconstituted by dialysis. J Biol Chem 272:859–866

Livshits VA, Zakataeva NP, Aleshin VV, Vitushkina MV (2003) Identification and characterization of the new gene *rhtA* involved in threonine and homoserine efflux in *Escherichia coli*. Res Microbiol 154:123–135

Lo TC, Bewick MA (1978) The molecular mechanisms of dicarboxylic acid transport in *Escherichia coli* K12. The role and orientation of the two membrane-bound dicarboxylate binding proteins. J Biol Chem 253:7826–7831

Locher KP, Lee AT, Rees DC (2002) The *E. coli* BtuCD structure: a framework for ABC transporter architecture and mechanism. Science 296:1091–1098

MacLeod PR, MacLeod RA (1992) Identification and sequence of a Na(+)-linked gene from the marine bacterium *Alteromonas haloplanktis* which functionally complements the *dagA* gene of *Escherichia coli*. Mol Microbiol 6:2673–2681

Marcus M, Halpern YS (1969) Genetic analysis of the glutamate permease in *Escherichia coli* K-12. J Bacteriol 97:1118–1128

May G, Faatz E, Villarejo M, Bremer E (1986) Binding protein dependent transport of glycine betaine and its osmotic regulation in *Escherichia coli* K12. Mol Gen Genet 205:225–233

Meng SY, Bennett GN (1992) Nucleotide sequence of the *Escherichia coli cad* operon: a system for neutralization of low extracellular pH. J Bacteriol 174:2659–2669

Merkens H, Beckers G, Wirtz A, Burkovski A (2005) Vanillate metabolism in *Corynebacterium glutamicum*. Curr Microbiol 51:59–65

Merlin C, Gardiner G, Durand S, Masters M (2002) The *Escherichia coli metD* locus encodes an ABC transporter which includes Abc (MetN), YaeE (MetI), and YaeC (MetQ). J Bacteriol 184:5513–5517

Metzer E, Halpern YS (1990) In vivo cloning and characterization of the *gabCTDP* gene cluster of *Escherichia coli* K-12. J Bacteriol 172:3250–3256

Moore BC, Leigh JA (2005) Markerless mutagenesis in *Methanococcus maripaludis* demonstrates roles for alanine dehydrogenase, alanine racemase, and alanine permease. J Bacteriol 187:972–979

Morbach S, Krämer R (2003) Impact of transport processes in the osmotic response of *Corynebacterium glutamicum*. J Biotechnol 104:69–75

Morrow JA, Collie IT, Dunbar DR, Walker GB, Shahid M, Hill DR (1998) Molecular cloning and functional expression of the human glycine transporter GlyT2 and chromosomal localisation of the gene in the human genome. FEBS Lett 439:334–340

Müller A, Thomas GH, Horler R, Brannigan JA, Blagova E, Levdikov VM, Fogg MJ, Wilson KS, Wilkinson AJ (2005) An ATP-binding cassette-type cysteine transporter in *Campylobacter jejuni* inferred from the structure of an extracytoplasmic solute receptor protein. Mol Microbiol 57:143–155

Nakamura J, Hirano S, Ito H (2006) US 2006/0141588

Nakayama K (1985) Lysine. In: Moo-Young M (eds) Comprehensive Biotechnology, vol 3. Pergamon Press, New York, pp 607–620

Nandineni MR, Gowrishankar J (2004) Evidence for an arginine exporter encoded by *yggA* (*argO*) that is regulated by the LysR-type transcriptional regulator ArgP in *Escherichia coli*. J Bacteriol 186:3539–3546

Niebisch A, Kabus A, Schultz C, Weil B, Bott M (2006) Corynebacterial protein kinase G controls 2-oxoglutarate dehydrogenase activity via the phosphorylation status of the OdhI protein. J Biol Chem 281:12300–12307

Nohno T, Saito T, Hong JS (1986) Cloning and complete nucleotide sequence of the *Escherichia coli* glutamine permease operon (*glnHPQ*). Mol Gen Genet 205:260–269

Nottebrock D, Meyer U, Krämer R, Morbach S (2003) Molecular and biochemical characterization of mechanosensitive channels in *Corynebacterium glutamicum*. FEMS Microbiol Lett 218:305–309

Ogawa W, Kayahara T, Tsuda M, Mizushima T, Tsuchiya T (1997) Isolation and characterization of an *Escherichia coli* mutant lacking the major serine transporter, and cloning of a serine transporter gene. J Biochem (Tokyo) 122:1241–1245

Oh BH, Pandit J, Kang CH, Nikaido K, Gokcen S, Ames GF, Kim SH (1993) Three-dimensional structures of the periplasmic lysine/arginine/ornithine-binding protein with and without a ligand. J Biol Chem 268:11348–11355

Okamoto K, Kino K, Ikeda M (1997) Hyperproduction of L-threonine by an *Escherichia coli* mutant with impaired L-threonine uptake. Biosci Biotechnol Biochem 61:1877–1882

Ovchinnikov YA, Aldanova NA, Grinkevich VA, Arzamazova NM, Moroz IN (1977) The primary structure of a Leu, Ile and Val (LIV)-binding protein from *Escherichia coli*. FEBS Lett 78:313–316

Oxender DL, Anderson JJ, Daniels CJ, Landick R, Gunsalus RP, Zurawski G, Selker E, Yanofsky C (1980) Structural and functional analysis of cloned DNA containing genes responsible for branched-chain amino acid transport in *Escherichia coli*. Proc Natl Acad Sci USA 77:1412–1416

Palmieri L, Berns D, Krämer R, Eikmanns B (1996) Threonine diffusion and threonine transport in *Corynebacterium glutamicum* and their role in threonine production. Arch Microbiol 165:48–54

Peter H, Bader A, Burkovski A, Lambert C, Krämer R (1997) Isolation of the *putP* gene of *Corynebacterium glutamicum* and characterization of a low-affinity uptake system for compatible solutes. Arch Microbiol 168:143–151

Peter H, Burkovski A, Krämer R (1996) Isolation, characterization, and expression of the *Corynebacterium glutamicum betP* gene, encoding the transport system for the compatible solute glycine betaine. J Bacteriol 178:5229–5234

Peter H, Weil B, Burkovski A, Krämer R, Morbach S (1998) *Corynebacterium glutamicum* is equipped with four secondary carriers for compatible solutes: identification, sequencing, and characterization of the proline/ectoine uptake system, ProP, and the ectoine/proline/glycine betaine carrier, EctP. J Bacteriol 180:6005–6012

Pi J, Wookey PJ, Pittard AJ (1991) Cloning and sequencing of the *pheP* gene, which encodes the phenylalanine-specific transport system of *Escherichia coli*. J Bacteriol 173:3622–3629

Pittman MS, Corker H, Wu G, Binet MB, Moir AJ, Poole RK (2002) Cysteine is exported from the *Escherichia coli* cytoplasm by CydDC, an ATP-binding cassette-type transporter required for cytochrome assembly. J Biol Chem 277:49841–49849

Ren Q, Paulsen IT (2005) Comparative analyses of fundamental differences in membrane transport capabilities in prokaryotes and eukaryotes. PLoS Comput Biol 1:e27

Rübenhagen R, Rönsch H, Jung H, Krämer R, Morbach S (2000) Osmosensor and osmoregulator properties of the betaine carrier BetP from *Corynebacterium glutamicum* in proteoliposomes. J Biol Chem 275:735–741

Ruhrmann J, Sprenger GA, Krämer R (1994) Mechanism of alanine excretion in recombinant strains of *Zymomonas mobilis*. Biochim Biophys Acta 1196:14–20

Ryback KV, Slivinskaja EA, Kozlov YI, Suzuki T (2006) WO 2006-068273

Sack JS, Saper MA, Quiocho FA (1989) Periplasmic binding protein structure and function. Refined X-ray structures of the leucine/isoleucine/valine-binding protein and its complex with leucine. J Mol Biol 206:171–191

Saier MH Jr (2000) Families of transmembrane transporters selective for amino acids and their derivatives. Microbiology 146:1775–1795

Saier MH Jr, Tran CV, Barabote RD (2006) TCDB: the Transporter Classification Database for membrane transport protein analyses and information. Nucleic Acids Res 34:D181–D186

Sarsero JP, Wookey PJ, Gollnick P, Yanofsky C, Pittard AJ (1991) A new family of integral membrane proteins involved in transport of aromatic amino acids in *Escherichia coli*. J Bacteriol 173:3231–3234

Schneider F, Krämer R, Burkovski A (2004) Identification and characterization of the main beta-alanine uptake system in *Escherichia coli*. Appl Microbiol Biotechnol 65:576–582

Seep-Feldhaus AH, Kalinowski J, Pühler A (1991) Molecular analysis of the *Corynebacterium glutamicum lysl* gene involved in lysine uptake. Mol Microbiol 5:2995–3005

Shiio I, Ohtska S, Takahashi M (1961) OGDH Significance of α-ketoglutarate dehydrogenase on the glutamic acid formation in *Brevibacterium flavum*. J Biochem 50:164–165

Shingu H, Terui G (1972) Studies on the process of glutamic acid fermentation at the enzyme level: I. On the changes of a-ketoglutaric acid dehydrogenase in the course of culture. J Ferment Technol 49:400–405

Simic P, Sahm H, Eggeling L (2001) L-threonine export: use of peptides to identify a new translocator from *Corynebacterium glutamicum*. J Bacteriol 183:5317–5324

Simic P, Willuhn J, Sahm H, Eggeling L (2002) Identification of *glyA* (encoding serine hydroxymethyltransferase) and its use together with the exporter ThrE to increase L-threonine accumulation by *Corynebacterium glutamicum*. Appl Environ Microbiol 68:3321–3327

Steffes C, Ellis J, Wu J, Rosen BP (1992) The *lysP* gene encodes the lysine-specific permease. J Bacteriol 174:3242–3249

Steger R, Weinand M, Krämer R, Morbach S (2004) LcoP, an osmoregulated betaine/ectoine uptake system from *Corynebacterium glutamicum*. FEBS Lett 573:155–160

Sumantran VN, Schweizer HP, Datta P (1990) A novel membrane-associated threonine permease encoded by the *tdcC* gene of *Escherichia coli*. J Bacteriol 172:4288–4294

Sun YJ, Rose J, Wang BC, Hsiao CD (1998) The structure of glutamine-binding protein complexed with glutamine at 1.94 A resolution: comparisons with other amino acid binding proteins. J Mol Biol 278:219–229

Tauch A, Hermann T, Burkovski A, Krämer R, Pühler A, Kalinowski J (1998) Isoleucine uptake in *Corynebacterium glutamicum* ATCC 13032 is directed by the *brnQ* gene product. Arch Microbiol 169:303–312

Tempest DW, Neijssel OM (1992) Physiological and energetic aspects of bacterial metabolite overproduction. FEMS Microbiol Lett 79:169–176

Thanbichler M, Neuhierl B, Böck A (1999) S-methylmethionine metabolism in *Escherichia coli*. J Bacteriol 181:662–665

Trötschel C, Deutenberg D, Bathe B, Burkovski A, Krämer R (2005) Characterization of methionine export in *Corynebacterium glutamicum*. J Bacteriol 187:3786–3794

Trötschel C, Kandirali S, Diaz-Achirica P, Meinhardt A, Morbach S, Krämer R, Burkovski A (2003) GltS, the sodium-coupled L-glutamate uptake system of *Corynebacterium glutamicum*: identification of the corresponding gene and impact on L-glutamate production. Appl Microbiol Biotechnol 60:738–742

van der Heide T, Poolman B (2002) ABC transporters: one, two or four extracytoplasmic substrate-binding sites? EMBO Rep 3:938–943

van der Ploeg JR, Weiss MA, Saller E, Nashimoto H, Saito N, Kertesz MA, Leisinger T (1996) Identification of sulfate starvation-regulated genes in *Escherichia coli*: a gene cluster involved in the utilization of taurine as a sulfur source. J Bacteriol 178:5438–5446

Vardy E, Steiner-Mordoch S, Schuldiner S (2005) Characterization of bacterial drug antiporters homologous to mammalian neurotransmitter transporters. J Bacteriol 187:7518–7525

Vrljic M, Sahm H, Eggeling L (1996) A new type of transporter with a new type of cellular function: L-lysine export from *Corynebacterium glutamicum*. Mol Microbiol 22:815–826

Wehrmann A, Morakkabati S, Krämer R, Sahm H, Eggeling L (1995) Functional analysis of sequences adjacent to *dapE* of *Corynebacterium glutamicum* reveals the presence of *aroP*, which encodes the aromatic amino acid transporter. J Bacteriol 177:5991–5993

Whipp MJ, Halsall DM, Pittard AJ (1980) Isolation and characterization of an *Escherichia coli* K-12 mutant defective in tyrosine- and phenylalanine-specific transport systems. J Bacteriol 143:1–7

Willis RC, Furlong CE (1975) Interactions of a glutamate-aspartate binding protein with the glutamate transport system of *Escherichia coli*. J Biol Chem 250:2581–2586

Willis RC, Woolfolk CA (1975) L-asparagine uptake in *Escherichia coli*. J Bacteriol 123:937–945

Winnen B, Felce J, Saier MH Jr (2005) Genome analyses of transporter proteins in *Corynebacterium glutamicum* and *Corynebacterium efficiens*. In: Eggeling L, Bott M (eds) Handbook of *Corynebacterium glutamicum*. Taylor & Francis, Boca Raton, pp 149–186

Wissenbach U, Keck B, Unden G (1993) Physical map location of the new *artPIQMJ* genes of *Escherichia coli*, encoding a periplasmic arginine transport system. J Bacteriol 175:3687–3688

Wissenbach U, Six S, Bongaerts J, Ternes D, Steinwachs S, Unden G (1995) A third periplasmic transport system for L-arginine in *Escherichia coli*: molecular characterization of the *artPIQMJ* genes, arginine binding and transport. Mol Microbiol 17:675–686

Yamada S, Awano N, Inubushi K, Maeda E, Nakamori S, Nishino K, Yamaguchi A, Takagi H (2006) Effect of drug transporter genes on cysteine export and overproduction in *Escherichia coli*. Appl Environ Microbiol 72:4735–4742

Yamashita A, Singh SK, Kawate T, Jin Y, Gouaux E (2005) Crystal structure of a bacterial homologue of Na+/Cl–dependent neurotransmitter transporters. Nature 437:215–223

Yao N, Trakhanov S, Quiocho FA (1994) Refined 1.89-A structure of the histidine-binding protein complexed with histidine and its relationship with many other active transport/chemosensory proteins. Biochemistry 33:4769–4779

Yernool D, Boudker O, Jin Y, Gouaux E (2004) Structure of a glutamate transporter homologue from *Pyrococcus horikoshii*. Nature 431:811–818

Zakataeva NP, Aleshin VV, Tokmakova IL, Troshin PV, Livshits VA (1999) The novel transmembrane *Escherichia coli* proteins involved in the amino acid efflux. FEBS Lett 452:228–232

Zittrich S, Krämer R (1994) Quantitative discrimination of carrier-mediated excretion of isoleucine from uptake and diffusion in *Corynebacterium glutamicum*. J Bacteriol 176:6892–6899

Microbiol Monogr (5)
V. F. Wendisch: Amino Acid Biosynthesis
DOI 10.1007/7171_2006_063/Published online: 5 December 2006

Occurrence, Biosynthesis, and Biotechnological Production of Dipeptides

Shin-ichi Hashimoto

Kyowa Hakko Kogyo Co., Ltd., Technical Research Laboratories, 1-1 Kyowa-cho, 747-8522 Hofu-shi, Japan
shashimoto@kyowa.co.jp

Abstract l-α-Dipeptides are not mere mixtures of amino acids. Some of them have their own properties which are not found in the corresponding amino acids. Despite such versatility, dipeptides have been poorly recognized, mainly due to the lack of an efficient manufacturing method. Though a variety of dipeptides, many of which contain unusual amino acids or have a cyclic form, have been found as metabolites of microorganisms, their biosynthetic routes had remained unclear. However, recent studies have revealed the existence of several ribosome-independent machineries capable of synthesizing dipeptides, such as nonribosomal peptide synthetases or the new enzyme, l-amino acid α-ligase. By using these activities, highly efficient methods for producing dipeptides have been devised. In this review, known functions and occurrence of dipeptides are reviewed first since many readers may not be familiar with dipeptides. Then a few important studies on the biosynthesis of dipeptides are summarized followed by a description of the emerging technologies for dipeptide manufacturing based on the recent findings.

1 Introduction

l-α-Dipeptide is the simplest peptide consisting of two molecules of l-amino acids. Although biosynthesis of amino acids and their regulation have been revealed in detail (see other sections of this volume), there are only limited insights into the formation of dipeptides in vivo as well as their physiological functions. This may be attributed to the general idea that dipeptides are intermediates of protein degradation to amino acids. Every living cellular organism possesses some dipeptide degrading activity as a necessary step in protein turnover. The importance of this function may be reflected by the fact that every organism is equipped with many peptidases (eg., *Escherichia coli* has more than 25 peptidase genes).

Dipeptides are, however, not mere intermediates of protein degradation. Some dipeptides have their own properties, which overcome drawbacks of amino acids or which are not found in the corresponding amino acids. Enzymes that specifically synthesize dipeptides have recently also been found. In this review, a brief introduction to the functions of dipeptides, their occurrence and biosynthesis in microorganisms, and recent progress in the manufacturing of dipeptides will be described.

2 Function of Dipeptides

Apparently, dipeptides hold the physiological functions of the amino acids, which constitute the dipeptide, since they are broken into the amino acids in organisms. But the physicochemical properties are different between dipeptides and amino acids. For example, the solubility of Gln (amino acids are noted in three letters and mentioned as l-form unless otherwise stated) and Ala is 0.41 M and 1.75 M, respectively, whereas that of alanylglutamine (Ala-Gln: dipeptides are noted in this manner throughout this review) is much higher, 2.71 M. Also their stability in solution is different. These features led to the use of some dipeptides as substitutes for amino acids since they have better properties. For instance, Gln, a conditionally essential amino acid for humans, is heat labile in water while Ala-Gln is stable and acts as a Gln source (Stehle et al. 1984; Abumrad et al. 1989). Tyr is practically insoluble, but the dipeptide Gly-Tyr is soluble (Daabees and Stegink 1978). On the basis of these characteristics, these dipeptides are used as ingredients of infusions for patients.

Some dipeptides have specific physiological functions (Table 1). For example, kyotorphin (Arg-Tyr), isolated from bovine brain, has analgesic effects; carnosine (β-Ala-His) and anserine (β-Ala-methyl-His), widely distributed in muscle and brain of fish, bird, animals, and humans, have been

Table 1 Physiological activities of dipeptides

Compound	Source	Activity	Refs.
Carnosine (β-Ala-His), Anserine (β-Ala-metyl His)	muscle and brain of many vertebrates	maintenace of cellular pH, antioxidation, etc.	Gulewitsch and Amiradzibi 1900 Hines and Sutfin 1956, Guiotto et al. 2005, Begum et al. 2005, Bauer 2005
Various dipeptides	synthetic	tastes	Mazur et al. 1969, Schiffman 1976, de Armas et al. 2004
Aspartame (Asp-Phe methyl ester)	synthetic	sweetener	Cloninger and Baldwin 1970
Kyotorphin (Tyr-Arg)	mammalian brain	analgesic	Takagi et al.1979, Ueda et al. 1980
Ser-Tyr, Gly-Tyr, Phe-Tyr, Asn-Tyr,	extract of garlic,	ACE inhibitor	Suetsunna 1998,
Ser-Phe, Gly-Phe, Asn-Phe, etc.	synthetic		Opris and Diudea 2001
Lys-Glu	synthetic	antitumor	Khavinson and Anisimov 2000
Val-Tyr	synthetic hydrolysate of sardine muscle	antihypertensive	Matsui et al. 2003

reported to have specific functions such as maintenance of cellular pH and antioxidation. Furthermore, some dipeptides derived from foods are known to have antihypertensive effects. The taste of dipeptides has also been examined (Mazur et al. 1969; Schiffman 1976; de Armas et al. 2004). The most famous tasty dipeptide is a sweetener, aspartame (Asp-Phe methyl ester). In addition to these, several dipeptides derived from microorganisms are known to possess pharmacological activities (see the next section).

3 Dipeptides and Dipeptide Derivatives from Microorganisms

There are many reports on the production of dipeptide(s) or its derivative(s) by microorganisms. Linear dipeptides that originate from microorganisms are listed in Table 2. Many of them were found as antibiotics and contain unusual amino acids such as bacilysin (Ala-anticapsin) produced by *Bacillus subtilis* (Rogers et al. 1965a,b) and alahopsin (Ala-[(2S,3R)-2-amino-4-

Table 2 Linear dipeptides from microorganisms

Compound	Producer	Activity	Refs.
Pro-Leu, Pro-Val	*Aspergillus ochraceus*, *Oopsra destructor*	anti insect	Kodaira, 1961
Bacilycin (Ala-anticapsin)	*Bacillus subtilis*	antibiotic	Rogers et al. 1965a,b
Bestatin ([(2S,3R)-3-amino-2-hydroxyl-4-phenylbutanoic acid]-Leu)	*Streptomyces olivo-reticuli*	antitumor, amino-peptidase inhibitor	Umezawa et al. 1976, Suda et al. 1976, Umezawa, 1977
(2,6-diamino-6-hydroxymethyl pimeric acid)-Ala	*Micromonospora chalcea*	antibiotic	Shoji et al. 1981
Bu-2743E ([2,3-dihydroxy-benzoyl-l-Ala]-Thr)	*Bacillus circulans*	aminopeptidase inhibitor	Kobaru et al. 1983
FR900148 (Val-[3-chloro-3,4-dehydro-Glu])	*Streptomyces xanthocidicus*	antibiotic	Kuroda et al. 1983, Yasuda and Sakane, 1991
Acetyl-Leu-Argininal	Bacterium BMG520-yF2	antiplasmin	Nishikiori et al. 1984
Alahopcin (Ala-[(2S,3R)-2-amino-4-formyl-3-(hydroxy-amiocarbonyl) butyric acid])	*Streptomyces albulus* subsp. *Ochragerus* subsp. nov.	antibiotic	Higashide et al. 1985a,b, Horii et al. 1985
Ile-amiclenomycin, Methyl-Ile-amiclenomycin, Methyl-Val-amiclenomycin	*Streptomyces venezuelae* Tu 2460	antibiotic	Poetsch and Zahner, 1985
Ala-azaserine	*Glycomyces harbinensis*	antibiotic	Lee et al. 1987
Rhizocticin A (Arg-[l-2-amino-5-phosphono-3-pentenoic acid])	*Bacillus subtilis*	antibiotic	Rapp et al. 1988, Fredenhagen et al. 1995
Acetyl-Gln-Gln	*Rhizobium meliloti*	osmoprotectant	Smith and Smith 1989
[(2S,3R)-3-amino-2-hydroxyl-4-phenylbutanoic acid]-Val	*Streptomyces neyagawaensis*	antibiotic	Chung et al. 1996

Table 3 Cyclic dipeptides from microorganisms

Compound	Producer	Activity	Refs.
Cyclo(Pro-Leu)	*Rosellinia necatrix*, *Aspergillus fumigatus*	attenuation of morphine dependence	Chen, 1960, Prasad, 1995
Cyclo(Pro-Val)	*Rosellinia necatrix*, *Aspergillus ochraceus*, *Metharrhizum ansiopha*, *Streptomyces sp.*	inhibition of SV40-transformed cell growth	Chen, 1960, Munekata and Tamura, 1981, Prasad, 1995
Cyclo(Pro-Phe)	*Rosellinia necatrix*, *Lactobacillus plantarum*, *Lactobacillus coryniformis*	antifungal	Chen, 1960, Prasad, 1995, Schnuer and Magnusson, 2005
Cyclo(Phe-Phe)	*Penicillium nigricans*, *Streptomyces noursei*	no description on the activity	Birkinshaw and Mohammed, 1962, Prasad, 1995
Albonoursin [derivative of cyclo(Phe-Leu)]	*Streptomyces noursei*, *S. alburus*	antibiotic	Khokhlov and Lokshin, 1963, Prasad, 1995
Echinulin [derivative of cyclo(Ala-Trp)]	*Aspergillus chevalieri*	alkaloid	MacDonard and Slater, 1966, Allen, 1972
Cyclo(Ala-Leu)	*Aspergillus niger*	alkaloid	Caesar et al. 1969, Prasad, 1995
Rhodotorulic acid (cyclo dimmer of δ-*N*-acetyl-δ-*N*-hydroxyornitine)	*Rhodotorula pilimanae*, *Leucosporidium*, *Rhodosporidium*, *Sporidiobolus*	siderophore	Atkin et al. 1970, Prasad, 1995
Bicyclomycin	*Streptomyces sapporonennsis*	antibiotics	Miyoshi et al. 1972, Kamiya et al. 1972
Cyclo(Pro-Tyr)	*Alternaria alternate*, *Streptomyces* sp.	Phytotoxin, inhibition of SV40-transformed cell growth	Wilson et al. 1973, Chen, 1960, Prasad, 1995
Cyclo(Trp-[dehydro-α-amino butyric acid])	*Streptomyces spectabilis*	no description on the activity	Kakinuma and Rinehart, 1974
Derivative of cyclo(Ala-Trp)	*Aspergillus chevalieri*	no description on the activity	Hamasaki et al. 1976
Verruculogen, Tryprostatin	*Penicillium simplicissimum*, *Aspergillus fumigatus*	mycotoxin	Day and Mantle, 1982, Cui et al. 1996
Dimerumic acid (cyclo dimmer of *trans*-fusarinine)	*Histoplasma capsulatum*	siderophore	Burt, 1982, Prasad, 1995

Table 3 (continued)

Compound	Producer	Activity	Refs.
Thaxtomin [derivative of cyclo(Trp-Phe)]	*Streptomyces acidiscabies*	phytotoxin	King et al. 1989, 1992
Cyclo(Pro-Trp)	*Penicillium brevicompactum*	phytotoxin	Prasad, 1995
Cyclo(*N*-metyl-Trp)$_2$	*Streptomyces griseus*	calpain inhibitor	Alvarez et al. 1994
Cyclo(Phe-hydroxy Pro)	*Lactobacillus plantarum*, *Lactobacillus coryniformis*	antifungal activity	Schnurer and Magnusson, 2005
Cyclo(Gly-Leu)	*Lactobacillus plantarum*	antifungal activity	Schnurer and Magnusson, 2005
Cyclo(Leu-Pro)	*Achromobacter xylosoxidans*	inhibition of aflatoxin production	Yan et al. 2004

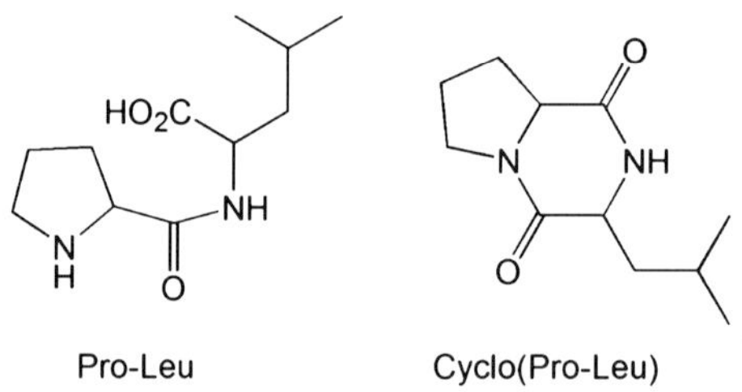

Fig. 1 The structures of linear and cyclic Pro-Leu

formyl-3-(hydroxyamiocarbonyl)butyric acid]) by a *Streptomyses* strain (Higashide et al. 1985ab; Horii et al. 1985). Also a dipeptide derivative, acetyl-Gln-Gln, has been described to play a role in osmoprotection in *Rhizobium meliloti* (Smith and Smith 1989).

Two molecules of amino acids can form a circular dipeptide (diketopiperazine) with two peptide bonds (for a reference, structures of Pro-Leu in the linear form and the cyclic form are shown in Fig. 1). Various cyclic dipeptides and their derivatives have been found in nature and are reported to exert physiological effects (for reviews see Prasad 1995, Schnurer and Magnusson 2005). The microorganism-originated ones are listed in Table 3. It should be noted that cyclodipeptides can be formed by heat treatment of linear dipeptides or by protease digestion of proteins (Prasada 1995), suggesting that some of the cyclodipeptides may not be direct metabolites of the organisms.

4 Biosynthesis of Dipeptides

Despite the relative abundance of reports on the production of dipeptide(s) by microorganisms, as shown in Tables 2 and 3, knowledge about their biosynthesis remains very limited. In this section, peptide bond-forming activities are reviewed first before examples of dipeptide synthesizing activities are summarized.

4.1 Activities and Mechanisms of Peptide-Bond Formation in Nature

Whereas the majority of peptide bond-forming reactions in organisms are conducted by the ribosome, which is present in all cellular organisms, there are many other enzymes that work on peptide-bond formation: (1) Nonribosomal peptide synthetases (NRPS) constitute a large enzyme family essential for the biosynthesis of many peptidyl secondary metabolites, such as penicillins, vancomycine, and gramicidin S (see review Byford et al. 1997; Marahiel et al. 1997; Finking and Marahiel 2004). (2) D-Alanine-D-alanine ligase is ubiquitously distributed among eubacteria and plays a role in the synthesis of the cell wall as well as other enzymes participating in peptidoglycan synthesis (Walsh 1989). (3) A tripeptide containing one γ-peptide bond, glutathione, is also known to be synthesized by a specific enzyme, glutathione synthetase (Li et al. 2004). (4) Several polymers are produced by specific enzymes, such as polyglutamate in polyglutamated folate by folylpolyglutamate synthetase (Sun et al. 1998) or the intracellular polymer of cyanobacteria, cyanophycin, by cyanophycin synthetase (Aboulmagd et al. 2000).

Although these activities are very diverse in their specificity and physiological function, all, including ribosome system, share one common feature: the requirement of ATP to catalyze the peptide bond-forming reaction. Theoretically, peptide bonds can be formed through the reverse reaction of peptidases or proteases as employed in aspartame production, but, in this review, these possibilities are not considered since unusual circumstances are required to precede the reverse reaction practically. While the requirement of ATP is common to the peptide-forming enzymes, they are divided into two categories based on their reaction intermediates, which can be either aminoacyl AMP or aminoacyl phosphate.

The representatives of the former are the ribosome (Fig. 2) and the NRPS (Fig. 3). NRPS is a large enzyme complex which is composed of catalytically independent modules and each module consists of a set of catalytic domains for the incorporation of one substrate amino acid into the nascent peptide chain. The adenylation domain (A domain) recognizes a specific amino acid and activates it by adenylation using ATP. The activated amino

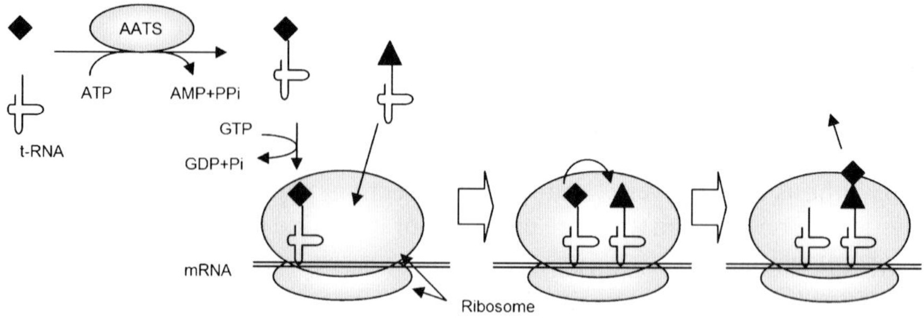

Fig. 2 Schematic diagram of peptide synthesis by the ribosomal system. *Closed diamonds and triangles* indicate amino acids. AATS: aminoacyl-t-RNA synthetase

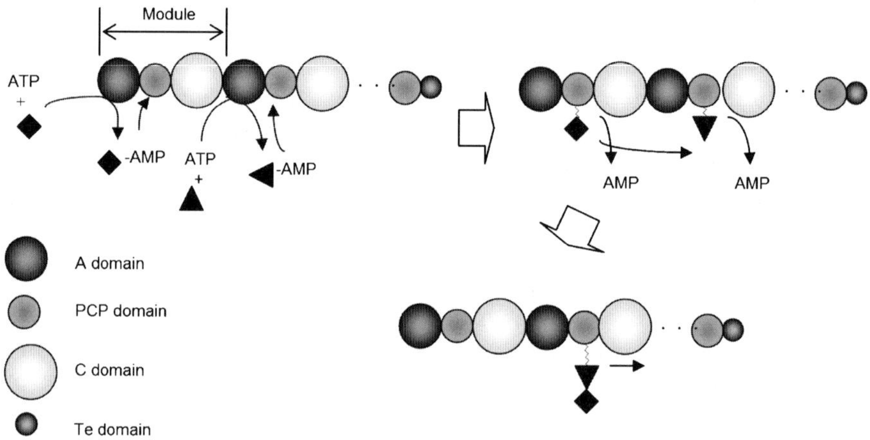

Fig. 3 Schematic diagram of peptide synthesis by a nonribosomal peptide synthetase. *Closed diamonds and triangles* indicate amino acids. A domain: adenylation domain, PCP domain: peptidyl carrier protein, C domain: condensation domain, Te domain: thioesterase-like domain. A, PCP, and C domains together make a module

acid is transferred to the peptidyl carrier protein (PCP) domain which mediates the transfer of the activated amino acid to the condensation domain. The condensation (C) domain catalyzes the peptide bond formation, and the thioesterase-like domain (Te domain) releases the synthesized peptide from the enzyme protein.

Enzymes generating aminoacyl phosphate as a reaction intermediate have a variety of activities; glutathione synthetase, d-alanine-d-alanine ligase, etc. These enzymes share a characteristic motif (ATP-grasp motif) in their amino-acid sequence and belong to ATP-dependent carboxylate-amine/thiol ligase superfamily (Evers et al. 1996; Galperin and Koonin 1997).

4.2
Dipeptide Synthesizing Activities

Knowledge on dipeptide biosynthesis has been scant and fragmental. However, recent studies have revealed that there are several ways to synthesize dipeptides that include novel enzymes.

Bacilysin This linear dipeptide produced by *B. sutilis* is the one whose biosynthesis has been most thoroughly investigated although the conclusive synthesis details are still absent. Biochemical analyses with cell free extracts revealed that bacilysin is synthesized from Ala and anticapsin in an ATP-dependent manner (Sakajoh et al. 1987). An Ala-dependent ATP-PPi exchange reaction was observed with a partially purified enzyme fraction, suggesting a NRPS-type mechanism (Yazgan et al. 2001). Genetic studies clarified that the *bac* operon is responsible for bacilysin synthesis and a gene in the operon (*bacABCDE* = *ywfBCDEF*), *bacD* (= *ywfE*), is essential for connecting the two amino acids (Inaoka et al. 2003; Steinborn et al. 2005). However, whether the *bacD* gene product alone is enough for the activity or whether other protein(s) are required remains unclear. The expression of the *bac* operon was reported to be under a dual regulation composed of stringent control mediated by guanosine 5′-diphosphate 3′-diphosphate (ppGpp) and repression control mediated by GTP (Inaoka et al. 2003).

Interestingly, a completely different line of study found that YwfE alone has an activity forming various dipeptides. Tabata et al. (2005) conducted in silico screening for a gene coding for α-dipeptide forming enzyme and

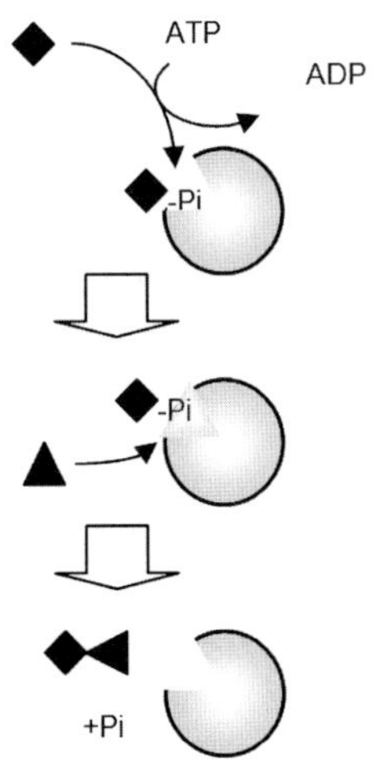

Fig. 4 Schematic diagram of peptide synthesis by l-amino acid α-ligase. *Closed diamonds and triangles* indicate amino acids

found *ywfE* as the only candidate. Purified YwfE was revealed to synthesize 44 kinds of dipeptide[1] coupled with dephosphorylation of ATP to ADP. It was also found that the enzyme did not synthesize tri- or longer peptide and that d-amino acids were inert. From these characteristics, the enzyme was tentatively named l-amino acid α-ligase (Lal, EC 6.3.2.28) (Fig. 4). Unfortunately, examinations on the substrate specificity of YwfE did not contain anticapsin due to its unavailability, leaving it unclear if Lal alone catalyzes bacilysin synthesis. Although Lal seems to be the bacilysin synthesizing enzyme, further examinations are needed to confirm this since the reaction mechanism suggested from the earlier study is of the NRPS-like type, which is different from the results obtained by the Lal study. At present, no obvious homologue of Lal can be found in public databases except *bacD* in *Bacillus licheniformis*, which codes for a protein 89% identical with Lal of *B. subtilis*. Unfortunately, whether the BacD protein of *B. licheniformis* possesses Lal activity has not been looked at.

Carnosine and Kyotorphin While both dipeptides are not microbial products, their biosynthetic activities have been elucidated to some extent. It was demonstrated that the extracts of pectoral muscles from chicks had the activity for catalyzing the following reaction; β-Ala + His + ATP$\rightarrow$carnosine + AMP + pyrophosphate (Kalyankar and Meister 1959). Similarly, a partially purified enzyme from rat brain synaptosomes was reported to catalyze the formation of kyotorphin, AMP and pyrophosphate from Tyr, Arg and ATP (Ueda et al. 1987). Though further biochemical information and genetic analyses are absent, these results suggest that both synthetic activities depend on NRPS-type enzymes.

Thaxtomins A plant phasogen, *Streptomyces scabies*, produces phytotoxins called thaxtomins [cyclo(4-nitro-Trp-Phe) and its related congeners] (King et al. 1989, 1992). Disruption of the *txtAB* genes coding for NRPS-type proteins abolished the phytotoxin production, clearly indicating that NRPS was responsible for the synthesis of thaxtomins (Healy et al. 2000). Since biochemical studies on other NRPSs showed that dimodular NRPS often form a cyclo-dipeptide, TxtAB is expected to catalyze the formation of the cyclo-dipeptide.

Albonoursin *Streptomyces noursei* and *S. alburus* have been known to produce an antibiotic compound, albonoursin, a dehydrated derivative of cyclo(Phe-Leu) (Khokhlov and Lokshin 1963, Kanzaki et al. 1999). A recombinant *S. lividans* expressing a gene from *S. noursei*, named *albC*, was found to synthesize cyclo(Phe-Leu). The amino acid sequence of AlbC has no homology with any

[1] The enzyme did not accept highly charged amino acids such as Lys and Glu and secondary amines such as Pro; the N-terminal residue of the dipeptide formed was limited to Ala, Gly, Ser, Thr, and Met whereas the C-terminal seemed to allow for a wider array of amino acids.

known NRPSs and dose not contain an ATP grasp motif, thus, the reaction mechanism of the enzyme remains unclear (Lautru et al. 2002).

5 Biotechnology for Dipeptide Manufacturing

The finding of activities synthesizing dipeptides in ribosome-independent manners prompted to devise new methods for manufacturing dipeptides. In this section, pre-existed methods are reviewed first to clarify their limitations, and then, newly developed or developing methods are introduced.

5.1 Current Technology for Dipeptide Manufacturing

Since Emil Fischer's work in 1901 (Fischer and Fourneau 1901), chemical synthetic methods of dipeptide manufacture have been successfully developed (for a review see Nilsson et al. 2005). Establishment of solid-phase techniques (Merrifield 1986) and other advancements have made the methods rapid and easily automated. Any kind of dipeptide can be synthesized by chemical methods, fulfilling the researcher's demands. In general, chemical methods require multiple reaction steps; protection of amino acids, condensation of them, and deprotection of the resulting product to get a dipeptide. Because of the multiple steps and the necessity for protection, production costs for these methods are so high that they cannot be applied in the wider commercial arena. Furthermore, harmful reagents, such as phosgene and organic solvents, are used in some cases.

There have been many attempts to apply an enzyme, such as a protease or peptidase, as a catalyst for the condensation reaction (for reviews see Morihara 1987; Sinisterra and Alcantara 1993; Bongers and Heimer 1994; Kumar and Bhalla 2005). Because these enzymes catalyze the reverse reaction, hydrolysis of peptide bonds, under normal conditions, protection of the amino- or carboxy-group of the substrate amino acids is needed to forward the synthetic reaction and to direct the order of the amino acids. The most successful example is the application of a protease, thermolysin, in manufacturing aspartame, a low calorie sweetener (Bornscheuer 2000). The enzyme condenses *N*-(benzyloxycarbonyl)-Asp and Phe methyl ester to form an insoluble product, *N*-(benzyloxycarbonyl)-Asp-Phe metyl ester. Precipitation of the product pulls the reversible reaction towards the synthesizing direction. The product is easily converted to Asp-Phe methyl ester, aspartame. This is a fortunate case that an insolubility of the product drives the thermodynamic control of the reaction. Thus, application of enzymatic methods also has its limitations.

The ideal way to synthesize dipeptides is by connecting unmodified amino acids in an irreversible reaction. This technique inevitably utilizes the peptide bond-forming activities mentioned in the previous section. Three lines of studies pursuing these possibilities have been reported.

5.1.1 Dipeptide Production using the Ribosome System

Since the ribosome system is the most versatile procedure for peptide synthesis, it is not surprising that the system is used for dipeptide production. Attempts aimed at the production of Asp-Phe for the precursor of aspartame have been reported. In these studies, synthetic genes encoding poly(Asp-Phe) were expressed in *Escherichia coli*, but the productivities were quite low (Doel et al. 1980; Murata et al. 1993). Fusing the synthetic gene with the prochymosin gene increased the productivity (Murata et al. 1993). Apart from the productivity, this method is laden with other difficulties, in particular, on efficient procedures to separate the desired dipeptide from the polymer. When the polymer was treated with a couple of proteases, not only Asp-Phe but also Phe-Asp was formed (Doel et al. 1980), suggesting that an efficient downstream process should be invented to make the method more viable.

5.1.2 Dipeptide Production using the NRPS Method

The modular architecture of NRPS and the finding that the substrate specificity of the enzyme largely depends on the adenylation domains (A domain) prompted researchers to engineer the enzyme to synthesize a designed dipeptide. Doeckel and Marahiel (Doeckel and Marahiel 2000) designed synthetic templates for dipeptide formation by combining parts of *bac* (coding for bacitracin-biosynthetic NRPS in *B. licheniformis*) and *tyc* (coding for tyrocidine-biosynthetic NRPS in *B. brevis*) genes. The first A domain of BacA1, which recognizes and activates Ile, was fused with the last part of TycC containing C-PCP-A (recognizes and activates Leu) – PCP-Te domains. The resulting artificial dimodular NRPS was co-produced in *E. coli* with 4′-phosphopantetheinyl transferase to make NRPS holo-enzyme by modifying with CoA in vivo. The enzyme was purified and incubated with Ile, Leu, and ATP. As expected, Ile-Leu was formed. They devised similar dimodular NRPSs with different combinations of the parts, which synthesized Phe-Leu or Ile-Phe. These pioneering experiments clearly demonstrated the effectiveness of a rationally designed NRPS for dipeptide synthesis. The similar strategy combining A (Phe activating), PCP, E (epimerization) domains of tyrocidine synthetase and C, A (Pro activating), PCP domains of gramicidin S led to the formation of a dipeptide, D-Phe-Pro, but in the cyclic form probably because of the absence of a Te domain (Keller and Schauwecker 2003).

A more detailed examination aimed at creating an enzyme synthesizing Asp-Phe was performed. Asp-Phe synthetases were constructed by fusing A-PCP or A-PCP-C domains of Asp-activating modules from surfactin synthetase of *B. subtilis* and A-PCP, C-A, A-PCP, or A domains of Phe activating modules from tyrocidine synthetase (Duerfaht et al. 2003). These artificial synthetases had Asp-Phe synthesizing activity, but their activities were significantly different, indicating the importance of the strategy to construct a hybrid NRPS.

Reflecting the increasing attention on NRPS, knowledge about characteristics of the enzyme and know-how for engineering it has been accumulated rapidly (Sieber and Marahiel 2005). The substrate specificity of the A domain has been explored in at least 397 domains and it covers all of the usual 20 kinds of amino acids (Rausch et al. 2005). Sequence comparison and structure-function mutagenesis have defined the specificity-conferring code of the A domain (Stachelhaus et al. 1999). Therefore, any dipeptide consisting of the usual amino acids can theoretically be synthesized by a customized NRPS. Because C and Te domains have been found to have some preference (Belshaw et al. 1999; Shwarzer et al. 2001), careful design plus some trial and error shall be needed to get a highly active enzyme.

5.1.3
Dipeptide Production using the Lal Method

Considering the broad substrate specificity and simplicity of Lal, this enzyme was expected to realize a novel process, which would overcome the drawbacks of the current dipeptide manufacturing processes (chemical synthesis or protease/peptidase dependent methods). Two types of the process have been invented; the resting cell reaction process and the direct fermentation process.

Resting Cell Reaction Process: The principle of the process is a coupling reaction between Lal and ATP regeneration, which enables the use of an inexpensive substrate instead of the expensive ATP (Fig. 5). Among several ATP-regeneration systems already known (Hashimoto and Ozaki 1999; Ishige et al. 2001), one of the simplest, polyphosphate kinase, which catalyzes degradation of polyphosphate coupled with phosphorylation of ADP to ATP, was chosen (Shiba et al. 2000). A recombinant *E. coli* strain producing Lal of *B. subtilis* and polyphosphate kinase (Ppk) of *Rhodobacter spaeroides*, whose optimal pH is relatively close to that of Lal, was constructed. When the cells were treated with a detergent for permeabilization and mixed with the substrate amino acids and polyphosphate, the corresponding dipeptides were accumulated in the reaction mixture. Ala-Met, Ala-Val, Ala-Leu, Ala-Ile, Gly-Met, and Gly-Phe were formed from the corresponding substrates. The highest titer was obtained for Ala-Met; from 200 mM each of Ala and Met, 127.9 mM of

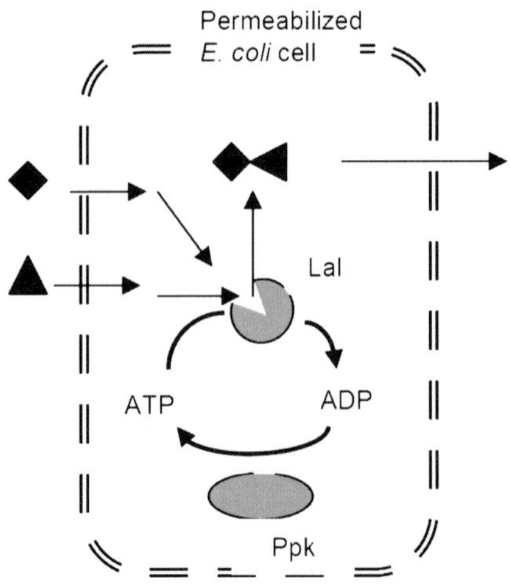

Fig. 5 Outline of the resting cell reaction system using Lal. *Closed diamonds and triangles* indicate amino acids. *Dashed lines* indicate the permeabilized cell envelope

Ala-Met (28 g/L) was produced in a 27 h reaction (Ikeda et al. 2006). This system is suitable for practical use by the ability to synthesize a dipeptide from unprotected amino acids in high yield. Furthermore, different dipeptides are easily produced with the same cells only by changing the substrate amino acids, suggesting the wide applicability of the system.

Direct Fermentation: As amino acids can be produced by fermentation, the next step to be pursued was dipeptide production through fermentation from the primary substrates, glucose and ammonia (Tabata and Hashimoto 2005, 2006). The fact that no dipeptide was overproduced by simply expressing the Lal gene in wild-type *E. coli* suggested two hurdles to be overcome; low affinity of the enzyme for the substrate amino acids and degradation of dipeptides by the producing cell itself.

As for the former issue, increasing the flux to the substrate amino acids was effective. To do this, many ways are known as described in other sections of this volume and many of them were found to be useful probably except for improving amino acid export. The most simplified version was overproducing the relevant biosynthetic enzyme(s). For instance, overproduction of Ald, alanine dehydrogenase from *B. subtilis*, with Lal evoked Ala-Ala accumulation. Deregulation of the biosynthetic enzyme was also effective; when the genes for Ald and Lal were coexpressed in a mutant deficient in *metJ*, coding for a repressor for Met biosynthesis, Ala-Met was produced. Furthermore, an amino-acid producing strain already obtained was able to be used as a host; overexpression of the genes for a deregulated PheA and for Lal in a Thr-producing *E. coli* strain obtained by classical mutagenesis resulted in the production of Thr-Phe. These examples show that dipeptide fermentation can be achieved by increasing the flux to the substrate amino acids and that the arts of metabolic engineering for amino acid production accumulated over several decades can be directly applied. It is worth noting that the necessity to enhance the flux is advantageous for avoiding undesired formation of other dipeptides.

Degradation was found to be attenuated by a combined disruption of the genes for several peptidases and for the dipeptide uptake system. While every single disruption of *pepA, pepB, pepD*, or *pepN*, which encode dipeptidases or aminopeptidases in *E. coli*, exerted little influence on the degradation, double, triple, or quadruple inactivation additively reduced it. Combination with the inactivation of the dipeptide uptake system (*dpp* genes) doubled the effect. Though destroying all the peptidases is apparently impossible, these genetic manipulations were sufficient for practical use.

These strategies were integrated to construct a producer strain for Ala-Gln, a useful ingredient of infusions. The *E. coli* host constructed was deficient in *pepA, pepB, pepD, pepN, dpp, glnB*, and *glnE*. Inactivation of *glnB* and *glnE*, the regulators of Gln biosynthesis, conferred Gln productivity on the cell. Ald and Lal were overproduced in the host cells. The resulting strain produced 100 mM (24.6 g/L) of Ala-Gln during 47 h cultivation on glucose and ammonia, demonstrating that the technical level of dipeptide fermentation reached was almost comparable to that of amino acid fermentation. The general scheme for dipeptide fermentation is depicted in Fig. 6.

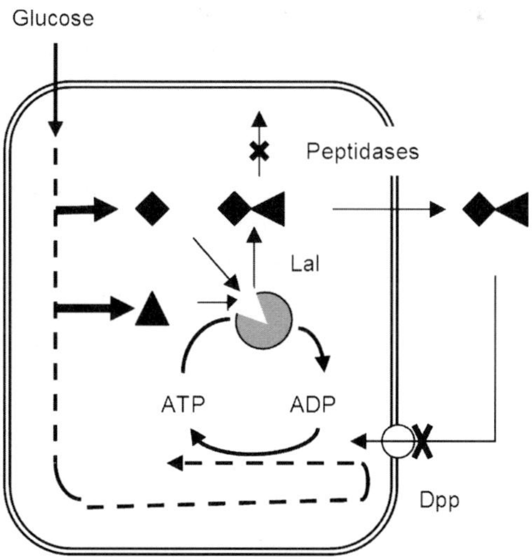

Fig. 6 Outline of dipeptide fermentation using Lal. *Closed diamonds and triangles* indicate amino acids. *Bold arrows*: the strengthened pathways, X: the disrupted or weakened pathways

While the direct fermentation is obviously the most cost effective way, strain construction requires a lot of effort. Balancing the two fluxes will be needed to increase the productivity, but remains an unexplored problem. The mechanism of dipeptide efflux (including whether an efflux system exists or not) also remains to be discovered. Accounting for these difficulties, the resting cell system is a worthwhile alternative to the fermentation system for easily obtaining many kinds of dipeptides.

5.2 Comparison between the NRPS Method and the Lal Method

Although the demonstration of dipeptide production by the NRPS method remains at the in vitro level, it will be applied to whole cell or living cells in the near feature. The technique to attenuate dipeptide degradation, described above, should also be integrated. Such NRPS-based manufacturing processes shall seemingly be the same as Lal-based ones though the intracellular events should be different. To get a perspective of the future technology it is worth comparing the two systems (Table 4).

One characteristic difference is the complexity of the enzyme. While Lal consists of a single, medium-size protein (472 amino acids) and requires no co-factor, NRPS is huge and is often a complex of several proteins and needs 4′-phosphopanteteine as a co-factor. The complexity of NRPS is a barrier for engineering the enzyme to some extent. The other outstanding difference is the product spectrum. A domains of NRPS corresponding to each usual amino acid are known and the "specificity conferring code" of the domain is also known, whereas Lal covers only 44 kinds of dipeptides and no obvious homologues have been found, indicating a much wider application space for designed NRPS. The energy requirement is also different between the two systems. To form a dipeptide from two molecules of amino acids, NRPS needs four high-energy phosphate bonds, whereas Lal uses one. As the energy supply is a burden for the cell, a lower need for energy is preferable for fermentation. However, how much effect the above difference will

Table 4 Comparison between NRPS and Lal on dipeptide synthesis

Characteristics	NRPS	Lal
Enzyme constitution	complex	simple
Co-factor	4′-phosphopantetheine	none
Energy requirement	(ATP → AMP + PPi) × 2	ATP → ADP + Pi
Product spectrum	20 × 20 (theoretically)	44 (confirmed)
Affinity for the substrates	high	low
Reaction rate	slow	moderate

have on cellular physiology remains to be explored. Differences in the affinity for the substrate amino acids should also be considered. Lal's Km for the amino acid for the C-terminus of the dipeptide produced has been reported to be quite high (for example, Km for Gln in Ala-Gln synthesis was 105 mM), which caused the necessity for the enhancement of the amino-acid biosynthesis. Designed NRPSs may not need such metabolic manipulation, but to avoid the formation of byproducts due to the ambiguity in the specificity of A domains (Stachelhaus et al. 1999), some manipulation might be required.

Both the designed NRPS and Lal systems have drawbacks and advantages. The current situation can be summarized that the former is advanced in science and the latter is advanced in practical use.

6 Concluding Remarks

While there have been several reports on the production of dipeptides by organisms, the synthetic machinery remains equivocal. However, findings of Lal and AlbC clearly indicate the existence of specific dipeptide synthesizing enzymes though their distribution in nature has yet to be confirmed. Furthermore, great progress has been made in the rational design of NRPS in the last decade. Irrespective of relying on NRPS or on Lal, many dipeptides can be produced by fermentation in the future. This technology represents a new generation of amino acid fermentation, but will also be the beginning of a new research field on the export of dipeptides, accurate metabolic control of biosynthetic fluxes to the amino acids, improved specificity of peptidases, etc. Considering that the discovery of glutamic acid fermentation in 1957 (Kinoshita et al. 1957) triggered the expansion of amino-acid applications, dipeptide fermentation will open the door to new frontiers of dipeptide research and application.

Acknowledgements The author sincerely thanks Drs. Akio Ozaki and Makoto Yagasaki for their kind and helpful discussion.

References

Abumrad NN, Morse EL, Lochs H, Williams PE, Adibi SA (1989) Possible sources of glutamine for parental nutrition: Impact on glutamine metabolism. Am J Physiol 257: E228–E234

Aboulmagd E, Oppermann-Sanio FB, Steinbuchel A (2000) Molecular characterization of the cyanophycin synthetase from *Synechocystis* sp. strainPCC6308. Arch Microbiol 174:297–306

Allen CM Jr (1972) Biosynthesis of echinulin. Isoprenylation of cyclo-l-alanyl-l-tryptophanyl. Biochemistry 23:2154–2160

Alvarez ME, Houck DR, White CB, Brownell JE, Bobko MA, Rodger CA, Stawicki MB, Sun HH, Gillum AM, Cooper R (1994) Isolation and structure elucidation of two new calpain inhibitors from *Streptomyces griseus*. J Antibiot (Tokyo) 47:1195–1201

de Armas RR, Diaz HG, Molina R, Conzalez MP, Uriarte E (2004) Stochastic-based descriptiors studying peptides biological properties: Modeling the bitter tasting threshold of dipeptides. Bioorg Med Chem 12:4815–4822

Atkin CL, Neilands JB, Phaff HJ (1970) Rhodotorulic acid from species of *Leucosporidium, Rhodosporidium, Rhodotorula, Sporidiobolus*, and *Sporobolomyces*, and a new alanine-containing ferrichrome from *Cryptococcus melibiosum*. J Bacteriol 103:722–733

Bauer K (2005) Carnosine and homocarnosine, the forgotten, enigmatic peptides of the brain. Neurochem Res 30:1339–1345

Begum G, Cunliffe A, Leveritt M (2005) Physiological role of carnosine in contracting muscle. Int J Sport Nutr Exerc Metab 15:493–514

Belshaw PJ, Walsh CT, Stachelhaus T (1999) Aminoacyl-CoAs as probes of condensation domain selectivity in nonribosomal peptide synthesis. Science 284:486–489

Birkinshaw JH, Mohammed YS (1962) Studies in the biochemistry of micro-organisms. Biochem J 85:523–527

Bongers J, Heimer EP (1994) Recent applications of enzymatic peptide synthesis. Peptides 15:183–193

Bornscheuer UT (2000) Industrial biotransformations. In: Rehm H-J, Reed G, Puhler A, Stadler P (eds) Biotechnology second, completely revised edition, vol 8. Wiley, Weinheim, p 277–294

Burt WR (1982) Identification of coprogen B and its breakdown products from *Histoplasma capsulatum*. Infect Immun 35:990–996

Byford MF, Baldwin JE, Shiau C, Schofield CJ (1997) The mechanism of ACV synthetase. Chem Rev 97:2631–2649

Caesar Von F, Jansson K, Mutschler E (1969) Uber nigragillin, ein neues alkaloid aus der *Aspergillus niger*-gruppe. Pharm Acta Helvet 44:676–690

Chen Y (1960) Studies on the metabolic products of *Rosellinia necatrix* Berlese, Part I. Isolation and characterization of several physiologically active neutral substances. Bull Agric Chem Soc Jpn 24:372–381

Chung M, Lee H, Chu H, Lee C, Kim S, Kho Y (1996) Bestatin analogue from *Streptomyces neyagawaensis* SL-387. Biosci Biotechnol Biochem 60:898–900

Cloninger MR, Baldwin RE (1970) Aspartylphenylalanine methyl ester: A low-calorie sweetner. Science 170:81–82

Cui C, Kakeya H, Okada G, Onose R, Osada H (1996) Novel mammalian cell cycle inhibitor, tryprostatins A, B and other diketopiperazines produced by *Aspergillus fumigatus*, vol I. Taxonomy, fermentation, isolation and biological properties. J Antibiot (Tokyo) 49:527–533

Daabees TT, Stegink LD (1978) l-Alanyl-l-tyrosine as a tyrosine source during intravenous nutrition of the rat. J Nutr 108:1104–1113

Day JB, Mantle PG (1982) Biosynthesis of radiolabeled verruculogen by *Penicillium simplicissimum*. Appl Environ Microbiol 43:514–516

Doekel S, Marahiel M (2000) Dipeptide formation on engineered hybrid peptide synthetases. Chem Biol 7:373–384

Doel MT, Eaton M, Cook EA, Lewis H, Patel T, Carey NH (1980) The expression in *E. coli* of synthetic repeating polymeric genes coding for poly(l-aspartyl-l-phenylalanine). Nucleic Acids Res 8:4575–4592

Duerfahrt T, Doekel S, Sonke T, Quaedflieg PJLM, Marahiel MA (2003) Construction of hybrid peptide synthetases for the production of α-l-aspartyl-l-phenylalanine, a precursor for the high-intensity sweetener aspartame. Eur J Biochem 270:4555–4563

Evers S, Casadwall B, Charles M, Dutka-Malen S, Galimand M, Courvalin P (1996) Evolution of structure and substrate specificity in d-alanine:d-alanine ligase and related enzymes. J Mol Evol 42:706–712

Finking R, Marahiel MA (2004) Biosynthesis of nonribosomal peptides. Annu Rev Microbiol 58:453–488

Fischer E, Fourneau E (1901) Ueber einige Derivate des Glykocolls. Ber Deutsch Chem Ges 34:2868–2877

Fredenhagen A, Angst C, Peter HH (1995) Digestion of rhizocticins to (Z)-l-2-amino-5-phophono-3-pentenoic acid: Revision of the absolute configuration of plumbemycins A and B. J Antibiot (Tokyo) 48:1043–1045

Galperin MY, Koonin EV (1997) A diverse superfamily of enzymes with ATP-dependent carboxylate-amine/thiol ligase activity. Protein Sci 6:2639–2643

Guiotto A, Calderan A, Ruzza P, Borin G (2005) Carnosine and carnosine-related antioxidants: a review. Curr Med Chem 12:2293–2315

Gulewitsch W, Amiradzibi S (1900) Ueber das Carnosin, eine neue organishe Base des Fleischextractes. Ber Deutsch Chem Ges 33:1902–1903

Hamasaki T, Nagayama K, Matsuda Y (1976) Structure of a new metabolite from *Aspergillus chevalieri*. Agric Biol Chem 40:203–205

Hashimoto S, Ozaki A (1999) Whole microbial cell processes for manufacturing amino acids, vitamins or ribonucleotides. Curr Opin Biotechnol 10:604–608

Healy FG, Wach M, Krasnoff SB, Gibson DM, Loria R (2000) The *txtAB* genes of the plant pathogen *Streptomyces acidiscabies* encode a peptide synthetase required for phytotoxin thaxtomin A production and pathogenicity. Mol Microbiol 38:794–804

Higashide E, Horii S, Ono H, Mizokami N, Yamazaki T, Shibata M, Yoneda M (1985a) Alahopcin, a new dipeptide antibiotic produced by *Streptomyces albulus* subsp. *ochragerus* subsp. nov. J Antibiot (Tokyo) 38:285–295

Higashide E, Kanamaru T, Fukase H, Horii S (1985b) Isolation of dealanylalahopcin, a new amino acid, and its biological activity. J Antibiot (Tokyo) 38:296–301

Hines HM, Sutfin DC (1956) Physiologic properties of anserine and carnosine. Am J Physiol 186:286–288

Horii S, Fukase H, Higashide E, Yoneda M (1985) Structure of alahopcin (nourseimycin), a new dipeptide antibiotic. J Antibiot (Tokyo) 38:302–311

Ikeda H, Yagasaki M, Hashimoto S (2006) Methods for manufacturing dipeptides or their derivatives. WO 2006/001382

Inaoka T, Takahashi K, Ohnishi-Kameyama M, Yoshida M, Ochi K (2003) Guanine nucleotides guanosine 5′-diphosphate 3′-diphosphate and GTP co-operatively regulate the production of an antibiotic bacilysin in *Bacillus subtilis*. J Biol Chem 278:2169–2176

Ishige K, Hamamoto T, Shiba T, Noguchi T (2001) Novel method for enzymatic synthesis of CMP-NeuAc. Biosci Biotechnol Biochem 65:1736–1740

Kakinuma K, Rinehart KL Jr (1974) Tryptophan-dehydrobutyrine diketopiperazine, a metabolite of *Streptomyces spectabilis*. J Antibiot (Tokyo) 27:733–737

Kalyankar G, Meister A (1959) Enzymatic synthesis of carnosine and related β-alanyl and γ-aminobutyryl peptides. J Biol Chem 234:3210–3218

Kamiya T, Maeno S, Hashimoto M, Mine Y (1972) Bicyclomycin, a new antibiotic, vol II. Structural elicidation and acyl derivatives. J Antibiot 25:576–581

Kanzaki H, Imura D, Sashida R, Kobayashi A, Kawazu K (1999) Effective production of dehydro cyclic dipeptide albonoursin exhibiting pronuclear fusion inhibitory activity, vol I. Taxonomy and fermentation. J Antibiot (Tokyo) 52:1017–1022

Keller U, Schauweeker F (2003) Combinatorial biosynthesis of non-ribosomal peptides. Comb Chem High Throughput Screen 6:527–540

Khavinson VK, Anisimov VN (2000) Synthetic dipeptide vilon (l-Lys-l-Glu) increases life span and inhibits a development of spontaneous tumors in mice. Doklady Akad Nauk 372:421–423

Khokhlov AS, Lokshin GB (1963) The structure of albonoursin. Tetrahedron Lett 27:1881–1885

King RR, Lawrence CH, Calhoun LA (1992) Chemistry of phytotoxins associated with *Streptomyces scabies*, the causal organism of potato common scab. J Agric Food Chem 40:834–837

King RR, Lawrence CH, Clark MC, Calhoun LA (1989) Isolation and characterization of phytotoxins associated with *Streptomyces scabies*. J Chem Soc Chem Commun 13:849–850

Kinoshita S, Udaka S, Simono M (1957) Studies on the amino acid fermentation. J Gen Appl Microbiol 3:193–205

Kobaru S, Tsunakawa M, Hanada M, Konishi M, Tomita K, Kawaguchi H (1983) Bu-2743E, a leucine aminopeptidase inhibitor, produced by *Bacillus circulans*. J Antibiot (Tokyo) 36:1396–1398

Kodaira Y (1961) Toxic substances to insects, produced by *Aspergillus ochraceus* and *Oopra destructor*. Agric Biol Chem 25:261–262

Kumar D, Bhalla TC (2005) Microbial proteases in peptide synthesis: Approaches and applications. Appl Microbiol Biotechnol 68:726–736

Kuroda Y, Okuhara M, Goto T, Yamashita M, Iguchi E, Kohsaka M, Aoki H, Imanaka H (1980) FR-900148, a new antibiotic I. Taxonomy, fermentation, isolation and characterization. J Antibiot (Tokyo) 33:259–266

Lautru S, Gondry M, Genet R, Pernodet J-L (2002) The albonoursin gene cluster of *S. noursei*: Biosynthesis of diketopiperazine metabolites independent of nonribosomal peptide synthetases. Chem Biol 9:1355–1364

Lee MD, Fantini AA, Kuck NA, Greenstein M, Testa RT, Borders DB (1987) New antitumor antibiotic, LL-D05139β fermentation, isolation, structure determination and biological activities. J Antibiot (Tokyo) 40:1657–1663

Li Y, Wei G, Chen J (2004) Glutathione: A review on biotechnological production. Appl Microbiol Biotechnol 66:233–242

MacDonald JC, Slater GP (1966) The utilization of tryptophan in the biosynthesis of echinulin. Can J Microbiol 12:455–463

Marahiel MA, Stachelhaus T, Mootz HD (1997) Modular peptide synthetase involved in nonribosomal peptide synthesis. Chem Rev 97:2651–2673

Matsui T, Hayashi A, Tamaya K, Matsumoto K, Kawasaki T, Murakami K, Kimoto K (2003) Depressor effect induced by dipeptide, Val-Tyr, in hypertensive transgenic mice is due, in part, to the suppression of human circulating rennin-angiotensin system. Clin Exp Pharmacol Physiol 30:262–265

Mazur RH, Schlatter JM, Goldkamp AH (1969) Structure-taste relationships of some dipeptides. J Am Chem Soc 91:2684–2691

Merrifield B (1986) Solid phase synthesis. Science 232:341–347

Miyoshi T, Miyairi N, Aoki H, Kohsaka M, Sakai H, Imanaka H (1972) Bicyclomycin, a new antibiotic, I. Taxonomy, isolation and characterization. J Antibiot (Tokyo) 25:569–575

Morihara K (1987) Using proteases in peptide synthesis. Trends Biotechnol 5:164–170
Munekata M, Tamura G (1981) Selective inhibition of SV40-transformed cell growth by diketopiperazines. Agric Biol Chem 45:2613–2618
Murata T, Horinouchi S, Beppu T (1993) Production of poly(l-aspartyl-l-phenylalanine) in *Escherichia coli*. J Biotechnol 28:301–312
Nilsson BL, Soellner MB, Raines RT (2005) Chemical synthesis of proteins. Annu Rev Biophys Biomol Struct 34:91–118
Nishikiori T, Kawahara F, Naganawa H, Muraoka Y, Aoyagi T, Umezawa H (1984) Production of acetyl-l-leucyl-l-argininal, inhibitor of dipeptidyl aminopeptidase III by bacteria. J Antibiot (Tokyo) 37:680–681
Opris D, Diudea MV (2001) Peptide property modeling by Cluj indices. SAR QASR Environ Res 12:159–179
Poetsch M, Zahaner H (1985) Metabolic products from microorganisms. 230 amiclenomycin-peptides, new antimetabolites of biotin. Taxonomy, fermentation and biological properties. J Antibiot (Tokyo) 38:312–320
Prasad C (1995) Bioactive cyclic dipeptides. Peptides 16:151–164
Rapp C, Jung G, Kugler M, Loeffler W (1988) Rhizocticins – new phosphono-oligopeptides with antifungal activity. Liebigs Annu Chem 1988:655–661
Rausch C, Weber T, Kohlbacher O, Wohleben W, Huson DH (2005) Specificity prediction of adenylation domains in nonribosomal peptide synthetase (NRPS) using transductive support vector machines (TSVMs). Nucleic Acids Res 33:5799–5808
Rogers HJ, Newton GGF, Abraham EP (1965a) Production and purification of bacilysin. Biochem J 97:573–577
Rogers HJ, Lomakina N, Abraham EP (1965b) Observation on the structure of bacilysin. Biochem J 97:579–586
Sakajoh M, Solomon NA, Demain AL (1987) Cell-free synthesis of the dipeptide antibiotic bacilysin. J Ind Microbiol 2:201–208
Schiffman SS (1976) Taste of dipeptides. Physiol Behav 17:523–525
Schnurer J, Magunsson J (2005) Antifungal lactic acid bacteria as biopreservatives. Trends Food Sci Technol 16:70–78
Schwarzer D, Mootz HD, Marahiel MA (2001) Exploring the impact of different thioesterase domain for the design of hybrid peptide synthetases. Chem Biol 8:997–1010
Shiba T, Tsutsumi K, Ishige K, Noguchi T (2000) Inorganic polyphosphate and polyphosphate kinase: Their novel biological functions and applications. Biochemistry (Mosc) 65:315–323
Shoji J, Hinoo H, Kato T, Nakauchi K, Matsuura S, Mayama M, Yasuda Y, Kawamura Y (1981) Isolation of *N*-(2,6-diamino-6-hydroxyetylpimelyl)-l-alanine from *Micromonospora chlcea*. J Antibiot (Tokyo) 34:374–380
Sieber SA, Marahiel MA (2005) Molecular mechanisms underlying nonribosomal peptide synthesis: Approaches to new antibiotics. Chem Rev 105:715–738
Sinisterra JV, Alcantara AR (1993) Synthesis of peptide catalyzed by enzyme: A practical overview. J Mol Catal 84:327–364
Smith LT, Smith GM (1989) An osmoregulated dipeptide in stressed *Rhizobium meliloti*. J Bacteriol 171:4714–4717
Stachelhaus T, Mootz H, Marahiel MA (1999) The specificity-conferring code of adenylation domains in nonribosomal peptide synthetases. Chem Biol 6:493–505
Stehle P, Pfaender P, Frust P (1984) Isotachophoretic analysis of a synthetic dipeptide l-alanyl-l-glutamine. Evidence for stability during heat sterilization. J Chromatogr 294:507–512

Steinborn G, Hajirezaei M-R, Hofemeister J (2005) *bac* Genes for recombinant bacilysin and anticapsin production in Bacillus host strains. Arch Microbiol 183:71–79

Suda H, Takita T, Aoyagi T, Umezawa H (1976) The structure of bestatin. J Antibiot (Tokyo) 29:100–101

Suetsuna K (1998) Isolation and characterization of angiotensin I-converting enzyme inhibitor dipeptides derived from *Allium sativum* L (garlic). J Nutr Biochem 9:415–419

Sun X, Bognar AL, Baker EN, Smith CA (1998) Structural homologies with ATP- and folate-binding enzymes in the crystal structure of folylpolyglutamate synthetase. Proc Natl Acad Sci USA 95:6647–6652

Tabata K, Ikeda H, Hashimoto S (2005) *ywfE* in *Bacillus subtilis* code for a novel enzyme, l-amino acid ligase. J Bacteriol 187:5195–5202

Tabata K, Hashimoto S (2005) Microorganisms producing dipeptides and process for producing dipeptides using the microorganisms. WO 2005/045006

Tabata K, Hashimoto S (2006) Methods for manufacturing dipeptides. WO 2006/001379

Takagi H, Shiomi H, Ueda H, Amano H (1979) Morphine-like analgesia by a new dipeptide, l-tyrosyl-l-arginie (Kyotorphin) and its analogue. Eur J Pharmacol 55:109–111

Ueda H, Shiomi H, Takagi H (1980) Regional distribution of a novel analgesic dipeptide kyotorphin (Tyr-Arg) in the rat brain and spinal cord. Brain Res 6:460–464

Ueda H, Yoshihara Y, Fukushima N, Shiomi H, Nakamura A, Takagi H (1987) Kyotorphin (tyrosine-arginine) synthetase in rat brain synaptosomes. J Biol Chem 262:8165–8178

Umezawa H, Aoyagi T, Suda H, Hamada M, Takeuchi T (1976) Bestatin, an inhibitor of aminopeptidase B, produced by actinomycetes. J Antibiot (Tokyo) 29:97–99

Umezawa H (1977) Microbial secondary metabolites with potential use in cancer treatment. Biomedicine 26:236–249

Walsh CT (1989) Enzymes in the D-alanine branch of bacterial cell wall peptidoglycan assembly. J Biol Chem 264:2393–2396

Wilson BJ, Yang DTC, Harris TM (1973) Production, isolation, and preliminary toxicity studies of brevianamide A from cultures of *Penicillium viridicatum*. Appl Microbiol 26:633–635

Yan P, Song Y, Sakuno E, Nakajima H, Nakagawa H, Yabe K (2004) Cyclo(l-leucyl-l-prolyl) produced by *Achromobacter xylosoxidans* inhibits aflatoxin production by *Aspergillus parasiticus*. Appl Environ Microbiol 70:7466–7473

Yasuda N, Sakane K (1991) Revised structure and the chemical transformations of FR900148. J Antibiot (Tokyo) 44:801–802

Yazgan A, Ozcengiz G, Ozcengiz E, Kilinc K, Marahiel MA, Alaeddinoglu NG (2001) Bacilysin biosynthesis by a partially-purified enzyme fraction from *Bacillus subtilis*. Enz Microb Technol 29:400–406

Microbiol Monogr (5)
V. F. Wendisch: Amino Acid Biosynthesis
DOI 10.1007/7171_2006_066/Published online: 9 December 2006

Genomes and Genome-Level Engineering of Amino Acid-Producing Bacteria

Hideaki Yukawa (✉) · Masayuki Inui · Alain A. Vertès

Microbiology Research Group,
Research Institute of Innovative Technology for the Earth (RITE), Soraku,
619-0292 Kyoto, Japan
mmg-lab@rite.or.jp

Abstract The complete nucleotide sequence of the genomes of several strains of *Escherichia coli* and *Corynebacterium glutamicum* reveal the genetic blueprint of these industrial organisms including their structural genetic organization and their metabolic networks and conversion capabilities, refine the understanding of their phylogenetic positions, and open the possibility to assess the expected size of their pan-genomes in order to harness their diversity. The genome of *C. glutamicum* R codes for approximately 3000 genes, a minimum of 5.3% of which are related to amino acid transport and metabolism and 4.6% to carbohydrate transport and metabolism. The genome of *E. coli* K-12 encodes approximately 4450 genes, 7.5% of which are involved in amino acid transport and metabolism and 6.1% in carbohydrate transport and metabolism. Global techniques were enabled by these complete genomic sequences, including analyses by global transcription profiling, proteomics and metabolomics to gather biological data, and megabase molecular biology tools to engineer at will these organisms at various scales, from the level of single base pairs to that of chromosomes. Systems biology represents the next technological paradigm necessary on the one hand to efficiently integrate and process the large volume of global biological information thus attained, in order to understand bacterial physiology and organization at a higher level; and on the other hand to enable in silico models useful for generating optimization strategies of increasing complexity and relevance, in the hope to lead faster towards improved metabolic engineering solutions with the aim of attaining expanded industrial process scope and superior economics.

Keywords *Corynebacterium glutamicum* · *Escherichia coli* · Genes and proteins repertoires · Genomics · Transcriptomics · Proteomics · Systems biology

1 Introduction

Corynebacterium glutamicum and *Escherichia coli* are the two primary biotechnological workhorses used for the production of amino acids. Other genera that have also been used for this purpose include *Arthrobacter*, *Bacillus*, *Pseudomonas*, and *Serratia* (Leuchtenberger 1996). The utilization of amino acids as a food and feed supplement takes its root in Japan from the traditional use of kelp as a food seasoning. While kelp is still a routine ingredient for Japanese cooking that uses simple infusion techniques for extracting the flavor of this seaweed, the molecule that gives kelp its main flavor was identified in 1908 as monosodium glutamate. It is this original discovery that virtually gave birth to a new industry that initially used hydrolysis of wheat, soybean, or other plant proteins to produce this amino acid. However, the economics of the production process improved in order to allow the large-scale use of this flavoring agent only after the isolation of *C. glutamicum* and concurrent identification of industrializable conditions for this organism to secrete large amounts of glutamate (Kumagai 2000). More recently, between 1990 and 2001, the worldwide annual market for amino acids grew from 600 000 t to more than 2 million t (Hermann 2003) and is poised to grow in the coming years at an average annual rate of 5 to 7% (Leuchtenberger et al. 2005) to represent a market in excess of 1 billion USD by 2009 (Brown 2005). As a result, exquisite operational expertise at an industrial scale up to 5000 hL has been acquired by manufacturers worldwide (Hermann 2003), thus contributing to make *E. coli*- and *C. glutamicum*-based processes robust, cost-effective, and relatively easy to scale.

C. glutamicum is a Gram-positive non-sporulating bacterium with a rapid growth rate that grows to high cell densities under aerobic conditions and that has a moderate to high G+C content. Corynebacteria typically exhibit limited extracellular protease activity. This species comprises numerous strains that were formerly classified as members of the genus *Brevibacterium*, such as *Brevibacterium flavum* or *Brevibacterium lactofermentum* (Liebl et al.

1991). Corynebacteria belong to the Actinomycetales order of the eubacteria and form a tight phylogenetic cluster with *Nocardia* and *Mycobacterium* species (Stackebrandt and Woese 1984; Ruimy et al. 1995; Stackebrandt et al. 1997; Liebl 2004). A structural feature that characterizes these organisms, with the notable exception of *Corynebacterium amycolatum* (Liebl et al. 1991) or *Corynebacterium kroppenstedtii* (Collins et al. 1998), is the presence of a mycolic acid permeability barrier that surrounds the peptidoglycan layer and particularly acts by conferring additional resistance to chemical injury (Barry et al. 1998). Non-medical Corynebacteria occupy a variety of ecological niches, including soil, dairy products, and plant materials (Liebl 1992). *C. glutamicum* strains have typically been recovered from soil isolates. This perhaps forms the basis of their industrial hardiness, as some strains have been demonstrated to not undergo autolysis and to actively maintain their main metabolic machinery under starvation or oxygen-deprivation conditions (Terasawa et al. 1985; Inui et al. 2004b). The manipulation of Corynebacteria is facilitated by the availability of a complete toolbox of molecular biology tools (Vertès et al. 2005) and a stable genome (Nakamura et al. 2003). The most commonly referenced industrial strains of Corynebacteria include *C. glutamicum* ATCC 13032 (Kinoshita 1985; Ikeda and Nakagawa 2003; Kalinowski et al. 2003), R (Kotrba et al. 2003), and ATCC 13869 (Kinoshita 1985). In addition, *Corynebacterium efficiens* YS-314 has been shown to grow and produce glutamate at temperatures significantly higher than the permissible temperature of typical *C. glutamicum* strains (Nishio et al. 2003).

The species *E. coli* includes both commensals and pathogens. *E. coli* is a facultative anaerobe that is commonly part of the microbial flora of the gastrointestinal tract of humans and animals. *E. coli* can be found in soil and water as a result of fecal contamination and its detection has been used as a marker of poor water or food quality (for example, Bettelheim 1992). Research on *E. coli* genetics and physiology has been at the forefront of biotechnology. As a result, this emphasis has enabled the production from a variety of *E. coli* strains of numerous compounds at various manufacturing scales, ranging from primary metabolites (Demain 1999) to large molecules such as human insulin (Graumann and Premstaller 2006). *E. coli* K-12 derivatives, for example protease deficient mutants, are most widely used. However, *E. coli* B strains exhibit particularly efficient recombinant protein expression capabilities and are increasingly being used for manufacturing purposes (Graumann and Premstaller 2006).

The objective of this chapter is to provide a brief description of the genomes of *C. glutamicum* and *E. coli* to serve as a basic reference of the genetic blueprint of these industrial species, as well as to provide a brief review of the main genetic tools that are available to date to enable their genetic manipulation from the scale of an individual gene to that of megabase DNA segments. In turn, these global analysis and manipulation techniques represent prerequisites for implementing global metabolic engineering and directed strain evolution rationales to further improve the economics of amino acid-production processes.

2 The *Corynebacterium glutamicum* Genome

The complete genomic sequence of at least six different Corynebacteria have been deciphered, including *C. glutamicum* R (Yukawa et al. manuscript in preparation), two isolates of the type strain of the *C. glutamicum* species, *C. glutamicum* ATCC 13032 (Ikeda and Nakagawa 2003; Kalinowski et al. 2003), and *C. efficiens* YS-314 (Nishio et al. 2003). In addition, the genomic sequences of *C. diphtheriae* NTCT 13129 (Cerdeño-Tárraga et al. 2003a) and *C. jeikeium* K411 (Tauch et al. 2005a), two pathogenic Corynebacteria, have also been determined.

2.1 Structure

The genomes of the saprophytic Corynebacteria used for amino acid production are larger in size by approximately 1 megabase than those of the two pathogenic species sequenced to date (Table 1). *C. glutamicum* R comprises 3 314 179 bp predicted, using the softwares GLIMMER (Delcher et al. 1999) and GeneMarkS (Besemer et al. 2001), to encode 2990 putative open reading frames (Yukawa et al. manuscript in preparation). The two isolates of the *C. glutamicum*-type strain ATCC 13032 have genomes that differ in size, 3 282 708 bp (Kalinowski et al. 2003) and 3 309 401 bp (Ikeda and Nakagawa 2003), assumed to encode 3002 and 2993 genes, respectively. This difference is mostly due to the presence of additional copies of insertion sequences and of an additional putative prophage present in the larger genome (Kalinowski 2005). The *C. glutamicum* chromosome is circular (Fig. 1). The G + C content of the *C. glutamicum* R and *C. glutamicum* ATCC 13032 genomes are, respectively, 54.1% and 53.8% overall (Table 1), 55.2% and 54.8% for the protein coding regions, and 47.4% and 47.5% for the non-coding regions. While *C. glutamicum* R harbors an episome 49 120 bp in size that exhibits a G + C content of 53.9%, *C. glutamicum* ATCC 13032 is devoid of any such element. GC skew analysis (Grigoriev 1998) of the genomic sequences of the two *C. glutamicum* ATCC 13032 strains reveals a bidirectional replication, also observed in *C. glutamicum* R, that is initiated at the *oriC* sequence near the *dnaA* gene and ends near the calculated replication terminus at around 1.6 Mb (Kalinowski 2005) (Fig. 1). *C. glutamicum* R encodes six 16S *r*RNA genes, six 23S *r*RNA genes, six 5S *r*RNA genes, and 57 *t*RNA genes; this is in good agreement with what is observed in *C. glutamicum* ATCC 13032 (respectively, six,

Table 1 Features of *C. glutamicum* R and related bacteria. Abbreviations: *Cg*, *Corynebacterium glutamicum*; K, Kitasato University *C. glutamicum* ATCC 13032 isolate; B, Bielefeld University *C. glutamicum* ATCC 13032 isolate; *Ce*, *Corynebacterium efficiens*; *Cj*, *Corynebacterium jeikeium*

Species	*Cg R*	*Cg K*	*Cg B*	*Ce*	*Cj*	*Cd*
Accession number	AP009044	NC003450	BX927147	BA000035	CR931997	BX248353
Strain	R	ATCC 13032	ATCC 13032	YS-314	K411	NCTC 13129
Reference	This work	(Ikeda and Nakagawa 2003)	(Kalinowski et al. 2003)	(Nishio et al. 2003)	(Tauch et al. 2005b)	(Cerdeño-Tárraga et al. 2003b)
Total genome size (bp)	3 314 179	3 309 401	3 282 708	3 147 090	2 462 499	2 488 635
Genome G + C content	54.10%	53.80%	53.80%	63.40%	61.40%	53.50%
Number of ORFs	2990	2993	3002	2950	2104	2320
Average ORF (bp)	957	933	916	979	1030	964
Coding density	86.30%	86.80%	88.30%	90.40%	89.20%	89.60%
*t*RNA	57	60	60	56	50	54
16S *r*RNA	6	6	6	5	3	5
23S *r*RNA	6	6	6	5	3	5
5S *r*RNA	6	6	6	5	3	5
Accession number episome 1	AP009045			AP005225	AF401314	
Size (bp)	PCgR1: 49 120			pCE2: 23 743	pKW4: 14 323	
G + C content	53.9%			54.4%	53.8%	
Accession number episome 2				AP005226		
Size (bp)				pCE3: 48 672		
G + C concent (%)				56.4%		

seven, six, and 60 genes) (Kalinowski et al. 2003). Based on data from strain R (Yukawa et al. manuscript in preparation) and strain ATCC 13032 (Kalinowski et al. 2003), the length of the average *C. glutamicum* gene is 935 bp, and the coding density of the *C. glutamicum* genome is 87.1%. On the other hand, the genome of *C. efficiens* with a G+C content of 63.4% reflects the capability of this organism to grow at a maximum temperature that is 5 °C higher than the permissible temperature for *C. glutamicum* (Nishio et al. 2003). The length of the *C. efficiens* YS-314 genome (3 147 090 bp) is close to that of *C. glutamicum* and encodes a similar number of genes (2950). However, the ancestral genome structure of Corynebacteria has been calculated by G + C content window analysis and GC skew analysis, using *C. diphtheriae* as an outgroup, to be closer to that of *C. glutamicum* than to that of *C. efficiens* (Nishio et al. 2003). Consequently, it is probable that *C. efficiens* acquired its increased thermostability after its divergence from *C. glutamicum*. The codon usage appears more biased in *C. efficiens* than in *C. glutamicum*. For example, in *C. glutamicum*, only seven out of ten of the most common codons have a G or C in the third position, whereas the ten most common codons in *C. efficiens* end with a G or C. This is consistent with the observation that *C. efficiens* can

Fig. 1 **a** Circular representation of the *C. glutamicum* R genome (accession number: AP009044). The two outermost *circles* represent the predicted coding sequences on the forward (maroon) strand on the outside and reverse (green) strand on the inside. The third circle from the outside shows the location of the 11 strain-specific islands greater than 10 kb (SSI-1 through SSI-11) along the genome. A region containing eight genes homologous to CGP1-borne genes of *C. glutamicum* ATCC 13032 is shown in grey. The distribution of insertion sequences (IS*CgR1* through IS*CgR13*) and putative phage-derived proteins (PPCgR1 through PPCgR7) is depicted. The color of the labeled genes corresponds to the color of the strand on which they are located. The fourth and fifth circles represent the G + C content and the GC skew (G–C/G + C), respectively, each plotted using a 3000-bp window with a 1000-bp window overlap. Red regions (pointing outwards) in these two circles are those of higher GC content or higher GC skew, whereas blue regions (pointing inwards) are those regions of lower GC content or lower GC skew, as compared to the genome averages. Tandem copies of IS*CgR8* and IS*1628* (IS*CgR9*) on SSI-5 constitute putative composite transposons. **b** Circular representation of the *C. glutamicum* ATCC 13032 genome (accession number: BX927147). The distribution of insertion sequences, and the location of strain-specific regions greater than 10 kb (SSICg-1, SSICg-2), prophages (CGP1, CGP2, CGP3), and regions of G + C content significantly lower or higher than the genome average (respectively, LGC1 and HGC1) are shown. The location of genes on both strands and the G + C content and GC skew (G–C/G + C) are as described in part A. Tn*Cg1* is a 3.1 kb composite transposon comprising two terminal copies of IS*Cg15* (Kalinowski et al. 2003). The maps were created using original scripts and the CGview software (Stothard and Wishart 2005). The first base of the initiation codon of *dnaA* is used as the coordinate origin. Coordinates are given in bp. Contiguous open reading frames putatively coding via frameshifts for only one transposase are represented as a single element. Mobile elements are named using the nomenclature initially used in the literature to describe virtually identical isoforms ▶

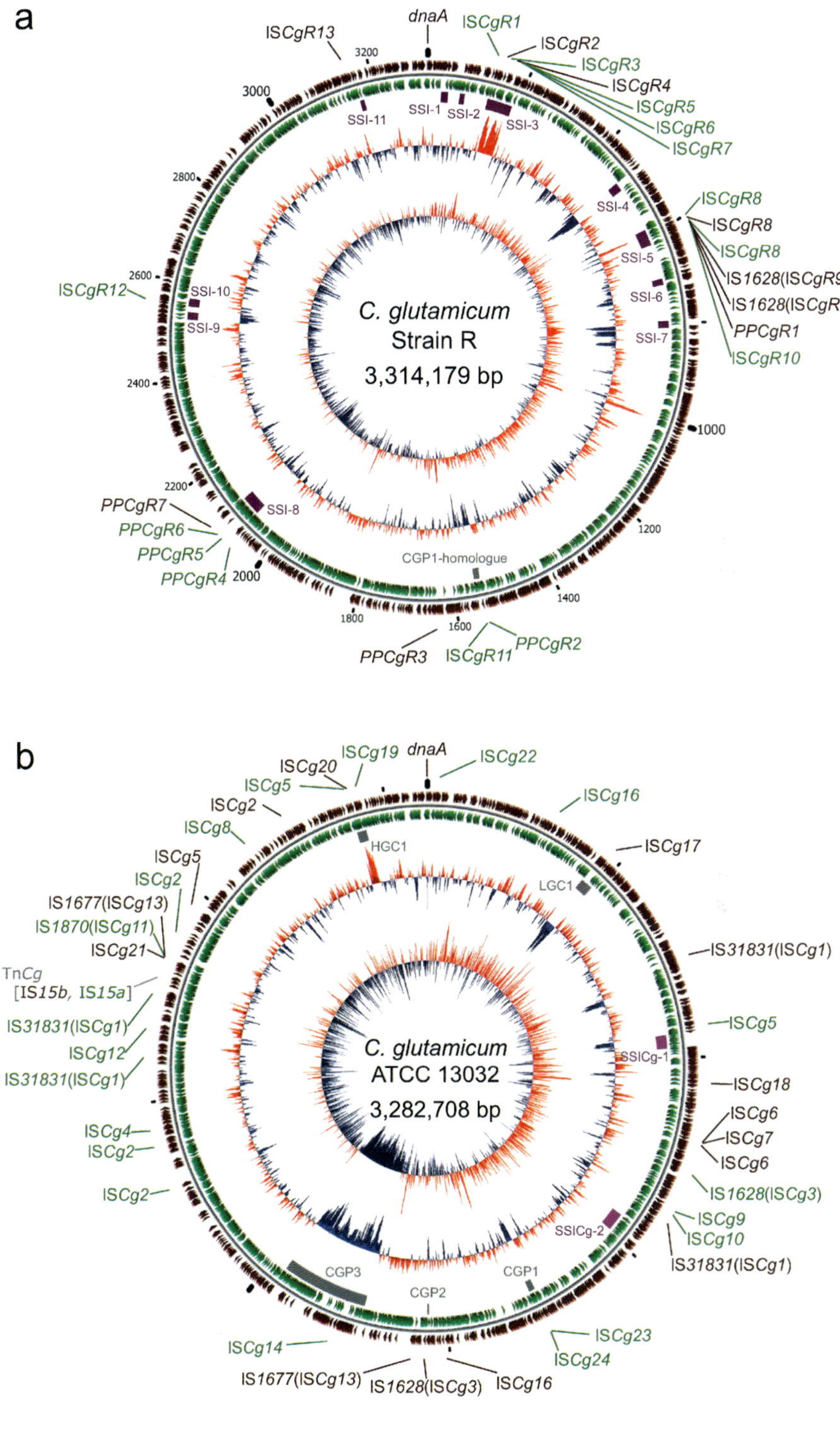
a
dnaA
ISCgR1
ISCgR13
ISCgR2
ISCgR3
ISCgR4
ISCgR5
ISCgR6
ISCgR7
ISCgR8
ISCgR8
ISCgR8
IS1628(ISCgR9)
IS1628(ISCgR9)
PPCgR1
ISCgR10
ISCgR12
PPCgR7
PPCgR6
PPCgR5
PPCgR4
PPCgR3
ISCgR11
PPCgR2
SSI-1
SSI-2
SSI-3
SSI-4
SSI-5
SSI-6
SSI-7
SSI-8
SSI-9
SSI-10
SSI-11
CGP1-homologue
C. glutamicum
Strain R
3,314,179 bp
b
dnaA
ISCg19
ISCg20
ISCg22
ISCg5
ISCg2
ISCg16
ISCg8
ISCg17
ISCg5
ISCg2
IS1677(ISCg13)
IS1870(ISCg11)
ISCg21
TnCg
[IS15b, IS15a]
IS31831(ISCg1)
ISCg12
IS31831(ISCg1)
IS31831(ISCg1)
ISCg5
ISCg18
ISCg6
ISCg7
ISCg6
IS1628(ISCg3)
ISCg9
ISCg10
IS31831(ISCg1)
ISCg4
ISCg2
ISCg2
ISCg14
IS1677(ISCg13)
IS1628(ISCg3)
ISCg16
ISCg23
ISCg24
HGC1
LGC1
SSICg-1
SSICg-2
CGP1
CGP2
CGP3
C. glutamicum
ATCC 13032
3,282,708 bp

grow at a higher temperature than *C. glutamicum*. On the other hand, notably 13 codons (GGG, ATA, CTA, TTA, AGA, AGG, CGA, CGG, AGT, TCA, TCG, ACA, GTA) appear to be rarely used in highly expressed genes of *C. glutamicum* (Malumbres et al. 1993). Such bias might represent a hurdle that would need to be addressed when manipulating Corynebacteria for high-level gene expression.

2.2 Mobile Elements, Prophages, and Strain-Specific Islands

Horizontal DNA transfer events are increasingly recognized as major forces of bacterial evolution (Ochman et al. 2000). In Corynebacteria, segments of foreign origin can be relatively easily identified by either G + C content differences or synteny breakpoints, since Corynebacteria exhibit a strikingly high level of synteny (Fig. 2). This latter property is particularly notable by the lack of major inversions in the genomes of these bacteria, with the

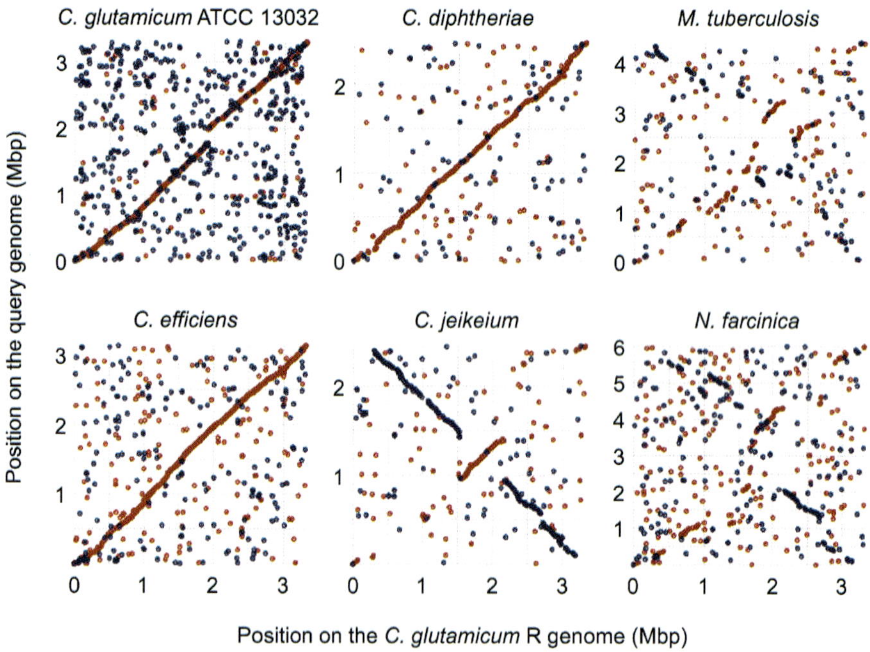

Fig. 2 Corynebacterial synteny. The dot plot of syntenic regions between the genomes of *C. glutamicum* R, *C. glutamicum* ATCC 13032 (Kalinowski et al. 2003), *C. efficiens* YS-314 (Nishio et al. 2003), *C. diphtheriae* NCTC 13129 (Cerdeño-Tárraga et al. 2003a), *C. jeikeium* K411 (Tauch et al. 2005a), *M. tuberculosis* H37Rv (Cole et al. 1998), and *N. farcinica* IFM 10152 (Ishikawa et al. 2004) catalogues each pair of orthologs present in the genomes being compared. Breakpoints in synteny reveal inversions and insertions

exception of *C. jeikeium* (Fig. 2). This striking genomic stability has been ascribed to the absence of a complete RecBCD recombination repair system in Corynebacteria (Nakamura et al. 2003), despite the presence of both *recA* and *recB* genes. This interesting characteristic could perhaps be useful to infer in these organisms, which conserved an ancestral genome structure, the evolution of metabolic pathways or species filiations, as exemplified by the comparison of the heme metabolic pathway in Corynebacteria and Mycobacteria (Fig. 3). Detailed examination of the complete genome sequences reveals that only a limited number of strain-specific islands are present in the genomes of Corynebacteria (Table 2). Specifically, the fine comparison between the genomes of *C. glutamicum* R and *C. glutamicum* ATCC 13032 reveals the presence of numerous small strain-specific islands (Fig. 4), but only a discrete number of these islands are greater than 10 kb in size. A particularly noteworthy insertion, 211 kb in size and referenced as CGP3, is present in the genome of *C. glutamicum* ATCC 13032, but is absent from the *C. glutamicum* R genome (Fig. 1, Table 2). This *C. glutamicum* ATCC 13032 DNA segment, characterized by a low G + C content, was demonstrated by cumulative GC profile analysis to have clear boundaries with its surrounding regions and to exhibit codon usage and amino acid usage biases that are signifi-

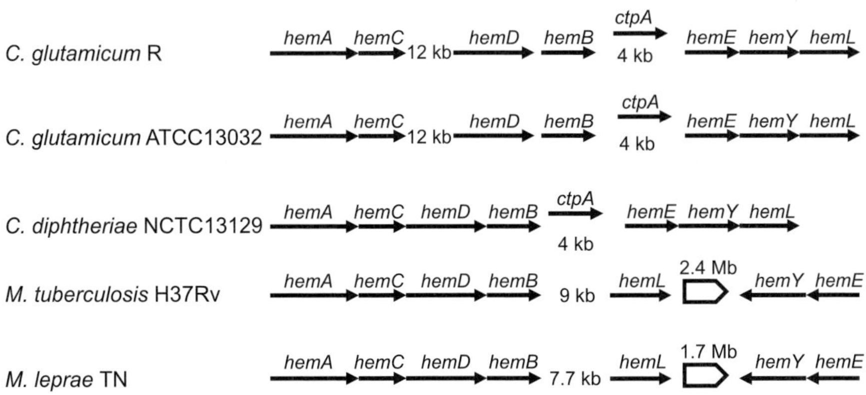

Fig. 3 The *hemACDBEYL* gene cluster of Corynebacteria and Mycobacteria. The physical organization of the *hemACDBEYL* genes is highly conserved in Corynebacteria and Mycobacteria. The dissociation and inversion of the *hemEY* genes observed in Mycobacteria suggests that this event might have occurred subsequently to the divergence of these two species. The presence of a 12-kb insertion in the *C. glutamicum* genomes, but neither in the *C. diphtheriae* genome nor in the mycobacterial genomes suggests that this insertion occurred after the divergence of *C. glutamicum* from *C. diphtheriae*. The intergenic region between the *hemB* and *hemE* genes observed in all five genomes suggests the presence of an intergenic region in the cluster that antedates the divergence of *Corynebacterium* from *Mycobacterium*. Genome sequence accession numbers are as follows: *C. glutamicum* R: AP009044, *C. glutamicum* ATCC 13032: AX114121, *C. diphtheriae* NCTC 13129: BX248353, *M. tuberculosis* H37Rv: AL123456, *M. leprae* TN: AL450380

Table 2 Strain-specific islands greater than 10 kb in the genomes of *C. glutamicum* R, *C. glutamicum* ATCC 13032, and *C. efficiens* YS-314. The complete genomic sequences of *C. glutamicum* R and *C. glutamicum* ATCC 13032 were analyzed with the software MUMer2.1 to identify DNA regions specific to each strain (Delcher et al. 2002). Only the SSIs greater than 10 kb in size are reported. The *C. glutamicum* ATCC 13032 nomenclature is from Kalinowski (Kalinowski 2005), the *C. glutamicum* R nomenclature is from Suzuki et al. (Suzuki et al. 2005b), the *C. efficiens* nomenclature is from Zhang and Zhang (Zhang and Zhang 2005). CGP4 is a prophage inserted into CGP3 and is only present in the largest (Ikeda and Nakagawa 2003) of the two *C. glutamicum* ATCC 13032 isolates that have been sequenced to date. CGP3 is a putative prophage inserted at a *t*RNA-Val gene carrying integrase, primase, restriction and modification, and lysin genes (Kalinowski 2005). CGP2 is a prophage remnant carrying integrase and lysin genes, and CGP1 is an incomplete prophage inserted at a *t*RNA-Leu gene carrying an integrase gene. LGC1 is a low G + C content region containing genes involved in cell wall formation and lipopolysaccharide synthesis, while HGC1 is a high G + C content region probably acquired from *C. diphtheriae* by horizontal transfer (Kalinowski 2005). CGP2, CGP3, and CGP4 are absent from the genome of *C. glutamicum* R, but an incomplete CGP1 homologue comprising eight genes is observed at the corresponding locus. These latter genes are 100% identical at the amino acid level and approximately 90% identical at the DNA level to those from the *C. glutamicum* ATCC 13032 CGP1. The *C. glutamicum* R SSI-3 and SSI-5 are rich in transposases, whereas SSI-8 contains a putative phage-associated protein, a protease from the ClpP family, and a putative single-stranded DNA binding protein (Yukawa et al., manuscript in preparation). SSI-3 comprises eight genes highly homologous to genes from the *C. glutamicum* ATCC 13032 HGC1 region (*cgR0120*: 99.8% identical at the amino acid level, *cgR0121*: 98.4%, *cgR0122*: 99.2%, *cgR0123*: 98.2%, *cgR0125*: 98.8%, *cgR0129*: 95.5%, *cgR0131*: 100%, *cgR0133*: 96.5%). In addition, the SSI-4 region comprises 5 genes homologous to the *C. glutamicum* ATCC 13032 LGC1 region (*cgR0432*: 94.8%, *cgR0433*: 95.7%, *cgR0439*: 96.8%, *cgR0440*: 96.4%, *cgR0441*: 97.6%). The *C. efficiens* YS-314-specific islands CEG-1, CEG-2, and CEG-4 contain transposase genes, whereas CEG-3 mostly encodes hypothetical proteins, and an integrase and a phage-associated protein (Zhang and Zhang 2005). The *C. efficiens* analysis is based on cumulative GC profiles, a method which was designed to detect G + C content changes (Zhang and Zhang 2005), and thus may not detect all strain-specific islands

SSI	G + C content (%)	Length (kb)	ORFs	Coordinates
C. glutamicum R (Suzuki et al. 2005b): G + C content is 54.1%				
1	52.3	15.6	11	29 081–44 644
2	52.5	11.1	8	70 366–81 562
3	60.7	56.2	59	131 880–188 128
4	45.7	22.6	6	472 908–492 060
5	55.2	32.8	35	590 539–625 656
6	53.1	16.2	15	701 786–715 258
7	49.9	16.5	3	792 926–808 305
8	52.1	45.1	58	2 068 313–2 111 033
9	49.3	16.3	17	2 524 162–2 540 644
10	53.0	14.6	17	2 552 809–2 570 013
11	50.6	12.0	4	3 165 697–3 175 508
CGP1 homologue	48.5	12.8	10	1 543 871–1 556 716

Table 2 (continued)

SSI	G + C content (%)	Length (kb)	ORFs	Coordinates
C. glutamicum ATCC 13032 (Ikeda and Nakagawa 2003): G + C content is 53.8%				
LGC1	46.4	26.9	23	360 111–390 733
CGP1	46.5	13.5	18	1 400 182–1 413 641
CGP2	48.7	3.9	3	1 637 081–1 641 004
CGP3	49.0	219.7	206	1 776 613–1 995 294
CGP4	51.7	23.5	34	1 963 136–1 986 590
HGC1	64.3	20.6	22	3 156 304–3 176 905
C. efficiens ATCC YS-314 (Zhang and Zhang 2005): G + C content is 63.5%				
CEG-1	59.5	70.5	60	225 152–295 658
CEG-2	55.5	54.6	51	371 109–425 727
CEG-3	59.5	40.4	50	873 060–913 441
CEG-4	56.6	32.3	34	1 270 647–1 302 930

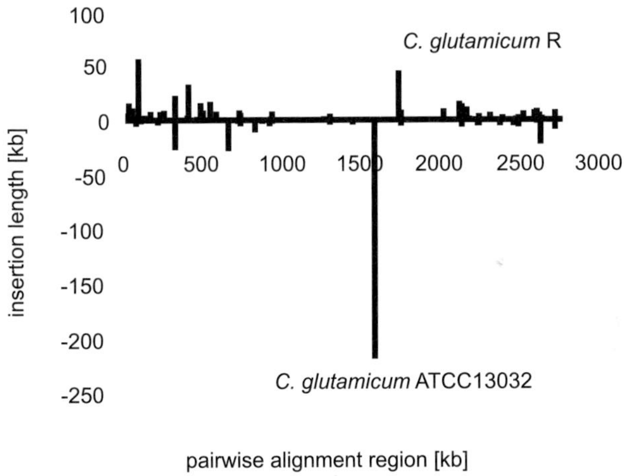

Fig. 4 Locations and sizes of strain-specific islands in the genomes of *C. glutamicum* strain ATCC 13032 and strain R. The origin of the horizontal axis coincides with the replication origins of both organisms. The island size is given on the vertical axis, the island location on the horizontal axis

cantly different from the rest of the *C. glutamicum* genome (Zhang and Zhang 2005). In addition, the horizontal transfer origin of this DNA segment, which comprises approximately 180 open reading frames, is corroborated by the similarity displayed by a few of these genes to transposase and known bacteriophage proteins such as a primase or an integrase (Kalinowski et al. 2003). The probable prophage nature of CGP3 is moreover confirmed by the integration in the CGP3 copy present in the largest of the *C. glutamicum* ATCC 13032 genomes, of CGP4, an additional prophage that carries genes coding for integrase, nuclease, single-stranded binding protein, and lysin genes (Kalinowski 2005). The important role that bacteriophages have played in the evolution of Corynebacteria by spreading genes horizontally is further stressed by the presence of phage-derived genes in all Corynebacteria sequenced to date. For example, the genomes of *C. glutamicum* R and *C. glutamicum* ATCC 13032 encode at least two and five putative phage integrases, respectively. It is interesting to note that only few native corynebacterial plasmids, such as plasmids of the pCG1 family, have been demonstrated to encode proteins putatively involved in bacterial conjugation (Tauch 2005).

Despite that no conjugative transposon has been observed in Corynebacteria to date, mobile elements constitute another group of genetic elements that are involved in species evolution by enabling DNA rearrangements, such as translocations, inversions, and deletions. For example, the genome of *C. glutamicum* ATCC 13032 encodes 27 mobile elements representing 24 different insertion sequences from nine different families, but 11 of these copies are partially deleted (Kalinowski et al. 2003). The best characterized of these mobile elements, IS*31831* (also sometimes referred to as IS*Cg1*) (Vertès et al. 1994) belongs to the IS*L3* family of insertion sequences and has been shown to be capable of spontaneous transposition (Garbe et al. 2004; Barreiro et al. 2005). Likewise, *C. glutamicum* R encodes a limited number of mobile elements, including six partially deleted elements and 16 complete and presumably functional insertion sequences, of which five could synthesize their transposases via the occurrence of frameshift mutations (Yukawa et al. manuscript in preparation). On the other hand, *C. efficiens* includes a much larger number (110) of open reading frames homologous to transposases. It is noteworthy that in *C. glutamicum* R, mobile elements are primarily located in large strain-specific islands, as exemplified by SSI-3 and SSI-5, which encode 12 and nine mobile element-derived open reading frames, respectively (Yukawa et al. manuscript in preparation). Similarly, the strain-specific islands CEG-1, CEG-2, and CEG-4 from *C. efficiens* encode numerous transposase genes (respectively, six, five, and 11) (Nishio et al. 2003). Nevertheless, while in both *C. glutamicum* R and *C. glutamicum* ATCC 13032 the strain-specific islands mostly code for hypothetical proteins, these genes may have important physiological functions, as exemplified by the presence in *C. glutamicum* R SSI-3 of genes coding for a transcriptional regulator of the LacI family and of a two-component system response regulator-sensory kinase couple. Furthermore, SSI-5 encodes a two-component system not found in any of the other Corynebacteria sequenced to date, but its component proteins are, respectively, 50% and 70% homologous to the CutS/CutR system that was observed to negatively regulate secondary metabolism in *Streptomyces lividans* (Chang et al. 1996). As a result, as is observed in *E. coli*, strain-specific islands constitute important sources of genetic diversity for

Corynebacteria that may confer numerous industrially useful genetic traits. Consequently, the specific sequencing of these islands from a large number of corynebacterial strains could provide a library of genetic properties that could significantly contribute to the optimization of *C. glutamicum* for industrial application, and thus could provide deeper insight into the biology of these bacteria by defining the *C. glutamicum* pan-genome.

2.3
Genomic Repertoires

An important attribute of *C. glutamicum* as a production organism is that it is able to synthesize (from a simple mineral medium) all the compounds that are necessary to sustain its life, with the important exception of biotin, since its biotin synthesis pathway is incomplete (Hatakeyama et al. 1993a, 1993b). An additional desirable trait displayed by *C. glutamicum* is its limited number of secreted proteins, as exemplified by its limited or absent proteolytic, nucleolytic, lipolytic, cellulolytic, and amylotic activities (Fig. 5). Notably, *C. glutamicum* can be successfully manipulated to express those catabolic properties when needed for specific biotechnological applications, as exemplified by the introduction in this organism of pathways to catabolize xylose (Kawaguchi et al. 2006) or soluble starch (Seibold et al. 2006). A functional classification in COGs (clusters of orthologous groups) (Tatusov et al. 2003) of the genes of the Corynebacteria sequenced to date is presented in Table 3. It is particularly worth noting that saprophytic Corynebacteria have more complete amino acid transport and metabolism, and carbohydrate transport and metabolism machineries than pathogenic Corynebacteria. In addition, a significant number of strain-specific genes are present in strain R and ATCC 13032, which provide a basis to understand the main physiological differences that can be observed between the two strains.

Based on the nucleotide sequences of their genomes, the Corynebacteria sequenced to date contain virtually all the genes that have been identified to be indispensable for the growth of the Gram-positive bacterium *Bacillus subtilis* in rich medium under standard laboratory conditions, including genes whose products are involved in DNA and RNA metabolism, protein synthesis, cell envelope, cell shape and division, glycolysis, respiratory pathways, nucleotides, and cofactors (Kobayashi et al. 2003).

Amino acid excretion by bacteria was modelled to occur via three possible mechanisms or any combination thereof: overflow metabolism (e.g., glutamate), limited oligopeptide catabolism (lysine, threonine, isoleucine), and deregulated anabolism (e.g., threonine production in *E. coli*) (Burkovski and Krämer 2002). The mechanisms of amino acid transport in amino acid-producing bacteria have been reviewed in detail by R. Krämer in a different section of this monograph. The most common excretion mechanism involves specific active export systems, even though diffusion of hydrophobic

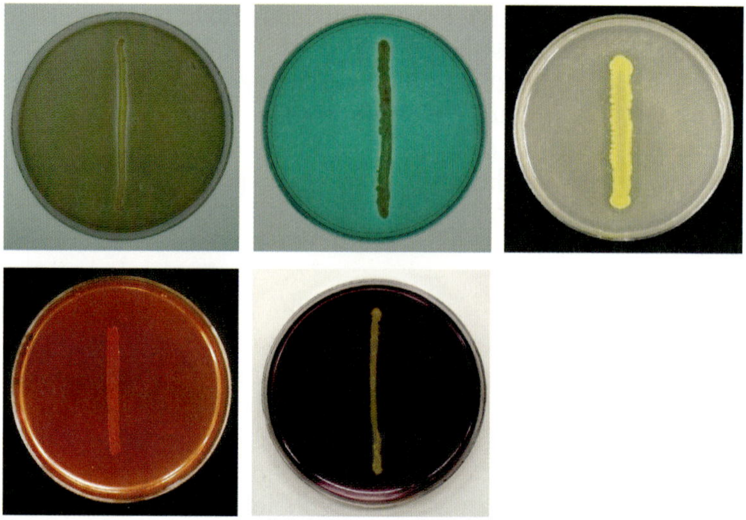

Fig. 5 Extracellular enzymatic activities of *C. glutamicum* R. Unless otherwise mentioned, assays were performed at 33 °C in rich A medium or mineral BT agar plates (Inui et al. 2004a), plates were incubated for 48 h. The listed reagents were all sourced from Difco except Avicel PH-101 which was sourced from Sigma and tributyrin which was sourced from Wako. The amylase test was performed in rich medium supplemented with 4% soluble starch, and were stained using 1 ml of an indole solution (1.3 mM I_2, 40.2 mM KI) at the end of the incubation period; the nuclease test was performed in mineral medium supplemented with DNase test agar with methyl green; the protease test was performed in rich medium in the presence of 4% skim milk, plates were incubated for 6 days; the cellulose test was performed in 2% mineral medium containing 0.5% Avicel PH-101, a microcrystalline cellulose, and 1 ml of a 5 mg/ml solution of Congo Red was used for staining at the end of the incubation period; the lipase test was performed on Nutrient Agar plates containing 1% (vol/vol) tributyrin which was emulsified by sonication after autoclaving, lipolytic activity was recorded following a 72-h incubation period

branched-chain amino acids has been observed in *C. glutamicum* (Burkovski and Krämer 2002). Notably, the various genes involved in the synthesis and export of the amino acids L-aspartate, L-lysine, L-threonine, L-methionine, L-isoleucine, and L-valine, and the vitamin D-pantothenate have all been identified. Moreover, the various metabolic pathways leading to the synthesis of these compounds have been reconstructed from the genome sequence of *C. glutamicum* ATCC 13032, including the steps of sugar uptake, central metabolism, biosynthesis, and transport (Kobayashi et al. 2003). *C. glutamicum* has been previously demonstrated to uptake fructose, glucose, mannose, and sucrose via the phosphotransferase system (Dominguez and Lindley 1996; Kotrba et al. 2001; Moon et al. 2005). However, the transport of pentoses remains still to this date ill-defined in these microorganisms despite the physiology of the corynebacterial pentose phosphate pathway has been studied in detail, and particularly the enzymes transketolase and transal-

Table 3 Clusters of orthologous genes in Corynebacteria. Upon BLASTP analysis of each of the genomes of the Corynebacteria sequenced to date against the NCBI COGs database, results were parsed for hits with an e-value cutoff of $1 \times e^{-25}$. After elimination of duplicates, the resultant hits, in accordance with their COGs identities, were grouped into the following categories: *CgR*, total hits of *Corynebacterium glutamicum* R; Core, hits with orthologs in all the Corynebacteria sequenced to date; Sap, hits with orthologs only in the saprophytic Corynebacteria sequenced to date; *Cglut*, hits with orthologs only in the two *C. glutamicum* R and Kitasato University *C. glutamicum* ATCC13032 isolate; R spec, non-core, non-*glutamicum* and non-saprophytic hits of strain R; K spec, non-core, non-*glutamicum*, and non-saprophytic hits of ATCC 13032 Kitasato strain; *Ce*, total hits of *Corynebacterium efficiens* YS-314; *Ce* spec, non-core and non-saprophytic hits of *C. efficiens*. Path, non-core hits of the pathogenic Corynebacteria sequenced to date; *Cd*, total hits of *Corynebacterium diphtheriae* NTCT 13129; *Cd* spec, non-core and non-pathogen-specific hits of *C. diphtheriae*; *Cj*, total hits of *Corynebacterium jeikeium* K411; *Cj* spec, non-core, non-pathogen-specific hits of *C. jeikeium*

	NCBI COG	*CgR*	Core	Sap	*Cglut*	R spec	*CgK*	K spec	*Ce*	*Ce* spec	*Cd*	Path	*Cd* spec	*Cj*	*Cj* spec
Information storage and processing															
J	Translation, ribosomal structure and biogenesis	**139**	117	13	6	3	**140**	**4**	**135**	**5**	**131**	**3**	**11**	**129**	**9**
A	RNA processing and modification	1	1	0	0	0	1	0	1	0	**1**	0	0	**1**	0
K	Transcription	**145**	42	58	35	10	**143**	**8**	**107**	7	**69**	**1**	**26**	**55**	**12**
L	Replication, recombination and repair	**96**	53	24	9	10	**113**	**27**	**102**	**25**	**79**	**9**	**17**	**77**	**15**
Cellular processes and signaling															
D	Cell cycle control, cell division, chromosome partitioning	**17**	13	2	2	0	**18**	**1**	**16**	**1**	**17**	**1**	**3**	**22**	**8**
V	Defense mechanisms	**42**	12	13	15	2	**43**	**3**	**39**	**14**	**35**	**4**	**19**	**43**	**14**
T	Signal transduction mechanisms	**37**	18	13	5	1	**40**	**4**	**34**	3	**22**	**1**	3	**25**	**6**
M	Cell wall/membrane/envelope biogenesis	**84**	49	16	8	11	**89**	**16**	**81**	**16**	**71**	**3**	**20**	**65**	**13**
U	Intracellular trafficking, secretion, and vesicular transport	16	11	3	1	1	16	1	15	1	15	**1**	3	12	0
O	Posttranslational modification, protein turnover, chaperones	**67**	35	21	9	2	**71**	**6**	**61**	**5**	**56**	**5**	**16**	**45**	**5**

Table 3 (continued)

NCBI COG	*CgR*	Core	Sap	*Cglut*	R spec	*CgK*	K spec	*Ce*	*Ce* spec	*Cd*	Path	*Cd* spec	*Cj*	*Cj* spec
Metabolism														
C Energy production and conversion	**117**	42	44	20	11	**113**	7	**99**	**13**	**78**	3	33	**68**	**23**
G Carbohydrate transport and metabolism	**136**	43	48	27	18	**124**	**6**	**107**	**16**	**83**	4	**36**	**64**	**17**
E Amino acid transport and metabolism	**161**	64	67	19	11	**160**	**10**	**145**	**14**	**123**	9	50	**112**	**39**
F Nucleotide transport and metabolism	**60**	35	20	4	1	**62**	**3**	**58**	**3**	**55**	4	16	**51**	**12**
H Coenzyme transport and metabolism	**95**	45	40	6	3	**95**	**4**	**95**	**10**	**92**	8	39	**66**	**13**
I Lipid transport and metabolism	**59**	22	18	7	12	**58**	**11**	**63**	**23**	**42**	5	15	**56**	**29**
P Inorganic ion transport and metabolism	**159**	23	75	43	18	**147**	**6**	**109**	11	**70**	16	31	**81**	**42**
Q Secondary metabolites biosynthesis, transport and catabolism	28	3	16	5	4	26	**2**	25	6	10	2	5	14	**9**
Poorly characterized														
R General function prediction only	223	80	100	38	5	238	20	200	20	133	6	47	123	37
S Function unknown	165	48	70	33	14	176	25	139	21	104	5	51	76	23
In Other COGs	251	79	101	48	23	253	25	214	34	150	10	61	141	52
Total	**2098**	**835**	**762**	**340**	**160**	**2126**	**189**	**1845**	**248**	**1436**	**100**	**502**	**1326**	**378**

dolase (Kamada et al. 2001; Krömer et al. 2004). Examination of the *C. glutamicum* ATCC 13032 genomic sequence reveals the presence of 153 open reading frames coding for subunits of transporters of the ATP-binding cassette (ABC) family, but, for most of these, specificities remain unclear. On the other hand, both *C. glutamicum* and *C. efficiens* have fewer secondary carriers for sugars than bacteria from the *Bacillus-Clostridium* branch of the eubacteria. For example, *C. glutamicum* ATCC 13032 displays only three genes coding for transporters of the major facilitator superfamily (MFS), while none was observed in the *C. efficiens* genome (Winnen et al. 2005). The understanding of primary and secondary sugar transport in amino acid-producing bacteria thus still to this date represents an important area of development for the generation of optimized strains of Corynebacteria. This view is particularly promoted by the observation that amino acid yields attained using Corynebacteria-based processes are linked to the nature of the carbon source utilized (Shiio et al. 1990; Dominguez et al. 1998; Kiefer et al. 2002; Wittmann et al. 2004; Georgi et al. 2005). In addition, at high growth rate, glucose transport via a secondary carrier has been evidenced to represent up to 15% of the total glucose transport (Cocaign-Bousquet et al. 1996). Similarly, the genomic sequence provides additional insights of the solute secretion machinery of these organisms and numerous novel carriers have thus been identified (Vrljic et al. 1996; Burkovski and Krämer 2002; Trotschel et al. 2005). However, the efflux properties of these novel carriers require additional experimental work to be understood. This is an area of research of high biotechnological significance, as transporters could in principle be modulated to further improve industrial fermentation yields (Bröer et al. 1993).

Likewise, the *C. glutamicum* genes coding for putative aminotransferases or pyridoxal-5′-phosphate-dependent enzymes have been catalogued from the genome sequence of the type strain (McHardy et al. 2003; Marienhagen et al. 2005). At least 20 of these enzymes putatively involved in amino acid synthesis or interconversion could be inferred by genome mining and their biological functions characterized (McHardy et al. 2003; Marienhagen et al. 2005). The enzymes identified in the *C. glutamicum* ATCC 13032 genome are all present in *C. glutamicum* R and *C. efficiens* YS-314, with the notable exception of two enzymes putatively involved in butanoate and pyridoxal-5′ phosphate synthesis that are absent in *C. efficiens*. On the other hand, *C. glutamicum* R encodes one additional putative aminotransferase (CgR2801), which is 64% and 63% homologous to the *C. glutamicum* ATCC 13032 and *C. glutamicum* R *aecD* gene products, respectively.

2.4 Comparative Genomics

The combination of bioinformatics techniques and comparative genomics is a powerful tool to characterize the function of proteins encoded by putative genes. Such an approach was used by Brune et al. (2005) to identify the repertoire of DNA-binding transcriptional regulators found in the genomes of *C. glutamicum* ATCC 13032 (Kalinowski et al. 2003), *C. efficiens* YS-314 (Nishio et al. 2003), *C. diphtheriae* NCTC 13129 (Cerdeño-Tárraga et al. 2003a), and *C. jeikeium* K411 (Tauch et al. 2005a). *C. glutamicum* appears to encode more DNA-binding regulators than *C. efficiens*, respectively 127 and 103, whereas the pathogenic Corynebacteria encode significantly fewer such proteins (*C. diphtheriae*: 63; *C. jeikeium*: 55). The corynebacterial transcriptional regulators thus identified correspond to 25 regulatory protein families. A total of 28 of these regulators, putatively involved in key cellular processes such as cell division and septation regulation, SOS and stress response, carbohydrate metabolism, macroelement and metal homeostasis, are present in all Corynebacteria sequenced to date (Brune et al. 2005). Understanding the regulation of genetic expression in Corynebacteria remains an issue of significant biotechnological importance (Wendisch 2006). Particularly, six conserved genes (*glxR*, *ramB*, *acnR*, *deoR* family, *iclR* family, *lacI* family) seem to play a central role in regulating carbohydrate metabolism in *C. glutamicum* (Brune et al. 2005), with for example the AcnR repressor being involved in the control of the tricarboxylic acid cycle (Krug et al. 2005). The higher number of regulators relative to genome size observed in saprophytic Corynebacteria, as compared to pathogenic Corynebacteria, is consistent with their lifestyles as soil organisms. As previously mentioned, saprophytic Corynebacteria flourish on the skin of a variety of plants including fruits and vegetables, and are thus exposed to a variety of fast changing conditions, including nutrient limitations, oxygen gradients, temperature variations, and water activity changes. This need is also reflected in the greater versatility of the metabolism exhibited by saprophytic Corynebacteria compared to pathogenic Corynebacteria and in their larger number of genes (Table 1, Table 3).

The complete genomic sequence of several strains of Corynebacteria also gave indications on the basis of transcriptional regulation mechanisms in these organisms as operated by specialized sigma factors. Sigma factors are RNA polymerase subunits that act by increasing the affinity of this enzyme to a promoter (Wosten 1998). In addition to σ^A, which directs the transcription of most genes in growing cells, and σ^B, which plays a crucial role in the maintenance of the stationary phase, Corynebacteria exhibit a limited number of sigma factors of the extracytoplasmic family (σ^C, σ^D, σ^E, σ^H, σ^M) and two hypothetical proteins (*pvds1*, *pvds2*) that show a limited homology to sigma factors (Oguiza et al. 1996; Kim et al. 2005). The sigma factor σ^D is interrupted by a spacer region in *C. glutamicum* R (Yukawa et al. manuscript in preparation). Moreover, except σ^A and σ^C, all the other sigma factor genes are dispensable as demonstrated by their disruption in *C. glutamicum* R by transposon mutagenesis (Suzuki et al. 2006). Nevertheless, the transcriptional machinery of *C. glutamicum* is versatile as demonstrated by the functionality of numerous promoters of this microorganism in other genera such as *E. coli*, *Streptomyces lividans*, or *B. subtilis*, and by the functionality of *E. coli* or *B. subtilis* promoters in *C. glutamicum* (Zupancic et al. 1995; Pátek et al. 2003; Vertès et al. 2005).

The corynebacterial core genome comprises approximately a third of the genes present in the saprophytic Corynebacteria and half of those present in pathogenic Corynebacteria (Tauch et al. 2005a). Of the genes that have been assigned to a particular COG category, 835 appear to be shared by all Corynebacteria sequenced to date (Table 3). It has been suggested that gene decay and gene reduction have been the major evolutionary response of *C. diphtheriae* to adapt to a clinical ecological niche (Nishio et al. 2004). Particularly, the common ancestor of *C. glutamicum*, *C. efficiens*, and *C. diphtheriae* was deduced by evolutionary phylogeny techniques to have been equipped with almost all of the subset of genes necessary for amino acid production, but *C. diphtheriae* had lost many of those genes during the course of its evolution. However, horizontal transfer events also have played a critical role in conferring to both *C. efficiens* and *C. glutamicum* some amino acid metabolic genes (Nishio et al. 2004).

2.5 Essential Genes

The subset of genes that are essential for *C. glutamicum* have been investigated by global transposon mutagenesis of strain R yielding 18 000 mutants (Suzuki et al. 2006). A minimum of 2,330 genes proved to be dispensable in rich medium under standard laboratory conditions. These included 30 open reading frames that show a high homology to candidate essential genes of *E. coli* (Gerdes et al. 2003; Kang et al. 2004) or *B. subtilis* (Kobayashi et al. 2003). Among the 658 candidate essential genes of *C. glutamicum* R, 116 code for hypothetical proteins and 21 for putative transcriptional regulators (Suzuki et al. 2006). Nonetheless, this number probably represents an overestimate since some genes may have been missed by chance or due to polar effects of some transposon insertions, for, when using optimum growth conditions and several knock-out approaches, the subset of genes that are essential to various microorganisms has been determined to comprise no more than 206 genes (Gil et al. 2004). Nevertheless, detailed analysis of these candidate essential genes might generate important insights for further improving amino acid manufacturing yields. Similarly, an alternative method involving the use of transposon site hybridization has been described in Mycobacteria to isolate subsets of genes whose disruption results in auxotrophic mutations under chosen incubation conditions (Sassetti et al. 2001). It is particularly noteworthy that numerous genes of the respiratory chain appear essential in *C. glutamicum* growing in rich medium under standard laboratory conditions, including heme biosynthesis genes (Fig. 3) (e.g., uroporphyrinogen III synthase, δ-aminolevulinic acid dehydratase, uroporphyrinogen decarboxylase, protoporphyrinogen oxidase, ferrochelatase), cytochrome synthesis genes, and the F_1F_0 ATP synthase operon (Suzuki et al. 2006). Interestingly, a reduced activity of the ATP synthase of *C. glutamicum* ATCC 14067 was observed to result in the abolition of glutamic acid production concomitant to the enhancement of the glucose consumption rate and the rate of lactic acid production (Sekine et al. 2001). In addition, the F_1F_0 ATP synthase operon appears to be induced at alkaline pH, and particularly at pH 9, which is the optimal pH for *C. glutamicum* growth, perhaps via a mechanism that senses pH changes to mediate a higher rate of ATP synthesis and increased growth rate near the optimal pH (Barriuso-Iglesias et al. 2006). More generally, changes in the efficiency of oxidative phosphorylation either via defects in the ATP synthase or at the level of the respiratory chain have strong effects on both general metabolism and amino acid production (Bott and Niebisch 2003). This result is not unexpected since *C. glutamicum* generates most of its ATP by oxidative phosphorylation. Therefore, the respiratory chain of *C. glutamicum*, which comprises various dehydrogenases, menaquinone, cytochrome bc_1-aa_3 supercomplex, cytochrome *bd* oxidase, and nitrate reductase (Bott and Niebisch 2003), represents a promising target for rational design that has been largely neglected to date, where both oxygen and nitrate can serve as terminal electron acceptor (Vertès et al. unpublished). This concept is perhaps best exemplified by the overexpression of *glbO*, a *C. glutamicum* gene coding for a hemoglobin-like protein, which resulted in improved lysine production yields (Möckel et al. 2002). This gene is present in all Corynebacteria sequenced to date, and was shown in *C. glutamicum* R to be dispensable (Suzuki et al. 2006).

2.6 Transcriptional Profiling Analyses

Complete genomic sequences have enabled the construction of DNA microarrays to measure the modulation of genetic expression under a variety of environmental conditions. Transcriptional profiling is a technique that is particularly useful to optimize manufacturing strains by rational design approaches, as in principle it allows to identify the subset of genes that are particularly important under given incubation conditions such as amino acid production under high oxygen tensions, moderate growth rate or carbohydrate rationing, including when various membrane weakening compounds such as penicillin G are added, or when heat treatment or biotin removal are performed. Furthermore, the co-regulation of hypothetical proteins with well characterized proteins provides a clue to their functions. However, a limitation of this technique is that it provides no information regarding constitutively expressed genes or those metabolic nodes that are regulated at the protein modification level, and it does not take into account differential transcript stability, such that the relationship between *m*RNA abundance and enzyme activity is not always straightforward (Glanemann et al. 2003). As a result, transcriptional profiling analyses need to be complemented with proteomics and biochemistry studies in order to gain a deeper insight of the physiology of a cell suspension under process conditions (Wendisch 2003; Polen and Wendisch 2004). Such an approach has been successfully used to define the physiological response of wild-type strains when submitted to various stresses such as ammonium limitation (Silberbach et al. 2005), phosphate starvation (Ishige et al. 2003), or heat shock (Muffler et al. 2002), or to better characterize the effect of specific modifications on the general metabolism of production strains (Hüser et al. 2005) as well as to better understand the physiological events occurring in *C. glutamicum* cells under amino acid production conditions (Lange et al. 2003; Krömer et al. 2004; Radmacher et al. 2005; Stansen et al. 2005; Hayashi et al. 2006b; Kataoka et al. 2006),

For example, under glutamate overproduction conditions, most genes involved in glycolysis, the pentose phosphate pathway, and the tricarboxylic acid pathway are down-regulated. This observation promotes the view that L-glutamate overproduction can be ascribed to pre-existing enzymes, rather than to enzymes induced upon the switch to glutamate production conditions. Glutamate production is typically triggered by a variety of environmental changes including temperature shift, biotin limitation, addition of Tween 40, Tween 80, or penicillin (Stansen et al. 2005; Kataoka et al. 2006). Under these conditions, the pentose phosphate pathway flux is low, as well as the rate of conversion of 2-oxoglutarate to oxaloacetate; anaplerotic demands are fulfilled by the action of pyruvate carboxylase (Stansen et al. 2005). Transcriptome analyses of *C. glutamicum* cells subjected to either biotin limitation, Tween or penicillin addition revealed minor down-regulation of the pyruvate carboxylase gene and severe reduction of the expression of the phosphoenolpyruvate carboxykinase gene thus enabling the efficient production of oxaloacetate (Kataoka et al. 2006). On the other hand, acetyl-CoA, which is with oxaloacetate an intermediate in L-glutamate synthesis, is not affected by either of these latter triggers, as only one of the two isocitrate lyase genes of *C. glutamicum* is regulated upon these changes. Nonetheless, five genes appear to be more than eightfold up-regulated under either of the glutamate overexpression conditions tested (Kataoka et al. 2006). The products of all these five genes are uncharacterized, though three of these appear to be highly homologous to each other (*VIMSS 377000*, *VIMSS 377001*, *VIMSS 377002*) and one is a putative copper chaperone (*VIMSS 377031*) (Kataoka et al. 2006). A similar observation was made upon addition of the cell wall inhibitor ethambutol as a means to promote L-glutamate efflux from *C. glutamicum* cells: 58 genes are down-regulated upon addition of this antibiotic, while 18 genes are more than eightfold up-regulated. Among the up-regulated genes, *VIMSS 374631* encoding the RNA polymerase subunit SigD, is perhaps the most outstanding since this sigma factor is involved in

Mycobacterium tuberculosis in regulating gene expression for optimal growth in the absence of oxygen limitation, with its expression linked to the stringent response by being increased upon starvation (Raman et al. 2004). Other notable up-regulated genes include the cell wall-related genes *ftsE* and *mepA*, respectively, coding for a cell division ATP-binding protein and a putative secreted metalloprotease. In addition, a putative ABC-type lipid transport permease (*VIMSS 374450*), and *VIMSS 377000* and *VIMSS 377002* are also overexpressed upon biotin limitation or Tween or penicillin addition (Radmacher et al. 2005; Kataoka et al. 2006). Differential expression of these numerous genes is likely to result in multiple molecular events that ultimately result in increased L-glutamate efflux via alteration of the lipid or peptidoglycan composition of the cellular envelope (Radmacher et al. 2005). Likewise, global transcription analyses of *C. glutamicum* 2262 cells revealed that this strain overexpresses an operon coding for a lactate permease and a quinone-dependent lactate dehydrogenase. Strain 2262 exhibits the useful property of being triggered to produce glutamate via temperature shift. Interestingly, uptake and utilization of lactate secreted into the medium during glutamate production constitutes a lead to improved glutamate yields (Stansen et al. 2005).

Numerous *C. glutamicum* strains producing other amino acids, such as lysine or threonine, have been derived mostly by classical random mutagenesis and screening techniques (Vertès et al. 2005). For example, the L-lysine producer *C. glutamicum* B-6 was derived from the *C. glutamicum*-type strain ATCC 13032 to produce industrial levels of L-lysine greater than 100 g/l (Hayashi et al. 2006b). The mutations in strain B-6 that enable lysine overproduction lead not only to high-level expression of the pentose phosphate pathway but also to increased global expression of amino acid biosynthetic genes including *lysC-asd*, *serAB*, and *aroEBKC*, and low levels of expression of genes coding for enzymes of the tricarboxylic acid pathway (Hayashi et al. 2006b). This coordinated up-regulation of numerous amino acid biosynthesis genes suggest changes in a global regulatory mechanism in strain B-6 (Hayashi et al. 2006b). A similar study was performed using the lysine-producing strain *C. glutamicum* ATCC 13287, a homoserine auxotroph derived from the type strain *C. glutamicum* ATCC 13032 by UV mutagenesis and selection (Krömer et al. 2006). Parallel analysis of intracellular metabolite concentrations, metabolic fluxes, and gene expression patterns revealed that the shift of strain ATCC 13287 from vegetative growth to lysine production is characterized by a decrease in glucose uptake flux, a redirection of the tricarboxylic acid flux towards anaplerotic carboxylation and synthesis of lysine, and complex changes in the genetic expression of central metabolism genes as well as transient dynamics of intracellular metabolite pools and strong induction of the lysine exporter. Notably, the observed decrease in glucose uptake flux corresponds to changes occurring at the expression level of glucose phosphotransferase genes. Interestingly, under the conditions tested, expression of lysine biosynthesis genes is relatively constant, in spite of significant changes in metabolic fluxes, thus demonstrating regulations at the metabolic level (Krömer et al. 2006).

Data gathered to this date clearly demonstrate the usefulness of combining various post-genomic profiling techniques to provide quantitative and qualitative information on the various biological networks that mediate amino acid production. These techniques thus offer the possibility to design improved bacterial amino acid producers bearing an optimal set of rationally selected mutations (Krömer et al. 2004). Moreover, combined post-genomic techniques enable to identify the beneficial mutations in strains developed using classical random mutagenesis and screening schemes and to transfer these mutations in a strain carrying an optimal number of mutations (Ikeda and Nakagawa 2003).

2.7 Proteomic Repertoire

The *C. glutamicum* ATCC 13032 genome encodes approximately 3000 open reading frames, 27% of which are predicted to be membrane proteins (Ikeda and Nakagawa 2003; Kalinowski et al. 2003). Techniques for the identification of the proteins expressed by *C. glutamicum* cells and present in either the soluble fraction or the membrane fraction have recently been developed. For example, Schaffer et al. used two-dimensional gel electrophoresis techniques of *C. glutamicum* ATCC 13032 protein extracts to yield 970 protein spots from the soluble protein fraction at the 4–6 p*I* range (Schaffer et al. 2001). They identified 169 protein spots by matrix-associated laser desorption ionization time of flight mass spectrometry (MALDI-TOF-MS) from tryptic digest of proteins recovered from excised spots of interest. These spots represented 152 different proteins, including central metabolism enzymes (18 proteins), amino acid biosynthesis enzymes (24 proteins), nucleotide biosynthesis enzymes (11 proteins), and proteins of unknown functions (35 proteins). Identifications were performed by matching a digest database of 3312 putative *C. glutamicum* proteins. Similarly, membrane-bound proteins were also submitted to two-dimensional gel electrophoresis but could be analyzed in the larger p*I* range of 4–7 (Schaffer et al. 2001). A total of 660 protein spots were observed, 22 of which were further characterized and identified to represent 13 different proteins. Of the identified proteins, four were present in both the soluble and the membrane-bound fraction (Schaffer et al. 2001). These results were confirmed in a separate study where the proteome of both *C. efficiens* YS-314 and *C. glutamicum* ATCC 13032 were analyzed by two-dimensional gel electrophoresis combined to tryptic peptide fingerprinting by MALDI-TOF-MS to cover a 3–7 p*I* range with a molecular range of 10–130 kDa (Hansmeier et al. 2006). Using cells grown at 37 °C in Luria–Bertani medium, of the 2950 coding regions identified in the genome of *C. efficiens* (Nishio et al. 2003),

635 protein spots where observed in the cytosolic fraction, 76 in the cell surface fraction, and 102 in the extracellular fraction. Mass spectrometry data enabled the identification of, respectively, 164, 49, and 89 protein spots representing 177 different proteins. Comparison with the proteome of *C. glutamicum* ATCC 13032 grown in the same medium but at 30 °C revealed not only the extensive similarity of the two organisms but also species-specific patterns. For example, *C. efficiens* shows a different cell surface structure. This is particularly manifested by the presence of fimbrial proteins that are absent from *C. glutamicum*. In addition, both of these organisms also exhibit significant genetic expression differences, as demonstrated by variations in protein abundance (Hansmeier et al. 2006). Particularly, *C. efficiens* expresses proteins of amino acid biosynthesis in its cytosol (e.g., IlvE, Asd, and Hom) in higher abundance than does *C. glutamicum*. Could this property possibly be exploited to confer to *C. efficiens* important industrial advantages for amino acid production?

Nevertheless, anion-exchange chromatography followed by sodium dodecyl sulfate polyacrylamide gel electrophoresis represents an improved technique to access intact integral membrane proteins. Using this procedure that appears to be suitable for proteins 10–120 kDa and in the 3.7–10.6 p*I* range, Schluesener et al. (2005) identified 50 integral membrane proteins from a wide functional space, including energy metabolism, transport, signal transduction, protein translocation, and proteolysis. On the other hand, separation strategies based on multi-dimensional protein identification technology (MudPIT) and improved digestion conditions enabled the identification of 326 integral membrane proteins, that is, more than 50% of the predicted number of such proteins (Fischer and Poetsch 2006; Fischer et al. 2006). MudPIT combines two-dimensional capillary chromatography of peptides with in-line ion trap mass spectrometry, and thus allows the high-throughput identification of integral membrane proteins in complex mixtures. Applied to a sample of a lysine-producing *C. glutamicum* culture, this technique particularly allowed the identification of 142 out of 417 of the predicted hypothetical proteins that are integral membrane proteins, as well as 82 out of 153 of the transport and binding integral membrane proteins, and 24 out of 39 integral membrane proteins related to energy metabolism (Fischer et al. 2006). It is worth noting that various components of the respiratory chain also could be identified in this sample using MudPIT, including numerous dehydrogenases (Ndh, Mqo, Poxb, Dld, Ldh, Glpd, PutA, SdhA, SdhB, and SdhCD), the nitrate reductase (Narg, NarH, NarI) and cytochrome-c-aaa$_3$ oxidase complex (QcrB, QcrA, QcrC, CtaF, CtaD, Ctaf), and the complete ATP synthase complex (Fischer et al. 2006). Notably, phosphorylated proteins of *C. glutamicum* also were mapped by two-dimensional gel electrophoresis, in an effort to better understand the mechanism of regulation of gene expression and protein activity in these bacteria (Bendt et al. 2003). A total of 119 phosphorylated proteins have been observed, and 41 of these proteins have been identified using MALDI-TOF-MS and database searches. Most of the proteins thus identified are relatively abundant and act at the level of the central metabolism: glycolysis (enolase, fructose-1,6-biphosphate aldolase, Gap dehydrogenase) and tricarboxylic acid cycle (aconitase, citrate synthase, isocitrate dehydrogenase), and fatty acid metabolism (acyl CoA synthetase, acyl CoA carboxylase, succinyl CoA:CoA transferase) (Bendt et al. 2003). However, regulatory proteins such as proteins from two-component systems could not be detected in global proteomic analyses to date, perhaps due to their relatively low abundance in the protein extracts. Similarly, protein phosphorylation stoichiometry of the cytoplasmic proteome of *C. glutamicum* was analyzed by either a combination of one-dimensional gel electrophoresis, in-gel digestion, and final capillary liquid chromatography and inductively coupled mass spectrometry (ICPMS), or a combination of one-dimensional gel electrophoresis, protein blotting, and imaging laser ablation and ICPMS (Krüger et al. 2006). Both methods yield consistent results suggesting that approximately 3% of the *C. glutamicum* protein is phosphorylated at least at one site, but the degree of phosphorylation remains undefined (Krüger et al. 2006). Moreover, proteins secreted by *C. glutamicum* were determined for cells incubated in minimal medium. A total of 40 spots in the 4.0–5.0 p*I* range were observed in the supernatant of late exponential phase culture. One of these proteins was demonstrated using an antiserum to be glutamine synthetase (Hermann et al. 2001).

In conclusion, while a leap in the analytical power of global proteomic techniques is still needed, these techniques already offer important complementary analyses to confirm results obtained using global transcriptional profiling (Lange et al. 2003; Krömer et al. 2004; Silberbach et al. 2005).

3
The *Escherichia coli* Genome

The Gram-negative *E. coli*, and particularly strain K-12, is, with the Gram-positive *B. subtilis* and the yeast *Saccharomyces cerevisiae*, one of the best characterized microorganisms. A full review of the current knowledge of its biology is available elsewhere (Böck et al. 2006).

3.1
Structure

To date, the genomic sequences of at least six different *E. coli* strains have been published including the benign strains K-12 MG1655 (Blattner et al. 1997) and K-12 W3110 (Fukiya et al. 2004; Hayashi et al. 2006a), and the clinical species CFT073 (Welch et al. 2002), UTI89 (Chen et al. 2006), and O157:H7 (Hayashi et al. 2001; Perna et al. 2001). The versatility of the *E. coli* group of

Table 4 Genomic features of *E. coli* strains. Values reported in the table are from the NCBI Entrez database, except the estimate of the number of genes present in strains MG1655 and W31130, which has been updated elsewhere based on sequence and annotation verifications (Baba et al. 2006). Native plasmids harbored by these strains are not indicated. Strains K-12, MG1655 and K-12 W3110 are commensal organisms, whereas UTI89 and CFT073 are uropathogenic and O157:H7 Sakai and O157:H7 EDL 933 are potent enterohaemorrhagic pathogens. *Shigella flexneri* strains serotype 2a 2457T and 301 that have also been sequenced (Yang et al. 2005) are considered to be *E. coli* strains but are not included in this comparative Table

Strain	K-12 MG1655	K-12 W3110	UTI89	CFT073	O157:H7 Sakai	O157:H7 EDL933
Accession #	U00096	AP009048	CP000243	AE014075	BA000007	AE005174
Reference	(Blattner et al. 1997)	(Hayashi et al. 2006a)	(Chen et al. 2006)	(Welch et al. 2002)	(Hayashi et al. 2001)	(Perna et al. 2001)
Total genome size (bp)	4 639 675	4 636 552	5 065 741	5 231 428	5 498 450	5 528 445
Genome G + C content	51%	50%	51%	50%	51%	50%
Number of ORFs	4464	4474	5176	5589	5395	5453
Coding density	86%	79%	89%	87%	85%	87%
*t*RNA	86	86	88	89	105	98
*r*RNA	22	22	22	21	22	22

organisms, which comprises commensal bacteria as well as enteropathogenic, enterohemorrhagic, enteroinvasive, enterotoxigenic, and enteroaggregative pathogens, is reflected in the diversity of the *E. coli* genetic blueprint and the complexity of its pan-genome.

The size of the *E. coli* genome varies from 4.6 to 5.5 megabases (Table 4). In contrast to what is observed in Corynebacteria, the genomes of the benign strains are approximately 400 to 900 kb smaller than those of the pathogenic strains. The *E. coli* genome is circular with a G + C content of 50 to 51% (Table 4). The genome of *E. coli* K-12 W3110 has a large inversion as compared to the original *E. coli* K-12 isolate and strain K-12 MG1655 (Itoh et al. 1999). Protein coding regions of the *E. coli* K-12 MG1655 genome represent 87.8% of the total sequence and the average open reading frame is 951 bp long (Blattner et al. 1997); 381 open reading frames are smaller than 100 amino acids (Blattner et al. 1997). Structural maps of the *E. coli* genome, revealing a large region comprising the replication terminus that displays, as expected, a high level of curvature, low level of stability, and low degree of helix stability, have been published elsewhere (Pedersen et al. 2000). In this organism also, chromosome replication is bidirectional; it starts at *oriC* and terminates when the two replication forks meet at the chromosome region opposite to *oriC*, where progress of the clockwise and counterclockwise replication forks is inhibited by *ter* regions acting in a direction-specific manner. The origin, terminus, and other genomic DNA regions exhibit marked migration patterns during active partitioning of daughter chromosomes (Niki and Hiraga 1998). Strain K-12 exhibits seven 16S *r*RNA, seven 23S *r*RNA, 8 5S *r*RNA, and 86 *t*RNA genes. All the seven *r*RNA operons, 53 of the 86 *t*RNA genes, and 55% of protein coding genes are expressed in the direction of replication (Blattner et al. 1997). The evolutionary importance of the direction of replication is furthermore emphasized by the statistically significant greater abundance of G (26.22%) over C (24.58%), A (24.52%) or T (24.69%) in the replication leading strand (Blattner et al. 1997). It is also noteworthy that Chi sites, which are recombination hot spots, are likewise clearly skewed towards the leading strand (Blattner et al. 1997). *E. coli* exhibits codon usage bias (Ikemura 1981), with codons CUA, CGA, AUA, AGA, and AGG being the least frequent codons found in *E. coli* proteins. Codon usage-related biotechnological manufacturing problems can be circumvented by site-directed mutagenesis of genes coding for proteins of interest or by increasing the gene dosage of specific *t*RNAs (Baneyx 1999; Jana and Deb 2005).

3.2 Mobile Elements, Prophages, and Strain-Specific Islands

Notably, the genomes of the two K-12 isolates sequenced to date, strain MG1655 and W3110, are not identical. They differ in size due to a higher number of insertion sequences and the absence of a defective phage in the

W3110 genome (Riley et al. 2006). The integrase of this prophage shares homology with the integrase of phiR-73, Sf6, and the CP4 cryptic prophage family, but its other genes are markedly different from those of this family of prophages (Blattner et al. 1997). The genome of *E. coli* K-12 comprises numerous other prophages and phage remnants, including phage lambda and the defective lambdoid prophages DLP12, Rac, Qin, the element e14, and CP4-57, CP4-6, CP4-44 (Blattner et al. 1997). *E. coli* K-12 includes also numerous transposase genes corresponding to a variety of elements from a variety of insertion sequence family as defined elsewhere (Siguier et al. 2006): IS*1* (13 copies), IS*3* family [IS*2* (12 copies), IS*3* (five copies), IS*150* (one copy), IS*911* (seven copies)], IS*4* family [IS*4* (nine copies), IS*186* (two copies)], IS*5* (ten copies), IS*30* (three copies), IS*605* (one copy), IS*609* (one copy), other elements (16 copies). Interestingly, prophage CP4-6 includes the IS*911A* complex, a partial copy of the insertion sequence IS*30*, two complete copies of IS*1* and one copy of IS*5* (Blattner et al. 1997). Moreover, these strains also exhibit numerous mutations reflecting changes that have presumably occurred during their maintenance in different laboratories (Riley et al. 2006).

E. coli can take up exogenous DNA by a variety of mechanisms including transduction, conjugation, and transformation (Matic 1995). Bacteriophage Mu for example is an important vector. It is a phage that replicates via transposition; it produces on average 100 phage particles per cell at the time of cell lysis and packages into virion particles approximately 2 kb of its host's DNA covalently attached to the prophage right end (Manna et al. 2004). The importance of horizontal transfer in shaping the genetic blueprint of *E. coli* during the course of the evolution of this organism has been examined by analyzing intraspecific recombination events in six complete *E. coli* genomes. Mosaic operons and genes from 187 genomic DNA clusters of single nucleotide differences, comprising a 100-kb-long region, were revealed that are likely to have resulted from horizontal transfer between lineages (Mau et al. 2006). As mentioned by Perna et al. (2001) and Wirth et al. (2006), results of various similar genome-scale analyses all promote the view that enterobacterial genomes are particularly prone to recombinational evolution, and moreover that the evolution of virulence is strongly linked to horizontal DNA transfer. Specific pathogen types appear to have arisen independently in several lineages, while other lineages contain mostly commensal strains and few pathogens (Wirth et al. 2006). This conclusion is validated particularly by the observation that large genomic islands appear to be frequent in this genus, as perhaps best exemplified by the uropathogenic *E. coli* strain UTI89, which exhibits four large pathogenicity islands, the largest of which is 119 kb in size, and numerous smaller genomic islands that are likely to also contribute to the virulence of this strain (Chen et al. 2006). The view that genomic islands contribute to a great extent to the diversity of *E. coli* isolates is further promoted by the observation that the general structure of the *E. coli* genome is largely composite in all strains studied, as exemplified by the genomes of the pathogenic strains CFT073 and EDL933 that are as different from each other as they are from the benign strain K-12 (Welch et al. 2002). Furthermore, a total of 1387 new genes encoded in strain-specific islands of various sizes were catalogued from the genome of strain EDL933, including putative virulence factors, alternative metabolic capacities, and prophages (Perna et al. 2001). Detailed comparison between *E. coli* K-12 MG1655 and *E. coli* O157:H7 EDL933 revealed that 1.34 megabases of the EDL933 genome, representing 26% of the total open reading frames, are not present in strain MG1655, and reciprocally that 0.53 megabases of the MG1655 genome, representing 12% of the total open reading frames, are not present in strain EDL933 (Perna et al. 2001). In addition, it was demonstrated that 755 open reading frames representing 548 kb have been introduced into the *E. coli* MG1655 genome via at least 234 horizontal transfer events since the divergence of *E. coli* from *Salmonella enterica* (Lawrence and Ochman 1998). As a result, similar to what is observed in Corynebacteria, strain-specific islands represent regions of high diversity that deserve to be studied in details in numerous isolates as a means to better define the enterobacterial pan-genome and to discover novel useful catabolic or anabolic properties.

3.3 Genomic Repertoires

The variety of ecological niches that *E. coli* strains are able to occupy, as well as the relatively large size of their genomes, suggest that these organisms encode a wide set of metabolic properties. *E. coli* strains are Gram-negative, non-sporulating, rod-shaped mesophilic bacteria that are often motile via peritrichously arranged flagella. Most strains grow well on both rich and mineral media under aerobic and anaerobic conditions. Noteworthily, *E. coli* typically can reduce nitrate to nitrite to conduct anaerobic respiration (Bettelheim 1992). Classification of the *E. coli* genes in COGs (Tatusov et al. 2003) reveals that the fraction of genes corresponding to each COG does not markedly differ between pathogenic and commensal strains (Table 5). Interestingly, the number of genes involved in defense mechanisms also remains approximately equal to 0.01% in both benign and pathogenic *E. coli* strains. Moreover, all *E. coli* strains sequenced to date encode at least 325 genes involved in amino acid transport and metabolism. Furthermore, *E. coli* has an exceptionally broad transport capability enabled by a relatively high number of transporters per unit of genome size reaching 66 transporters/megabase, as calculated on the basis of 304 transporters in the *E. coli* genome, comprising per megabase 14 transporters of the major facilitator superfamily and 15 transporters of the ABC superfamily (Paulsen et al. 2000). *E. coli* primarily transports sugars via the phosphotransferase system (Kotrba et al. 2001; Saier 2001) and can use a large variety of carbon sources, including glucose, arabinose, lactose, maltose, mannitol, mannose, rhamnose, sorbitol, trehalose,

Table 5 Classification of *E. coli* genes in clusters of orthologous genes. Values reported in the table are from the NCBI database. Some proteins, such as multi-functional domain proteins, may be listed in more than one COG as duplicates have not been excluded. Strains K-12 MG1655 and K-12 W3110 are commensal organisms, whereas UTI89 and CFT073 are uropathogenic and O157:H7 Sakai and O157:H7 EDL 933 are potent enterohaemorrhagic pathogens

NCBI COG		*E. coli* strains K-12 MG1655	K-12 W3110	UTI89	CFT073	O157:H7 Sakai	O157:H7 EDL933
Information storage and processing							
J	Translation, ribosomal structure and biogenesis	152	152	148	145	150	150
A	RNA processing and modification	1	1	1	1	1	1
K	Transcription	269	269	277	289	281	287
L	Replication, recombination and repair	212	225	178	251	273	293
Cellular processes and signaling							
D	Cell cycle control, cell division, chromosome partitioning	30	29	29	27	29	29
V	Defense mechanisms	46	46	52	53	51	54
T	Signal transduction mechanisms	158	158	157	159	168	171
M	Cell wall/membrane/envelope biogenesis	218	216	237	241	242	246
N	Cell motility	104	103	113	109	120	122
W	Extracellular structures	0	0	2	1	2	2
U	Intracellular trafficking, secretion, and vesicular transport	118	117	133	136	146	153
O	Posttranslational modification, protein turnover, chaperones	125	125	136	131	134	137

Table 5 (continued)

NCBI COG		*E. coli* strains K-12 MG1655	K-12 W3110	UTI89	CFT073	O157:H7 Sakai	O157:H7 EDL933
Metabolism							
C	Energy production and conversion	276	277	278	270	274	277
G	Carbohydrate transport and metabolism	268	269	294	298	261	268
E	Amino acid transport and metabolism	328	329	334	331	325	326
F	Nucleotide transport and metabolism	81	81	84	82	77	78
H	Coenzyme transport and metabolism	128	128	138	139	135	135
I	Lipid transport and metabolism	98	97	101	116	106	107
P	Inorganic ion transport and metabolism	200	200	215	222	220	227
Q	Secondary metabolites biosynthesis, transport and catabolism	67	65	84	93	75	75
Poorly characterized							
R	General function prediction only	397	397	432	433	469	459
S	Function unknown	210	209	249	234	273	282
Not in COG		1215	1192	1888	2127	1947	1962
Total proteins predicted		4243	4227	5066	5379	5253	5324

Table 6 Overview of the aminotransferases and pyridoxal-5′-phosphate-dependent enzymes of *E. coli* K-12. Annotations and gene identification numbers were compiled from the MicrobesOnline (Alm et al. 2005) and BioCyc databases. AT: aminotransferase. The list is not exhaustive and does not include for example pyridoxal-5′-phosphate binding racemases

Gene ID	Gene	Class	Enzyme	Probable Cellular Function
VIMSS 14149	*thrC*		Threonine synthase	Threonine synthesis from homoserine
VIMSS 14300	*hemL*	III	Glutamate-1-semialdehyde AT	Uroporphyrinogen synthesis
VIMSS 14340	*proS*		Gamma-glutamylphosphate reductase	Proline and citrulline synthesis
VIMSS 14737	*ybdl*	I, II	Putative aspartate/tyrosine aromatic AT	
VIMSS 14899	*bioA*	III	7,8-diaminopelargonic acid synthetase	Biotin synthesis
VIMSS 14901	*bioF*	II	8-amino-7-oxononanoate synthase	Biotin synthesis
VIMSS 15031	*serC*	V	3-phosphoserine AT	Serine synthesis
VIMSS 15052	*aspC*	I, II	Aspartate AT	Aspartate synthesis
VIMSS 15218	*pabC*	IV	4-amino-4-deoxychorismate lyase	Tetra-hydrofolate synthesis
VIMSS 15381	*trpB*		Tryptophan synthase, beta protein	Tryptophan synthesis
VIMSS 15422	*goaG*	III	4-aminobutyrate AT	
VIMSS 15614	*gadB*		Glutamate decarboxylase isozyme	Glutamate catabolism
VIMSS 15799	*b1680*	V	Putative protein with AT activity	
VIMSS 15866	*cstC*	III	Acetyl-ornithine delta AT	Amino acid catabolism
VIMSS 16129	*hisC*	I, II	Histidinol-phosphate AT	Histidine synthesis
VIMSS 16360	*b2253*		Putative protein with AT activity	
VIMSS 16397	*b2290*	I, II	Putative aspartate / tyrosine aromatic AT	
VIMSS 16471	*dsdA*		Serine dehydratase	Conversion of serine to pyruvate and ammonia
VIMSS 16484	*b2379*	I, II	Putative aspartate/tyrosine aromatic AT	
VIMSS 16513	*cysK*		Cysteine synthase A	Cysteine synthesis
VIMSS 16520	*cysM*		Cysteine synthase B	Cysteine synthesis

Table 6 (continued)

Gene ID	Gene	Class	Enzyme	Probable cellular function
VIMSS 16629	*yfhO*	V	Putative cysteine sulfate desulfinase/ cysteine desulfurase	
VIMSS 16752	*gabT*	III	4-aminobutyrate AT	Utilization of γ-aminobutyrate as nitrogen source
VIMSS 16894	*b2810*	V	Putative protein with AT activity	
VIMSS 16951	*ygeX*		2,3-diaminopropionate ammonia-lyase	α, β-elimination of both the D and L stereoisomer of 2,3-diaminopropionate, weak activity on D-serine
VIMSS 16983	*gcvP*		Glycine decarboxylase	Glycine cleavage, formyl tetra-hydrofolate synthesis
VIMSS 17085	*metC*		Cystathione beta-lyase	Methionine synthesis
VIMSS 17030	*yggS*		Putative pyridoxal binding protein	
VIMSS 17148	*ygjG*	III	Putrescine transaminase	Pathway for utilization of arginine as nitrogen source
VIMSS 17191	*tdcB*		Threonine dehydratase	Threonine degradation pathway
VIMSS 17422	*argD*	III	acetyl-ornithine AT	Arginine synthesis
VIMSS 17578	*gadA*		Glutamate decarboxylase isozyme	Glutamate catabolism
VIMSS 17633	*avtA*	I, II	Alanine-valine transaminase	Interconversion of alanine, valine and α-aminobutyrate into their alpha-keto acids
VIMSS 17678	*kbl*	II	2-amino-3-ketobutyrate CoA ligase	Threonine degradation pathway
VIMSS 17789	*glmS*		L-glutamine:D-fructose-6-phosphate AT	Hexosamine synthesis
VIMSS 17822	*ilvE*	IV	Branched-chain amino acid AT	Branched-chain amino acid synthesis
VIMSS 17824	*ilvA*		Threonine deaminase	Isoleucine biosynthesis from threonine
VIMSS 17978	*metB*		Cystathione γ-synthase	Methionine synthesis
VIMSS 18082	*tyrB*	I, II	Tyrosine AT	Phenylalanine and tyrosine synthesis

and xylose (Brenner and Farmer 2005). On the other hand, genes coding for proteins involved in secondary metabolites, transport and synthesis only represent a small proportion, approximately 0.02% of the total number of open reading frames of the *E. coli* genomes sequenced to date, including those of the commensal strains (Table 5).

In contrast, *E. coli* K-12 encodes a relatively large number of aminotransferases and pyridoxal-5′-phosphate-binding proteins. As reported in Table 6, 14 of these proteins are putatively directly involved in amino acid synthesis. While the biochemical activities of several of the products of these genes have been well documented, a significant number of those enzymes remain hypothetical and the regulation of their genetic expression and biochemical functions would need to be characterized.

3.4 Comparative Genomics

As mentioned previously, despite exhibiting similar genomic sizes and gene compositions if measured by the number of genes present in each COG category, *E. coli* shows a remarkable diversity in both genome and specific ecological niches (Dobrindt et al. 2003; Fukiya et al. 2004). The core genome has been calculated to comprise approximately 2800 genes. This result was attained using both the complete genomic sequences of seven *E. coli* strains (Chen et al. 2006), including the sequence of the phylogenetically identical *Shigella flexneri*, and genomic hybridization of *E. coli* K-12 W3110 microarrays with genomic DNA from 22 pathogenic *E. coli* and *Shigella* strains (Fukiya et al. 2004). Many of these genes from the 2800 genes constituting the ancestral *E. coli* backbone appear to have accumulated vertically transmitted sequence changes (Welch et al. 2002) while, as observed by the synteny plot between strain K-12 and O157:H7 (Hayashi et al. 2001; Perna et al. 2001) (Fig. 6), to a large extent no large rearrangement has occurred in the ancestral genomic regions comprising these conserved genes. However, the genes that are not part of this core mostly represent genes that have been, in evolutionary terms, relatively newly introduced. Strong evidence thus favors the hypothesis that the dynamic *E. coli* genome is mostly shaped by a combination of horizontal DNA transfer events mediated by a variety of systems including bacteriophages, conjugative plasmids, and natural competence, and by recombination and deletion events operated by the cell's own DNA recombination and repair system or by mobile genetic elements. As pointed out earlier, integration of these genes in the ancestral genetic blueprint of *E. coli* constituted important events that have enabled *E. coli* to adapt to novel ecological niches.

As a result, it is this subset of genes that define the industrial usefulness or pathogenic potential of *E. coli* strains. For example, the benign *E. coli* K-12 MG1655 encodes 567 open reading frames that are absent from the en-

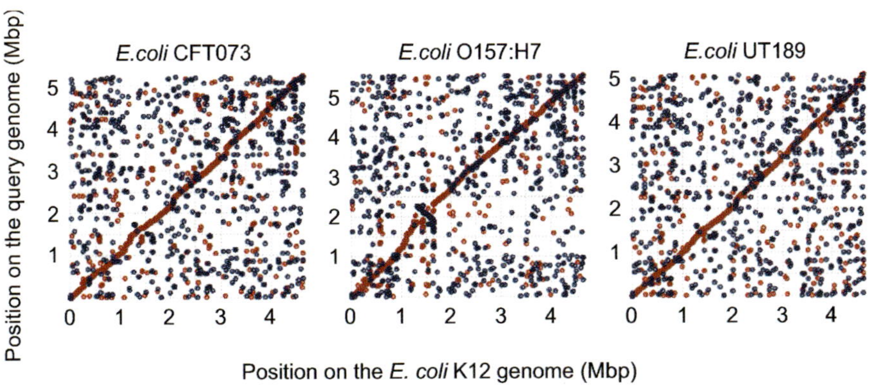

Fig. 6 *E. coli* synteny. The dot plot of syntenic regions between the genomes of *E. coli* catalogues each pair of ortholog present in the genomes being compared. Breakpoints in synteny reveal inversions and insertions. The ancestral *E. coli* backbone is revealed by the conservation of gene order and orientation of many of the genes constituting the core enterobacterial genome

terohaemorrhagic pathogen *E. coli* O157:H7 Sakai. These K-12-specific genes confer various specific properties such as a general secretion pathway or the ability to rapidly ferment D-sorbitol or conjugated glucuronides (Hayashi et al. 2001). Likewise, comparison of the genetic blueprint of *E. coli* K-12 W3110 with 22 pathogenic enteric bacteria of the *E. coli* and *Shigella* group revealed that 1424 of the *E. coli* K-12 genes, frequently organized in clusters, were missing in at least one of the pathogens (Fukiya et al. 2004). These latter genes chiefly corresponded to a limited number of functional categories: cell envelope such as lipopolysaccharide synthesis or fimbriae and pili synthesis genes, central intermediary metabolism, energy metabolism, transport and binding proteins encoding genes, regulatory genes (a total of 140), and hypothetical proteins (Fukiya et al. 2004). In addition, 96 *E. coli* K-12 W3110 genes appeared to be absent from all 22 pathogenic species tested, 75 of which were carried by prophage regions (Fukiya et al. 2004). A similar study was performed using DNA arrays containing the translatable genes of *E. coli* K-12 MG1655 to compare this genome with those of 26 commensal, and extraintestinal or intestinal pathogenic *E. coli* strains (Dobrindt et al. 2003). In agreement with the former observation, the presence of a similar proportion of the *E. coli* K-12 MG1655 genes, a total of 1165, was found to be variable (Dobrindt et al. 2003). Particularly, the number of *E. coli* K-12 MG1655 genes that were absent in each individual isolate tested varied from 112 to 427 (Dobrindt et al. 2003). These differences were essentially scattered along the chromosome (Dobrindt et al. 2003). Furthermore, for uropathogenic *E. coli*, strain-specific genomic DNA island acquisitions resulted in the capability to evade host defenses without compromising the ability to colonize the intestine, and in the capability to infect the urinary tract and bloodstream. On the other hand, for intestinal pathogens, genes issued from horizontal transfer events conferred new modes of interactions with host tissues and the capability to colonize specific regions of the intestine (Welch et al. 2002).

Notably, the number of new genes discovered from the complete sequence of a new *E. coli* strain has been calculated to be 441 (Chen et al. 2006), suggesting that the *E. coli* pan-genome is open and of great diversity. It thus becomes increasingly compelling that the lifestyle of each particular *E. coli* strain has had a profound influence on the evolution of its genetic blueprint, as further demonstrated by the identification of 29 genes that are under positive selection in uropathogenic *E. coli* strains and involved in cell surface structure, DNA metabolism, nutrient acquisition, and urinary tract infection (Chen et al. 2006).

3.5 Essential Genes

As determined from the analysis of various different bacterial genera, the subset of genes that is necessary to ensure bacterial survival under optimum laboratory conditions in rich medium comprises a minimum of 206 genes encoding an essentially complete DNA replication and transcription and translation machineries; rudimentary cell division, DNA repair, and protein processing systems; limited lipid biosynthesis; basic substrate transport; and functional substrate-level phosphorylation and non-oxidative branch of the pentose phosphate pathway (Gil et al. 2004). The essentiality of genes of these classes has been verified also in *E. coli*. Systematic single-gene deletion knockout of the genes of *E. coli* K-12 BW25113 was conducted to assemble a library of 7970 mutants representing 3985 genes. However, 303 genes could not be deleted, including 37 genes of unknown functions (Baba et al. 2006). Among these candidate essential genes, 282 are universally present (Baba et al. 2006) and preferentially located on the leading replication strand (Rocha and Danchin 2003). A total of 67 genes that appear to be essential to *E. coli* are not essential to *B. subtilis*, and 86 *E. coli* essential genes do not have any ortholog in *B. subtilis* (Baba et al. 2006). In addition, as observed with Corynebacteria, numerous genes from the respiratory chain appear to be essential, including for example the ATPase operon and heme biosynthesis genes (*hemL, hemB, hemH, hemA, hemD, hemC, hemG*).

3.6 Transcriptional Profiling Analyses

Numerous transcriptional studies have been performed with *E. coli* to determine at steady state the patterns of gene expression when this organism is faced with a variety of incubation conditions. Particularly, it was demonstrated that during the exponential phase of growth, the expression of pyrimidine and amino acid biosynthesis genes is significantly increased when cells are incubated in minimum medium as opposed to rich medium (Wei et al. 2001). Likewise, expression of the ATP and proton motive force machinery increases two-fold (Wei et al. 2001). Moreover, it was shown that when *E. coli* cells are incubated in growth-limiting chemostat, such as ammonia limitation, most of the genes coding for enzymes of the amino acid metabolism are up-regulated, and particularly *glnA* coding for glutamine synthetase (Hua et al. 2004). Interestingly, in *E. coli* cells incubated under conditions of ammonia limitation, as opposed to glucose limitation, *asnB* coding for the glutamine-dependent asparagine synthetase as well as numerous other genes whose products are involved in amino acid biosynthesis are overexpressed, including *thrA* and *asd* (involved in the biosynthesis of amino acids of the aspartate family); *lysC*, *dapB*, and *dapD* (lysine), *hisG* (histidine), *ilvIH* (valine and isoleucine), and *argECBH* (arginine) (Hua et al. 2004). However, except *astCADBE*, which code for the main enzymes of arginine catabolism, the transcript levels of other amino acid catabolic enzymes do not seem to be significantly altered between glucose- and ammonia-limiting chemostats (Hua et al. 2004). Similarly, most genes coding for enzymes of the glycolysis pathway and the pentose phosphate pathway are overexpressed under ammonia-limitation, while those corresponding to most genes of the tricarboxylic acid cycle are expressed at levels lower than those observed under glucose limitations (Hua et al. 2004). This observation notwithstanding, under glucose limitation, transcription profiling analyses revealed that numerous genes coding for proteins involved in oxidative phosphorylation are overexpressed (Hua et al. 2004). This is coherent with the view that *E. coli* cells respond to carbon source limitations by maintaining its overall energy balance via increasing its level of oxidative phosphorylation, probably to compensate the reduced substrate-level phosphorylation activity that results from reduced glucose availability. In addition, global transcription profiling allows the precise and relatively exhaustive identification of genes differentially expressed due to direct or indirect effects mediated by particular regulators, using comparisons of transcription levels in wild-type *E. coli* and in regulator mutants. This is perhaps best exemplified by the survey of the genes and operons controlled by the leucine-responsive regulatory protein (Lrp), a regulator that in most cases represses the expression of genes coding for catabolic enzymes and activates the expression of genes coding for biosynthetic enzymes (Hung et al. 2002). The nitrogen limitation response mediated by the nitrogen regulatory protein (NtrC) was investigated in a similar manner (Zimmer et al. 2000). NtrC appears to activate transcription of numerous genes including genes coding for regulatory proteins (*ntrB* or *glnL*, *ntrC* or *glnG*, *glnK*, and *nac*). NtrC directly activates transcription of σ^{54}-dependent genes, and indirectly activates transcription of σ^{70}-dependent genes via the nitrogen assimilation control protein Nac, which serves as an adapter between NtrC and σ^{70}-dependent operons. Transcription profile analyses particularly revealed that the NtrC/Nac regulon activates 45 transport operons, thereby conferring to nitrogen starved *E. coli* the capability to scavenge for compounds, including for compounds recycled during cell wall synthesis. Such compounds include D-alanine and D-alanine-D-alanine that *E. coli* cells released during growth and division in both the external medium and the periplasmic space (Zimmer et al. 2000). In *E. coli* cells grown in Luria–Bertani medium, the onset of the NtrC-mediated response coincides with the depletion of the medium in glycine, serine, aspartate, and glutamine. Serine and aspartate are used by *E. coli* to synthesize other amino acids and nucleotides, while glycine serves in the synthesis of purine, thymine, and methionine, and glycine and serine both stimulate growth as auxiliary carbon sources (Baev et al. 2006). Genes coding for biosynthetic enzymes of these amino acids are expressed first. Other genes coding for amino acid biosynthetic enzymes are expressed later, with the exception of those for threonine and alanine whose biosynthesis does not exhibit a clear trend, and that of glutamine synthetase which returns to basal level following a transient peak (Baev et al. 2006).

Global transcriptional responses of various *E. coli* amino acid producers were studied in conditions of high extracellular amino acid concentrations. For example, the genetic expression response of *E. coli* in the presence of 0.5 g/l of phenylalanine reflects a perturbed aromatic amino acid modulated by TyrR regulation, as compared with a mutant in which the gene coding for this σ^{70} transcription factor, which controls the expression of a group of genes involved in aromatic amino acid biosynthesis and transport, had been deleted. Addition of 5 g/l of phenylalanine reduces the growth rate of this mutant by half, but supplementation with shikimate, tyrosine, or tryptophan relieves the inhibition (Polen et al. 2005). These straightforward experiments clearly illustrate the value of such global analyses, since for example a TyrR-defective strain could prove to be a more efficient phenylalanine producer (Polen et al. 2005). The influence of TyR on aromatic amino acid biosynthesis was particularly demonstrated by the observation of global changes in *m*RNA abundance that result from perturbations of tryptophan metabolism (Khodursky et al. 2000). Interestingly, only one two-gene operon (*tnaAB*), involved in tryptophan degradation, is up-regulated by the addition of tryptophan. Shikimate production is another economically important biotechnological conversion that is limited by by-product formation. Comparison of the transcription profiles of *E. coli* W3110 and its *aroL* deletion derivative W3110.shik1 that has been engineered to overexpresses *aroF* provide clues to this phenomenon. The *aroL* deletion restricts the carbon flow towards the synthesis of tryptophan, phenylalanine, and tyrosine. The cellular response to carbon limitation includes the up-regulation of numerous genes coupled to the shikimate pathway, including *ydiB*, *aroD*, and *ydiN* that are strongly induced. It is this up-regulation that leads to by-products formation under carbon source limitation conditions (Johansson and Lidén 2006). Notably, there is a larger difference in expression patterns under carbon-limitation as opposed to phosphate limitation. This could be ascribed to starvation for aromatic acids under carbon-limitation which the cell circumvents under phosphate limitation by up-regulation of *aroK* and *aroA* (Johansson and Lidén 2006). Similarly, clues to the improvement of the production process of the essential amino acid threonine were attained by global proteomic and transcription profile analyses. On the one hand, accumulation of the main threonine precursors (oxaloacetate, aspartate, or homoserine), mediated via significant changes in expression of enzymes of threonine biosynthesis and product degradation, was shown to be critical to reach high threonine yields from an *rpoS*$^-$ *E. coli* mutant (auxotroph for methionine and isoleucine, and resistant to threonine, lysine, and proline analogues). On the other hand, only 1.3% of the genes of a threonine overproducing *rpoS*$^+$ *E. coli* mutant (also auxotroph for methionine and isoleucine, and resistant to threonine, lysine, and proline analogues) exhibit significantly changed expression profiles. These latter changes in genetic expression result in increased expression of genes of the glyoxylate shunt, tricarboxylic acid cycle and amino acid biosynthesis including particularly the *thr* operon, and in down-regulation of many ribosomal protein genes and *dadAX*, *hdeA*, *hdeB*, *ompF*, *oppA*, *oppB*, *oppF*, *yfiD* (Lee et al. 2003; Kim et al. 2004). Moreover, the two specific mutations *thrA345* and *ilvA97* were found to be essential in the latter strain to reach high threonine yields (Lee et al. 2003).

Transcription profiling thus proves very useful in studying pleiotropic effects of single-gene changes affecting a regulatory protein for example (Salmon et al. 2003, 2005), or of the adaptive response exhibited by bacteria to specific growth conditions, such as pH of the incubation medium (Tucker et al. 2002), osmostress (Weber and Jung 2002), or carbon sources poorer than glucose such as acetate or propionate (Oh et al. 2002; Polen et al. 2003; Liu et al. 2005). Combined, these various data provide the basis to reconstruct in silico the various dynamic regulatory networks that act upon the *E. coli* cell's biology (Gutiérrez-Ríos et al. 2003; Covert et al. 2004). Reconstruction of regulatory network structures including both allosteric and genetic controls, coupled with metabolic flux and proteomic analyses, thus paves the way to develop predictive virtual cells with which hypotheses could first be tested in silico and verified in vitro or in vivo, and that could be applied in a cost- and time-effective manner to facilitate the conduct of industrial fermentations or contribute to research projects effectiveness.

3.7 Proteomic Repertoire

The p*I* values displayed by *E. coli* K-12 proteins range from 3.38 to 13.0 and the molecular masses from 1.59 to 248 kDa (Lopez-Campistrous et al. 2005). In addition to proteins that are extracellular, *E. coli* proteins are predicted to be located in four cellular compartments: cytosol (2885 known and putative proteins), inner membrane (670 known and putative proteins), periplasm (138 known and putative proteins), and outer membrane (87 known and putative proteins). The *E. coli* K-12 proteome has been analyzed by a variety of techniques using cells that have been grown under different conditions. For example, high-pressure liquid chromatography-tandem mass spectrometry was successful in identifying from crude extracts and membrane extracts 1147 different proteins that are present in approximately at least 100 copies per cell (Corbin et al. 2003). Moreover, analysis of proteins from each isolated cellular fraction led to the identification of 575 different proteins forming 2160 spots. In addition, it was shown that 42% of these exist in multiple forms and 459 partition into only one of these compartments (Lopez-Campistrous et al. 2005). The inner *E. coli* membrane proteome was further defined by global topology analysis using C-terminal tagging with alkaline phosphate and green fluorescent proteins to define the periplasmic or cytoplasmic location of the C-termini of 601 proteins of the inner membrane (Daley et al. 2005). Interestingly, only a small subset of the *E. coli* K-12 genes originating from horizontal transfer seem to be expressed, despite being transcribed as efficiently as other genes, suggesting that gene decay has played an important part in the evolution of the *E. coli* genome (Taoka et al. 2004) since genes that are not expressed, and thus that do not confer to the cell any competitive advantage, are typically lost during the course of evolution.

Notably, proteomic profiling was performed also on recombinant *E. coli* K-12 cells during high-cell density fermentations, for instance on cells incubated under phosphate limitation to produce a biopharmaceutical antibody fragment (Aldor et al. 2005). Significant quantitative and qualitative differences in protein spots were observed between the samples originating from production experiments as compared to controls, with a strong correlation existing between production of the antibody fragment and synthesis of the stress protein phage shock protein A (PspA). This analytical work has had immediate biotechnological applications as, in this particular system, manipulation of PspA levels led to improved fermentation yields (Aldor et al. 2005). Exhaustive understanding of the wild-type proteomic repertoire is a prerequisite to attaining the highest possible levels of amino acid productivities, as exemplified by the process optimizations made possible by the comparison of the wild-type proteome with the proteomes of threonine overexpressing mutants (Lee et al. 2003; Kim et al. 2004).

Notwithstanding the useful information that can be derived from proteomics data in isolation, and as suggested in the preceding section, it is the integration of these data with other global analyses, such as complete genome sequences, transcription profiling, and *m*RNA half-lives data that has the greatest potential to generate novel paradigms in the understanding of microbial physiology, and thus lead to more efficient industrial production strains (Allen et al. 2003; Vertès et al. 2005).

4 Genetic Engineering at the Megabase Level

Genetic techniques to engineer *C. glutamicum* or *E. coli* have been reviewed elsewhere (Baneyx 1999; Vertès et al. 2005; Wendisch et al. 2006a). Of particular interest are genome-level recombinant DNA techniques that enable the scarless deletion of large DNA fragments from the *E. coli* and the corynebacterial genomes. These methods are based on the Cre/*loxP* excision system (Kuhn and Torres 2002) or the λ-Red system (Hashimoto et al. 2005). Similar to other site-specific recombination systems of *E. coli*, such as Flp/*FRT* (Martinez-Morales et al. 1999) or λ-Red (Sung et al. 2006), the Cre recombinase promotes the site-specific deletion of sequences that are present between two *loxP* sequences (Yu et al. 2002; Suzuki et al. 2005b). Sequential deletions of strain-specific islands greater than 10 kb of the *C. glutamicum* R genome were performed and resulted in a multiple deletion strain that lacked 190 kb encoding a total of 188 open reading frames. Despite this large deletion, the mutant did not exhibit any obvious abnormal growth pattern in minimal medium with glucose as the sole carbon source under standard laboratory conditions though competition experiments were not performed (Suzuki et al. 2005a). Likewise, the *E. coli* genome was reduced in size by deleting 313 kb of genomic DNA representing a total of 287 open reading frames. This was achieved by iterations combining phage P1 transduction (to tag with two *loxP* sites the regions to be deleted) and site-specific Cre recombinase mediated recombination (Yu et al. 2002). Again, in spite of the fact that a large number of genes had been lost, the cumulative deletion strain thus obtained exhibited normal growth patterns under standard laboratory conditions (Yu et al. 2002). However, as expected, upon larger-scale deletion of non essential genes, representing up to 29.7% of the wild-type genome, mutant *E. coli* cells have been observed to grow more slowly than the parental strain. In addition, some of these mutants are also morphologically affected in their cell lengths and widths, perhaps caused by synergistic effects of multiple mutations (Hashimoto et al. 2005). Nonetheless, precise selection of genes to be eliminated such as recombinogenic or mobile DNA and cryptic virulence genes including fimbriae-encoding genes can lead not only to improved biotechnological converters (Sung et al. 2006) but also to improved laboratory strains, as demonstrated by the high electroporation efficiency and accurate propagation of recombinant genes and plasmids in a multiple deletion derivative of *E. coli* K-12 in which up to 15% of genes had been deleted (Pósfai et al. 2006).

On the other hand, the feasibility to integrate large fragments of DNA into bacterial chromosomes was demonstrated in *B. subtilis* where cycles of overlap-elongation, Campbell-type integration, and gap-filling were used to integrate a 48.5-kb-long prophage lambda into the genome of strain 168 Marburg (Itaya 1995). Techniques that rely on iterative assembly of DNA fragments can in essence be applied without obvious limitation of size, as demonstrated by the stable cloning of the complete 3.5-megabase genome of the photosynthetic bacterium *Synechocystis* PCC6803 into the chromosome of *B. subtilis* 168 (Itaya et al. 2005). The recombinant strain contained a chimerical 7.7-megabase genome that proved to be structurally stable (Itaya et al. 2005). Megacloning techniques thus complete the set of cloning tools for manipulating bacterial chromosomes, such as bacterial artificial chromosomes, which have proven to be stable and offer large but limited carrying capacities in the 100 to 300 kb range (Shizuya et al. 1992).

These various technical developments, together with the inherent plasticity and mosaic structure of bacterial genomes, and particularly of the genomes of *E. coli* and *C. glutamicum*, make thus possible the manipulation of important industrial organisms at any scale, from single-nucleotide changes to global chromosomal rearrangements, in order to achieve always optimized productivity levels.

5 Perspectives

Technical platforms and the vast amount of knowledge accumulated on the biology and genetics of *E. coli* and *C. glutamicum* enable to combine directed strain evolution techniques and rational design. Specifically, this can be achieved by integrating all the data gathered on the various dynamic interacting networks, which characterize a biological system (Vertès et al. 2005). Rational strain development by metabolic engineering has already proven to be invaluable for generating *E. coli* and *C. glutamicum* industrial strains with largely improved industrial performances (Choi et al. 2006; Wendisch et al. 2006a). Similar approaches are already tested for the development of novel pharmaceutical compounds. These techniques consolidate biological and biochemical knowledge recorded in the scientific literature, information management techniques via the use of ontologies, and data processing via the reconstruction in silico of virtual biological pathways. The increasing completeness of the biological information that is available in the scientific literature synergistically reinforces the increasing ability to process it, such that it is now possible to envision the creation of virtual patients. These simulations have not only the potential to facilitate the design of novel pharmaceutical entities but also to significantly decrease the cost of clinical trials (Willis 2003). Systems biology of whole microorganisms (Schilling et al. 2002; Arita et al. 2005), which has largely inspired the developments achieved in the pharmaceutical sciences, is also being increasingly applied to industrial microbiology, though more slowly than in the pharmaceutical industry as it is driven by a lower pace of investment. For example, systems biology approaches that rely on models aiming at improving either energy efficiency, or carbon conversion efficiency of the production route, proved to constitute a useful optimization framework to formulate gene addition and deletion strategies. Using such a scheme, a genetic optimization program could be designed that in theory would lead to the improvement of arginine production yield in *E. coli* by up to 9% (Burgard and Maranas 2001). Notably, systems biology techniques are only in the initial phase of their development for both *C. glutamicum* (Wendisch et al. 2006b) and *E. coli*. However, their value has already been demonstrated in these organisms also, for example by defining the control of *r*RNA synthesis in *E. coli* (Dennis et al. 2004) or by predicting metabolic capabilities of *E. coli* (Edwards et al. 2001; Ibarra et al. 2002). As a result, metabolic engineering of amino acid-producing bacteria can thus both benefit from these integrated techniques and contribute to their further development as discovery tools.

References

Aldor IS, Krawitz DC, Forrest W, Chen C, Nishihara JC, Joly JC, Champion KM (2005) Proteomic profiling of recombinant *Escherichia coli* in high-cell-density fermentations for improved production of an antibody fragment biopharmaceutical. Appl Environ Microbiol 71:1717–1728

Allen TE, Herrgard MJ, Liu M, Qiu Y, Glasner JD, Blattner FR, Palsson BO (2003) Genome-scale analysis of the uses of the *Escherichia coli* genome: model-driven analysis of heterogeneous data sets. J Bacteriol 185:6392–6399

Alm EJ, Huang KH, Price MN, Koche RP, Keller K, Dubchak IL, Arkin AP (2005) The Microbes Online Web site for comparative genomics. Genome Res 15:1015–1022

Arita M, Robert M, Tomita M (2005) All systems go: launching cell simulation fueled by integrated experimental biology data. Curr Opin Biotechnol 16:344–349

Baba T, Ara T, Hasegawa M, Takai Y, Okumura Y, Baba M, Datsenko KA, Tomita M, Wanner BL, Mori H (2006) Construction of *Escherichia coli* K-12 in-frame, single-gene knockout mutants: the Keio collection. Mol Syst Biol 2:2006 0008

Baev MV, Baev D, Radek AJ, Campbell JW (2006) Growth of *Escherichia coli* MG1655 on LB medium: monitoring utilization of amino acids, peptides, and nucleotides with transcriptional microarrays. Appl Microbiol Biotechnol 71:317–322

Baneyx F (1999) Recombinant protein expression in *Escherichia coli*. Curr Opin Biotechnol 10:411–421

Barreiro C, Gonzalez-Lavado E, Brand S, Tauch A, Martin JF (2005) Heat shock proteome analysis of wild-type *Corynebacterium glutamicum* ATCC 13032 and a spontaneous mutant lacking GroEL1, a dispensable chaperone. J Bacteriol 187:884–889

Barriuso-Iglesias M, Barreiro C, Flechoso F, Martin JF (2006) Transcriptional analysis of the F_0F_1 ATPase operon of *Corynebacterium glutamicum* ATCC 13032 reveals strong induction by alkaline pH. Microbiology 152:11–21

Barry CE, Lee RE, Mdluli K, Sampson AE, Schroeder BG, Slayden RA, Yuan Y (1998) Mycolic acids: structure, biosynthesis and physiological functions. Prog Lipid Res 37:143–179

Bendt AK, Burkovski A, Schaffer S, Bott M, Farwick M, Hermann T (2003) Towards a phosphoproteome map of *Corynebacterium glutamicum*. Proteomics 3:1637–1646

Besemer J, Lomsadze A, Borodovsky M (2001) GeneMarkS: a self-training method for prediction of gene starts in microbial genomes. Implications for finding sequence motifs in regulatory regions. Nucleic Acids Res 29:2607–2618

Bettelheim KA (1992) The genus *Escherichia*. In: Balows A, Trüper HG, Dworkin M, Harder W, Schleifer KH (eds) The Prokaryotes. Springer, Berlin Heidelberg New York, pp 2696–2736

Blattner FR, Plunkett G III, Bloch CA, Perna NT, Burland V, Riley M, Collado-Vides J, Glasner JD, Rode CK, Mayhew GF, Gregor J, Davis NW, Kirkpatrick HA, Goeden MA, Rose DJ, Mau B, Shao Y (1997) The complete genome sequence of *Escherichia coli* K-12. Science 277:1453–1474

Böck A, Curtiss RI, Kaper JB, Neidhardt FC, Nyström T, Rudd KE, Squires CL (2006) *Escherichia coli* and *Salmonella typhimurium*: cellular and molecular biology. ASM Press, Washington, DC, http://www.ecosal.org, (Last visited on: Nov 23, 2006)

Bott M, Niebisch A (2003) The respiratory chain of *Corynebacterium glutamicum*. J Biotechnol 104:129–153

Brenner DJ, Farmer JJ III (2005) Enterobacteriaceae. In: Brenner DJ, Krieg NR, Staley JT, Garrity GM (eds) Bergey's Manual of Systematic Bacteriology. Springer, Berlin Heidelberg New York, pp 587–607

Bröer S, Eggeling L, Krämer R (1993) Strains of *Corynebacterium glutamicum* with different lysine productivities may have different lysine excretion systems. Appl Environ Microbiol 59:316–321

Brown K (2005) Amino acids: highlighting synthesis applications. Business Communications Company, Inc., Norwalk, CT, USA

Brune I, Brinkrolf K, Kalinowski J, Pühler A, Tauch A (2005) The individual and common repertoire of DNA-binding transcriptional regulators of *Corynebacterium glutamicum*, *Corynebacterium efficiens*, *Corynebacterium diphtheriae* and *Corynebacterium jeikeium* deduced from the complete genome sequences. BMC Genomics 6:86

Burgard AP, Maranas CD (2001) Probing the performance limits of the *Escherichia coli* metabolic network subject to gene additions or deletions. Biotechnol Bioeng 74:364–375

Burkovski A, Krämer R (2002) Bacterial amino acid transport proteins: occurrence, functions, and significance for biotechnological applications. Appl Microbiol Biotechnol 58:265–274

Cerdeño-Tárraga AM, Efstratiou A, Dover LG, Holden MT, Pallen M, Bentley SD, Besra GS, et al (2003a) The complete genome sequence and analysis of *Corynebacterium diphtheriae* NCTC13129. Nucleic Acids Res 31:6516–6523

Cerdeño-Tárraga AM, Efstratiou A, Dover LG, Holden MT, Pallen M, Bentley SD, Besra GS, et al (2003b) The complete genome sequence and analysis of *Corynebacterium diphtheriae* NCTC 13129. Nucleic Acids Res 31:6516–6523

Chang HM, Chen MY, Shieh YT, Bibb MJ, Chen CW (1996) The CutRS signal transduction system of *Streptomyces lividans* represses the biosynthesis of the polyketide antibiotic actinorhodin. Mol Microbiol 21:1075–1085

Chen SL, Hung CS, Xu J, Reigstad CS, Magrini V, Sabo A, Blasiar D, Bieri T, Meyer RR, Ozersky P, Armstrong JR, Fulton RS, Latreille JP, Spieth J, Hooton TM, Mardis ER, Hultgren SJ, Gordon JI (2006) Identification of genes subject to positive selection in uropathogenic strains of *Escherichia coli*: a comparative genomics approach. Proc Natl Acad Sci USA 103:5977–5982

Choi JH, Keum KC, Lee SY (2006) Production of recombinant proteins by high cell density culture of *Escherichia coli*. Chem Eng Sci 61:876–885

Cocaign-Bousquet M, Guyonvarch A, Lindley ND (1996) Growth rate-dependent modulation of carbon flux through central metabolism and the kinetic consequences for glucose-limited chemostat cultures of *Corynebacterium glutamicum*. Appl Environ Microbiol 62:429–436

Cole ST, Brosch R, Parkhill J, Garnier T, Churcher C, Harris D, Gordon SV, et al. (1998) Deciphering the biology of *Mycobacterium tuberculosis* from the complete genome sequence. Nature 393:537–544

Collins MD, Falsen E, Akervall E, Sjoden B, Alvarez A (1998) *Corynebacterium kroppenstedtii* sp. nov., a novel *Corynebacterium* that does not contain mycolic acids. Int J Syst Bacteriol 48(Pt4):1449–1454

Corbin RW, Paliy O, Yang F, Shabanowitz J, Platt M, Lyons CE Jr, Root K, McAuliffe J, Jordan MI, Kustu S, Soupene E, Hunt DF (2003) Toward a protein profile of *Escherichia coli*: comparison to its transcription profile. Proc Natl Acad Sci USA 100:9232–9237

Covert MW, Knight EM, Reed JL, Herrgård MJ, Palsson BO (2004) Integrating high-throughput and computational data elucidates bacterial networks. Nature 429:92–96

Daley DO, Rapp M, Granseth E, Melen K, Drew D, von Heijne G (2005) Global topology analysis of the *Escherichia coli* inner membrane proteome. Science 308:1321–1323

Delcher AL, Harmon D, Kasif S, White O, Salzberg SL (1999) Improved microbial gene identification using GLIMMER. Nucleic Acids Res 27:4636–4641

Delcher AL, Phillippy A, Carlton J, Salzberg SL (2002) Fast algorithms for large-scale genome alignment and comparison. Nucleic Acids Res 30:2478–2483

Demain AL (1999) Metabolites, primary and secondary. In: Flickinger MC, Drew SW (eds) Bioprocess Technology: Fermentation, Biocatalysis, and Bioseparation. Wiley, New York, pp 1713–1732

Dennis PP, Ehrenberg M, Bremer H (2004) Control of *r*RNA synthesis in *Escherichia coli*: a systems biology approach. Microbiol Mol Biol Rev 68:639–668

Dobrindt U, Agerer F, Michaelis K, Janka A, Buchrieser C, Samuelson M, Svanborg C, Gottschalk G, Karch H, Hacker J (2003) Analysis of genome plasticity in pathogenic and commensal *Escherichia coli* isolates by use of DNA arrays. J Bacteriol 185:1831–1840

Dominguez H, Lindley ND (1996) Complete sucrose metabolism requires fructose phosphotransferase activity in *Corynebacterium glutamicum* to ensure phosphorylation of liberated fructose. Appl Environ Microbiol 62:3878–3880

Dominguez H, Rollin C, Guyonvarch A, Guerquin-Kern JL, Cocaign-Bousquet M, Lindley ND (1998) Carbon-flux distribution in the central metabolic pathways of *Corynebacterium glutamicum* during growth on fructose. Eur J Biochem 254:96–102

Edwards JS, Ibarra RU, Palsson BO (2001) In silico predictions of *Escherichia coli* metabolic capabilities are consistent with experimental data. Nat Biotechnol 19:125–130

Fischer F, Poetsch A (2006) Protein cleavage strategies for an improved analysis of the membrane proteome. Proteome Sci 4:2

Fischer F, Wolters D, Rögner M, Poetsch A (2006) Toward the complete membrane proteome: high coverage of integral membrane proteins through transmembrane peptide detection. Mol Cell Proteomics 5:444–453

Fukiya S, Mizoguchi H, Tobe T, Mori H (2004) Extensive genomic diversity in pathogenic *Escherichia coli* and *Shigella* Strains revealed by comparative genomic hybridization microarray. J Bacteriol 186:3911–3921

Garbe TR, Suzuki N, Inui M, Yukawa H (2004) Inhibitor-associated transposition events in *Corynebacterium glutamicum*. Mol Genet Genomics 271:729–741

Georgi T, Rittmann D, Wendisch VF (2005) Lysine and glutamate production by *Corynebacterium glutamicum* on glucose, fructose and sucrose: roles of malic enzyme and fructose-1,6-bisphosphatase. Metab Eng 7:291–301

Gerdes SY, Scholle MD, Campbell JW, Balázsi G, Ravasz E, Daugherty MD, Somera AL, Kyrpides NC, Anderson I, Gelfand MS, Bhattacharya A, Kapatral V, D'Souza M, Baev MV, Grechkin Y, Mseeh F, Fonstein MY, Overbeek R, Barabási AL, Oltvai ZN, Osterman AL (2003) Experimental determination and system level analysis of essential genes in *Escherichia coli* MG1655. J Bacteriol 185:5673–5684

Gil R, Silva FJ, Pereto J, Moya A (2004) Determination of the core of a minimal bacterial gene set. Microbiol Mol Biol Rev 68:518–537

Glanemann C, Loos A, Gorret N, Willis LB, O'Brien XM, Lessard PA, Sinskey AJ (2003) Disparity between changes in *m*RNA abundance and enzyme activity in *Corynebacterium glutamicum*: implications for DNA microarray analysis. Appl Microbiol Biotechnol 61:61–68

Graumann K, Premstaller A (2006) Manufacturing of recombinant therapeutic proteins in microbial systems. Biotechnol J 1:164–186

Grigoriev A (1998) Analyzing genomes with cumulative skew diagrams. Bioinformatics 14:252–258

Gutiérrez-Ríos RM, Rosenblueth DA, Loza JA, Huerta AM, Glasner JD, Blattner FR, Collado-Vides J (2003) Regulatory network of *Escherichia coli*: consistency between literature knowledge and microarray profiles. Genome Res 13:2435–2443

Hansmeier N, Chao TC, Pühler A, Tauch A, Kalinowski J (2006) The cytosolic, cell surface and extracellular proteomes of the biotechnologically important soil bacterium *Corynebacterium efficiens* YS-314 in comparison to those of *Corynebacterium glutamicum* ATCC 13032. Proteomics 6:233–250

Hashimoto M, Ichimura T, Mizoguchi H, Tanaka K, Fujimitsu K, Keyamura K, Ote T, Yamakawa T, Yamazaki Y, Mori H, Katayama T, Kato J (2005) Cell size and nucleoid organization of engineered *Escherichia coli* cells with a reduced genome. Mol Microbiol 55:137–149

Hatakeyama K, Hohama K, Vertès AA, Kobayashi M, Kurusu Y, Yukawa H (1993a) Genomic organization of the biotin biosynthetic genes of coryneform bacteria: cloning and sequencing of the *bioA-bioD* genes from *Brevibacterium flavum*. DNA Seq 4:177–184

Hatakeyama K, Kohama K, Vertès AA, Kobayashi M, Kurusu Y, Yukawa H (1993b) Analysis of the biotin biosynthesis pathway in coryneform bacteria: cloning and sequencing of the *bioB* gene from *Brevibacterium flavum*. DNA Seq 4:87–93

Hayashi K, Morooka N, Yamamoto Y, Fujita K, Isono K, Choi S, Ohtsubo E, Baba T, Wanner BL, Mori H, Horiuchi T (2006a) Highly accurate genome sequences of *Escherichia coli* K-12 strains MG1655 and W3110. Mol Syst Biol 2:2006 0007

Hayashi M, Ohnishi J, Mitsuhashi S, Yonetani Y, Hashimoto S, Ikeda M (2006b) Transcriptome analysis reveals global expression changes in an industrial L-lysine producer of *Corynebacterium glutamicum*. Biosci Biotechnol Biochem 70:546–550

Hayashi T, Makino K, Ohnishi M, Kurokawa K, Ishii K, Yokoyama K, Han CG, Ohtsubo E, Nakayama K, Murata T, Tanaka M, Tobe T, Iida T, Takami H, Honda T, Sasakawa C, Ogasawara N, Yasunaga T, Kuhara S, Shiba T, Hattori M, Shinagawa H (2001) Complete genome sequence of enterohemorrhagic *Escherichia coli* O157:H7 and genomic comparison with a laboratory strain K-12. DNA Res 8:11–22

Hermann T (2003) Industrial production of amino acids by coryneform bacteria. J Biotechnol 104:155–172

Hermann T, Pfefferle W, Baumann C, Busker E, Schaffer S, Bott M, Sahm H, Dusch N, Kalinowski J, Pühler A, Bendt AK, Krämer R, Burkovski A (2001) Proteome analysis of *Corynebacterium glutamicum*. Electrophoresis 22:1712–1723

Hua Q, Yang C, Oshima T, Mori H, Shimizu K (2004) Analysis of gene expression in *Escherichia coli* in response to changes of growth-limiting nutrient in chemostat cultures. Appl Environ Microbiol 70:2354–2366

Hung SP, Baldi P, Hatfield GW (2002) Global gene expression profiling in *Escherichia coli* K-12. The effects of leucine-responsive regulatory protein. J Biol Chem 277:40309–40323

Hüser AT, Chassagnole C, Lindley ND, Merkamm M, Guyonvarch A, Elisakova V, Pátek M, Kalinowski J, Brune I, Pühler A, Tauch A (2005) Rational design of a *Corynebacterium glutamicum* pantothenate production strain and its characterization by metabolic flux analysis and genome-wide transcriptional profiling. Appl Environ Microbiol 71:3255–3268

Ibarra RU, Edwards JS, Palsson BO (2002) *Escherichia coli* K-12 undergoes adaptive evolution to achieve in silico predicted optimal growth. Nature 420:186–189

Ikeda M, Nakagawa S (2003) The *Corynebacterium glutamicum* genome: features and impacts on biotechnological processes. Appl Microbiol Biotechnol 62:99–109

Ikemura T (1981) Correlation between the abundance of *Escherichia coli* transfer RNAs and the occurrence of the respective codons in its protein genes: a proposal for a synonymous codon choice that is optimal for the *E. coli* translational system. J Mol Biol 151:389–409

Inui M, Kawaguchi H, Murakami S, Vertès AA, Yukawa H (2004a) Metabolic engineering of *Corynebacterium glutamicum* for fuel ethanol production under oxygen-deprivation conditions. J Mol Microbiol Biotechnol 8:243–254

Inui M, Murakami S, Okino S, Kawaguchi H, Vertès AA, Yukawa H (2004b) Metabolic analysis of *Corynebacterium glutamicum* during lactate and succinate productions under oxygen deprivation conditions. J Mol Microbiol Biotechnol 7:182–196

Ishige T, Krause M, Bott M, Wendisch VF, Sahm H (2003) The phosphate starvation stimulon of *Corynebacterium glutamicum* determined by DNA microarray analyses. J Bacteriol 185:4519–4529

Ishikawa J, Yamashita A, Mikami Y, Hoshino Y, Kurita H, Hotta K, Shiba T, Hattori M (2004) The complete genomic sequence of *Nocardia farcinica* IFM 10152. Proc Natl Acad Sci USA 101:14925–14930

Itaya M (1995) Toward a bacterial genome technology: integration of the *Escherichia coli* prophage lambda genome into the *Bacillus subtilis* 168 chromosome. Mol Gen Genet 248:9–16

Itaya M, Tsuge K, Koizumi M, Fujita K (2005) Combining two genomes in one cell: stable cloning of the Synechocystis PCC6803 genome in the *Bacillus subtilis* 168 genome. Proc Natl Acad Sci USA 102:15971–15976

Itoh T, Okayama T, Hashimoto H, Takeda J, Davis RW, Mori H, Gojobori T (1999) A low rate of nucleotide changes in *Escherichia coli* K-12 estimated from a comparison of the genome sequences between two different substrains. FEBS Lett 450:72–76

Jana S, Deb JK (2005) Strategies for efficient production of heterologous proteins in *Escherichia coli*. Appl Microbiol Biotechnol 67:289–298

Johansson L, Lidén G (2006) Transcriptome analysis of a shikimic acid-producing strain of *Escherichia coli* W3110 grown under carbon- and phosphate-limited conditions. J Biotechnol 126:528–545

Kalinowski J (2005) The genomes of amino acid-producing Corynebacteria. In: Eggeling L, Bott M (eds) Handbook of *Corynebacterium glutamicum*. CRC Press, Boca Raton, FL, USA, pp 37–56

Kalinowski J, Bathe B, Bartels D, Bischoff N, Bott M, Burkovski A, Dusch N, et al. (2003) The complete *Corynebacterium glutamicum* ATCC 13032 genome sequence and its impact on the production of L-aspartate-derived amino acids and vitamins. J Biotechnol 104:5–25

Kamada N, Yasuhara A, Takano Y, Nakano T, Ikeda M (2001) Effect of transketolase modifications on carbon flow to the purine-nucleotide pathway in *Corynebacterium ammoniagenes*. Appl Microbiol Biotechnol 56:710–717

Kang Y, Durfee T, Glasner JD, Qiu Y, Frisch D, Winterberg KM, Blattner FR (2004) Systematic mutagenesis of the *Escherichia coli* genome. J Bacteriol 186:4921–4930

Kataoka M, Hashimoto KI, Yoshida M, Nakamatsu T, Horinouchi S, Kawasaki H (2006) Gene expression of *Corynebacterium glutamicum* in response to the conditions inducing glutamate overproduction. Lett Appl Microbiol 42:471–476

Kawaguchi H, Vertès AA, Okino S, Inui M, Yukawa H (2006) Engineering of a xylose metabolic pathway in *Corynebacterium glutamicum*. Appl Environ Microbiol 72:3418–3428

Khodursky AB, Peter BJ, Cozzarelli NR, Botstein D, Brown PO, Yanofsky C (2000) DNA microarray analysis of gene expression in response to physiological and genetic changes that affect tryptophan metabolism in *Escherichia coli*. Proc Natl Acad Sci USA 97:12170–12175

Kiefer P, Heinzle E, Wittmann C (2002) Influence of glucose, fructose and sucrose as carbon sources on kinetics and stoichiometry of lysine production by *Corynebacterium glutamicum*. J Ind Microbiol Biotechnol 28:338–343

Kim TH, Kim HJ, Park JS, Kim Y, Kim P, Lee HS (2005) Functional analysis of *sigH* expression in *Corynebacterium glutamicum*. Biochem Biophys Res Commun 331:1542–1547

Kim YH, Park JS, Cho JY, Cho KM, Park YH, Lee J (2004) Proteomic response analysis of a threonine-overproducing mutant of *Escherichia coli*. Biochem J 381:823–829

Kinoshita S (1985) Glutamic acid bacteria. In: Demain AL, Solomon NA (eds) Biology of Industrial Microorganisms. Benjamin Cummings, London, pp 115–146

Kobayashi K, Ehrlich SD, Albertini A, Amati G, Andersen KK, Arnaud M, Asai K, et al (2003) Essential *Bacillus subtilis* genes. Proc Natl Acad Sci USA 100:4678–4683

Kotrba P, Inui M, Yukawa H (2001) Bacterial phosphotransferase system (PTS) in carbohydrate uptake and control of carbon metabolism. J Biosci Bioeng 92:502–517

Kotrba P, Inui M, Yukawa H (2003) A single V317A or V317M substitution in Enzyme II of a newly identified beta-glucoside phosphotransferase and utilization system of *Corynebacterium glutamicum* R extends its specificity towards cellobiose. Microbiology 149:1569–1580

Krömer JO, Sorgenfrei O, Klopprogge K, Heinzle E, Wittmann C (2004) In-depth profiling of lysine-producing *Corynebacterium glutamicum* by combined analysis of the transcriptome, metabolome, and fluxome. J Bacteriol 186:1769–1784

Krömer JO, Wittmann C, Schröder H, Heinzle E (2006) Metabolic pathway analysis for rational design of L-methionine production by *Escherichia coli* and *Corynebacterium glutamicum*. Metab Eng

Krug A, Wendisch VF, Bott M (2005) Identification of AcnR, a TetR-type repressor of the aconitase gene *acn* in *Corynebacterium glutamicum*. J Biol Chem 280:585–595

Krüger R, Kübler D, Pallissé R, Burkovski A, Lehmann WD (2006) Protein and proteome phosphorylation stoichiometry analysis by element mass spectrometry. Anal Chem 78:1987–1994

Kuhn R, Torres RM (2002) *Cre/loxP* recombination system and gene targeting. Methods Mol Biol 180:175–204

Kumagai H (2000) Microbial production of amino acids in Japan. Adv Biochem Eng Biotechnol 69:71–85

Lange C, Rittmann D, Wendisch VF, Bott M, Sahm H (2003) Global expression profiling and physiological characterization of *Corynebacterium glutamicum* grown in the presence of L-valine. Appl Environ Microbiol 69:2521–2532

Lawrence JG, Ochman H (1998) Molecular archaeology of the *Escherichia coli* genome. Proc Natl Acad Sci USA 95:9413–9417

Lee JH, Lee DE, Lee BU, Kim HS (2003) Global analyses of transcriptions and proteomes of a parent strain and an L-threonine-overproducing mutant strain. J Bacteriol 185:5442–5451

Leuchtenberger W (1996) Amino acids - technical production and use. In: Roehr M (ed) Biotechnology, Vol VI. VCH, Weinheim, Germany, pp 452–454

Leuchtenberger W, Huthmacher K, Drauz K (2005) Biotechnological production of amino acids and derivatives: current status and prospects. Appl Microbiol Biotechnol 69:1–8

Liebl W (1992) The genus *Corynebacterium*-Nonmedical. In: Balows A, Trüper HG, Dworkin M, Harder W, Schleifer KH (eds) The Prokaryotes. Springer, Berlin Heidelberg New York, pp 1157–1171

Liebl W (2004) *Corynebacterium* taxonomy. In: Eggeling L, Bott M (eds) Handbook of *Corynebacterium glutamicum*. CRC Press, Boca Raton, pp 9–34

Liebl W, Ehrmann M, Ludwig W, Schleifer KH (1991) Transfer of *Brevibacterium divaricatum* DSM 20297T, "*Brevibacterium flavum*" DSM 20411, "*Brevibacterium lactofermentum*" DSM 20412 and DSM 1412, and *Corynebacterium glutamicum* and their distinction by *r*RNA gene restriction patterns. Int J Syst Bacteriol 41:255–260

Liu M, Durfee T, Cabrera JE, Zhao K, Jin DJ, Blattner FR (2005) Global transcriptional programs reveal a carbon source foraging strategy by *Escherichia coli*. J Biol Chem 280:15921–15927

Lopez-Campistrous A, Semchuk P, Burke L, Palmer-Stone T, Brokx SJ, Broderick G, Bottorff D, Bolch S, Weiner JH, Ellison MJ (2005) Localization, annotation, and comparison of the *Escherichia coli* K-12 proteome under two states of growth. Mol Cell Proteomics 4:1205–1209

Malumbres M, Gil JA, Martin JF (1993) Codon preference in corynebacteria. Gene 134:15–24

Manna D, Breier AM, Higgins NP (2004) Microarray analysis of transposition targets in *Escherichia coli*: the impact of transcription. Proc Natl Acad Sci USA 101:9780–9785

Marienhagen J, Kennerknecht N, Sahm H, Eggeling L (2005) Functional analysis of all aminotransferase proteins inferred from the genome sequence of *Corynebacterium glutamicum*. J Bacteriol 187:7639–7646

Martinez-Morales F, Borges AC, Martinez A, Shanmugam KT, Ingram LO (1999) Chromosomal integration of heterologous DNA in *Escherichia coli* with precise removal of markers and replicons used during construction. J Bacteriol 181:7143–7148

Matic I (1995) Les mécanismes du contrôle des échanges génétiques interspécifiques et de la variabilité génétique chez les bactéries. Bulletin Institut Pasteur 93:187–219

Mau B, Glasner JD, Darling AE, Perna NT (2006) Genome-wide detection and analysis of homologous recombination among sequenced strains of *Escherichia coli*. Genome Biol 7:R44

McHardy AC, Tauch A, Rückert C, Pühler A, Kalinowski J (2003) Genome-based analysis of biosynthetic aminotransferase genes of *Corynebacterium glutamicum*. J Biotechnol 104:229–240

Möckel B, Marx A, Pfefferle W (2002) Novel nucleotide sequences coding for the *glbO* gene, pp WO/2001/094569, publication date: Dec 094513, 092001

Moon MW, Kim HJ, Oh TK, Shin CS, Lee JS, Kim SJ, Lee JK (2005) Analyses of enzyme II gene mutants for sugar transport and heterologous expression of fructokinase gene in *Corynebacterium glutamicum* ATCC 13032. FEMS Microbiol Lett 244:259–266

Muffler A, Bettermann S, Haushalter M, Horlein A, Neveling U, Schramm M, Sorgenfrei O (2002) Genome-wide transcription profiling of *Corynebacterium glutamicum* after heat shock and during growth on acetate and glucose. J Biotechnol 98:255–268

Nakamura Y, Nishio Y, Ikeo K, Gojobori T (2003) The genome stability in *Corynebacterium* species due to lack of the recombinational repair system. Gene 317:149–155

Niki H, Hiraga S (1998) Polar localization of the replication origin and terminus in *Escherichia coli* nucleoids during chromosome partitioning. Genes Dev 12:1036–1045

Nishio Y, Nakamura Y, Kawarabayasi Y, Usuda Y, Kimura E, Sugimoto S, Matsui K, Yamagishi A, Kikuchi H, Ikeo K, Gojobori T (2003) Comparative complete genome sequence analysis of the amino acid replacements responsible for the thermostability of *Corynebacterium efficiens*. Genome Res 13:1572–1579

Nishio Y, Nakamura Y, Usuda Y, Sugimoto S, Matsui K, Kawarabayasi Y, Kikuchi H, Gojobori T, Ikeo K (2004) Evolutionary process of amino acid biosynthesis in *Corynebacterium* at the whole genome level. Mol Biol Evol 21:1683–1691

Ochman H, Lawrence JG, Groisman EA (2000) Lateral gene transfer and the nature of bacterial innovation. Nature 405:299–304

Oguiza JA, Marcos AT, Malumbres M, Martin JF (1996) Multiple sigma factor genes in *Brevibacterium lactofermentum*: characterization of *sigA* and *sigB*. J Bacteriol 178:550–553

Oh MK, Rohlin L, Kao KC, Liao JC (2002) Global expression profiling of acetate-grown *Escherichia coli*. J Biol Chem 277:13175–13183

Pátek M, Muth G, Wohlleben W (2003) Function of *Corynebacterium glutamicum* promoters in *Escherichia coli*, *Streptomyces lividans*, and *Bacillus subtilis*. J Biotechnol 104:325–334

Paulsen IT, Nguyen L, Sliwinski MK, Rabus R, Saier MH Jr (2000) Microbial genome analyses: comparative transport capabilities in eighteen prokaryotes. J Mol Biol 301:75–100

Pedersen AG, Jensen LJ, Brunak S, Staerfeldt HH, Ussery DW (2000) A DNA structural atlas for *Escherichia coli*. J Mol Biol 299:907–930

Perna NT, Plunkett G 3rd, Burland V, Mau B, Glasner JD, Rose DJ, Mayhew GF, et al (2001) Genome sequence of enterohaemorrhagic *Escherichia coli* O157:H7. Nature 409:529–533

Polen T, Krämer M, Bongaerts J, Wubbolts M, Wendisch VF (2005) The global gene expression response of *Escherichia coli* to L-phenylalanine. J Biotechnol 115:221–237

Polen T, Rittmann D, Wendisch VF, Sahm H (2003) DNA microarray analyses of the long-term adaptive response of *Escherichia coli* to acetate and propionate. Appl Environ Microbiol 69:1759–1774

Polen T, Wendisch VF (2004) Genomewide expression analysis in amino acid-producing bacteria using DNA microarrays. Appl Biochem Biotechnol 118:215–232

Pósfai G, Plunkett G 3rd, Feher T, Frisch D, Keil GM, Umenhoffer K, Kolisnychenko V, Stahl B, Sharma SS, de Arruda M, Burland V, Harcum SW, Blattner FR (2006) Emergent properties of reduced-genome *Escherichia coli*. Science 312:1044–1046

Radmacher E, Stansen KC, Besra GS, Alderwick LJ, Maughan WN, Hollweg G, Sahm H, Wendisch VF, Eggeling L (2005) Ethambutol, a cell wall inhibitor of *Mycobacterium tuberculosis*, elicits L-glutamate efflux of *Corynebacterium glutamicum*. Microbiology 151:1359–1368

Raman S, Hazra R, Dascher CC, Husson RN (2004) Transcription regulation by the *Mycobacterium tuberculosis* alternative sigma factor SigD and its role in virulence. J Bacteriol 186:6605–6616

Riley M, Abe T, Arnaud MB, Berlyn MK, Blattner FR, Chaudhuri RR, Glasner JD, Horiuchi T, Keseler IM, Kosuge T, Mori H, Perna NT, Plunkett G 3rd, Rudd KE, Serres MH, Thomas GH, Thomson NR, Wishart D, Wanner BL (2006) *Escherichia coli* K-12: a cooperatively developed annotation snapshot–2005. Nucleic Acids Res 34:1–9

Rocha EP, Danchin A (2003) Gene essentiality determines chromosome organisation in bacteria. Nucleic Acids Res 31:6570–6577

Ruimy R, Riegel P, Boiron P, Monteil H, Christen R (1995) Phylogeny of the genus *Corynebacterium* deduced from analyses of small-subunit ribosomal DNA sequences. Int J Syst Bacteriol 45:740–746

Saier MH Jr (2001) The bacterial phosphotransferase system: structure, function, regulation and evolution. J Mol Microbiol Biotechnol 3:325–327

Salmon K, Hung SP, Mekjian K, Baldi P, Hatfield GW, Gunsalus RP (2003) Global gene expression profiling in *Escherichia coli* K-12. The effects of oxygen availability and FNR. J Biol Chem 278:29837–29855

Salmon KA, Hung SP, Steffen NR, Krupp R, Baldi P, Hatfield GW, Gunsalus RP (2005) Global gene expression profiling in *Escherichia coli* K-12: effects of oxygen availability and ArcA. J Biol Chem 280:15084–15096

Sassetti CM, Boyd DH, Rubin EJ (2001) Comprehensive identification of conditionally essential genes in mycobacteria. Proc Natl Acad Sci USA 98:12712–12717

Schaffer S, Weil B, Nguyen VD, Dongmann G, Günther K, Nickolaus M, Hermann T, Bott M (2001) A high-resolution reference map for cytoplasmic and membrane-associated proteins of *Corynebacterium glutamicum*. Electrophoresis 22:4404–4422

Schilling CH, Covert MW, Famili I, Church GM, Edwards JS, Palsson BO (2002) Genome-scale metabolic model of *Helicobacter pylori* 26695. J Bacteriol 184:4582–4593

Schluesener D, Fischer F, Kruip J, Rogner M, Poetsch A (2005) Mapping the membrane proteome of *Corynebacterium glutamicum*. Proteomics 5:1317–1330

Seibold G, Auchter M, Berens S, Kalinowski J, Eikmanns BJ (2006) Utilization of soluble starch by a recombinant *Corynebacterium glutamicum* strain: growth and lysine production. J Biotechnol 124:381–391

Sekine H, Shimada T, Hayashi C, Ishiguro A, Tomita F, Yokota A (2001) H^+-ATPase defect in *Corynebacterium glutamicum* abolishes glutamic acid production with enhancement of glucose consumption rate. Appl Microbiol Biotechnol 57:534–540

Shiio I, Sugimoto S, Kawamura K (1990) Effects of carbon source sugars on the yield of amino acid production and sucrose metabolism in *Brevibacterium flavum*. Agric Biol Chem 54:1513–1519

Shizuya H, Birren B, Kim UJ, Mancino V, Slepak T, Tachiiri Y, Simon M (1992) Cloning and stable maintenance of 300-kilobase-pair fragments of human DNA in *Escherichia coli* using an F-factor-based vector. Proc Natl Acad Sci USA 89:8794–8797

Siguier P, Pérochon J, Lestrade L, Mahillon J, Chandler M (2006) ISfinder: the reference centre for bacterial insertion sequences. Nucleic Acids Res 34:D32–36

Silberbach M, Schäfer M, Huser AT, Kalinowski J, Pühler A, Krämer R, Burkovski A (2005) Adaptation of *Corynebacterium glutamicum* to ammonium limitation: a global analysis using transcriptome and proteome techniques. Appl Environ Microbiol 71:2391–2402

Stackebrandt E, Rainey FA, Ward-Rainey NL (1997) Proposal for a new hierarchical classification system, *Actinobacteria* classis nov. Int J Syst Bacteriol 47:479–491

Stackebrandt E, Woese CR (1984) The phylogeny of prokaryotes. Microbiol Sci 1:117–122

Stansen C, Uy D, Delaunay S, Eggeling L, Goergen JL, Wendisch VF (2005) Characterization of a *Corynebacterium glutamicum* lactate utilization operon induced during temperature-triggered glutamate production. Appl Environ Microbiol 71:5920–5928

Stothard P, Wishart DS (2005) Circular genome visualization and exploration using CGView. Bioinformatics 21:537–539

Sung BH, Lee CH, Yu BJ, Lee JH, Lee JY, Kim MS, Blattner FR, Kim SC (2006) Development of a biofilm production-deficient *Escherichia coli* strain as a host for biotechnological applications. Appl Environ Microbiol 72:3336–3342

Suzuki N, Nonaka H, Tsuge Y, Inui M, Yukawa H (2005a) New multiple-deletion method for the *Corynebacterium glutamicum* genome, using a mutant lox sequence. Appl Environ Microbiol 71:8472–8480

Suzuki N, Okai N, Nonaka H, Tsuge Y, Inui M, Yukawa H (2006) High-throughput transposon mutagenesis and construction of mutant library of *Corynebacterium glutamicum*. Appl Environ Microbiol 72:3750–3755

Suzuki N, Okayama S, Nonaka H, Tsuge Y, Inui M, Yukawa H (2005b) Large-scale engineering of the *Corynebacterium glutamicum* genome. Appl Environ Microbiol 71:3369–3372

Taoka M, Yamauchi Y, Shinkawa T, Kaji H, Motohashi W, Nakayama H, Takahashi N, Isobe T (2004) Only a small subset of the horizontally transferred chromosomal genes in *Escherichia coli* are translated into proteins. Mol Cell Proteomics 3:780–787

Tatusov RL, Fedorova ND, Jackson JD, Jacobs AR, Kiryutin B, Koonin EV, Krylov DM, Mazumder R, Mekhedov SL, Nikolskaya AN, Rao BS, Smirnov S, Sverdlov AV, Vasudevan S, Wolf YI, Yin JJ, Natale DA (2003) The COG database: an updated version includes eukaryotes. BMC Bioinformatics 4:41

Tauch A (2005) Native plasmids of amino acid-producing Corynebacteria. In: Eggeling L, Bott M (eds) Handbook of *Corynebacterium glutamicum*. CRC Press, Boca Raton, FL, USA, pp 57–80

Tauch A, Kaiser O, Hain T, Goesmann A, Weisshaar B, Albersmeier A, Bekel T, Bischoff N, Brune I, Chakraborty T, Kalinowski J, Meyer F, Rupp O, Schneiker S, Viehoever P, Pühler A (2005a) Complete genome sequence and analysis of the multiresistant nosocomial pathogen *Corynebacterium jeikeium* K411, a lipid-requiring bacterium of the human skin flora. J Bacteriol 187:4671–4482

Tauch A, Kaiser O, Hain T, Goesmann A, Weisshaar B, Albersmeier A, Bekel T, Bischoff N, Brune I, Chakraborty T, Kalinowski J, Meyer F, Rupp O, Schneiker S, Viehoever P, Pühler A (2005b) Complete genome sequence and analysis of the multiresistant nosocomial pathogen *Corynebacterium jeikeium* K411, a lipid-requiring bacterium of the human skin flora. J Bacteriol 187:4671–4682

Terasawa M, Yukawa H, Takayama Y (1985) Production of L-aspartic acid from *Brevibacterium* by the cell reusing process. Process Biochem 20:124–128

Trotschel C, Deutenberg D, Bathe B, Burkovski A, Krämer R (2005) Characterization of methionine export in *Corynebacterium glutamicum*. J Bacteriol 187:3786–3794

Tucker DL, Tucker N, Conway T (2002) Gene expression profiling of the pH response in *Escherichia coli*. J Bacteriol 184:6551–6558

Vertès AA, Inui M, Kobayashi M, Kurusu Y, Yukawa H (1994) Isolation and characterization of IS*31831*, a transposable element from *Corynebacterium glutamicum*. Mol Microbiol 11:739–746

Vertès AA, Inui M, Yukawa H (2005) Manipulating corynebacteria, from individual genes to chromosomes. Appl Environ Microbiol 71:7633–7642

Vrljic M, Sahm H, Eggeling L (1996) A new type of transporter with a new type of cellular function: L-lysine export from *Corynebacterium glutamicum*. Mol Microbiol 22:815–826

Weber A, Jung K (2002) Profiling early osmostress-dependent gene expression in *Escherichia coli* using DNA macroarrays. J Bacteriol 184:5502–5507

Wei Y, Lee JM, Richmond C, Blattner FR, Rafalski JA, LaRossa RA (2001) High-density microarray-mediated gene expression profiling of *Escherichia coli*. J Bacteriol 183:545–556

Welch RA, Burland V, Plunkett G 3rd, Redford P, Roesch P, Rasko D, Buckles EL, Liou SR, Boutin A, Hackett J, Stroud D, Mayhew GF, Rose DJ, Zhou S, Schwartz DC, Perna NT, Mobley HL, Donnenberg MS, Blattner FR (2002) Extensive mosaic structure revealed by the complete genome sequence of uropathogenic *Escherichia coli*. Proc Natl Acad Sci USA 99:17020–17024

Wendisch VF (2003) Genome-wide expression analysis in *Corynebacterium glutamicum* using DNA microarrays. J Biotechnol 104:273–285

Wendisch VF (2006) Genetic regulation of *Corynebacterium glutamicum* metabolism. J Microbiol Biotechnol 16:999–1009

Wendisch VF, Bott M, Eikmanns BJ (2006a) Metabolic engineering of *Escherichia coli* and *Corynebacterium glutamicum* for biotechnological production of organic acids and amino acids. Curr Opin Microbiol 9:268–274

Wendisch VF, Bott M, Kalinowski J, Oldiges M, Wiechert W (2006b) Emerging *Corynebacterium glutamicum* systems biology. J Biotechnol 124:74–92

Willis RC (2003) In silico medicine. Today's Chemist at Work, October, pp 15–18

Winnen B, Felce J, Saier MH Jr (2005) Genomic analyses of transporter proteins in *Corynebacterium glutamicum* and *Corynebacterium efficiens*. In: Eggeling L, Bott M (eds) Handbook of *Corynebacterium glutamicum*. CRC Press, Boca Raton, FL, USA

Wirth T, Falush D, Lan R, Colles F, Mensa P, Wieler LH, Karch H, Reeves PR, Maiden MC, Ochman H, Achtman M (2006) Sex and virulence in *Escherichia coli*: an evolutionary perspective. Mol Microbiol 60:1136–1151

Wittmann C, Kiefer P, Zelder O (2004) Metabolic fluxes in *Corynebacterium glutamicum* during lysine production with sucrose as carbon source. Appl Environ Microbiol 70:7277–7287

Wosten MM (1998) Eubacterial sigma-factors. FEMS Microbiol Rev 22:127–150

Yang F, Yang J, Zhang X, Chen L, Jiang Y, Yan Y, Tang X, et al (2005) Genome dynamics and diversity of *Shigella* species, the etiologic agents of bacillary dysentery. Nucleic Acids Res 33:6445–6458

Yu BJ, Sung BH, Koob MD, Lee CH, Lee JH, Lee WS, Kim MS, Kim SC (2002) Minimization of the *Escherichia coli* genome using a Tn5-targeted Cre/*loxP* excision system. Nat Biotechnol 20:1018–1023

Zhang R, Zhang CT (2005) Genomic islands in the *Corynebacterium efficiens* genome. Appl Environ Microbiol 71:3126–3130

Zimmer DP, Soupene E, Lee HL, Wendisch VF, Khodursky AB, Peter BJ, Bender RA, Kustu S (2000) Nitrogen regulatory protein C-controlled genes of *Escherichia coli*: scavenging as a defense against nitrogen limitation. Proc Natl Acad Sci USA 97:14674–14679

Zupancic TJ, Kittle JD, Baker BD, Miller CJ, Palmer DT, Asai Y, Inui M, Vertès AA, Kobayashi M, Kurusu Y, Yukawa H (1995) Isolation of promoters from *Brevibacterium flavum* strain MJ233C and comparison of their gene expression levels in *B. flavum* and *Escherichia coli*. FEMS Microbiol Lett 131:121–126

Subject Index

Printing: Krips bv, Meppel
Binding: Stürtz, Würzburg